SVEC
2012
12

Jean-Jacques Rousseau and botany

the salutary science

SVEC (formerly known as *Studies on Voltaire and the Eighteenth Century*) is dedicated to eighteenth-century research. *SVEC* welcomes work across a broad range of disciplines and critical methodologies.

www.voltaire.ox.ac.uk

Jean-Jacques Rousseau and botany

the salutary science

ALEXANDRA COOK

VOLTAIRE FOUNDATION
OXFORD
2012

ISBN 978 0 7294 1055 7
ISSN 0435-2866

The Voltaire Foundation is a department of the University of Oxford. It furthers the University's objective of excellence in research, scholarship and education by publishing worldwide.

Voltaire Foundation
99 Banbury Road
Oxford OX2 6JX, UK
www.voltaire.ox.ac.uk

A catalogue record for this book is available from the British Library

Rousseau / history of science / philosophy
Rousseau / histoire de la science / philosophie

Cover illustration: Left-hand image: Jean-Jacques Rousseau, 'Signes ou abréviations pour les descriptions et caractères des plantes', MsR80, f.135, by permission of the Bibliothèque publique et universitaire de Neuchâtel, Neuchâtel, Switzerland; right-hand image: *Euphorbia exigua* L., Herbarium for Madelon Delessert, by permission of the Musée Jean-Jacques Rousseau – Montmorency – France. Photography: Laurent Juillard.

MIX
Paper from responsible sources
FSC® C023114

FSC® (the Forest Stewardship Council) is an independent organization established to promote responsible management of the world's forests.

This book is printed on acid-free paper
Printed in the UK by Page Bros (Norwich) Ltd

Contents

List of illustrations and tables

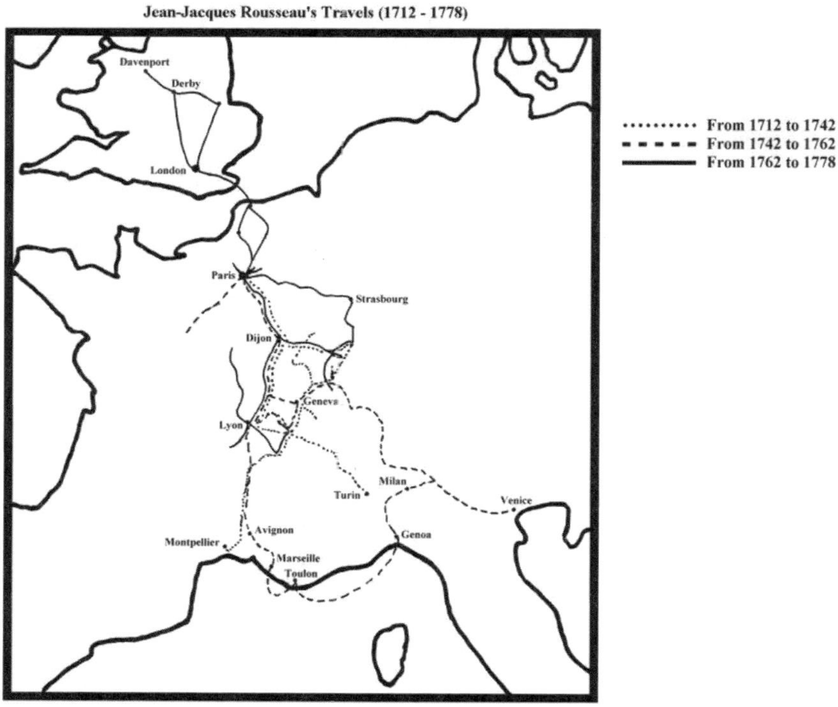

Frontispiece: Kenneth Yung, *Map of Rousseau's travels 1712-1778*, 2011, Hong Kong.

Rousseau: chronology

1712: b. 28 June, Geneva, Switzerland

1728: leaves Geneva and apprenticeship; meets Mme de Warens ('Maman'); converts to Catholicism

1729-1742: with Maman intermittently at Annecy, Chambéry and Les Charmettes in Savoy; introduced to botany by Maman's valet, Claude Anet

1742, July: relocates to Paris; meets Daniel Roguin; presents his system of musical notation to the Académie royale des sciences

1743, Spring: becomes acquainted with the Dupin family; begins to study chemistry with Dupin de Francueil

1743-1744: secretary to the comte de Montaigu, French ambassador in Venice

1745: forms lifelong relationship with Thérèse Levasseur

1749: writes articles on music for the *Encyclopédie* of Diderot and d'Alembert, later part of *Dictionnaire de musique* (1767)

1750: *Discours sur les sciences et les arts* wins the prize of the Académie de Dijon, making Rousseau an overnight sensation

1752: composes *Le Devin du village*, performed before Louis XV at Fontainbleau

1754, June-September: visits Geneva; citizenship and membership in Reformed Church reinstated, tour of Lake Geneva with the geologist Jean-André Deluc and his family

1755: publication of the *Discours sur les origines de l'inégalité*, dedicated to the Republic of Geneva

1756: resides at the Hermitage (valley of Montmorency) as guest of Mme d'Epinay

1757: *Lettre à M. d'Alembert sur les spectacles*, defends Geneva's Calvinist traditions against Voltaire's reform proposals

1759: moves to the Mont-Louis (town of Montmorency) at the invitation of the maréchal and maréchale de Luxembourg; writes *Emile* and *Julie, ou La Nouvelle Héloïse*

1762, April-May: publication of *Emile* and *Du contrat social*; they are condemned in Paris, Geneva, and elsewhere; arrest warrants issued in France and Geneva

1762, June: flees to Daniel Roguin's home in Yverdon, a Bernese territory, from which he is expelled by the authorities

July: finds refuge in Môtiers (Prussian principality of Neuchâtel) with assistance of Roguin's niece, Mme Boy de la Tour

1763, 12 May: surrenders Genevan citizenship; registered 16 May by the Petit Conseil

1764: studies botany in Neuchâtel and begins acquiring a botanical library

1765, June: studies botany with Abraham Gagnebin in La Ferrière (Swiss Jura)

1765, 12 September: forced by the stoning of his house to leave Môtiers; settles on the Ile St Pierre (Bernese territory) in the Lake of Bienne, botanical studies (cinquième promenade)

1765, 25 October: expelled from the Ile St Pierre by the Bernese authorities

1765, November-December: traverses France; fêted as a celebrity

1766, January: accepts David Hume's invitation to go to England; mobbed by visitors; settles at Wootton Hall; begins the *Confessions* and botanises with the Duchess of Portland and Daniel Malthus (father of the economist, Thomas Malthus)

1767, May-June: returns to France; spends two weeks at Fleury-sous-Meudon, end of June moves to Trye under the protection of the Prince de Conti; assumes the name 'Renou'

1768, June: goes to Lyon with his herbarium and library; botanises with Dombey, who gives him a valuable herbarium

1768, July: botanical trip to the Grande Chartreuse with Latourrette and abbé Rozier

1768, August: marries Thérèse

1769, January: settles at the farm of Monquin, above Bourgoin

1769, August: botanical trip to Mont Pilat with Dr Meynier, abbé Baurin and Donin de Champagneux

1769-1770: considers giving up botany; makes it known that his botanical library is for sale; possibly returns Dombey's herbarium?

1770, January: drops pseudonym

1770, June: returns to Paris; copies music, botanises, receives constant stream of visitors

1770-1771: gives private readings of the *Confessions*; Mme d'Epinay appeals to the police to stop them

1771, June: completes *Considérations sur le gouvernement de Pologne*; botanises with Thouin and Antoine Laurent de Jussieu in Montmorency

1771, July: becomes acquainted with Bernardin de St Pierre

1771, August: begins eight letters on botany to Madeleine-Catherine Delessert, née Boy de la Tour

1772, April: proposes herbiers portatifs to the Duchess of Portland; sends herbier for Julie Emélie Boy de la Tour (Zürich) to her sister, Mme Delessert

1773: sends herbiers portatifs to Malesherbes and the Duchess of Portland

1774: concludes eight letters on botany to Mme Delessert, but they continue to correspond about botany; gives herbarium to her daughter Madelon ('Montmorency')

1774, Spring: Antoine Laurent de Jussieu introduces Linnaean names to the Jardin du roi and arranges the plants according to his natural method

1775: sells botanical library to Daniel Malthus

1776: botanises with Lamarck; completes *Dialogues: Rousseau juge de Jean-Jacques*; begins *Rêveries du promeneur solitaire*

1777: continues *Rêveries*; takes notes from borrowed botanical works

1778, 2 July: dies at Ermenonville, the estate of the marquis de Girardin; leaves a small botanical library of fourteen works

1781-1782: *Collection complète des œuvres de J.-J. Rousseau* published by Du Peyrou and Moultou in Geneva

1781: German translation of *Lettres élémentaires* published in Mannheim

1789: Danish translation of *Lettres élémentaires* published in Copenhagen

1785: Martyn's *Letters on the elements of botany* published in London; eight editions follow

1794, 9-11 October: transfer of Rousseau's remains to the Panthéon in Paris, to join those of Voltaire, his adversary in life

1798: Dutch translation of Martyn published in London and Rotterdam

1801: Portuguese translation of Martyn published in Lisbon

1810: Russian translation of *Lettres élémentaires* published in Moscow

Acknowledgements

This has not been an easy book to research, given the dispersion and disappearance of virtually all of Rousseau's botany books (described in Appendix 1), and some of his herbaria (described in Appendix 3, Table 2). I am therefore deeply grateful to the many librarians and archivists who assisted me in my research, and gave me invaluable leads to pursue.

At the Bibliothèque publique et universitaire of Neuchâtel I was graciously received by Michel Schlup, Maryse Surdez-Schmidt, Sylvie Béguelin and Anne-Lise Grobéty; François Felber of the Institut de botanique of the University of Neuchâtel granted me permission to see the herbarium of Rousseau's teacher, Abraham Gagnebin. On that occasion I was accompanied by a very knowledgeable guide in Marcel Jacquat of La Chaux-des-fonds, Switzerland. Claire Jaquier Kaempfer of the University of Neuchâtel gave me valuable information regarding the ongoing mystery concerning the herbarium of another teacher of Rousseau, Jean-Antoine d'Ivernois. Crystal Smith and Stephen Greenberg of the History of Medicine Division of the National Library of Medicine, Bethesda, MD, introduced me to works from their magnificent collection. J. M. Vasseur of the Musée Jacquemart-André, Fontaine-Chaalis showed me artefacts from Rousseau's last days in Ermenonville, and told me that Rousseau's system of musical notation had travelled all the way to China. I have also been graciously assisted at various times by Chantal Mustel and Laurence Feo-Robin of the Musée Jean-Jacques Rousseau, Montmorency, the Rare Book and Special Collections Reading Room of the Library of Congress, Washington, DC, the Bibliothèque centrale of the Muséum national d'histoire naturelle (Paris) and the Library of the Academy of Natural Sciences, Philadelphia, PA.

Special thanks are due to Gina Douglas and Janet Ashdown for their generous reception at the Linnean Society, London, where they showed me the magnificent herbaria of Carolus Linnaeus and Sir James Edward Smith, and brought me up to speed on herbarium conservation over cups of tea. Patrick Bungener, my collaborator on the Rousseau tercentenary exhibit, 'Je raffole de la botanique', in 2012 at the Conservatoire et jardin botaniques of Geneva, Switzerland, made me aware of Rousseau's legacy for Geneva's botanists.

I owe enormous thanks to several research assistants, in particular, Rinske Kuiper, who prepared the images for the book and spent long hours on the complex task of assembling and verifying Rousseau's botanical sources in Appendix 1 and Michelle Ha Ming Wai, who researched Rousseau's correspondents and botanical library using old-fashioned methods before we had access to the electronic version of the *Correspondance complète* in *Electronic Enlightenment*; her work forms the core for Appendices 1 and 2. Kenneth Yung Cai Chung, Macy Yip, Brian Ho Kin Ting, Irwin Chan Yu Shing and Justin Li Ying Kit provided patient help with various aspects of this project.

I am grateful for comments on various aspects of this work by Peter Stevens, Mark Elvin, Robert Peckham, Frank James, John C. O'Neal, Marianne Klemun, Leo Damrosch, Gerard G. Aymonin, Cécile Aupic, Gerald Sullivan, Raymond Birn and my colleagues in the Department of Philosophy, University of Hong Kong. I would like to give belated thanks to Ralph Lerner, who took me on my first botanical field trip to the Morton Arboretum in Milton Township, Illinois, those many years ago.

The research for this book was substantially supported by a General Research Fund grant from the Research Grants Council of the Hong Kong Special Administrative Region, China (Project No., HKU 743307H). Work on maps, images and appendices was partially supported by a General Research Fund grant (Project No. HKU 743711H), by University of Hong Kong Small Project Funding grants (Project Nos. 201007176310, 200507176184), and by the HKU SPACE Research Fund.

Abbreviations

The correspondence and complete works of Rousseau are abbreviated as:

CC	*Correspondance complète*, ed. Ralph A. Leigh, 52 vols (Geneva, Madison, WI, Banbury, Oxford, 1965-1998).
OC	*Œuvres complètes*, ed. Bernard Gagnebin and Marcel Raymond, 5 vols (Paris, 1959-1995).

Note: For the above abbreviations, Roman numerals indicate volume numbers, Arabic numerals indicate page numbers, for example: *CC*, xv.53, *OC*, i.999.

BdAN	Bibliothèque de l'Assemblée nationale, Paris
BPUN	Bibliothèque publique et universitaire, Neuchâtel
MdAD	Musée des arts décoratifs, Paris

Introduction

i

Jean-Jacques Rousseau's fame as a philosopher, a novelist, a music theorist and all-round provocateur has caused his works to be studied, dissected, translated and debated in innumerable publications. With the exception of the decades immediately after his death, Rousseau's botanical works have received considerably less attention, however.[1] Rousseau's excursion into botany commenced circa 1764 and lasted – with intermissions – until his death in 1778; it has generally been viewed as the eccentric and irrelevant affectation or hobby of a bitter, isolated and ageing philosopher who dared not write about his real concerns in the human realm: politics, education, religion and morals.[2]

Considered as a hobby, Rousseau's botany has been seen as irrelevant to his philosophy; the fad for botany having passed in the early nineteenth century, his enthusiasm for the subject seems remote even to those who study his other works with avid interest. Since Rousseau made no great discovery in the manner of a Copernicus or a Newton, there seems to be no reason to read his presumably outdated scientific texts; the allegedly progressive nature of science precludes these works from teaching us anything.[3] Furthermore, outdated science may appear even more inaccessible than contemporary science.[4] So it has been considered

1. I discuss these works in the third section of this Introduction, p.6-8.
2. '[N]obody was further removed from the scientific frame of mind than Rousseau [...] Living exclusively in a subjective world of his own imagination, he could not bear the cold objectivity of men of science [...] He provided indeed the cardinal example of the terribly dangerous result of the working of an untrained mind [...] without logical critical judgment or a sound basis of objective knowledge': Gavin de Beer, 'Jean-Jacques Rousseau: botanist', *Annals of science* 10:3 (1954), p.189-223 (189-90). This article has recently been relied upon in Sam George, *Botany, sexuality and women's writing 1760-1830: from modest shoot to forward plant* (Manchester, 2007), p.72, n.26.
3. This, in effect, is the stance of Roger de Vilmorin, who edited these writings in Jean-Jacques Rousseau, *Œuvres complètes*, 5 vols (Paris, 1959-1995), ed. Bernard Gagnebin and Marcel Raymond. All subsequent references to Rousseau's works will refer to this edition, hereafter abbreviated as *OC*. References to Rousseau's correspondence will be to the *Correspondance complète*, ed. Ralph A. Leigh, 52 vols (Geneva, 1965-1998), abbreviated as *CC*.
4. It is anachronistic to use the term 'science' before the nineteenth century. The term serves here only as a convenient shorthand. Andrew Cunningham and Perry Williams, 'De-centring the "big picture": *The Origins of modern science* and the modern origins of science', *British journal for the history of science* 26 (1993), p.407-32 (420).

acceptable to ignore these works, even when addressing Rousseau's relationship with the scientific culture and ideas of his era.[5]

A more fruitful series of interpretations has been offered by Jean Starobinski, who identified the medical and therapeutic logic at work in Rousseau's interest in plants.[6] Elsewhere, Starobinski pointed to Rousseau's adoption of a homeopathic therapy in his writings on society.[7] These two moments are connected, although Starobinski does not explicitly relate them; furthermore, Starobinski denied that Rousseau did anything more in botany than perform activities that are 'demi-machinales'; this interpretation, as I have argued elsewhere, misrepresents Rousseau's botanical knowledge, skills and practice.[8]

In this work, I take issue with this marginalisation of Rousseau's botanical enterprise, showing how it supports his central philosophical premise of the primacy of nature. I accept Starobinski's argument that botany served Rousseau as therapy – a point that Rousseau himself never disputed – but I argue that botany came to form part of an askesis: a practice dating back to antiquity that tends to the soul and its maladies.[9] Today we would call this use of nature to treat the soul 'environmental therapy'.

Askesis involves regular exercises, usually directed toward oneself: meditation, dialogue and/or writing and privation.[10] For example, in the *Rêveries*, Rousseau meditated on his past failings and conducted a rigorous self-examination to prepare the account he was to give of himself.[11]

5. See for example Mark Hulliung, *The Autocritique of Enlightenment: Rousseau and the philosophes* (Cambridge, MA, 1994).
6. Jean Starobinski, *Jean-Jacques Rousseau: la transparence et l'obstacle* (Paris, 1971), p.279-80.
7. Jean Starobinski, *Le Remède dans le mal: critique et légitimation de l'artifice à l'âge des Lumières* (Paris, 1989), p.166ff. Starobinski's translator, Arthur Goldhammer, chose the term 'homeopathic'. It is technically anachronistic since the founder of homeopathy, Samuel Hahnemann (1755-1843), developed his theories starting in the late eighteenth century.
8. P. Tolila, 'Rousseau et le matérialisme des fleurs', *La Pensée* 202 (1978), p.116-31 (122); A. Cook, 'Jean-Jacques Rousseau et les réseaux d'échange botanique', in *Rousseau et les sciences*, ed. Bernadette Bensaude-Vincent and Bruno Bernardi (Paris, 2003), p.93-114 (108, n.17); A. Cook, 'Idées et pratiques scientifiques dans la correspondance botanique de Jean-Jacques Rousseau', *Annales de la société Jean-Jacques Rousseau* 47 (2007), p.265-85 (274-75, n. 21). See especially A. Cook, 'The septième promenade of the *Rêveries*: a peculiar account of Rousseau's botany?', in *The Nature of Rousseau's Rêveries: physical, human, aesthetic*, ed. John C. O'Neal *SVEC* 2008:03, p.11-36.
9. Davis argues that Rousseau 'did not seek nature as a cure'. Michael Davis, *The Autobiography of philosophy: Rousseau's Rêveries of the solitary walker* (Lanham, MD, 1999), p.215. While Rousseau may not have initially sought nature as a cure, he immediately saw botany as 'salutaire'. Rousseau to Malesherbes, 11 November 1764; *CC*, xxii.43-45.
10. Michel Foucault, *L'Herméneutique du sujet: cours au Collège de France (1981-1982)*, edition established by Frédéric Gros under the direction of François Ewald and Alessandro Fontana (Paris, 2001), p.476.
11. Rousseau, *Rêveries*, *OC*, i.999.

However, this askesis was directed outward as well inward: an 'âme expansive cherche[...] à étendre ses sentiments et son existence sur d'autres êtres'.[12] Thus, Rousseau's askesis entailed not only the usual self-examination,[13] but also the attentive study of something external to himself that raised his sights 'vers le haut':[14] 'L'étude de la nature nous détache de nous-mêmes et nous *élève* à son Auteur. C'est en ce sens qu'on devient vraiment *philosophe*; c'est ainsi que l'histoire naturelle et la botanique ont un usage pour la *sagesse* et pour la *vertu*.'[15] Botany's interest to him derived therefore not from its material benefits, but from the wisdom and virtue that arise from detachment and edification.

Botany now assumed the pharmacological function that Rousseau denied to plants; he used botany not to treat the body's ailments, but rather, to cleanse his *soul*. Botany thereby enabled him to experience tranquility – ataraxia: 'Si l'étude des plantes me purge l'âme, c'est assez pour moi, je ne veux point d'autre pharmacie'.[16] Botany itself was for him therefore a medicine, a 'pharmakon': the incessant study of the physical world that Epicurus recommended to his disciples for their serenity of soul.[17] Practised in this way, natural philosophy operates as a 'pratique de soi'.[18]

My central claim is that in pursuing botanical study as therapy, Rousseau is not functioning as an automaton, but rather, as an engaged *natural philosopher* who studied the same book of nature that he extolled to his correspondents.[19] In the mid-eighteenth century, taxonomy was the dominant episteme of natural philosophy, providing an access point from which to approach nature. I show that Rousseau progressed from an artificial classification system, the sexual system of Carolus Linnaeus (1707-1778), to one that placed his botany on a footing compatible with his philosophy in general: the natural method. This method was supposed to be based on the real, underlying relations among plants, understood by contemporary botanical thought as continuous, an idea based on the 'chain of being'. Linnaeus and other botanists developed

12. Rousseau, *Rêveries*, *OC*, i.1066. See also John C. O'Neal, 'Nature as refuge in Rousseau's *Rêveries du promeneur solitaire*', in *The Nature of Rousseau's Rêveries*, *SVEC* 2008:03, p.59-69.

13. Rousseau, *Rêveries*, *OC*, i.999.

14. Foucault, *L'Herméneutique du sujet*, p.476.

15. My emphasis; Rousseau to the Duchess of Portland, 3 September 1766; *CC*, xxx.314.

16. The Greek term 'pharmakon' was used to describe the 'discours vrais'. Foucault, *L'Herméneutique du sujet*, p.480.

17. Foucault, *L'Herméneutique du sujet*, p.232-33. See also Robert Mauzi, *L'Idée du bonheur dans la littérature et la pensée françaises au XVIIIe siècle* (Paris, 1979), p.16. Rousseau often refers to the 'sérénité' that botany and being in nature give him.

18. Foucault, *L'Herméneutique du sujet*, p.197.

19. On natural philosophy, see Cunningham and Williams, 'De-centring the "big picture"', p.421.

this method during the eighteenth century, focusing on the 'habit' (habitus) of the plant.[20]

The teaching of the habit was esoteric for Linnaeus,[21] and Rousseau never mentioned it in his more public texts such as the *Confessions* and the *Rêveries*.[22] These well-known works offer limited and conventional accounts of his botanical ideas; he presented a more complete account in the *Lettres élémentaires*, originally a private communication that arose from his familial relations with the Delessert and Boy de la Tour families.[23] Although Rousseau revised the *Lettres élémentaires*, there is no suggestion on his part that he wished to publish them.

My account stresses, moreover, that Rousseau's botany is not merely theoretical or contemplative; it constituted a practice, a botany of *doing*, of producing: letter-writing, exchanging books and specimens, annotating books with Linnaean binomial names, collecting plants, describing their environments and preserving them with great aesthetic value. This was not a mechanical activity, 'une admiration stupide et monotone' of nature,[24] but rather, the conscious activity of a true philosopher of nature.

ii

This work revolves around three key questions:

What were the origins of Rousseau's interest in botany?
How does botany relate to his philosophy, which he asserts in the *Dialogues* forms an integrated whole;[25] and
What is the significance of Rousseau's excursion into botany, and what is its legacy?

Chapter 1 examines how Rousseau treated both social and personal problems with a derivative of the disease itself on the model of like curing like. Botany is 'salutaire', a medicine that purges the soul, rather than the body; at the same time it is natural philosophy, the study of the divine plan of Creation that affords unmediated contact with nature, not

20. I elaborate on this issue in Chapter 6.
21. Carolus Linnaeus, *Philosophia botanica*, trans. Stephen Freer (Oxford, 2003), p.130-32, section 168.
22. Rousseau read the *Confessions* aloud in various gatherings in 1771, and the *Rêveries* is the last part of a trilogy of autobiographies and self-justifications. While Rousseau claimed the *Rêveries* were for himself alone, there might be reason to contest this. *Rêveries*, *OC*, i.999. On Rousseau's intentions in writing his last work, see Davis, *The Autobiography of philosophy*, p.91, 93.
23. He suggested projects of botanical dissemination to some of his correspondents, most notably, Claret de Latourrette.
24. Rousseau, *Confessions*, *OC*, i.641; see also Rousseau, *Rêveries*, *OC*, i,1062.
25. Rousseau, *Dialogues*, *OC*, i.930.

forced by the hand of man. Horticulture and gardens, on the other hand, offer the soul no solace; Rousseau's botany is therefore a completely distinct enterprise from gardening or garden theory.

Chapter 2 shows how Rousseau's study of chemistry predisposes him to look beyond chemistry for a salutary science because chemistry studies what is dead, while botany studies what is living; chemistry destroys, while botany preserves; chemistry is for the rich, who can afford costly apparatus, while botany is for anyone who can afford a few simple tools. Chemistry produces medicines of dubious value and poisons the body with heavy metals; it attempts to transmute base metals into gold, which is associated with cupidity and commerce. Botany, if practised independently of material concerns, soothes the troubled soul.

Chapter 3 examines the conditions for learning and doing botany in eighteenth-century Switzerland, where decentralisation, emigration for education and careers, and the lack of a court or national centre enabled a plurality of scientific viewpoints to flourish. This pluralism gave Rousseau options in botanical classification from which his work benefited.

Chapter 4 opens with Rousseau's return to Switzerland, where he commenced his study of botany in the Prussian principality of Neuchâtel. I argue that for Rousseau, botany and politics met on the soil of his one true homeland ('pays'): Switzerland. This chapter examines the pluralistic antecedents of Rousseau's botany in Neuchâtel, where new scientific ideas were well-received.

Chapter 5 takes a historical view of the development of plant taxonomy in the west up to Linnaeus. While Rousseau inveighed against systems and methods of classification in the seventh promenade of the *Rêveries*, he did so as an adept of those systems and methods. In 1764 he may have used the artificial sexual system of Linnaeus, but by 1765 he was already exploring other classification systems, including the natural methods being proposed at the time; his reception of these initiatives has been ignored by recent interpretations of his botanical thought, which depict him as wedded to the *artificial* sexual system.

Chapter 6 demonstrates that rather than remaining frozen in adherence to the sexual system, Rousseau's botanical thought, like that of Linnaeus, evolved over time. Rousseau was influenced by direct contact with the progenitors of natural methods of classification, such as the Jussieus and Michel Adanson; he offered his own version of this method in his *Lettres élémentaires* to his friend, Mme Delessert. His taxonomic choices were consistent with his long-held presumption of nature's primacy; in this domain as in others, he fitted his prescriptions to circumstances rather than prescribing one abstract principle for all situations.[26]

26. See for example Rousseau, *Préface à Narcisse*, *OC*, ii.971-72; *Du contrat social*, *OC*, iii.414-19.

Chapter 7, 'Modes of mediation: botanical books and binomial names', addresses the problem of mediation as it presents itself to Rousseau in two important aspects of botany: books and nomenclature. Books lacked pictures, treated plants only discursively, and were too advanced for the beginner. The Latin binomial names invented by Linnaeus improved on pre-Linnaean polynomials, but nevertheless constituted a mediation between humanity and nature that posed theoretical, if not practical, problems for Rousseau.

Chapter 8 examines Rousseau's herbaria: why and how he made them, who his teachers were, what techniques he employed and how his approach to this central practice of botany compares with that of other practitioners. I show that Rousseau aligned his work with the best practice of the time, displaying sensitivity to both aesthetic and scientific norms. His herbaria connected Rousseau to the wider world of European botany and the republic of science.

Chapter 9, 'A forged legacy: the strange fates of Rousseau's botanical works', exposes how Rousseau's literary executors manipulated Rousseau's botanical *Nachlass*, especially the texts published posthumously as *Fragmens pour un dictionnaire des termes d'usage en botanique*. I also assess the effects of another important revision of Rousseau's botanical work: Thomas Martyn's 1785 English translation of the 'Introduction' and *Lettres* accompanied by explanations of the twenty-four classes of the Linnaean sexual system. Martyn's Linnaean framing of the eight letters influenced generations of botanical authors, especially women. In France and Germany, on the other hand, Rousseau's botanical practice was appreciated largely on its own terms. The German poet and naturalist, Goethe, saw in Rousseau the botanical teacher par excellence, who taught the natural method of classification.

Chapter 10, 'Reading Rousseau on botany at the end of nature', considers the significance and implications of Rousseau's botanical works, taking up their relevance for current debates in environmental ethics.

iii

This does not purport to be a definitive, exhaustive work on Rousseau and botany, filling all the gaps in our knowledge of this topic. The one previous attempt of this kind, Albert Jansen's *Jean-Jacques Rousseau als Botaniker* (1885), while useful and detailed, is flawed and unreliable.[27] A

27. A more recent work by Guy Ducourthial, *La Botanique selon Jean-Jacques Rousseau* (Paris, 2009), attempts to provide a comprehensive treatment of Rousseau's botanical works, but offers no account of the relationship between his botanising and his philosophy. The question of Rousseau's taxonomic choices, which I consider the keystone of my argument, is barely broached.

definitive, modern work on this subject has long been awaited, yet there are still too many lacunae in the history of Rousseau's botany to provide a complete, exhaustive account of it. For example, we have no manuscript of his purported botanical dictionary; nor, for the reasons given in Chapter 9, 'A forged legacy', do I believe that we are likely to find one. Nor do we have a complete catalogue of Rousseau's botanical library, although Appendix 1 is the best effort to date to provide one.

Similarly, Rousseau's relations with a number of important figures in the botanical world of the time are somewhat obscure; these have to be inferred and reconstructed from a variety of sources, especially letters and sometimes second- or third-hand accounts. Rousseau's face-to-face relations with some of the era's most important botanists was not science at a distance entailing the writing of letters, as in the case of his correspondence with Claret de Latourrette and the Duchess of Portland. Teasing out these relations helps establish my claim that Rousseau followed the natural method.

My approach is by necessity historical and biographical in order to understand the origins of Rousseau's botanical thought, situate it within its context and assess its significance. In addition to Rousseau's own works (discussed below), I establish the context for his botanical thought using core botanical texts such as Linnaeus's *Philosophia botanica*, contemporary Swiss publications such as the *Journal helvétique*, unpublished manuscripts, correspondence[28] and material culture, most notably Rousseau's extant herbaria.[29]

A word of caution is in order concerning Rousseau's most well-known comments about botany in his *Rêveries* and *Confessions*: these accounts provide an inadequate basis for interpreting Rousseau's botanical thought.[30] Of the works on botany that Rousseau produced, or allegedly produced, I focus primarily on two categories: (1) his famous eight *Lettres élémentaires sur la botanique* and (2) three of his extant herbaria. The former were a kind of free-standing botany course that was widely translated, disseminated and imitated in the decades immediately following his death. The *Lettres* assume a central role in my account as the work in which Rousseau revealed his most thorough understanding of the plant kingdom in the taxonomic mode of the day. I also consult the various *Fragments de botanique*, including the essay on nomenclature, usually

28. My work makes ample use of the Voltaire Foundation's *Electronic Enlightenment* database; before this invaluable tool came online, it was extremely difficult to mine the wealth of this very rich correspondence. Terms such as 'botanique' were not found in the original index.

29. See Chapter 8 and Appendix 3.

30. See Paul A. Cantor, 'The metaphysics of botany: Rousseau and the new criticism of plants', *Southwest review* 70:3 (1985), p.362-80.

termed 'Introduction', that editors have placed at the head of either the *Lettres* or the *Fragmens pour un dictionnaire des termes d'usage en botanique*. The *Fragmens pour un dictionnaire* receives less attention in my account because of serious doubts concerning its authenticity, discussed in chapter 9.

In addition to the *Lettres*, I ground my analysis in a thorough study of Rousseau's herbaria – dried plant collections arranged according to a system, and preserved for future consultation in teaching and research. When I started this project I did not know what a herbarium was; Rousseau changed that. Because of his dedication to the herbarium, I have spent time in herbaria in France, Switzerland and Great Britain, not only examining specimens and comparing various ways of presenting them, but also talking to curators and botanists in order to understand how and why they make and keep herbaria. I now appreciate the many purposes to which this versatile boundary object can be put; the herbarium is a scientific tool that, as Rousseau discovered, can serve many purposes – not least of which is the advertisement of nature study.

The herbarium is hardly a quaint or useless artefact of yesterday; its scientific importance today perhaps even exceeds that of the past, as the human race struggles to document the planet's biodiversity. In his own humble way, Rousseau contributed to this process, noting down ecological circumstances of his specimens, and preparing them with such exquisite care that Pierre Prévost could write: 'Jamais herboriste n'a poussé plus loin la délicatesse et la propreté dans l'arrangement des plantes sur le papier; sa diligence à ce travail n'était moins remarquable'.[31] Rousseau's were herbaria with a difference – some were miniature, or at least portable for use in the field. In them he attached the plants with tiny paper bands that kept them free of adhesives used at the time, such as glue and wax. His work was even praised by experts who were otherwise far from sharing Rousseau's views. Herbarium specimens continue to arouse wonder to this day.[32]

This book argues that the recent trend of invoking Rousseau as the champion of the sexual system of Linnaeus and thus, it is often suggested, of a hidebound sexual hierarchy, has steered interpretation of his botanical work in the wrong direction. My work instead stresses Rousseau's *flexibility* in this domain as in others; instead of dictating one solution or rule to apply in all circumstances, he proposed different answers to apply to different cases: just as he insisted that what suited Geneva did not suit Paris, and vice versa.

This same flexibility characterised Rousseau's approach to botany.[33]

31. Pierre Prévost, A659. 'Souvenirs de Pierre Prévost', *CC*, xl.267.

32. See for example Nick Knight, *Flora* (New York, 2001).

33. In chapter 5 I give a short historical review of botany, but space constraints preclude a

As early as 1769 when he was writing the *Confessions*, he saw the sexual system as 'vide', but he could continue to admire Linnaeus because the Swedish botanist had more than one perspective to offer: in addition to the sexual system, Linnaeus outlined 'natural orders' in his *Philosophia botanica* of 1751, a work that for Rousseau surpassed all the works of morality. In the last quarter of the eighteenth century, Antoine Laurent de Jussieu acknowledged the value of Linnaeus's orders to his own 'méthode naturelle'. The influence of these important developments on his thinking is palpable in Rousseau's botanical work: he sometimes used the artificial sexual system of classification to order large amounts of information (such as plant specimens, pictures) while using the natural method to teach botany.

This line of interpretation takes up the insights of the earliest readers of the *Lettres*, who understood Rousseau as a practitioner of the natural method, as opposed to the sexual or other *artificial* system: the German poet, Goethe, the French revolutionary, Mme Roland and Augustin Pyramus de Candolle, a Genevan and one of the fathers of modern botany.[34]

fuller account. For the best recent work on Linnaeus's taxonomy, see Staffan Müller-Wille, 'Collection and collation: theory and practice of Linnaean botany', *Studies in history and philosophy of biological and biomedical sciences* 38 (2007), p.541-62. For an explication of the natural method, see Peter F. Stevens, *The Development of biological systematics: Antoine-Laurent de Jussieu, nature, and the natural system* (New York, 1994). An accessible, well-illustrated account is Joëlle Magnin-Gonze, *Histoire de la botanique*, revised edn (Paris, 2009).

34. These viewpoints are discussed further in chapter 6.

1. The 'remède dans le mal'[1]

> [C]ette étude si conforme à ma vie ambulante m'amusera beaucoup et me sera *salutaire*.[2]
>
> Il y a dans cette oiseuse occupation un charme qu'on ne sent que dans le plein calme des passions.[3]

i. The therapeutic project

At both its beginning and at its end, Rousseau's philosophical project had a therapeutic character. In the first *Discours* Rousseau assumed the role of a physician of society, arguing that society is *sick* and, by implication, that it requires a *cure* or treatment. What is the nature of this illness? In brief: moral degeneration caused by the sciences: 'la culture des sciences corrompe les mœurs d'une nation' by enervating the strong, and preserving the weak;[4] '[L]a science convient à quelques grands génies; mais [...] elle est toujours nuisible aux peuples qui la cultivent'.[5] The sciences give rise to luxury, vanity and the neglect of our duties: 'Qui voudrait [...] passer à de stériles contemplations si chacun ne consultant que les devoirs de l'homme et les besoins de la nature, n'avait de temps que pour la patrie, pour les malheureux, et pour ses amis?'[6] Inauthenticity and lack of transparency follow hard on the heels of the sciences and the arts: we come to live only in the opinion of others; 'il règne dans nos mœurs une vile et trompeuse uniformité [...] On n'ose plus paraître ce qu'on est'.[7]

Having painted this grim portrait of a sick society in the first *Discours*, Rousseau indicated that all was not hopeless: while the human race may be corrupted, enervated, and alienated, hope lies in the 'remède dans le mal', which has been characterised as a 'homeopathic cure' (lit. 'similar suffering').[8] Rousseau's 'remède dans le mal' derives from two folk

1. Jean Starobinski, *Le Remède dans le mal: critique et légitimation de l'artifice à l'âge des Lumières* (Paris, 1989), p.166ff.
2. My emphasis; Rousseau to Malesherbes, 11 November 1764; *CC*, xxii.43-45.
3. Rousseau, *Rêveries*, *OC*, i.1069.
4. Rousseau, *Observations de J.-J. Rousseau, sur la réponse à son 'Discours' (Réponse à Stanislas)*, *OC*, iii.39.
5. Rousseau, *Lettre à Grimm*, *OC*, iii.64.
6. Rousseau, *Discours sur les sciences et les arts*, *OC*, iii.17-18.
7. Rousseau, *Discours sur les sciences et les arts*, *OC*, iii.8.
8. Starobinski, *Le Remède dans le mal*, p.167. Starobinski's translator, Arthur Goldhammer, chose the term 'homeopathic'. It is technically, anachronistic, since the founder of

notions: (1) that antidotes are to be found growing near poisonous plants,[9] and (2) that like cures like.[10]

Used by a physician of society in the right dose, some poisons, like Plato's noble lie, can treat sick societies: these salutary social poisons include academies of science, theater, spectacles, and novels. The royal founding of academies draws on both cures: 'La prévoyance éternelle, en plaçant à côté de diverses plantes nuisibles des simples salutaires[...] a enseigné aux Souverains qui sont ses ministres à imiter sa sagesse'. The second cure was applied when Louis XIV drew the academies 'du sein même des sciences et des arts, sources de mille dérèglements'.[11] The homeopathic metaphor appears again in the first version of *Du contrat social* (the so-called *Geneva manuscript*): 'efforçons nous de tirer du mal même le remède qui doit le guérir'.[12]

Corrupt peoples should therefore enjoy the benefit of institutions of science, including those critical to early-modern and Enlightenment science, the Academies:

> J'ai loué les Académies et leurs illustres fondateurs, et j'en répéterai volontiers l'éloge. Quand le mal est incurable, le *Médecin* applique les *palliatifs* et proportionne les remèdes, moins aux besoins qu'au *tempérament* du malade. C'est aux *sages législateurs* d'imiter sa prudence; et, ne pouvant plus approprier aux Peuples malades, la plus excellente police, de leur donner du moins, comme Solon, la meilleure qu'ils puissent comporter.[13]

In the preface to his play, *Narcisse*, Rousseau asks whether the sciences should be abolished; he responds to his own question: 'les mêmes causes

---

homeopathy, Samuel Hahnemann (1755-1843) developed his theories starting in the late eighteenth century.

9. The homeopathic doctrine had yet to be elaborated. Unlike the homeopath, Rousseau did not prescribe very small doses of the medicine. Although he did not refer to 'inoculer' in the *Discours*, he discussed vaccination in *Emile*, *OC*, iv.378-79. Vaccination, which uses the disease agent to stimulate an immune response, is isopathic rather than homeopathic. Vaccination was long practised by the Circassians, and introduced into Europe in the eighteenth century. The Church opposed it, and in France it was resisted, much to the dismay of Voltaire in his *Lettres philosophiques* (Amsterdam, Lucas, 1734), lettre xi.
10. The notion that 'like cures like' (*similia similibus curentur*) had a long history, expressed in the doctrine of signatures, which holds that curative properties are indicated by a resemblance to the organ to be treated. Homeopathy added to this notion the 'law of infinitesimals', i.e. that very small doses induce the disease state necessary to effective treatment. Vital energy is injected into the medicine through 'potenisation' or 'dynamisation'.
11. Rousseau, *Discours*, *OC*, iii.26. The metaphor of medical treatment permeates Plato's *Republic*; the philosopher-king, like a physician, must know how to administer dangerous substances in the correct doses as medicines; lies are such a medicine – beneficial if used in the right way, pernicious if not. Plato, *Republic*, section 414b.
12. Rousseau, *Du contrat social*, *OC*, iii.288.
13. My emphasis; Rousseau, *Observations de J.-J. Rousseau, de Genève, sur la réponse qui a été faite à son discours*, *OC*, iii.56.

qui ont corrompu les peuples servent quelquefois à prévenir une plus grande corruption [...] c'est ainsi comment les arts et les sciences, après avoir fait éclore les vices, sont nécessaires pour les empêcher de se détourner des crimes [...] Elles détruisent la vertu, mais elles en laissent le simulacre public'.[14] Hence,

> Mon avis est donc de laisser subsister et même d'entretenir avec soin les Académies, les Collèges, les Universités, les Bibliothèques, les Spectacles, et tous les autres amusements qui peuvent faire quelque diversion à la méchanceté des hommes, et les empêcher d'occuper leur oisiveté à des choses plus dangereuses. Car dans une contrée où il ne serait plus question d'honnêtes gens ni de bonnes mœurs, il vaudrait encore mieux vivre avec des fripons qu'avec des brigands.[15]

The philosopher-physician, following Plato's *Republic*, can administer the potentially poisonous sciences as medicine in clinically indicated doses.

Of all the sciences it was perhaps with medicine that Rousseau had the most fraught relationship; having consulted physicians in vain to treat his various maladies,[16] he became disillusioned with academic medicine, and decided to submit to 'les seules lois de la nature'.[17] It was his longstanding belief that 'l'homme dans l'état de nature n'a donc guère besoin de remèdes, moins encore de médecins'.[18]

ii. The unique 'pharmacie'

Rousseau's disillusionment with the academic medicine of his day did nothing to diminish his interest in the conditions for optimal health, especially the health of his soul; in throwing off academic medicine he sought the true principles for human well-being in treating the maladies of his soul, in conducting what Foucault has termed the 'care of the self'.[19] This he could do by applying the 'remède dans le mal', taking up a

14. Rousseau, *Préface à Narcisse*, *OC*, ii.972.
15. Rousseau, *Préface à Narcisse*, *OC*, ii.972.
16. These maladies included urinary and digestive problems.
17. Rousseau, *Rêveries*, *OC*, i.1065. Academic medicine was based on the ancient theory of balancing the humours; herbal and chemical medicines were popular, as was blood-letting, which could be fatal, as in the case of George Washington, who was bled to death. See Pierre Huard, 'L'enseignement médico-chirugical', in *Enseignement et diffusion des sciences en France au XVIIIe siècle*, ed. René Taton (Paris, 1964), p.171-236; Josef N. Neumann, 'Rousseaus Kritik an der Heilkunde seiner Zeit: zur Frage nach der handlungstheoretischen und ethischen Begründung medizinischen Handelns', *Medizinhistorisches journal* 26:3-4 (1991), p.195-213.
18. Rousseau, *Discours*, *OC*, iii.139.
19. In Hellenistic and Roman times '[t]he care of the self [...] has to become permanent medical care. Permanent medical care is one of the central features of the care of the self. One must become the doctor of oneself': Michel Foucault, 'Technologies of the self', in *Technologies of the self: a seminar with Michel Foucault*, ed. Luther H. Martin *et al.* (Amherst, MA 1988), p.31.

science to treat the ills promoted by the sciences. The science he chose was botany: 'Si l'*étude* des plantes me purge l'*âme*, c'est assez pour moi, je veux point d'autre pharmacie'.[20]

The application of this 'pharmacie' is the study of plants *qua* plants, rather than studying how plants can be transformed into medicines, foods, clothing or other materially useful things.[21] It thus avoids sullying the mind with thoughts of the body.[22] Plants had traditionally served many uses, including as sources of medicines; Rousseau considered pharmaceutical botany pernicious, as it had caused the neglect of the study of plants themselves in favour of the reductive search for pharmaceuticals, a direct result of the unhealthy conditions of civilised life.[23] By using the *study* of the plant rather than the plant itself as medicine, Rousseau turned traditional pharmaceutical botany on its head.

Rousseau's rejection of academic and pharmaceutical medicine did not mean he rejected all medicine. In fact, his works are replete with medical ideas. In *Emile* he offered a plethora of medical prescriptions ranging from breast-feeding and childhood diet and exercise to prevention of scurvy and the merits of inoculation. Of particular interest to him was the ancient idea of regimen, based on the correct prescription of diet, environment and exercise according to the patient's constitution.[24]

Some commentators have accordingly hailed Rousseau as a prophet of natural medicine or naturopathy, whose core assumption is that human beings are by nature healthy; they become unhealthy through habits contrary to nature, that is, in civilisation; the goal of the practitioner is to return the patient to health by aligning his or her regime with nature's principles.[25] Rousseau so valued this approach that he suspended his aversion to the medical profession in order to consult a famous Swiss

20. My emphasis; Rousseau, *Fragments de botanique*, *OC* iv.1251. He refers to purging his soul, but he derided using plants to purge the body: 'pharmaciens ne voient dans le riche émail des prairies que des herbes pour les lavements [...] Pour moi j'y vois des objets d'admiration qui me transportent et qui me font respecter l'organisation qui les produit': Rousseau, *Rêveries*, *OC*, i.1064.
21. Rousseau, 'Introduction', *OC* iv.1201; *Dialogues*, *OC*, i.833.
22. Rousseau, *Rêveries*, *OC*, i.1064.
23. Rousseau, 'Introduction', *OC* iv.1201; *Dialogues*, *OC*, i.833.
24. Rousseau prescribes constitutionally in *Observations de J.-J. Rousseau, de Genève, sur la réponse qui a été faite à son discours*, *OC*, iii.56; for the principles of Hippocratic medicine, see G. E. R. Lloyd, ed. *Hippocratic writings*, trans. J. Chadwick *et al.* (London, 1978).
25. Gerhard Rudolf, 'Jean-Jacques Rousseau (1712-1778) – Stammvater der Naturheilkunde?', *Hippokrates* 12 (1969), p.474-79, and 'Jean-Jacques Rousseau (1712-1778) und die Medizin', *Sudhoffs Archiv: Zeitschrift für Wissenschaftsgeschichte* 53:1 (1969), p.30-67; Serge A. Thériault, *Jean-Jacques Rousseau et la médicine naturelle* (Montreal, 1979); Carlos Bartissol, 'Sources des idées médicales de Rousseau', doctoral dissertation, Faculty of Medicine, Paris (Paris, 1914). However, these works show that Rousseau's medical ideas were often derivative, rather than original.

specialist, Dr Samuel Tissot, on the regimen of diet and exercise to follow.[26]

When Rousseau proposed botany as his unique 'pharmacie', what did he think needed to be purged from his soul? His is an Epicurean or a Stoic answer: the passions, especially the desire for revenge on his presumed tormentors.[27] The ancient Stoic countered such feelings with meditation, as did Rousseau: 'Quelquefois mes rêveries finissent par la *méditation*, mais plus souvent mes *méditations* finissent par la rêverie'.[28] In this meditation he forgot himself and blended into nature: 'Je médite, je ne rêve jamais plus délicieusement que quand je m'oublie moi-même. Je sens des extases, des ravissements inexprimables à me fondre pour ainsi dire dans le système des êtres, à m'identifier avec la nature entière'.[29] A particularly corrosive passion, the desire for vengeance, was thereby dampened:

> [C]'est le moyen de ne laisser germer dans mon cœur aucun levain de *vengeance* ou de *haine* et pour trouver encore dans ma destinée du goût à quelque amusement, il faut assurément avoir un naturel bien épuré de toutes passions irascibles. C'est me *venger* de mes persécuteurs à ma manière, je ne saurais les punir plus cruellement que d'être heureux malgré eux.[30]

Botany calms the emotions by focusing the mind on something outside itself: an 'âme expansive cherche [...] à étendre ses sentiments et son existence sur d'autres êtres'.[31] This is not just a matter of enjoying a beautiful landscape or 'spectacle', but rather, a contemplative activity based in natural philosophy.[32] It is impossible, Rousseau claims, to appreciate the ensemble of nature as a whole without 'quelque idée du système végétal' and 'cette chaine de rapports et de combinaisons qui accable de ses merveilles l'esprit de l'observateur'.[33]

The object of contemplation is plant structures, reproductive organs, analogies among the various parts of the plant and overall familial appearance – the habit or 'port'. Rousseau engaged in careful, unhurried observation, wandering from plant to plant 'nonchalamment', 'brouter [son] foin presque par hasard', plucking specimens here and there.[34]

26. Rousseau to Dr Samuel Auguste André David Tissot, 1 February 1769; *CC*, xxxvii.37-40.
27. He believed a malicious 'cabal' was acting against him. See Rousseau, *Rêveries*, *OC*, i.1066.
28. My emphasis; Rousseau, *Rêveries*, *OC*, i.1062.
29. Rousseau, *Rêveries*, *OC*, i.1065-66.
30. My emphasis; Rousseau, *Rêveries*, *OC*, i.1061.
31. Rousseau, *Rêveries*, *OC*, i.1066.
32. Foucault notes that in imperial Roman times 'one couldn't assimilate ethical principles without a theoretical framework such as science, as for example in Lucretius's *De rerum naturae*', Foucault, 'Technologies of the self', p.35.
33. Rousseau, *Confessions*, *OC*, i.641.
34. Rousseau, *Confessions*, *OC*, i.641.

Sometimes he lay down on the ground to contemplate his subject: 'là je me couchais auprès de la plante en question par terre pour l'examiner sur pied, tout à mon aise'.[35] Lest these comments be taken to imply that Rousseau was not doing anything, it should be noted that examining a plant in this fashion requires its own kind of practice and hard work.

Concentration on the details of plants soothes the soul: 'J'y trouverais cette précieuse *sérénité d'âme* que donne la contemplation des merveilles qui nous entourent, et, que j'en devinsse ou non meilleur botaniste, j'en deviendrais à coup sûr et plus *sage* et plus *heureux* [...] Plus l'esprit s'éclaire et s'instruit, plus le cœur demeure plus *paisible*'.[36] The Roman Stoic, Seneca, asked: 'What is a happy life? Peacefulness and constant tranquility'.[37] Rousseau universalises this environmental therapy: 'à tout âge l'étude de la nature émousse le goût des amusements frivoles, prévient le *tumulte des passions* et porte à *l'âme* une nourriture qui lui profite en la remplissant du plus digne objet de ses contemplations.'[38]

iii. A metaphysics of botany: paradise regained?[39]

> Le sage Wolmar [...] observait-il si mal la nature? Ignorait-il que si son Auteur est grand dans les grandes choses, il est très grand dans les petites?[40]
>
> [I]l n'est pas aisé d'imaginer un jardin mieux assorti de plantes que celui d'Eden[41]

Rousseau averred that 'le premier Apothicaire fut Adam'.[42] If Adam was the first apothecary, then the Author of nature must have been the first botanist; while the first may have known how to use the plants, the second knows the plants themselves because he made them. This distinction is critical to understanding how botany facilitates detachment from the passions; it points to the divine force at work in the world:

35. Rousseau, *Confessions*, *OC*, i.643.
36. My emphasis; Rousseau to the Duchess of Portland, 3 September 1766; *CC*, xxx.314.
37. Seneca, *Letter* 92.3, LS 63F, in *The Hellenistic philosophers: translations of the principal sources with philosophical commentary*, ed. A. A. Long and D. N. Sedley (Cambridge, 1987), vol.1, p.396. See also Robert Mauzi, *L'Idée du bonheur dans la littérature*, p.16.
38. My emphasis; Rousseau à Mme Delessert, 22 août 1771; *CC*, xxxviii.251.
39. Renaissance gardens such as that of Padua attempted to present a vision of nature encompassed in a garden. John Prest, *The Garden of Eden: the botanic garden and re-creation of paradise* (New Haven, CT, 1981), p.29ff, 38ff.
40. Rousseau, *Julie, ou La Nouvelle Héloïse*, *OC*, ii.482, note.
41. Rousseau, *Rêveries*, *OC*, i.1064.
42. Rousseau, *Rêveries*, *OC*, i.1064. Rousseau did on one occasion admire Adam's knowledge: 'Je suis tenté d'essayer de la botanique, non [...] en grand et comme une branche de l'histoire naturelle; mais tout au plus en garçon *apothicaire*, pour savoir faire ma tisane et mes bouillons', Rousseau to Malesherbes; 11 November 1764; *CC*, xxii.44.

'L'étude de la nature nous détache de nous-mêmes et nous élève à son *Auteur*. C'est en ce sens qu'on devient vraiment *philosophe*; c'est ainsi que l'histoire naturelle et la *botanique* ont un usage pour la *sagesse* et pour la *vertu*'.[43] The many benefits of botanical study, even those that seem secular or personal, derive from plants' truth status: 'la nature qui ne ment jamais'.[44] This truth status in turn derives from their creation by the Author of things, not by 'hommes qui sont menteurs'.

The ultimate justification for studying botany therapeutically relies upon Rousseau's metaphysical assumptions, which he outlined most notably in the *Profession de foi du vicaire savoyard*, and confirmed in the *Rêveries*.[45] He understood there to be a First Cause, the Author of nature who has set in place final ends.[46] Restating the argument from design, Rousseau attested to nature's divine origins:[47] 'chaque pièce est faite pour les autres, j'admire l'ouvrier dans le détail de son ouvrage'.[48] Studying nature therefore promises to reveal the munificence of the author of nature:

> Comparons les *fins* particulières, les moyens, les rapports ordonnés de toute espèce, puis écoutons le sentiment intérieur; quel esprit sain peut se refuser à son témoignage, à quels yeux non prévenus *l'ordre sensible* de l'univers n'annonce-t-il pas une suprême intelligence, et que de sophismes ne faut-il point entasser pour méconnaître l'harmonie des êtres et l'admirable concours de chaque pièce pour la conservation des autres?[49]

In botany Rousseau likewise invoked '[l]e suprême ouvrier, attentif à la conservation de tous les êtres, [qui] a mis de grands soins à garantir la fructification des plantes des atteintes qui lui peuvent nuire'.[50] Plants reveal 'l'art *divin*' et le 'goût exquis de *l'ouvrier* qui fabriqua la robe de

43. My emphasis; Rousseau to the Duchess of Portland, 3 September 1766; *CC*, xxx.314. One's 'regard est attiré "vers le haut"', Foucault, *L'Herméneutique du sujet*, p.476.
44. Rousseau, *Fragments de botanique*, *OC*, iv.1250; see also Rousseau, *Discours sur l'origine de l'inégalité*, *OC*, iii.133, Rousseau, *Dialogues*, *OC*, i.833.
45. Rousseau, *Rêveries*, *OC*, i.1018.
46. See Cantor, 'The metaphysics of botany'. This article is based solely on the *Rêveries*, and is therefore of limited use for an analysis of Rousseau's position on teleology, which has to be teased out of many sources, including his correspondence. See also Peter Emberley, 'Rousseau versus the Savoyard vicar: the profession of faith considered', *Interpretation* 14:2-3 (1986), p.299-329.
47. The argument from design can be broken down into the argument to complexity (the world has a special, complex nature) and the argument to design (complexity implies a creator); proponents of these positions have always found their firmest support in the 'organised complexity' of living nature. The argument from design has recently been revived by intelligent design theory. See Michael Ruse, *Darwin and design: does evolution have a purpose?* (Cambridge, 2003), p.15-17, 315-36.
48. Rousseau, *Emile*, *OC*, iv.578.
49. My emphasis; Rousseau, *Emile*, *OC*, iv.578; note also his teleological reference to 'la fin'. Rousseau, *Rêveries*, *OC*, i.1069.
50. Rousseau to Mme Delessert, 16 May 1772; *CC*, xxxix.53.

notre mère commune'.[51] Impressed by this divine skill, Rousseau rejected the banal exercise of identifying useful ends in nature in favour of 'la botanique [...] une étude de pure curiosité et qui n'a d'autre utilité réelle que celle que peut tirer un être pensant et sensible de l'observation de la nature et des *merveilles* de l'univers'.[52]

These marvels include analogy, symmetry and sheer variety surpassing anything the human mind can grasp. For those able to see it, a chain of relations holds these marvels together:

> Cette constante *analogie* et pourtant cette *variété* prodigieuse ne transporte que ceux qui ont déjà quelque idée du système végétal. Les autres [...] ne voient rien en *détail*, parce qu'ils ne savent pas même ce qu'il faut regarder, et ils ne voient pas non plus l'ensemble, parce qu'ils n'ont aucune idée de cette *chaîne* de rapports[53] et de combinaisons qui accable de ses merveilles l'esprit de l'observateur'.[54]

Similarly, 'c'est dans cette succession qu'on voit des *métamorphoses* et un *enchaînement* de *merveilles* qui tiennent tout esprit sain qui les observe dans une continuelle *admiration*'.[55] Hence, the therapeutic distraction provided by botany arises from the range, interconnectedness, and fascination of the wonders created by the Author of things.

Botany approached in this way requires 'productions spontanées que la terre non forcée par les hommes offre [...] de toutes parts'.[56] This means plants whole, alive, and unaltered by human intervention, not pulverised, and unrecognisable.

> Les végétaux dans nos bois et dans nos montagnes sont encore tels qu'ils Sortirent originairement de[s] mains [de l'Auteur de nature], et c'est là que j'aime à étudier la nature; car je vous avoue que je ne sens plus le même charme à herboriser dans un *jardin.* Je trouve qu'elle n'y est plus la même; elle y a plus d'éclat, mais elle n'y est pas si *touchante.* Les hommes disent qu'ils l'embellissent, et moi je trouve qu'ils la *défigurent.*[57]

51. My emphasis; Rousseau, *Fragments de botanique*; *OC*, iv.1249.
52. My emphasis; Rousseau to Mme Delessert, March/April 1774; *CC*, xxxix.234. '[T]he whole point of "natural philosophy" was to look at nature [...] as created by God, and thus as capable of being understood as embodying God's powers and purposes'; natural philosophy 'was an autonomous study separate from theology and natural theology, but whose practitioners had at the forefront of their minds [...] the same God whose attributes the theologians studied', Cunningham and Williams, 'De-centring the "big picture"': p.421.
53. The 'chaîne' was a common eighteenth-century way to portray the beings of the living world as a seamless hierarchy in which there could be no gaps. See Arthur O. Lovejoy, *The Great chain of being: a study of the history of an idea* (Cambridge, MA, 1936). I discuss this idea further in Chapter 6 in connection with the natural method of Antoine Laurent de Jussieu. My emphasis.
54. My emphasis; Rousseau, *Confessions*, *OC*, i.641; see also Rousseau, *Rêveries*, *OC*, i.1062.
55. My emphasis; Rousseau to Mme Delessert, 2 May 1773; *CC*, xxxix.157.
56. Rousseau, *Rêveries*; *OC*, i.1070.
57. My emphasis; Rousseau to the Duchess of Portland, 12 February 1767; *CC*, xxxii.135.

Nature's true traits emerge only in the wildest places: 'elle fuit les lieux fréquentés; c'est au sommet des montagnes, au fond des forêts, dans des îles désertes qu'elle étale ses charmes les plus touchants'.[58] In such deserted places a realm full of 'riants objets' is to be found,[59] untainted by human power relations.[60]

Horticulture, on the other hand, concerns itself with an unsmiling second nature shot through with social relations and amour propre;[61] horticulture produces monsters such as sterile double flowers:

> Tant que vous les trouverez doubles, ne vous attachez pas à leur examen; elles seront *défigurées*, ou si vous voulez, parées à notre *mode*, la nature ne s'y trouvera plus: elle refuse de se reproduire par des *monstres ainsi mutilés*; car si la partie la plus brillante, savoir la corolle, s'y multiplie, c'est aux dépends des parties plus essentielles, qui disparaissent sous cet éclat.[62]

Gardens, which had traditionally offered a space for contemplation and spiritual renewal, are for Rousseau human creations fraught with human concerns, and therefore play no role in his salutary botany. He accordingly expresses disdain for them: 'Ceux qui veulent embellir et parer la nature sont des gens sans âme et sans goût qui n'a jamais connu ses beautés';[63] 'Je connais quelque chose à l'ouvrage de la nature, mais rien à celui du jardinier';[64] '[Q]uand dans les œuvres de ses mains [l'homme] croit étudier vraiment la nature, il se trompe. Cette erreur a lieu surtout dans la société civile, elle a lieu de même dans les jardins'.[65] Rousseau rejected the typical European garden, proclaiming it a disfigurement of nature in the service of vanity, curiosity and the 'manie' which has 'quelque chose de petit et de vain qui le rend puéril et ridiculement coûteux'.[66] He even criticised the famous garden of Richard Temple, first Viscount Cobham (1675-1749) at Stowe, widely considered a model of the new naturalistic landscape garden (fig.1.1, p.20).[67]

58. Rousseau, *Julie*, *OC*, ii.479.
59. Rousseau, *Rêveries*, *OC*, i.1066.
60. Rousseau, *Rêveries*, *OC*, i.1073.
61. According to Wolmar, '[l]'air grand est toujours triste'. Rousseau, *Julie*; *OC*, ii.480.
62. My emphasis; Rousseau to Mme Delessert, 18 October 1771; *CC*, xxxviii.277. See also Chandra Mukerji, *Territorial ambitions and the gardens of Versailles* (Cambridge, 1997), p.171: 'In this period [...] Great gardens were like fine clothing for the countryside'.
63. Rousseau is commenting on the Swiss author Solomon Gessner's *Idyllen*, which Michael Huber translated into French. Rousseau to Huber, 24 December 1761; *CC*, ix.350.
64. Rousseau, *Confessions*, *OC*, i.643. This is an exaggeration since Rousseau related his gardening activities at Les Charmettes in *Confessions*, *OC*, i.231.
65. Rousseau to Mme Delessert, April 1774; *CC*, xxxix.234.
66. This is Wolmar's response to St Preux's description of London's 'fleuristes'. Rousseau, *Julie*, *OC*, ii.481.
67. Rousseau, *Julie*, *OC*, ii.484. Dates are supplied for lesser-known figures. Please also see Appendices 1 and 2.

Figure 1.1: Jacques Rigaud, *The Queen's Theatre from the Rotunda, Stowe House, Bucks, with Lord Cobham and Charles Bridgeman*, 18th century, private collection/© Christie's Images/The Bridgeman Art Library.

Figure 1.2: Constant Bourgeois (drawing) and Guyot (etching), *L'Etang du désert, à Ermenonville*, from Alexandre Louis Joseph Laborde, *Description des nouveaux jardins de la France et de ses anciens châteaux* (Paris, Delance, 1808) by permission of the Bibliothèque publique et universitaire de Neuchâtel.

The horticulture of exotic botany, then the rage, was particularly suspect, since exotics and horticultural monstrosities are contrary to nature, deformed and exiled like slaves. Europeans sought exotic plants for their uses as condiments (spices), beverages (coffee, tea), drugs (quinine, ginseng, nicotine), clothing (silk, hemp) and much else; the source plants or their seeds for these products were often brought to Europe for naturalisation which sometimes succeeded. In other cases, plants were naturalised in the colonies, as in the case of the breadfruit's infamous transfer from Tahiti to the Caribbean by the British.[68] Entire empires were built on the trade in these plant products.

Despite the economic interests at stake, contemporary botanical literature concurred that exotics were altered through transplantation;[69] on this basis Rousseau objected to exotic botany: 'dans ma plus grande passion pour la botanique, content du foin que je trouvais sous mes pas je n'eus jamais de goût pour les plantes étrangères qu'on ne trouve parmi nous qu'en exil et dénaturées, dans les jardins des curieux'.[70] Exotic botany takes on the degeneracy of its object of study, and 'dégénère pas moins que les plantes exotiques dans les jardins des curieux'.[71] This is not the divine botany of the Creator, but rather the vain science condemned in the first *Discours*, which brings only evils in its wake. The true botanist studies the plants themselves as did Theophrastus, the one true botanist of antiquity.[72]

iv. A realm of freedom?

In contrast with the gardens of the day, Rousseau depicted in Julie's Elysium a protected zone free of exotic transplants, containing only indigenous plants: 'si je ne trouvai point de plantes *exotiques* et de productions des *Indes*, je trouvai celles du pays disposés et réunies de manière à produire un effet plus riant et plus agréable'.[73] Elysium can arguably claim to be a new Eden,[74] the ultimate anti-garden which shows

68. The first attempt to do this was by William Bligh (1754-1817), captaining the ill-fated *Bounty* in 1787. Bligh succeeded on the second breadfruit voyage in 1791-1793.
69. See for example Nikolaus Joseph Freiherr von Jacquin, 'Prefatio', *Selectarum stirpium americanarum historia* (Vienna, ex officina Krausiana, 1763).
70. Rousseau to the Duchess of Portland, 11 July 1776, *CC*, xl.77. Nature also effects transplantations, for example by the movement of wind or birds, which Rousseau, to my knowledge, does not discuss.
71. My emphasis; Rousseau, *Rêveries*, *OC*, i.1070.
72. Rousseau, *Rêveries*, *OC*, i.1063.
73. My emphasis; Rousseau, *Julie*, *OC*, ii.472; see A. Cook, 'Jean-Jacques Rousseau and exotic botany', *Eighteenth-century life* special issue 26:3 (Fall 2002, distributed Spring 2003): *Exoticism and the culture of exploration*, p.181-201.
74. For the view that Elysium is an 'Eden retrouvé', see Jean-Michel Racault, 'L'utopie narrative en France et en Angleterre, 1675-1761', *SVEC* 280 (1991), p.722. See the view

no trace of human intervention: 'la nature a tout fait, mais sous [l]a direction' of Julie, displaying 'point du travail humain';[75] there 'le main du jardinier ne se montre point'.[76]

In fact, Elysium is not a garden, but rather an 'azile', a 'desert', and a 'verger' in which trees, their plant parasites and water abound (fig.1.2). In Elysium art imitates nature exactly. Julie has produced a simulacrum of indigenous nature by using plants 'du pays disposées et réunies de manière à produire un effet plus riant et plus agréable'.[77] Filled with trees and water, Elysium bears more than a passing resemblance to the description of Eden in the Bible.[78] The knowledge of how to stock Elysium comes from nature itself; this is the divine botany that created Paradise.

Elysium recreates a space in which birds, plants, and indeed all beings, flourish on their own terms, according to their *habit* which, as we shall see, is a key unifying principle of Rousseau's botanical teaching (discussed in Chapter 6, p.176-77, p.191-98). Each plant is free in the negative sense of being free from interference. The birds, too, are free to come and go; it is they who are the 'maîtres', tolerating human intruders 'quelquefois'.[79] Instead of nature encumbered, manipulated and confined, Elysium offers a refuge from social relations of domination and exploitation.

Yet it has to be acknowledged that Julie's Elysium passes the key test of garden-hood: it is bounded. The term 'garden' denotes a closed environment; the various European words for garden refer to encirclement and *enclosure*, hence to control and confinement.[80] Like all gardens, Elysium has a boundary, however difficult to discern.[81] The same was true for Eden, as Adam and Eve discovered when they found themselves outside the perimeter of the only home they had ever known.[82]

---

that Elysium 'n'offre guère les apparences de l'Eden'. Geneviève Goubier-Robert, 'L'Elysée: leçon de morale dans un parc', *Annales de la société Jean-Jacques Rousseau: Jean-Jacques Rousseau et les arts visuels* (2003), p.113-29 (124-25).

75. Rousseau, *Julie*, *OC*, ii.472.
76. Rousseau, *Julie*, *OC*, ii.478.
77. Rousseau, *Julie*, *OC*, ii.472.
78. Andrew Cunningham, 'The culture of gardens', in *Cultures of natural history*, ed. N. Jardine, J. A. Secord, and E. C. Spary (Cambridge, 1996), p.38-56 (39).
79. Rousseau, *Julie*, *OC*, ii.475-76.
80. These include *jardin* (Fr.), *Garten* (Ger.), and *gorod* (Russian). The root, **garda-* ('Haus als umzäunter Besitz' ['house as fenced property']), is related to other German words referring to enclosure such as Gürtel and Gurt (belt, girdle). See *Wahrig Deutsches Wörterbuch* (Munich, 1986), under 'Garten.'
81. Rousseau, *Julie*, *OC*, ii.471.
82. Cunningham, 'The culture of gardens', p.39.

v. Agriculture and the good society

> Linnaeus [...] a un peu tiré la botanique des écoles de pharmacie pour la rendre à l'histoire naturelle et aux usages économiques.[83]

In rejecting the garden, Rousseau did not reject agriculture and animal husbandry, which occupy a class by themselves. These activities support good societies such as Clarens, the Switzerland of yesteryear and the Corsica of the future; they are the primary concern of true citizens. Yet, at the same time, Rousseau did not see his study of botany as contributing to a Baconian nature-management system for medicine, food, raiment, shelter or profit; this role was reserved to agriculture and horticulture. Taken to its logical extreme, this Baconian agenda yields biotechnology and genetically modified organisms; Sir Francis Bacon's dream is Rousseau's nightmare.[84]

The mid-eighteenth-century economic school known as Physiocracy, and what was called 'Kameralistik' in German-speaking lands, epitomised Bacon's approach to nature. The Physiocrats, led by Dr François Quesnay (1694-1774) and Victor Riqueti, the Marquis de Mirabeau (1715-1789), held land to be the premier source of a nation's prosperity, and advocated improvements in agronomy, taxation and market liberalisation to benefit both farmers and the nation as a whole; while Rousseau disagreed with some of the Physiocrats' ideas, for example their desire to abolish the feudal labour tax, known as the 'corvée', he shared their respect for agriculture.[85]

Rousseau praised the agricultural life in *Julie, ou La Nouvelle Héloïse*, the *Projet de constitution pour la Corse* (mid-1760s) and *Considérations pour le gouvernement de Pologne* (1771-1772?).[86] He told the Corsicans, '[l]e seul moyen de maintenir un état dans l'indépendance des autres est l'agriculture'.[87] For Rousseau, agriculture is not only the key to a nation's prosperity (in contrast with the luxury trades, whose products should be

83. Rousseau, *Rêveries*, *OC*, i.1064.

84. A. Cook, 'Jean-Jacques Rousseau, "terminator" and telos in nature', *Pensée libre* 8 (2001), p.311-25 (312).

85. Rousseau praised the corvée in Switzerland for reducing the circulation of money, and advocated its adoption in Poland. He also praised the Swiss practice (except in the canton of Bern) of paying officials in kind. See *Considérations pour le gouvernement de Pologne*, *OC*, iii.1006.

86. On Rousseau's political economy, and the Swiss case, see Michael Sonenscher, 'French economists and Bernese agrarians: the marquis de Mirabeau and the economic society of Berne', *History of European ideas* 33:4 (December 2007), p.411-26 and Béla Kapossy, 'Republican political economy', *History of European ideas* 33:4 (December 2007), p.377-89. Kapossy's discussion of Rousseau's political economy ignores Rousseau's advice to the Corsicans and the Poles.

87. Rousseau, *Projet de constitution pour la Corse*, *OC*, iii.905.

heavily taxed), but likewise the source of a nation's morality: for example, the early Romans' 'mœurs simples' were correlated with their 'goût pour l'agriculture'.[88]

In small states, the practical common sense of peasants is an infallible guide–witness the wisdom of the inhabitants of the poorer Swiss cantons: 'troupes des paysans règle[nt] les affaires de l'Etat sous un chêne et se conduisent toujours sagement'.[89] The relative equality of farming peoples makes democracy, otherwise impossible, a feasible option. Pastoral and farming peoples rule themselves well because they are constantly occupied, like those who work on the estate of the Wolmars in *Julie*. They have little or no time for vice, gossip or trouble-making, while their outdoor life is inevitably a healthy one; they make good soldiers, and preserve the vigour that is indispensable to the *patrie*,[90] 'La culture de la terre forme les hommes patients et robustes tels qu'il les faut pour faire des bons soldats'.[91] Thus farming should be made honourable, the corvée retained and magistrates paid in kind from the land.

However, the cultivation of the land is not divine botany. The farmer and the horticulturalist take no philosophical interest in plant life, nor should they; they do not need to experience wonder when they watch their wheat or barley grow. Rather, they need to know how to produce the best and most abundant crop they can. Their agronomic studies should therefore not be confused with botany; these studies have their proper place in social policy, in constituting the bucolic life celebrated in *Julie* and the best-selling paean of praise for this life in the famous physiocratic text, *Le Socrate rustique*.[92] The author of this surprise best-seller, the city physician of Zürich and correspondent of Rousseau, Johann Kaspar Hirzel (1725-1803), favoured intensive farming practices that would increase the productivity of the land and addressed the question of whether it is more economical to till the soil using oxen or horses (in response to Quesnay's *Encyclopédie* article, 'Fermiers'). Rousseau regarded the hero of the work, the Swiss peasant, 'Kleinjogg' (Jacob

88. Rousseau, *Du contrat social*, Book 4, ch. 4, *OC*, iii.448.

89. Rousseau, *Du contrat social*, *OC*, iii.437. On the link between the poverty of the people and democracy, see *Projet de constitution pour la Corse*, *OC*, iii.906. This same relationship existed in the Swiss Alpine republics, which were poor by virtue of the harsh terrain.

90. Rousseau, *Considérations pour le gouvernement de Pologne*, *OC*, iii.963.

91. Rousseau, *Projet de constitution pour la Corse*, *OC*, iii.905.

92. Johann Kaspar Hirzel, *Le Socrate rustique, ou Description de la conduite économique et morale d'un paysan philosophe*, trans. Jean Rodolphe Frey des Landres (Zürich, Heidegger, 1762; Lausanne, François Grasset, 1767). Originally published in German in 1761 as *Die Wirtschaft eines philosophischen Bauers*, it was translated into English, Italian and Russian. Arthur Young incorporated it into his *Rural œconomy* (London, T. Becket, 1770), whence it enjoyed a huge success, including in the United States. See abbé Jean Baptiste François Rozier, *Cours complet d'agriculture*, vol.6 (Paris, Delalain fils, 1787), p.111-31, under 'Kiloogg' on Kleinjogg's principles of rural economy.

Gujer, 1716-1785), as 'un paysan plus sage, plus *vertueux*, plus sensé que tous les *philosophes* de l'univers', an exemplar of the agrarian life extolled in his own works.[93]

vi. Conclusion

Botany for Rousseau is neither horticultural nor agricultural. It is not an escape either; it is a *quest* – a quest for freedom and Divine munificence. Botany not only reveals truth; it reveals natural freedom: life according to one's species being, without being forced or dominated. Instead of the products of horticulture and agriculture, Rousseau sought 'nature entière, et son inconcevable *auteur* [...] où rien ne montrant la main des hommes n'annonçat la servitude ou domination'.[94] Rousseau's botany studied a domination-free realm: plants that come direct from 'celui qui [...] fit [les hommes] pour être bons, et dont ils ont si indignement dégradé l'ouvrage'.[95] These plants are liberated like the birds of the air in Julie's Elysium from the denaturing, crushing hand of man in the throes of amour-propre, keen to use nature to display his wealth or status.

Botany not only shows Rousseau free beings, it allows him to be one: 'Ne pouvant laisser ma tête vide je la veux empailler, c'est de foin qu'il faut l'avoir pleine pour vivre *libre* et *vrai*, sans crainte d'être décrété'.[96] The manner of studying this kind of botany is inherently free – it is the poor man's science, with no strings attached, accessible to anyone with some leisure and sufficient funds to purchase a few simple, inexpensive instruments of 'facile usage';[97] botany has thus long enjoyed a popular or democratic status, celebrated during the French Revolution. This marks it out from the costly, elite science of chemistry, which Rousseau studied, explicated, and eventually rejected. It is to this chemical episode that I turn in the next chapter.

93. My emphasis; Rousseau to Michael Huber, 24 December 1761; *CC*, ix.350. The Reformed minister, Leonhard Usteri, wrote Rousseau a long letter about Gujer, so Rousseau was relatively well-informed about the work before receiving his own copy of the second edition of the French translation (including a letter from Mirabeau) in 1764 from Hirzel. Usteri to Rousseau, 20 November 1761; *CC*, ix.254-64. See also Paul H. Johnstone, 'The rural Socrates', *Journal of the history of ideas* 5:2 (April 1944), p.151-75.

94. My emphasis; Rousseau to Malesherbes, 26 January 1762; *CC*, x.53-54.

95. Rousseau to the Duchess of Portland, 12 February 1767; *CC*, xxxii.135.

96. My emphasis; Rousseau to Charles Pinot Duclos, 2 December 1764; *CC*, xxii.148.

97. Rousseau, *Rêveries*, *OC*, i.1069.

2. Rousseau's botany: the chemical background

> [N]i le soleil, ni tous les astres, ni tout le feu, ni tout le mouvement qui existent dans l'univers ne sont pas capables de produire la moindre de toutes les plantes [...] cet abîme de la génération[1] dans lequel les Philosophes se sont si longtemps perdus est encore aujourd'hui le désespoir des incrédules [...] Ces observations suffisent pour me montrer le point ou doivent commencer mes recherches.[2]

i. Introduction

It is well-known that Rousseau studied chemistry; yet the significance, content and duration of his chemical pursuits have been largely overlooked for a variety of reasons. First, the manuscript of Rousseau's unpublished *Institutions chymiques* was not discovered until the 1880s, and has only recently been included in standard, edited version of his writings in French. Second, his writings on chemistry were early on marked out by the editor of the *Institutions chymiques*, Maurice Gautier, as derivative work, and not even in chemistry, but in alchemy – a term that Gautier no doubt applied pejoratively.[3] Théophile Dufour, who discovered the manuscript of the *Institutions chymiques*, called for experts in eighteenth-century chemistry to analyse the work's merits,[4] but this call was not answered until the late 1990s when a volume of the French

1. The two reigning theories of generation were preformationism and spontaneous generation; most Academicians were dubious about spontaneous generation, yet 'some of the best minds continued to accept that plants and animals could be generated spontaneously'. Alice Stroup, *A Company of scientists: botany, patronage, and community at the seventeenth-century Parisian Royal Academy of Sciences* (Berkeley, CA, 1990), p.147. Rousseau later attacked materialism, which was usually not inconsistent with mechanistic understandings of nature: 'Montrez-moi clairement et sensiblement la génération purement matérielle du premier être intelligent, je ne vous demande rien de plus', Rousseau, *Lettres morales* 3, *OC*, iv.1096.
2. Jean-Jacques Rousseau, *Institutions chimiques*, ed. Christophe van Staen (Paris, 2009), p.100-101.
3. However, 'alchemy' in the seventeenth century also referred to the normal activities and products of chemistry; most of the latter were found in standard pharmacopeias: Ursula Klein and Wolfgang Lefèvre, *Materials in eighteenth-century science: a historical ontology* (Cambridge, MA, 2007), p.17.
4. Théophile Dufour, *Les Institutions chymiques de Jean-Jacques Rousseau* (Geneva, 1904, 1905), p.22-23.

philosophical journal *Corpus* examined this neglected and, by some accounts not so early, work of Rousseau.[5]

If Rousseau's chemistry has been neglected, the investigation of the chemical roots of his botanical interests has been even more so. Yet, as I shall show in this chapter, the origins of Rousseau's botanical pursuits can be traced to his time with Maman (1728-1742), and his early years in Paris (1740s) when he undertook an intensive study of chemistry. Chemistry opened up a particular aspect of botany: the properties (especially medical) and components of plants such as oils and salts. Rousseau was most certainly au fait with these investigations. Furthermore, while he was certainly unaware of it at the time, Rousseau's approach to chemistry contained within it the seeds of his eventual rejection of it in favour of botany. Thus, from being a youthful undertaking enthusiastically engaged in, chemistry eventually provided a straw man for Rousseau's later reflections on botany as the best science, which he defines *against* chemistry. This choice can only be fully understood when viewed against the background of eighteenth-century chemistry – its goals, traditions, assumptions, taxonomies, practices and instruments.

This chapter accordingly makes three arguments: (1) Rousseau's study of chemistry, especially plant chemistry, set the stage for his plant studies; (2) in studying chemistry Rousseau developed a critique of mechanics (physics) as all-explanatory; some phenomena are not susceptible to explanation by the laws of motion or by reductionist laboratory procedures; and (3) chemistry became a straw man for Rousseau's advocacy of botany as an ideal science.

ii. Pharmaco-herbalism *chez* Mme de Warens

Rousseau first became acquainted with plant-based pharmacology, then a branch of chemistry, when he entered the home of Mme de Warens, his 'Maman', at the age of sixteen. Much has been written about the significance of Rousseau's fateful encounter in 1728 with Françoise-Louise de la Tour, Mme de Warens (1699-1762), a highly intelligent, non-conforming aristocrat, who undertook to convert him to Catholicism, educate him and set him up in life (fig. 2.1, p.29).

Maman encouraged Rousseau's efforts at self-education and he assisted in her herbal-pharmacological business, which she conducted with her valet, Claude Anet (1697-1734). Anet was 'un paysan de Moutreux qui

5. Marco Beretta, 'Sensiblerie vs. mécanisme: Jean-Jacques Rousseau et la chimie', *Corpus* 36 (1999), p.103-20; B. Bensaude-Vincent and B. Bernardi, 'Pour situer les *Institutions chymiques*', *Corpus* 36 (1999), p.5-38; see Dufour, *Les Institutions chymiques*, p.6-8. Dufour did not believe that the *Institutions chymiques* could be the same 'manuscrit de Chimie' that Rousseau discussed with baron d'Holbach in 1757. See Rousseau, *Confessions*, *OC*, i.461.

Figure 2.1: Achille Jacques Jean Marie Deveria (artist) and Antoine-Joseph Chollet (engraver), *Louise-Eleonore de Warens*, coloured copper engraving, by permission of the Bibliothèque publique et universitaire de Neuchâtel.

dans son enfance herborisait dans le Jura pour faire du thé de Suisse [...] Il se passiona si bien pour l'étude des plantes [...] qu'il devint un vrai botaniste, et qu'il ne fut mort jeune il se serait fait un nom dans cette science'.[6] Rousseau's relationship with Anet was complicated, however; when Rousseau arrived at Maman's Anet was also her lover and seems to have perceived Rousseau as a rival. Presumably for that reason he tried to commit suicide by swallowing laudanum. Maman's concern to save Anet's life revealed the intimate relationship between mistress and servant. For Rousseau; Anet served as 'une espéce de gouverneur', who saved him from his 'folies', but Rousseau was, like Anet, jealous and could never wholly accept his rival's intimate relations with Maman.[7] Rousseau claimed that those who loved Maman loved each other: 'Une des preuves de l'excellence du caractère de cette aimable femme est que tous ceux qui l'aimait s'aimaient entre eux', but the claim has a disingenuous ring, given Rousseau's insensitive conduct after Anet's premature demise.[8]

With his introduction into Maman's household, which doubled as a pharmaceutical laboratory, Rousseau was immediately immersed in the 'activité médico-chimique' of commercial production of herbal remedies in which she and Anet were engaged.[9] During this period (1728-1742) Rousseau became involved with practical chemistry, its instruments, vocabulary and operations.[10] That the site of this introduction was a private household, involved in a cottage industry, makes the chemical nature of the activity no less valid, for '[t]he bulk of the processed substances studied by eighteenth-century chemists – in particular in the first half of the century – were neither found in nature nor were they the outcome of disinterested academic investigations. Rather, they were products long known in mining and metallurgy, in distilleries and apothecary's shops.'[11] Furthermore, 'chemistry itself was not restricted

6. Rousseau, *Confessions*, *OC*, i.177.
7. Rousseau, *Confessions*, *OC*, i.177.
8. Rousseau, *Confessions*, *OC*, i.178.
9. Rousseau esteemed a leading home remedy that he probably knew from his time with Maman: 'L'eau de la reine de Hongrie m'a fait grand plaisir', Rousseau to Davenport, 9 February 1767; *CC*, xxxii.121. On 'Hungary water', see Elaine Leong, 'Making medicines in the early-modern household', *Bulletin of the history of medicine* 82 (2008), p.153. For the recipe, consisting of an infusion of rosemary and spirit of wine, see Nicolas Lémery, *Pharmacopée universelle* vol.2 (Paris, De Saint & Saillant, 1763-1764), p.839. It supposedly treated a wide range of maladies (p.123).
10. Bensaude-Vincent and Bernardi, 'Pour situer les *Institutions chymiques*', p.9, 14. On the artisanal dimension of Rousseau's chemistry, as evidenced in his treatment of glass, glassmakers' secrets and discourse, see Jonathan Simon, 'L'homme de verre? Les trois règnes et la promiscuité de la nature', *Corpus* 36 (1999), p.65-80.
11. Klein and Lefèvre, *Materials*, p.16.

to academies, medical faculties, places of public and private teaching, botanical gardens, museums and other institutions of learned inquiry'.[12]

Maman's pharmacological enterprise was not only chemical, but also botanical, because herbal remedies used plant materials: 'elle ne recherchoit que les plantes usuelles pour les appliquer à ses drogues'.[13] Rousseau was exposed to the typical eighteenth-century treatment of plants as sources of pharmaceutical substances, not as objects of interest in and of themselves: 'The plants and most plant materials studied by eighteenth-century chemists belonged to the world of commodities, production and trade.'[14]

In early-modern Europe it was not uncommon for women of Maman's social status to spend 'considerable time in collecting the know-how to produce medicaments, and in making up various medicines. Their interest in medicine production can be seen in the large numbers of manuscript recipe collections [...] in major libraries and county record offices' in the United Kingdom; such practices also flourished on the Continent;[15] 'The process of producing medicaments was time-consuming and labour-intensive. A glance at contemporary recipes shows that many of them require multiple steps to be performed over several days': 'Boiling was the most common process used in the production of household medicines [...] Other common processes include the pounding of herbs, and the melting and mixing of fats such as butter and grease with herbs for salves and plasters.'[16]

Another important feature of early-modern domestic medicine production was its close interaction with the commercial market for medicines, principally through the use of commercial products in home remedies.[17] Maman was exceptional, however, because she produced medicines not only for domestic consumption, but also for the market.

Hence, Rousseau became aware not only of chemistry, but also of pharmaceutical and economic botany. His resulting conception of botany was negative and confused, reflecting the actual status of botany as primarily pharmacological: 'la botanique, la chimie et l'anatomie confondues dans mon esprit sous le nom de médecine, ne servaient qu'à me fournir de sarcasmes plaisants toute la journée, et à m'attirer des

12. Klein and Lefèvre, *Materials*, p.18.
13. Rousseau, *Confessions*, *OC*, i.180
14. Ursula Klein, 'Shifting ontologies, changing classifications: plant materials from 1700 to 1830', *Studies in the history and philosophy of science* 36 (2005), p.261-329 (264).
15. Leong, 'Making medicines', p.145-68 (146, 148, n.5).
16. Leong, 'Making medicines', p.152.
17. Leong, 'Making medicines', p.157.

soufflets de temps en temps'.[18] Thus, while occasionally drawn to botany, Rousseau was repelled by its pharmacological aspect, regarding it 'avec une sorte de mépris et même de dégoût [...] que comme une étude d'Apothicaire'. This was understandable, given Maman's purely instrumental interest in plants. Anet certainly had a good knowledge of plant names and classification, but theoretical studies of taxonomy or plant structure were not his primary concern.[19]

Rousseau not only assisted Maman and Anet with their pharmacological enterprises, but likewise undertook his own chemical experimentation, temporarily blinding himself in an experiment to produce sympathetic (i.e. invisible) ink. This incident led Rousseau to learn 'ainsi à ne pas me mêler de physique expérimentale sans en savoir les éléments',[20] and probably at least in part as a result of this accident he set out to learn chemistry properly, consulting the leading treatises on the subject, and sitting the courses of Rouelle.[21] In *Emile* he shows a respect for the science, teaching its practical side in a lesson on the dangers of food adulteration.[22]

While the profit-oriented exploitation of plants led Rousseau to dislike botany, Anet showed him another side to botany: 'le contentement que je voyois dans les yeux d'Anet revenant chargé de plantes nouvelles me mit deux ou trois fois sur le point d'aller herboriser avec lui'.[23] From the outset, therefore, it was not the pharmacological or useful side that attracted Rousseau to plant study, but rather, the 'contentement' derived from exploring the out-of-doors, and encountering the beauty and symmetry of plant forms and their relation to the 'spectacle' of nature as a whole. Later he wrote, 'Je suis presque assuré que si j'y avois été une seule fois cela m'auroit gagné, et je serois peut-être aujourd'hui un grand botaniste: car je ne connais point d'étude au monde qui s'associe mieux avec mes goûts naturels que celle des plantes, et la vie que je mène depuis dix ans à la campagne n'est guère qu'une herborisation continuelle.'[24] In reaction to this encounter with botany as an adjunct to pharmacy, Rousseau later took a rigorously theoretical, anti-medical, and anti-instrumental approach to the subject: instead of searching for 'simples' and other plant uses, he would examine plant characteristics, their comparison, and classification.

18. Rousseau, *Confessions*, *OC*, i.180.
19. Rousseau, *Confessions*, *OC*, i.205.
20. Rousseau, *Confessions*, *OC*, i.218.
21. Rousseau, *Confessions*, *OC*, i.218.
22. Rousseau, *Emile*, *OC*, iv.451-53.
23. Rousseau, *Confessions*, *OC*, i.180.
24. Rousseau, *Confessions*, *OC*, i.180.

iii. Paris, the Jardin du roi and the Académie royale des sciences

Rousseau next encountered the chemical study of plants in Paris; in this second exposure he again experienced the instrumental approach to plants that subjected them to decomposition and analysis for medical and pharmaceutical purposes.[25] Shortly after his arrival in Paris in 1742, he became involved with chemistry as it was taught and practised in the capital – in the Jardin du roi, the Académie royale des sciences and in private venues provided by the well-heeled.[26] In fact, Paris and chemistry were virtually synonymous for Rousseau: 'Revenir à Paris ce fut aussi revenir à la chimie, mais cette fois pour en faire une étude sérieuse et suivie.'[27] This activity was fuelled by relations of patronage in which Rousseau was expected to make himself useful.[28] Rousseau pursued the 'étude sérieuse et suivie' of chemistry with Dupin de Francueil, the stepson of his patron, Mme Dupin; Rousseau perceived that he was perhaps to serve as ghost-writer for his wealthy patron: 'M. de Francueil suivait alors l'histoire naturelle et la chymie et faisait un cabinet. Je crois qu'il aspirait à l'Académie des Sciences: il voulait pour cela faire un livre, et il jugeait que je pouvais lui être utile dans ce travail.'[29] On 11 March 1743 the two men began the renowned chemistry course of Guillaume-François Rouelle (1703-1770) at the Jardin du roi: 'M. de Francueil me prenait en amitié, je travaillais avec lui; nous commençâmes ensemble un cours de Chymie chez Rouelle.'[30] Rousseau and Francueil likewise set up

25. 'Much of the practice of chemistry in the seventeenth century was ancillary to the preparation of drugs. The pedagogical framework within which chemistry emerged as a discipline during this century remained strongly attached to its medical objectives [...] Even as they sought to define chemistry as an independent art seeking knowledge of the composition of natural substances, their definitions tended to coalesce with the goals of identifying the medical virtues of these substances and of improving drugs through knowledge of their underlying principles,' Frederic Lawrence Holmes, *Eighteenth-century chemistry as an investigative enterprise* (Berkeley, CA, 1989), p.61.
26. The Jardin royal des plantes médicinales was founded in 1626 by Guy de la Brosse (1589-1641, a physician of Montpellier, to support the King's medical needs. It became an important teaching institution that provided medical education free of charge, and fostered innovative thinking, following Montpellier's lead. Lucie Roule, 'Les médecins du Jardin du Roi: origine médicale du Muséum national d'histoire naturelle', doctoral dissertation, Faculty of Medicine, Paris (Paris, 1942), p.15-58; Edouard Brygoo, 'Les médecins de Montpellier et le Jardin du roi à Paris', *Histoire et nature* 14 (1979), p.3-29. During the Revolution, it became the Jardin des plantes of the Muséum national d'histoire naturelle.
27. Bensaude-Vincent and Bernardi, 'Pour situer les *Institutions chymiques*', p.16.
28. He later rejected such patronage as a limitation on his freedom with the potential to bias his views; he therefore rejected the pension offered to him by Louis XV. Rousseau, *Confessions*, *OC*, i.380.
29. Rousseau, *Confessions*, *OC*, i.341.
30. Rousseau, *Confessions*, *OC*, i.293; Rhoda Rappaport, 'G.-F. Rouelle: an eighteenth-century chemist and teacher', *Chymia* 6 (1960), p.68-101.

a laboratory at the Dupin family home of Chenonceau.[31] According to Rousseau:

> Je m'attachai à la Chimie. J'en fis plusieurs cours avec M. de Francueil chez M. Rouelle, et nous nous mîmes a barbouiller du papier tant bien que mal sur cette science dont nous possédions à peine les éléments. En 1747 nous allâmes passer l'automne en Touraine au Château de Chenonceau, maison royale sur le Cher, bâtie par Henri II pour Diane de Poitiers [...] J'y composai d'autres petits ouvrages [...] et tout cela se fit sans discontinuer mon travail sur la Chimie [...].[32]

Rousseau was able to pursue chemistry owing only to the great wealth of Dupin, the tax farmer's son; for the expense entailed in setting up a well-appointed laboratory restricted the study of this science only to the wealthy or to well-heeled apothecaries such as Nicolas Lémery (1645-1715?), Simon Boulduc and Claude Bourdelin, who were elected to the Académie as chemists.

Rousseau's teacher, Rouelle, played a key role in mid-eighteenth-century French chemistry. While he never published a work on chemistry, he was an important teacher and Academician whose expertise was sought by the government. His work is mainly known through manuscripts of his lectures made by students, and from direct reports by students such as Denis Diderot (1713-1784) and Antoine Laurent de Lavoisier (1743-1794). He gave the nascent field of chemistry a much-needed boost through his charismatic and eloquent teaching; while his originality is open to question, he did a considerable service to the sciences, lecturing not only on chemistry, but also on geology, in which he developed 'a general theory of geological stratification' and made 'plusieurs voyages lithologiques' with Bernard de Jussieu (1699-1777?), one of the fathers of the natural method of plant classification.[33]

Rouelle's teaching method was grounded in experiment, and included important quantitative elements such as temperature regulation of experiments and sealing vessels to avoid loss of reagents and products.[34] He is particularly important for introducing his students to the ground-breaking work of Stephen Hales in gas theory (photosynthesis), and by popularising the phlogiston theory of the seventeenth- and eighteenth-century German chemists, Johann Joachim Becher (1628-1685), Georg Ernst Stahl, and Johann Juncker (1679-1744). This theory, which held that all substances contain 'combustible earth' or 'phlogiston' (Stahl's neologism), is now known as a discredited reject in the dustbin of science, but at the time it constituted an important stage in the devel-

31. Bensaude-Vincent and Bernardi, 'Pour situer les *Institutions chymiques*', p.17.
32. Rousseau, *Confessions*, *OC*, i.342.
33. Rappaport, 'G.-F. Rouelle', p.96-97.
34. Rappaport, 'G.-F. Rouelle', p.98.

opment of chemical theory, and facilitated Lavoisier's discovery of oxygen.

Rouelle is equally important for having trained the next generation of chemists, including Lavoisier, Pierre-Joseph Macquer (1718-1784), author of the well-known *Elémens de chymie théorique* (1749) and Gabriel-François Venel (1723-1775), neo-Paracelsian professor of medicine at Montpellier and author of many articles for the *Encyclopédie*.[35] Rouelle likewise taught several individuals who did not become professional chemists, but who nevertheless used his teaching and ideas: Rousseau, Chrétien Guillaume de Lamoignon de Malesherbes (1721-1794), Anne Robert Jacques Turgot (1727-1781) and Diderot; the last became a particular friend of Rouelle. Rousseau's attendance at Rouelle's course supported his collaboration with Francueil, while Diderot went on to edit the *Encyclopédie*, in which chemistry, as a science with extensive industrial applications, was amply discussed both in texts and illustrations by Venel.

Rousseau's pursuit of chemistry at the Jardin du roi is extremely significant, because the Jardin was one of Europe's leading scientific institutions, where it was possible to receive an entire medical education gratis, including botany, pharmacology, anatomy and chemistry. Many leaders in the sciences taught there, including not only Rouelle, but also Bernard de Jussieu (1699-1777), the father of the natural method of plant classification that Rousseau used in his famous *Lettres* on botany to Madeleine-Catherine Delessert. Rousseau later had contact with other botanists at the Jardin du roi in the 1770s – Antoine Laurent de Jussieu (1748-1836), André Thouin (1747-1824), Louis-Jean-Marie Daubenton (1716-1800) and possibly his second cousin, Edmé-Louis Daubenton (1732-1785).[36]

iv. Chemistry in the Académie: the natural history of plants and beyond

Another leading centre of chemical work in Paris was the Académie royale des sciences ('Académie'), founded by Colbert in 1669 and reformed under new statutes in 1699.[37] Those Academicians working

35. Rousseau considered Venel an 'honnête homme' for whom he had 'la plus véritable estime', Rousseau to Moultou, 5 November 1768; *CC*, xxxvi.168-71; Rousseau to Moultou, 10 October 1768; *CC*, xxxvi.143 Venel would have been Rousseau's neighbour in Pézenas had the latter moved to the Domaine de Lavagnac in Languedoc, west of Montpellier, in 1768, as Moultou proposed. According to Rousseau's testimony, Venel assisted with this plan. Rousseau to Moultou, 12 December 1768; *CC*, xxxvi.202-203.

36. Rousseau, 'Botanical writings', in *Collected writings of Rousseau*, ed. A. Cook (Hanover, NH, 2000), vol.8, p.327, n.326.

37. Louis XIV, 'Règlement ordonné par le roi pour l'Académie royale des sciences', in Bernard le Bovier de Fontenelle, *Eloges des Académiciens, avec l'histoire de l'Académie royale des sciences*, facs. edn (Brussels, 1969 [The Hague, Isaac van der Kloot, 1740]), p.5-17.

in chemistry included apothecaries and botanists. Rousseau encountered the Académie in 1742 when he unsuccessfully sought official approval for his project to reform musical notation, *Projet concernant de nouveaux signes pour la musique*.[38] Despite his failure to win over the Academicians to his novel system of musical notation (which, he argued, they did not understand), Rousseau regarded the productions of this learned body as valuable in other fields. A manuscript in his hand lists sixty-six of the Académie's *Mémoires* and *Histoires* on chemistry: 'Catalogue de tous les Mémoires de la Chymie contenus dans tous les volumes de l'Histoire de l'Académie R[oyale] des sciences'.[39] These range from analysis by distillation, and solvents, to essential oils, volatile salts and the 'iron of plants'.[40] The most common topic (13 documents) is purgatives – a medical concern, followed by salts (9 documents), fermentation (another method of analysis: 5 documents) and essential oils (4 documents).[41]

Rousseau's list comprises fourteen authors, including many important figures in the history of chemistry: Guillaume (Wilhelm) Homberg (1652-1715), Simon (1652-1729) and Gilles-François Boulduc (1675-1741), Nicolas Lémery (1645-1715), René Antoine Ferchault de Réaumur (1683-1757), Etienne François (1672-1731) and Claude Joseph Geoffroy (1685-1752), Antoine de Jussieu (1686-1758), Claude Bourdelin (1621-1699) and Joseph Pitton de Tournefort (1656-1708). Why or when Rousseau drew up this list is not known; the first item listed is a memoire of 1699 by Guillaume Homberg on the salts of plants; the last entry is a 1734 memoire by Simon Boulduc's son, Gilles-François, entitled 'Essai d'Analyse des Plantes'.

The Académie played an important role in Rousseau's chemical and botanical studies; when he later speaks of 'l'analyse chimique encore plus fautive'[42] and 'travaux sédentaires de laboratoire',[43] it is most certainly to the practices of the Académie's chemists that he referred (figs 2.2 and 2.3). Claude Bourdelin, Simon Boulduc and Etienne François Geoffroy loomed large in plant chemistry. Much of the Académie's work in botany overlapped with chemistry, and vice versa; this was not due to a failure to distinguish these fields of endeavour from each other since these disciplines were already recognized as distinct in the Académie's statues

38. Rousseau, *Confessions*, *OC*, i.284ff.; Alexandra Cook, 'Rousseau and the languages of music and botany', *SVEC* 2004:08, p.75-87.
39. Bibliothèque publique universitaire de Neuchâtel, MsR 81.
40. [Bernard le Bovier de Fontenelle], 'Sur le fer des plantes', in *Histoire de l'Académie royale des sciences. Année MDCCVI. Avec les mémoires de mathématique & de physique, pour la même année, tirés des registres de cette académie* (Paris, Jean Boudot, 1707), p.38-40.
41. The complete list is found in BPUN MsR81, f.2-26.
42. Rousseau, *Fragments de botanique*, *OC*, iv.1249.
43. Rousseau, 'Introduction', *Fragmens pour un dictionnaire*, *OC*, iv.1201.

Figure 2.2: 'Laboratoire [et table des rapports]', drawing from Denis Diderot and Jean le Rond d'Alembert, *Encyclopédie ou Dictionnaire raisonné des sciences, des arts et des métiers*, The Project for American and French research on the treasury of the French language, 2012.

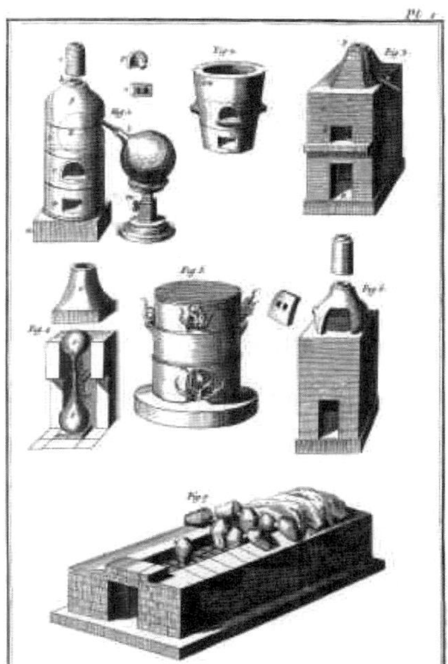

Figure 2.3: 'Planche 1, Des fourneaux, des vaisseaux & autres ustensiles du laboratoire de Chimie', including '1. Grand fourneau de réverbère pour la distillation des végétaux à feu nud. 2. Fourneau pour la distillation des substances végétales au bain-marie ou au degré de l'eau bouillante.' Engraving from Denis Diderot and Jean le Rond d'Alembert, *Encyclopédie ou Dictionnaire raisonné des sciences, des arts et des métiers*, The Project for American and French research on the treasury of the French Language, 2012.

of 1699.[44] The Académie's chemists understood the plant realm as a subject of chemical as well as botanical investigation: 'On voit assez combien il est important de connaître exactement le corps humain, et les remèdes que l'on peut tirer des minéraux, et des plantes.'[45] Hence it was necessary 'to identify the constituents of plants, to improve medicine, and to understand how plants nourish humans'.[46] Chemists interested in using chemistry in pharmacology or other medical applications saw the plant realm as an important source of potential medicines, especially the purgatives on which the medicine of the day depended.[47] Chemists such as Boulduc clearly demarcated their task of identifying the useful medicinal and other properties of plants from the botanist's task of description.[48]

> [E]arly eighteenth-century plant chemistry was not concerned with plants in general, but with a carefully selected group of plants. Chemists selected plants and plant materials that were applied as remedies, and, as comparison with the contemporary pharmacopoeias show, chemists carefully sorted out the chemical remedies they were interested in from the more traditional Galenic medicines and from the alchemical arcana considered as obscure or ineffective remedies.[49]

Not all chemists accepted this as their task, however: Venel, a famous student of Rouelle, wrote:

> Quelques autres, en bien plus grand nombre, restreignent l'idée de la *Chimie* à ses usages médicinaux: ce sont ceux qui demandent du produit d'une opération, de quoi cela guérit-t-il? Ils ne connaissent la *Chimie* que par les remèdes que lui doit la médecine pratique, ou tout au plus par ce côté et par les hypothèses qu'elle a fournies à la médecine théorique des écoles.[50]

44. Louis XIV, 'Règlement ordonné par le roi pour l'Académie royale des sciences', p.6.
45. 'Préface', *Histoire de l'Académie royale des sciences. Année MDCXCIX. Avec les mémoires de mathématique & de physique, pour la même année, tirés des registres de cette académie* 3rd edn (Paris, Gabriel Martin, Jean Baptiste Coignard fils and H. Louis Guerin, 1732), p.i-xix (vii).
46. Stroup, *Company*, p.98.
47. The Paracelsians considered minerals and chemicals such as mercury, bezoars and antimony to be therapeutic. From the sixteenth century onward, most of the king's physicians, including the founder of the Jardin du roi, Guy de la Brosse, came from the medical faculty of Montpellier, where Paracelsian ideas replaced the Galenic emphasis on balancing the four humours. The Paris medical faculty remained loyal to Galen, losing ground to Montpellier.
48. Simon Boulduc, 'Analise de l'ipecacuanha', in *Histoire de l'Académie royale des sciences. Année MDCC. Avec les mémoires de mathématique et de physique, pour la même année, tirés des registres de cette académie* (Paris, Charles Estienne Hochereau, 1719), p.1-5 (3).
49. Klein, 'Shifting ontologies', p.268.
50. Gabriel-François Venel, 'Chimie', in Denis Diderot and Jean Le Rond d'Alembert (eds), *Encyclopédie, ou Dictionnaire raisonné des sciences, des arts, et des métiers, par une société de gens de lettres*, CD-ROM (Paris, 2000). Stahl, a professor of medicine, denied the applicability of

A major project of the Académie in the late seventeenth century was a comprehensive natural history of all known plants; this included not only descriptive natural history in the Aristotelian and more recent Baconian modes,[51] but also chemical analysis, which was supposed to seek not only useful medicinal and nutritional substances, but also the causes of phenomena. The Académie's novel approach to plant research brought together old and new methods of investigation, moving beyond the merely descriptive; it was therefore a different kind of project from the sweeping taxonomic exercise that seventeenth-century botanists such as Caspar [Gaspard] Bauhin (1560-1624) had undertaken in his *Pinax*.[52]

The Académie's project was originally conceived as a 'grand botanical compendium' in response to pressures peculiar to late seventeenth-century botany, especially the quadrupling of the number of known plants, and the analogies being noted by researchers such as Nehemiah Grew (1641-1712) and Marcello Malpighi (1628-1694) between the plant and animal kingdoms, for example the circulation of the blood in animals and the sap in plants.[53] It was to include engravings of plants and other traditional botanical information, as well as an attempt, through chemical analysis, 'to explain what they were describing'.[54] Some of the participating Academicians believed in the importance of rigorous *chemical* analysis of plants – in the first instance through distillation – in order to identify useful plant constituents and properties or 'virtues'. Bourdelin was the premier practitioner of distillation for the Académie. The object was 'to improve medicine, and to understand how plants nourish humans'.[55] Academicians hoped to base practical applications on theories such as the corpuscular theory of matter: 'The main purpose of analyzing plants chemically was to discover their constituents', but this proved difficult through the reigning method of analysis, distillation, for reasons discussed below.[56]

After 1686 the direction of the Académie's investigation of plants became focused on results of immediate utility due to ministerial intervention, motivated by pecuniary considerations and a near-fatal illness of Louis XIV. Natural philosophy was rejected in favour of practical

chemistry to medicine, while acknowledging that alchemy might have some value. J. R. Partington, *A History of chemistry* (London, 1961), vol.2, p.657.

51. According to Bacon, comprehensive natural histories were to provide the basis for a new, practically useful science. Bacon, 'Great Instauration', in *Great instauration and New Atlantis*, ed. J. Weinberger (Arlington Heights, IL, 1989), p.26.

52. Caspar Bauhin, *Pinax theatri botanici* (Basel, Ludwig König, 1623).

53. Stroup, *Company*, p.67, 107, 119.

54. Stroup, *Company*, p.73.

55. Stroup, *Company*, p.98.

56. Stroup, *Company*, p.98.

results; the Academicians were advised to study medicine, improve and republish Duclos's book on mineral waters, try to desalinate sea water and analyse wines, 'efforts in which they were already engaged'.[57] 'Academicians had always sought medical applications of their work and planned to include the medical uses of plants in the natural history.'[58]

The medical focus of the work of the apothecary-chemist, Simon Boulduc, fitted this mandate closely; he took as his special brief the investigation of purgative plants:

> L'examen des vertus et différences des médicaments purgatifs est à mon sens, ce qu'il y a de plus mystérieux et de plus nécessaire dans toute la Physique expérimentale; ce sujet a fait jusqu'ici l'attention de tous nos Savants; je l'ay pris ici pour mon partage, heureux, si je puis en suivant leurs lumières, ajouter quelque chose à ce qu'ils nous en ont laissé, et pour donner quelque chose à la nouveauté, j'ay cru pouvoir d'abord commencer par la racine d'ipécacuanha.[59]

Boulduc developed new methods with which to examine purgative plants such as ipecachuana, jalapena, colocynth and black hellebore. The principal interest of many of the plants he investigated was their promise as purgatives, as purgation was one of the major methods of treating illness in the eighteenth century: references to the purgative effects of these plant extracts that 'M. Boulduc a éprouvé sur différents malades avec les précautions nécessaires' are frequent.[60]

For the duration of the project – which effectively ended unfinished in 1686 – much of this analysis was carried out through distillation: 'the process whereby plants were placed in a receptacle and heated to obtain liquid and solid products'.[61] Distillation had been in use for two centuries and was the most favoured tool at the disposal of seventeenth- and early eighteenth-century chemists. Despite its popularity, the shortcomings of distillation were already known by the end of the seventeenth century and evidence against the utility of distillation was mounting year by year.[62]

57. Stroup, *Company*, p.108.
58. Stroup, *Company*, p.109; on the medical goals of the Académie, see also p.171ff.
59. Boulduc, 'Analise de l'ipecacuanha', p.1-2.
60. [Bernard le Bovier de Fontenelle], 'Analise de l'ipecacuanha', in *Histoire de l'Académie royale des sciences. Année MDCC. Avec les mémoires de mathématique et de physique, pour la même année, tirés des registres de cette académie* (Paris, Charles Estienne Hochereau, 1719), p.46-48 (47); see also [Fontenelle], 'Analises de la coloquinthe, du jalap, de la gomme gutte, et de l'ellébore noir', in *Histoire de l'Académie royale des sciences. Année MDCCI. Avec les mémoires de mathématique et de physique, pour la même année, tirés des registres de cette Académie* (Paris, Charles Estienne Hochereau, 1719), p.58-62 (59).
61. Stroup, *Company*, p.90.
62. Academicians (1) doubted the purity of extracts; (2) they were concerned about the destructiveness of the procedure, which broke down or altered plant components that

Aware of the inadequacy of this approach, Boulduc expanded the range of methods of analysis beyond distillation. The Académie's 'Histoire' of Boulduc's 'Analise de l'ipecacuanha', notes, for example: 'La distillation ne lui a pas donné de grandes lumières, mais par l'Analyse qu'il appelle d'Extraction, il a vu que ce Mixte[63] contenait des parties salines, et sulfureuses ou résineuses, les salines en plus grande quantité'.[64] In his Mémoire on the same topic, Boulduc continues: 'Par la distillation je n'en ai tiré d'abord qu'un flegme, qu'un esprit acide et qu'un peu d'huile, et de la masse noire restée dans la cornue et calcinée à feu très violent, j'en ai retiré très peu de sel fixe.'[65]

Boulduc and Nicolas Lémery, another chemist-Academician, formulated their objections to distillation even more succinctly in subsequent publications. In 1719 Lémery wrote that fire reduces plants

> [E]n une espèce de poussière [...] il donne lieu à la formation de nouveaux composés qui diffèrent souvent très fort de ceux qui habitaient naturellement dans ce Mixte; c'est apparemment par les raisons déjà alléguées [...] que deux Plantes dont l'une est très salutaire et l'autre un poison, et dont la composition est par conséquent très différente, se ressemblent néanmoins [sic] très fort par les substances qu'on en a retiré, et par la quantité de ces substances; de manière que si on ne connaissait pas d'ailleurs leurs qualités, on serait tenté de croire, en vertu de l'Analyse, qu'elles seraient les mêmes.[66]

In 1734 Boulduc reiterated this conclusion, noting that it was necessary to abandon distillation since 'le feu que l'on étoit obligé d'employer pour ces analyses changent entiérement et dénaturant leurs principes, en sorte qu'ils n'étoient plus qu'en partie des créatures du feu, et non pas les

---

produced the curative effects in which academicians were interested; (3) they noted that uniformity of substances extracted suggested that distillation, 'could not account for diversity among plants'. Stroup, *Company*, p.91. Other pitfalls of distillation included conditions of heating that could make products of the same plants more different than those from different plants. Academicians tried to avoid these pitfalls by weighing all the products, using the least possible degree of heat and employing a standard procedure.

63. This term has to be taken in context and not translated into modern chemical terminology: 'A chemical union of two or more true "elements" or "principles" [...] Although the term had great philosophical complexities, it was roughly equivalent to our term "compound", but the latter is not to be considered a synonym.' Jon Eklund, *The Incompleat chymist: being an essay on the eighteenth-century chemist in his laboratory, with a dictionary of obsolete chemical terms of the period* (Washington, DC, 1975), p.32; see Rhoda Rappaport, 'Rouelle and Stahl: the phlogistic revolution in France', *Chymia* 7 (1961), p.73-102 (75-76).

64. [Fontenelle], 'Analise de l'ipecacuanha', p.46-47.

65. Boulduc, 'Analise de l'ipecacuanha', p.3.

66. Nicolas Lémery, 'Réflexions physiques: sur le défaut et le peu d'utilité des analises ordinaires de plantes et des animaux', in *Histoire de l'Académie royale des sciences. Année MDCCXIX. Avec les mémoires de mathématique et de physique, pour la même année, tirés des registres de cette Académie* (Paris, de l'Imprimerie royale, 1721), p.173-88 (174-75).

principes que la Nature avait employés à leur composition'.[67] These two statements occur in documents recorded on Rousseau's list of *Histoires* and *Mémoires* of the Académie.

While the history of plants never came to fruition, chemists continued to investigate plants and to report their findings on an annual basis in the *Histoires* and *Mémoires* of the Académie, a condition of their tenure in the Académie.[68] Hence, the Académie's *Mémoires* of 1701 contain further reports on purgative plants. In 'Observations Analytiques de la coloquinte' (likewise found on Rousseau's list), we read:

> M. Boulduc [...] a d'abord examiné la Coloquinte en la distillant dans la Cornuë par portion selon la methode commune; mais comme cette Méthode ne donne assez souvent qu'une connaissance générale et superficielle des Mixtes, M. Boulduc a eu recours à d'autres voies, dont on a déjà vu l'essai et le succès dans son 'Analyse de l'Ipecacuanha, Histoire de 1700'.[69]

Unlike the substances obtained by distillation, this extract was medically efficacious: 'Cet Extrait au poids de dix ou douze graines purge doucement et sans irritation'.[70] The role of quantitative methods such as weighing is also evident here.

Solvent analysis was a newer method of plant investigation alluded to in Boulduc's analysis of ipecacuanha that was being explored in the early eighteenth century; this method employed spirit of wine or water as a solvent:

> Soluble substances were often put into solution directly by 'extraction': the separation of a substance from a mixture by means of a solvent or 'menstruum'. If no heat were involved, the product of this process was called an infusion. If the solids and solvent were heated, the result was a 'decoction'. Other processes related to these included 'maceration,' which was essentially the same as infusion, and 'digestion,' which, though sometimes used in a way similar to decoction, usually implied a spontaneous reaction that needed the aid of prolonged moderate heat. Digestion, often carried out in the matrass, was felt to be 'useful to favour the action of certain substances upon each other.' It also was used as a preparative procedure 'to soften and open certain bodies,' especially in the animal and vegetable realms.[71]

67. Boulduc, 'Essai d'analise des plantes', in *Histoire de l'Académie royale des sciences. Année MDCCXXXIV. Avec les mémoires de mathématique et de physique, pour la même année, tirés des registres de cette académie* (Paris, de l'Imprimerie royale, 1736), p.101-106 (101).
68. Louis XIV, 'Règlement ordonné par le roi pour l'Académie royale des sciences', p.10.
69. [Fontenelle], 'Analises de la coloquinthe, du jalap, de la gomme gutte, et de l'ellébore noir', p.59.
70. Simon Boulduc, 'Observations analytiques de la coloquinthe', in *Histoire de l'Académie royale des sciences. Année MDCCI. Avec les mémoires de mathématique et de physique, pour la même année, tirés des registres de cette académie* (Paris, Charles Estienne Hochereau, 1719), p.12-17 (15).
71. Eklund, *The Incompleat chymist*, p.13.

Boulduc was a prime mover in pioneering this new method of analysis.[72] His application of these new methods achieved a breakthrough that, like many scientific innovations, was only adopted slowly; eminent chemists such as Nicolas Lémery continued to rely on distillation as the primary method of chemical analysis of plants.[73]

Rousseau was undoubtedly aware of the significance of these developments, since all the *Histoires* and *Mémoires* referred to above are inventoried in his 'Catalogue de tous les Mémoires de la Chymie contenus dans tous les volumes de l'Histoire de l'Academie R[oyale] des sciences'.[74] Perhaps Rousseau used this material in composing his *Institutions chymiques*, discussed in the next section. Despite our lack of information on these points, the list attests to Rousseau's extensive acquaintance with the Académie's experimental programme in plant chemistry, and it provides concrete support to Rousseau's allusions in his botanical writings to 'l'analyse encore plus fautive' of plants.[75] Some of these investigations aimed at purely practical outcomes – Boulduc sought to learn for example, about the efficacy of purgative plants, while others had theoretical goals, for example Homberg's investigation of the salts of plants[76] which sought to discover how plants reproduce themselves, an attempt to unravel the mystery of generation.

v. The *Institutions chymiques*[77]

At some unknown date Rousseau's studies in chemistry culminated in a lengthy treatise on chemical principles, the *Institutions chymiques*. This hitherto unknown text was first discovered by Théophile Dufour in 1882 in Trelex, Switzerland (Vaud), at the home of Suzanne Nicole, great-granddaughter of Rousseau's literary executor, Paul Moultou. Maurice Gautier published an edited version of the *Institutions chymiques* in 1918-1921.[78]

72. Thomas Southwell, *Medical essays and observations*, (London, J. Knox, 1764), p.207.
73. Holmes, *Eighteenth-century chemistry*, p.75.
74. BPUN MsR81, f.2-26.
75. Rousseau, *Fragments de botanique*, *OC*, iv.1249.
76. [Bernard le Bovier de Fontenelle], ' Sur les sels des plantes', in *Histoire de l'Académie royale des sciences. Année MDCXCIX. Avec les mémoires de mathématique et de physique, pour la même année, tirés des registres de cette Académie* (Paris, la veuve de Jean Boudot, 1718), p.63-65; Guillaume Homberg, 'Essais pour examiner les sels des plantes', in *Histoire de l'Académie royale des sciences. Année MDCXCIX. Avec les mémoires de mathématique et de physique, pour la même année, tirés des registres de cette académie* (Paris, la veuve de Jean Boudot, 1718), p.69-74.
77. It is not my object here to provide an analysis of Rousseau's chemical theory; rather I seek to locate in his chemical studies key influences on his later botanical activities.
78. Jean-Jacques Rousseau, 'Les Institutions chymiques de Jean-Jacques Rousseau', ed. Maurice Gautier *Annales de la société Jean-Jacques Rousseau* 12-13 (1918-1921). Dufour states that he found the MS twenty-two years before he published his study of the *Institutions*

Reconstructing the history of the *Institutions chymiques* is somewhat conjectural; scholars who have examined the few references in the *Confessions* and the correspondence as well as the manuscripts themselves, have come to different conclusions. Rousseau may have composed the work circa 1747 to assist Dupin in seeking election to the Académie royale des sciences (Paris); or it may have had an entirely distinct genesis as a work in its own right, perhaps at a later date; for example winter 1753-1754, as Bensaude-Vincent and Bernardi argue.[79] This work may have even been under discussion for publication as late as 1757, as implied by references in the *Confessions* and the correspondence to a chemistry manuscript in which baron d'Holbach (1723-1789) took an interest.[80] Dufour lends credence to this idea when he notes that Rousseau retained the MS of the *Institutions* until his death, not destroying it as he did many other manuscripts at various points during his lifetime, including most letters preceding 1757.[81] However, the manuscript material Rousseau left behind is of little help in dating when he learned what. For example, his list of the Académie's *Histoires* and *Memoires* on chemistry carries no date, and the date of the last entry (1734) tells us only that the list was drawn up during or after that year.

We cannot even be certain when or where he learned the practical side of chemistry; we can conjecture, however, that he may have already learned much of it with Maman, since the 'material culture of academic laboratories overlapped strongly with the realm of instruments, reagents, technologies and materials applied and produced in apothecary's laboratories, assaying shops and distilleries'.[82]

The *Institutions chymiques* has traditionally been considered merely 'un vaste travail de compilation'.[83] Gautier's edition (1918-1921) gave a biased view of Rousseau's treatise, suggesting that chemistry before Lavoisier was nothing more than alchemy, understood as the search for the philosopher's stone and the attempt to transform base metals into gold.[84] Gautier stresses Rousseau's reliance on so-called 'alchimistes'

chymiques, the first version of which appeared in *Semaine littéraire*, 17 December 1904; he therefore made the discovery in 1882, not in the early twentieth century, as the Pléiade editors state. *OC*, i.1414.

79. Bensaude-Vincent and Bernardi, 'Pour situer les *Institutions chymiques*', p.18-19, 23.
80. Rousseau, *Confessions*, *OC*, i.461.
81. Dufour, *Les Institutions chymiques*, p.8-9.
82. Klein and Lefèvre, *Materials*, p.34; see also Bensaude-Vincent and Bernardi, 'Pour situer les *Institutions chymiques*', p.11.
83. Gautier, 'Introduction', in Rousseau, 'Les *Institutions chymiques*', vol.12, p.xvii.
84. Gautier, 'Introduction', in Rousseau, 'Les Institutions chymiques', vol.12, p.xviii; Bensaude-Vincent and Bernardi, 'Pour situer les *Institutions chymiques*', p.25; Klein and Lefèvre, *Materials*, p.17.

such as J. J. Becher (1628-1685), Herman Boerhaave (1668-1738) and J. Juncker (1679-1744), ignoring their important contributions to the development of modern chemistry. Becher, for example, was a metallurgical chemist who promulgated the phlogiston theory, an important stage in chemical theory. Boerhaave, professor at Leiden, was an important physician and botanist, as well as chemist. Leading historians of science see Becher and Stahl as central to the history of chemistry.[85]

Moreover, contrary to what Gautier suggests, Rousseau rejected 'transmutations impossibles', a clear reference to one of alchemy's principal concerns.[86] He also condemned 'le langage favori des alchimistes': 'parler énigmatiquement de leurs sublimes travaux'.[87] He thus contrasted the obfuscation he takes to be characteristic of alchemical discourse with a clear, unambiguous scientific language, such as that favoured by Bacon and the Royal Society.[88]

It is equally mistaken to assume that Rousseau's interest in substances such as arsenic, mercury, antimony, bismuth, zinc and lead was alchemical in the limited sense invoked by Gautier.[89] This interest could just as well reflect the influence of the iatrochemists descended from Paracelsus, who dominated the medical faculty of Montpellier, populated the medical personnel of the royal court and wielded significant influence in the Académie. Paracelsian ideas, while linked to alchemy through a common interest in metals and minerals, should not be taken to be identical with alchemy, since Paracelsian physicians sought to treat illness, not transmute metals.

Gautier believed it was a sign of Rousseau's deficiencies as a chemical author that he lifted whole passages out of leading authorities such as Becher, Stahl and the Dutch chemist and botanist, Boerhaave. Rappaport concurs that the *Institutions* 'contains nothing of any originality; most passages have in fact been traced to the writings of Boerhaave, to the Stahlian chemists whose works were available to Rousseau, and to a manuscript of Rouelle's lectures owned by Rousseau'.[90]

Yet in borrowing from these authors, Rousseau was following contemporary practice, and adhering to the teachings of his teacher,

85. Partington, *A History*, vol. 2, p.643, 653.

86. Rousseau, *Institutions chimiques*, p.58.

87. Rousseau, *Institutions chimiques*, p.73; see also Lissa Roberts, 'Setting the table: the disciplinary development of eighteenth-century chemistry as read through the changing structure of its tables', in *The Literary structure of scientific argument*, ed. P. Dear (Philadelphia, PA, 1991), p.99-132; Holmes, *Eighteenth-century chemistry*; Eklund, *The Incompleat chymist*; and Klein, 'Shifting ontologies'.

88. Cook, 'Languages of music and botany', p.78.

89. Gautier, 'Introduction', in Rousseau, 'Les Institutions chymiques', vol.12, p.xviii.

90. Rappaport, 'G.-F. Rouelle', p.79.

Rouelle, who regarded Stahl and Becher as the founders of chemistry.[91] Despite his considerable borrowings from these major authors, which are evident on every page of the treatise, Rousseau does not accept their doctrines uncritically; for example, he rejects Becher's theory of principles, espousing instead the four elements theory of Boerhaave. He does not accept the theories of the great Boerhaave uncritically either, subjecting the latter's theses on fire and air to critique.[92]

However derivative the *Institutions chymiques* may be, it hints at the views that Rousseau would later espouse in botany. For example, in book 2, chapter 1 of his *Institutions chymiques*, entitled 'Du mécanisme de la nature', Rousseau set forth his view of nature not as a mere complex of mechanical forces, but as an awe-inspiring spectacle that can only be apprehended by those who know how to see it: 'J'entends tous les hommes vanter la magnificence du spectacle de la nature, mais j'en trouve fort peu qui la sachent voir.'[93] He continues with a lyrical invocation of this 'spectacle': 'Nous voyons le jour et la nuit, la neige et la verdure couvrir alternativement la face de la terre; nous voyons les eaux dont elle est mêlée s'élever en vapeurs par la chaleur du soleil retomber en rosée ou se condenser en nuages et se résoudre en une pluie.'[94]

The processes employed by nature include 'les digestions, les dissolutions, les filtrations, les fermentations, les calcinations';[95] while filtration was considered to be a mechanical process, digestion and fermentation were of interest to Paracelsian iatrochemists. These processes are studied in 'un Laboratoire artificiel sur le modèle de la nature'. In this laboratory 'il ne suffit pas de voir en gros les voies qu'elle emploie; il faut surtout connaître bien parfaitement les instruments dont elle se sert'.[96] These diverse instruments include 'le soleil, les vents, la pluie, les eaux, les sels, les terres, les parties mêmes des corps diversement mues, diversement figurées'.[97] Yet Rousseau objects that chemistry is unable to penetrate the mystery of life itself:

> ni le soleil, ni tous les astres, ni tout le feu, ni tout le mouvement qui existent dans l'univers ne sont pas capables de produire la moindre de toutes les

91. Bernadette Bensaude-Vincent and Bruno Bernardi, 'Rousseau chimiste', in *Rousseau et les sciences*, ed. B. Bensaude-Vincent and B. Bernardi (Paris, 2003), p.59-76 (69).
92. Bensaude-Vincent and Bernardi, 'Pour situer les *Institutions chymiques*', p.29, n.68. Rousseau criticised Boerhaave's thermometer, and his theory of the fixation of air; he was indignant that Boerhaave could have denied the reality of ice crystals, which had been attested by 'peuples des montagnes'.
93. Rousseau, *Institutions chimiques*, p.97.
94. Rousseau, *Institutions chimiques*, p.101.
95. Rousseau, *Institutions chimiques*, p.102.
96. Rousseau, *Institutions chimiques*, p.103.
97. Rousseau, *Institutions chimiques*, p.103.

> plantes [...]; cet abîme de la génération dans lequel les Philosophes se sont si longtemps perdus est encore aujourd'hui le désespoir des incrédules [...] Ces observations suffisent pour me montrer le point où doivent commencer mes recherches.[98]

For Rousseau in the 1740s, the study of matter by means of chemical analysis therefore held the key to understanding life itself. Chemistry gave him a ground within natural philosophy from which to refute the mechanist philosophers: 'la construction d'un corps organisé par les seules lois du mouvement est une chimère qu'on est contraint d'abandonner à ceux qui se payent de mots'.[99] This could be regarded as the opening salvo of Rousseau's anti-mechanist natural philosophy, an important step toward his later elevation of botany as the best kind of science.

vi. Demarcating the three realms of nature

In the early eighteenth century natural philosophy, including chemistry, was still organised according to the ancient division of nature into three realms – animal, vegetable, mineral; this division structured not only books, but also knowledge spaces such as natural history cabinets. However, there was no established order among the divisions; some authorities placed the vegetal first, while others privileged the mineral. Rouelle, Rousseau's teacher in chemistry at the Jardin du roi, placed the vegetal realm in first place, aligning himself with the chemists for whom pharmacological investigations of plants were primary. However, it was increasingly recognised that fundamental elements came from the mineral realm, so some treatises (such as those of the German chemists, Becher and Stahl, founders of the chemical tradition that Rousseau followed) considered the mineral realm the first division of nature. Rousseau likewise adopted this approach, calling the mineral realm 'le principal objet du Chimiste';[100] this decision may have been made under the influence of Becher's *Physica subterranea*, corrected and revised by Georg-Ernst Stahl (1703). The German chemists who influenced French chemistry after 1740 came from a metallurgical tradition based in the mining industries, clearly an influence on their privileging of the mineral realm.[101]

98. Rousseau, *Institutions chimiques*, p.100-101.

99. Rousseau, *Institutions chimiques*, p.100; see also Beretta, 'Sensiblerie vs mécanisme', p.11.

100. Rousseau, *Institutions chimiques*, p.92.

101. Klein and Lefèvre argue for the priority of the mineral kingdom around 1700 on the basis of the purity of the substances it offered for examination. However, these substances had not been isolated in academic settings and subjected to reversible experiments, but rather had long been 'commodities, applied in the apothecary trade and metallurgy', *Materials*, p.58.

However, the organisation of Rousseau's treatise did not follow the usual tripartite schemes:[102] 'Si les chimistes contemporains avaient eu l'occasion de lire les *Institutions chymiques* de Rousseau, ils auraient été frappés par une manque. On ne trouve pas dans les *Institutions chymiques* l'organisation caractéristique des traités et manuels de chimie pendant le dix-huitième siècle, comme dans le siècle qui le précéda: la division des textes selon les trois règnes – animal, végétal et minéral.' Yet '[s]i l'organisation générale des *Institutions chymiques* ne suit pas ce schéma tripartite traditionnel, il n'en reste pas moins que les similarités et différences entre les trois règnes constituent l'une des préoccupations majeures de Rousseau'.[103] This is evident, for example, in Rousseau's discussion of fire in Book 2; the first topic he investigates is fire in plants, pointing out how they embody fire in their combustible parts and how they can be analysed by means of distillation.

Rousseau's idiosyncratic organisation has a strongly functional character that focuses not so much on the realms of nature, but on the means ('instruments') by which chemical processes take place:

Livre 1: Des éléments des corps et de leur composition (De la mixtion et composition des corps, Du principe de la cohésion des corps, Des différents mixtes[104] and composés naturels)
Livre 2: Les instruments naturels (Du feu, De l'air, De l'eau, De la terre)
Livre 3: Les instruments artificiels (Des fourneaux et vaisseaux, Des autres instruments chimiques, Des dissolvants, Des précipitants et des rapports des substances)
Livre 4: Des opérations (De l'analyse ou syncrèse, De la diacrèse ou composition, De la combinaison).[105]

This focus on instruments may not have been as ephemeral a concern for Rousseau as we might think, because he continued to take great care in his selection and use of instruments when doing botany.

vii. Chemistry: the straw man?

In the *Institutions*, Rousseau presents himself as an advocate of the modernity and utility of chemistry, claiming that chemistry should be a leading science that can explain what physics could not:

> [L]a physique ne considère les corps que par leurs mouvements, leurs figures,

102. The Arab physician, Rhazes (860-925 CE), was cited by William Davison (1593-1669), first professor of chemistry at the Jardin du roi, as the source for this tripartite division. See Klein and Lefèvre, *Materials*, p.12, n.3; Klein, 'Shifting ontologies'.
103. Simon, 'L'homme de verre?', p.65; see also p.72; Rousseau, *Institutions chimiques*, p.111.
104. Eklund, 'Mixt', p, 32.
105. See the discussion of these terms in Dufour, *Les Institutions chymiques*, p.10, n.1.

et par d'autres semblables modifications [...] mais comme elle n'en examine, pour ainsi dire que l'écorce et la surface, elle ne parvient point seule à les connaître intérieurement et par leur propre substance; la chimie qui s'attache à cette recherche est donc la plus importante de toutes les parties de la physique, et il est certain que s'il y a quelque voie pour parvenir à la vraie connaissance de la nature, c'est à dire des corps qui la composent, c'est par leur analyse et la connaissance des éléments dont ils sont formés eux-mêmes qu'on y peut parvenir.[106]

Physics and chemistry complement each other and should work in tandem, rather than in opposition. Chemistry here appears as part of the master science of physics, the role later to be occupied by botany in Rousseau's revised economy of the sciences.

Indeed, chemistry was dynamic: it was developing quantitative methods of analysis (for example the use of weighing, as seen above) and creating a variety of classifications to serve myriad purposes, such as Geoffroy's 1718 'Table des différents rapports observés entre différentes substances'. Geoffroy's table of affinities gave chemistry legitimacy and organisation, raising it above a mere collection of facts or experimental events. In this respect chemistry exhibited similarities with the taxonomic discipline of botany.

His experiences with chemistry led Rousseau to condemn the chemical analysis of plants in the service of pharmacology:[107]

Le premier malheur de la Botanique est d'avoir été regardé dès sa naissance, comme une partie de la Médecine. Cela fit qu'on ne s'attacha qu'à trouver ou supposer des vertus aux plantes, et qu'on négligea la connaissance des plantes mêmes; car comment se livrer aux courses immenses et continuelles qu'exige cette recherche, et en même tems aux travaux sédentaires du laboratoire et aux traitements des malades, par lesquels on parvient à s'assurer de la nature des substances végétales, et de leurs effets dans le corps humain.[108]

In the *Rêveries* Rousseau analysed why he was attracted to botany, following the traditional tripartite division of nature. He first discussed the difficulties, dangers and costs of investigating the mineral realm, which had been the first realm he investigated in chemistry:

Pour *profiter* dans l'étude des minéraux, il faut être *chimiste* et *physicien*; il faut

106. Rousseau, *Institutions chimiques* p.58. 'La *figure* de toutes les parties de la vigne est du ressort du botaniste; le physicien tâche à force d'hypothèses d'expliquer quelques unes des lois de la végétation, c'est au chimiste à dire *tout le reste*.' My emphasis; Rousseau, *Institutions chimiques*, p.82.

107. My emphasis; Rousseau, *Fragments de botanique*, *OC*, iv.1249.

108. Rousseau, 'Introduction', *OC*, iv.1201. This document was long, but probably mistakenly, taken to be the introduction to Rousseau's unfinished botanical dictionary. Henry Cheyron, 'Ray et Sauvages annotés par Jean-Jacques Rousseau', *Littératures* 15 (Autumn 1986), p.83-99 (86).

> faire les expériences pénibles et coûteuses, travailler dans des laboratoires,[109] dépenser beaucoup d'argent et de temps parmi le charbon, les creusets, les fourneaux, les cornues, dans la fumée et les vapeurs étouffantes, toujours au risque de sa vie et souvent aux dépends de sa santé. De tout ce triste et fatigant travail résulte pour l'ordinaire beaucoup moins de savoir que d'orgueil, et où est le plus médiocre chimiste qui ne croit pas avoir pénétré toutes les grandes opérations de la nature pour avoir trouvé par hasard peut-être quelques petites combinaisons de l'art?[110]

For Rousseau the chemistry laboratory evoked the worst social motives of profit and fame – in other words – amour-propre; it represented economic inequality since chemistry was an expensive pursuit for the rich, such as the tax farmer, Lavoisier, and the tax farmer's son, Dupin de Francueil. The chemist is less attached to knowledge than to pride; he is occupied with dead nature, and even if he works hard, he produces little: just 'petites combinaisons de l'art'.[111]

The chemistry laboratory also symbolises the death of nature at the hands of man in its proto- or pre-industrial form: the artisanal workshop in which Bacon expected the workings of nature to be revealed.[112] Critiquing the unhealthy life of modern man, Rousseau depicts 'cette quantité de métiers mal-sains qui abrègent les jours ou détruisent le tempérament; tels que sont les travaux des mines, les diverses préparations des métaux, des minéraux, surtout du Plomb, du Cuivre, du Mercure, du Cobalt, de l'Arsenic, du Réalgar'.[113] These were substances of vital interest to chemists.

Studying the animal realm likewise has its limitations – while the animal realm is more accessible than the mineral, it is difficult and disgusting to practise anatomy: 'Le règne animal est plus à notre portée et certainement mérite encore mieux d'être étudié. Mais enfin cette étude n'a-t-elle pas aussi ses difficultés, ses embarras, des dégoûts et ses peines?'[114]

109. For a description of the typical laboratory as a kitchen, see Klein and Lefèvre, *Materials*, p.36. Proper ventilation and excess humidity were important concerns.
110. My emphasis; Rousseau, *Rêveries*, *OC*, i.1067.
111. See Venel, 'Chimie', in Diderot and d'Alembert (eds), *Encyclopédie*. 'C'est la nécessité de toutes ces connaissances pratiques, les longueurs des expériences chimiques, l'assiduité du travail et de l'observation qu'elles exigent, les dépenses qu'elles occasionnent, les dangers auxquels elles exposent, l'acharnement même à ce genre d'occupation qu'on risque toujours de contracter, qui ont fait dire aux Chimistes les plus sensés, que le goût de la *Chimie* était une passion de fou'.
112. Francis Bacon, *The New organon*, ed. L. Jardine and M. Silverthorne (Cambridge, 2000), p.70.
113. Rousseau, *Discours sur l'inégalité*, *OC*, ii.205, n.IX; see also Rousseau, *Institutions chimiques*, p.359-69. Realgar is arsenic disulfide.
114. Rousseau, *Rêveries*, *OC*, i.1067. For Rousseau's view of animals, see Jean-Luc Guichet, *Rousseau, l'animal et l'homme: l'animalité dans l'horizon anthropologique des Lumières* (Paris, 2006).

For Rousseau, the exemplary student of plants was not a chemist, apothecary or physician seeking 'simples', or useful 'properties' of plants, but rather a philosopher such as Theophrastus (c.370-285 BCE), who sought knowledge of plants themselves. The true botanist abandons the dead specimens, ground-up plant matter, distillates, apparatus and closed, smoke-filled space of the chemical laboratory for '[d]es près émaillés des fleurs [...] l'unique *laboratoire* du botaniste'. Unlike the chemist, with his elaborate, costly, and fixed apparatus, the botanist travels light, 'porte aisément tous ses outils dans sa poche'.[115] Note, however, that the laboratory and instruments play an unexpected role in botany, which is supposed to be an 'aimable' science in which no intermediary should come between oneself and nature.[116]

In his undated fragments on botany Rousseau returns to the division of labour between the emergent sciences of chemistry and botany, giving pride of place to the botanist rather than to the chemist. The botanist now takes the lead as student of the *living*, whereas the chemist's activity is limited to studying what is dead:[117]

> L'art d'étudier les plantes par la combinaison de leurs mixtes qui composent leur substance, l'art d'en déterminer les vertus médicinales vraies ou fausses soit par l'expérience et l'observation toujours imparfaite et trompeuse soit par l'analyse chimique encore plus fautive n'a rien de commun avec la botanique. Le botaniste étudie dans les végétaux leur tissu, leur figure, leur organisation, leur génération, leur naissance, leur croissance, leur vie et leur mort. Il les peut considérer aussi par leur couleur, par leur goût, leur odeur, leur saveur, par toutes les prises qu'elles peuvent offrir aux sens; mais ce n'est là qu'une étude analogique et secondaire, pour éclairer et confirmer celle des formes, les végétaux, ne sont aux yeux du botaniste que des êtres organiques, sitôt que le végétal est mort, qu'il cesse de végéter, que ses parties n'ont plus la mutuelle correspondance qui le faisait vivre et le constituait un, il n'est plus du ressort du botaniste, c'est une simple substance, une matière, un mixte, une terre morte, qui n'appartient plus dès lors au règne végétal mais au minéral. Les médecins, les charlatans, les empiriques peuvent lui parler des vertus merveilleuses des plantes mortes et décomposées, tout cela ne le regarde plus. Il faudrait pour l'admettre qu'il donnât à l'autorité des hommes qui sont menteurs, la foi qu'il n'a que dans la nature qui ne ment jamais et qui ne lui dit rien de pareil.[118]

115. My emphasis; Rousseau, *Fragments de botanique*, *OC*, iv.1250. These instruments included magnifying glasses (loupes), small scissors and tweezers for plucking small plant parts. See Rousseau to François Henri d'Ivernois, 20 July 1765; *CC*, xxvi.105-107.

116. Rousseau, *Fragments de botanique*, *OC*, iv.1250.

117. On the distinction between the living and the non-living, see Klein, 'Shifting ontologies'.

118. Rousseau, *Fragments de botanique*, *OC*, iv.1249.

viii. Conclusion

Chemistry's contribution to Rousseau's philosophical programme is most significant for the unfavourable view of this science, its concerns, its practice and its practitioners that he developed. Rousseau rejected chemistry for neglecting the living and being concerned only with the dead matter of the plant which thereby becomes 'une simple substance, une matière, un mixte, une terre morte, qui n'appartient plus dès lors au règne végétal mais au minéral'.[119] His encounter with chemistry may have therefore spurred the development of his anti-mechanist philosophy.

Rousseau espoused an approach to the natural world free of the medical motive on the one hand, and the chemist's destructive procedures (such as distillation) on the other. He came to see botany as the natural science par excellence, not only for its method, but also for its choice of subject matter: 'le plus riche et plus varié des trois' realms of nature.[120]

Yet chemistry's legacy for Rousseau's botanical project is not entirely negative. This can be seen in two instances: (1) he accepted experimental and technical knowledge, and is prepared to use instruments as aids to the senses in gaining knowledge of the external world; this stands in contrast to the visual, intuitive means of acquiring knowledge that are usually attributed to him; (2) he continued to assimilate nature to a laboratory – the site of the work – in which the boundary between nature and artifice is effaced.[121] It is therefore hardly surprising that Rousseau requested the best optical instruments from Geneva for his botanical studies, and depicted the fields as the 'unique laboratoire du botaniste'.[122]

119. Rousseau, *Fragments de botanique*, *OC*, iv.1250.

120. Rousseau, *Fragments de botanique*, *OC*, iv.1249.

121. 'La promenade est son unique travail', Rousseau, *Fragments de botanique*, *OC*, iv.1250.

122. I extend to botany an analysis offered by Bernadette Bensaude-Vincent, 'La nature laboratoire', *Rousseau et les sciences*, ed. Bernadette Bensaude-Vincent and Bruno Bernardi (Paris, 2003), p.155-74.

3. Helvetia mediatrix:[1] the atmosphere of eighteenth-century Swiss science

> Les Suisses ont cultivé, avec succès, toutes les sciences, depuis qu'elles ont recommencé à fleurir en *Europe*; mais on peut dire qu'ils se sont surtout distingués par leurs travaux et leurs progrès dans la botanique. Les grands Hommes que la *Suisse* a vu naître, et qu'elle produit encore en ce genre, sont un des ornements de la patrie et le sujet de l'admiration des curieux étrangers.[2]

i. Introduction

Rousseau received his introduction to botany in a Prussian principality on the western flank of the disaggregated Swiss Confederation of the mid-eighteenth century. This fact is replete with consequences, for in eighteenth-century Europe Switzerland played a critical role as 'médiatrice', where French ideas were germanised and German ideas were latinised.[3]

Neither French nor German, and rich in biodiversity, Switzerland offered its botanists important natural endowments and a variety of scientific practices and traditions. Swiss botanists, who usually had some foreign education or training, charted their own course in the minefield of eighteenth-century botanical nomenclature and classification; this created a rich and varied spectrum of epistemological and methodological choices, of which Rousseau made ample use.[4] I argue that the presence of foreign as well as local ideas on mid-eighteenth-century Swiss terrain played a hitherto unacknowledged role in Rousseau's botanical formation.

This important point, neglected in the literature, demands further

1. The phrase refers to 'la Suisse comme carrefour des cultures européennes', André Bandelier, *Des Suisses dans la république des lettres: un réseau savant au temps de Frédéric le Grand* (Geneva, 2007), p.26.
2. Original emphasis; 'Idée générale du grand ouvrage de botanique, que M. Haller, professeur à Gottingen, vient de publier sur les plantes naturelles de la Suisse', *Journal helvétique* (February 1743), p.174-83 (174).
3. François Jost, *Jean-Jacques Rousseau suisse, étude sur sa personnalité et sa pensée*, vol.2 (Fribourg, 1961), p.445.
4. Had Rousseau been introduced to botany in Paris it is highly unlikely, for example, that he would have studied the works of the Swiss botanist and polymath, Albrecht von Haller, discussed in this chapter.

development. This chapter examines the Swiss background of Rousseau's botany, beginning with an analysis of some of the principal factors that affected the development of the sciences in Switzerland, including: (1) the country's unusual political history as a confederation lacking a national centre, or court;[5] (2) its largely Calvinist Protestant identity, which connected it to other Protestant states and their scientific cultures, especially Holland and Great Britain; and (3) its large-scale recourse to foreign employment opportunities.

ii. Eighteenth-century Switzerland

Eighteenth-century Switzerland, known in French as the 'Corps helvétique', was a defensive confederation of thirteen cantons, their 'alliés protégés' such as Geneva[6] and Neuchâtel, and their 'alliés proches', for example St Gallen and Grisons. Rousseau famously analogised the Confederation to one large city:

> La Suisse entière est comme une grande ville divisée en treize quartiers, dont les uns sont sur les vallées, d'autres sur les coteaux, d'autres sur les montagnes. Genève, St-Gal, Neuchâtel sont comme les faubourgs: il y a des quartiers plus ou moins peuplés, mais tous le sont assez pour marquer qu'on est toujours dans la Ville [...] On ne croit plus parcourir des déserts quand on trouve des clochers parmi les sapins, des troupeaux sur des rochers, des manufactures dans des précipices, des ateliers sur des torrents.[7]

Despite this conurbation, the cantons and their allies were neither sufficiently coordinated nor focused on particular goals to organise or promote scientific activity directly on anything but a very limited basis. A city might, for example, have an official doctor (Stadtarzt) or other officials who might exercise scientific functions.[8] Examples include Albrecht von Haller (1707-1777) of Bern, Johann Kaspar Hirzel (1725-1803) of Zürich and Frédéric-Samuel Neuhaus (1733-1802) of Neuchâtel.

France, by contrast, was a centralised nation-state whose Académie royale des sciences was founded in 1669 by Louis XIV to foster a range of sciences, from physics to botany; similarly, the Prussian Akademie der

5. I shall not attempt to review the history of the Swiss Confederation (Confoederatio Helvetica, Corps helvétique, Eidgenossenschaft) here. See Oliver Zimmer, *A Contested nation: history, memory and nationalism in Switzerland, 1761-1891* (Cambridge, 2003); Ulrich Im Hof, *Mythos Schweiz: Identität, Nation, Geschichte, 1291-1991* (Zürich, 1991).
6. Jost, *Jean-Jacques Rousseau Suisse*, (vol.1, p.40ff), argues for the Helvetic orientation of Geneva from the sixteenth century onwards. Geneva's accession to the Helvetic Confederation was long blocked by the cantons that had remained Roman Catholic during the Reformation, such as Fribourg and Lucerne.
7. Rousseau to the maréchal de Luxembourg; 20 January 1763; *CC*, xv.49.
8. This post usually dated from the era of the Black Death, especially in Germany, when public health measures such as quarantine, supervised by the Stadtarzt, had a demonstrated benefit.

Wissenschaften, founded in 1700, made Berlin the focal point of scientific research in Prussia. The Imperial St Petersburg Academy of Sciences, founded in 1724-1725 by Peter I, aided by Leibniz, was modelled on the Paris Académie. The Royal Society of London served as the focal point of scientific research in Great Britain even though it was independent of the Crown. Learned societies and academies – which were usually founded by a royal patron (the Royal Society of London being the exception), and which played a key role in fostering the sciences elsewhere in Europe – appeared relatively late on the Swiss scene, at the end of the eighteenth century.[9]

Rousseau celebrated the creation of the academies in the first *Discours* as 'ces sociétés célèbres', '[c]es sages institutions' founded by Louis XIV to which all men of letters aspired, by producing 'ouvrages utiles' and exhibiting 'des mœurs irréprochables'.[10] Rousseau made a presentation of his system of musical notation to the Académie royale des sciences in 1742. Geneva, his city of origin, had to wait until 1790 for the founding of an indigenous equivalent.

These conditions of late institutionalisation obviously delayed the direct stimulation of scientific research and dissemination of its results (via the academies). So, unlike France or Great Britain, Switzerland had no central city or court to which men of letters could look for direction, patronage, or new initiatives; nor was there a central authority such as an enlightened monarch to serve as the nerve centre or stimulus for scientific research by creating institutions, providing subsidies or a system of rewards and incentives.

The absence of direct stimulation favoured emigration by Swiss savants who achieved fame and gained exposure to the newest scientific ideas abroad. Yet emigration also stimulated the sciences in Switzerland. The absence of an academy on the French or English model did not hold eighteenth-century Geneva back; the Protestant Rome 'fournit plus d'hommes aux sociétés savantes internationales [...] qu'aucun d'autre pays d'Europe, proportionnellement à sa population'.[11]

9. 'Auffallend war, wie spät alle Naturwissenschaftlichen Vereinigungen der Schweiz im Vergleich zu Italien, England, Frankreich, Deutschland, Holland und Skandanavien enstanden [It was striking how much later all natural science institutions arose in Switzerland compared with Italy, England, France, Germany, Holland and Scandanavia]', Eduard Fueter, 'Geschichte der exakten Wissenschaften in der schweizerischen Aufklärung (1680-1780)', doctoral dissertation, University of Zürich (Zürich, 1941), p.104; see also René Sigrist, *L'Essor de la science moderne à Genève* (Lausanne, 2004).

10. Rousseau, *Discours sur les sciences et les arts*, *OC*, iii.26.

11. Alain Perrot, *Jean Calvin et la science genevoise* (Geneva, 1996), p.57, n.22.

iii. Calvinist networks

This observation brings us to the second important characteristic that influenced the development of science in eighteenth-century Switzerland, namely, the largely Calvinist identity of the Swiss Confederation, with notable exceptions such as the Catholic cantons of St Gallen, Lucerne, Soleure and Fribourg. The important Swiss cities were Protestant: Geneva, Basel, Zürich and Bern. While only Basel boasted a university and a medical faculty, the other cities had theological academies for training future pastors; starting in the second half of the seventeenth century, these academies began to diversify their academic offerings to include history, philosophy and the sciences, most notably, Calvin's Académie in Geneva.

In the 'conservative' Swiss enlightenment, the sciences were introduced, practised and taught within the framework of natural theology. The Swiss accepted the argument from design: that is, the inductive proposition that the order of the natural world demonstrates the role of a designer. Important sources of this teaching were the English divine and Boyle lecturer, Samuel Clarke (1675-1729), author of *Evidences of natural and revealed religion* (1706) and the French cleric, the abbé Noël Antoine Pluche (1688-1761), in his famous *Spectacle de la nature* (1732). Newton's ideas became popular because he attributed the continued functioning of nature's laws to a continuous divine intervention. Swiss theologians and savants agreed that nature everywhere gives evidence of a beneficent Creator and they saw the sciences as the bulwark of religion, not as a threat to it.[12] As I argue in chapter 1, this theological teleology also guided Rousseau's thinking: 'L'étude de l'homme et de l'univers m'avait montré partout les causes finales et l'intelligence qui les dirigeait'.[13]

The emergence of modern science is inextricably linked with the development of the republic of letters, a multi-national community of researchers working both in institutions such as academies, as well as in informal correspondence networks: 'un rassemblement de tous les hommes attachés aux progrès des connaissances, et désireux de bâtir l'empire de la raison et de la libre critique'.[14] This republic relied on reciprocal *exchange* of people, correspondence and objects; epistolary exchange – in which important scientific findings were often reported –

12. Virginia P. Dawson, 'Foundations of natural history in eighteenth-century Geneva', in Patrick Coleman, Anne Hofmann and Simone Zurbuchen (eds), *Reconceptualizing nature, science, and aesthetics: contribution à une nouvelle approche des Lumières helvétiques* (Geneva, 1998), p.31-42 (33).
13. Rousseau, *Confessions*, *OC* i.392.
14. René Sigrist, *L'Essor*, p.16.

was the lifeblood of the republic of letters,[15] and correspondent members of learned academies played a key role in the advancement of science.[16] The republic was based on an ethic of free and reciprocal exchange of knowledge in academies, correspondence and exchange networks, and the periodical press.[17] The republic of science, which grew out of the republic of letters, was predicated upon peer review: witnessing experimental proofs of scientific claims was rigorously practised at the Royal Society of London and the Paris Académie royale des sciences. In Switzerland, the republic of science likewise prospered outside formal academic institutions because the majority of the Confederation's institutions of higher learning had a theological and humanistic focus grounded in Calvinism.

Exchanges of people, letters and things were facilitated by confessional ties among the Protestant centres of the Republic of Letters; the ties of the Protestant cantons to co-religionists in other countries facilitated the exposure of the Swiss to the main currents of European scientific culture, and their participation in the republic of science. Switzerland's prominence in science in the eighteenth century can be attributed in part to the role of the Protestant cantons as havens for French Huguenots after Louis XIV's revocation of the Edict of Nantes in 1685, when an influx of well-educated, technically-qualified Huguenots settled in the Suisse romande, transforming its economy. These refugees formed part of what one scholar has termed the 'Protestant International', the Huguenot diaspora.[18] Several Swiss botanists of note came from French Huguenot roots, notably Jean and Caspar [Gaspard] Bauhin of Geneva and Laurent Garcin (1683-1752) of Neuchâtel.

Owing to the connections of the Protestant cantons with Protestant states such as Holland, Prussia and Great Britain, Switzerland earned the epithet, *Helvetia mediatrix*, that is, a cultural crossroads. For example, thirty Swiss savants belonged to the Akademie der Wissenschaften in Berlin, and six served as division directors; many more corresponded with Jean Henri Samuel Formey (1711-1797), perpetual secretary of the Akademie, who was a second-generation Huguenot born in Prussia. One hundred and sixty Swiss correspondents exchanged two thousand letters

15. Rousseau's eight letters on botany to Mme Delessert exemplify epistolary exchange in the republic of science.

16. See for example Virginia P. Dawson, *Nature's enigma: the problem of the polyp in the letters of Bonnet, Trembley and Réaumur* (Philadelphia, PA, 1987), p.51ff.

17. See A. Cook, 'Jean-Jacques Rousseau et les réseaux d'échange botanique', p.93-114; Sigrist, *L'Essor*, p.18; Anne Goldgar, *Impolite learning: conduct and community in the Republic of Letters, 1680-1750* (New Haven, CT, 1995).

18. J. F. Bosher, 'Huguenot merchants and the Protestant International in the seventeenth century', *The William and Mary quarterly*, 3rd series, 52:1 (January 1995), p.77-102; see especially Anne Goldgar, *Impolite learning*, p.9-10; René Sigrist, *L'Essor*, p.15.

with Formey; only one of these was Catholic – a theologian from Fribourg.[19]

The Swiss Calvinists forged particularly strong links with their closest confessional brethren in the Dutch Reformed Church, and nature helped: the Rhine provided easy access from western Switzerland (Geneva, Basel and Schaffhausen) to the cosmopolitan cities of the Dutch Republic. Through this connection, Dutch science played an important role in shaping Swiss science, especially in physics, medicine and chemistry.[20]

Dutch science took its lead from the experimental, empirical focus of Sir Francis Bacon, on the one hand, and from Newtonian physics on the other.[21] Newton had the virtue of accommodating the theological preferences of conservative Calvinists in both Holland and Switzerland, since his theory of gravity could be taken as 'evidence of God's continuing, active presence in nature'.[22] The exigencies of Dutch global commerce likewise influenced the methods and research programmes of Dutch medical and botanical researchers in important ways.[23]

The Dutch Republic offered access to new ideas through exchanges of correspondence, scientific specimens and books, but also through movements of people. Education or a tour of duty in the wealthy, internationalised Dutch Republic exposed the Swiss to cosmopolitan cities, lively coffee house debates and a tolerant publishing regime. The vast Dutch colonial enterprise stretching from the Caribbean to southeast Asia offered unprecedented access to global natural history and the immense profits from the cultivation and trade in plant products – most notably spices; the acclimatisation of exotic species was an important element of the Dutch colonial venture, carried out in public and university botanical gardens at Amsterdam and Leiden, and private gardens such as that of the Anglo-Dutch merchant, George Clifford, whose rare plants are commemorated in Linnaeus's *Hortus Cliffortianus*.[24]

Examples of the influence of Dutch science on Swiss savants abound: for instance, the Neuchâtel physician-botanist Laurent Garcin (1683-

19. Bandelier, *Des Suisses dans la République des Lettres*, p.24.
20. 'L'influence des Pays-Bas, et particulièrement de Boerhaave, 's Gravesande et Muschenbroek, a été décisive dans l'implantation à Genève des traditions expérimentales en physique, en médecine, et en chimie'. Sigrist, *L'Essor*, p.132.
21. A good example was Bernard Nieuwentijt, who claimed that experiments should be used to demonstrate theological truths. Harold Cook, *Matters of exchange: commerce, medicine and science in the Dutch Golden Age* (New Haven, CT, 2007), p.382.
22. Dawson, *Nature's enigma*, p.71.
23. Cook, *Matters of exchange*, p.410ff.
24. Carolus Linnaeus, *Hortus cliffortianus* (Amsterdam, n.p., 1737); A. Cook, 'Jean-Jacques Rousseau and exotic botany'; Londa Schiebinger and Claudia Swan (eds), *Colonial botany: science, commerce and politics in the early modern world* (Philadelphia, PA, 2005).

1752), of whom more will be said in the next chapter, studied with Boerhaave at Leiden.[25] Boerhaave taught that the physician should proceed from the facts alone, free of all a priori prejudices or judgements, avoiding speculation on ultimate causes or God's will.[26] Garcin spent much of his medical career in the Vereenigde Oost-Indische Compagnie (VOC – Dutch East India Company). This was a hard life, carrying a high risk of mortality not only for the sailors, but also for the officers and ships' surgeons; those, like Garcin, who survived to return home, had in all likelihood endured many serious hardships and witnessed much human suffering.[27]

Garcin remained in contact with Boerhaave during his years with the VOC; he followed his teacher's instructions to learn about Asian medicine and returned home not only with Asian remedies in his pocket, but also with the new sexual classification system of Linnaeus, which was published in Holland in *Systema naturae* (Leiden, 1735).[28] The rapid introduction of this new system of classification to Neuchâtel demonstrates the speed with which foreign ideas could enter the Swiss scientific community. This episode is also a key element in our narrative, for it frames Rousseau's introduction to botany in Neuchâtel a quarter of a century later.

Garcin's son, Jean Laurent (1733-1781), likewise served in Holland as a pastor; some Swiss pastors took up posts in Protestant countries or in the chapels that operated in Catholic countries within the immunity of diplomatic missions. Like his father, Garcin fils was a keen botanist, and re-appears later in our story as the unacknowledged editor of the texts that were posthumously published in 1780-1782 as Rousseau's *Fragmens pour un dictionnaire des termes d'usage en botanique*.

iv. Demography

This brings us to the third important factor that crucially shaped the environment for doing science in eighteenth-century Switzerland: demographic increase. Like much of Europe, Switzerland's population increased substantially during the eighteenth century, from 1 million to 1.7 million, a net increase of 70%.[29] This increase was due in part to the

25. Boerhaave's accomplishments in botany, as contrasted with those in other sciences, were modest. See Cook, *Matters of exchange*, p.396.

26. Cook, *Matters of exchange*, p.390, 394.

27. Cook, *Matters of exchange*, p.178-80.

28. Gagnebin to Haller, 25 November 1739; quoted in Marcel S. Jacquat, 'Abraham Gagnebin, médecin, (1707-1800)', *Biographies neuchâteloises*, ed. Michel Schlup (Hauterive, 1996), vol.1, p.97-102 (101).

29. Carol Blum, '"Soyez toujours ce que vous êtes": J.-J. Rousseau, Genève, la Corse et le paradoxe de la population', in *Republikanische Tugend: Aussbildung [sic] eines schweizer*

continuation of the second French 'Réfuge', following Louis XIV's revocation of the Edict of Nantes in 1685.

This demographic increase occurred in tandem with proto-industrialisation in watch-making and textile production, especially in the Suisse romande. Stimulated by the migration of French Huguenots after the revocation of the Edict of Nantes in 1685, these industries flourished. Geneva was a major centre for textile production and watch-making, the latter industry employing 20% of males in 1725, 40% in 1788 and a large number of women as well; the city produced 100,000 watches per year: 'elle travaillera pour le monde entier'.[30] Watch-making in turn stimulated related technical fields, such as engraving and metallurgy. Swiss industries participated in global exchange networks, including the infamous triangular trade ('commerce triangulaire'); for example, Neuchâtel produced calico ('indiennages'), which was traded lucratively in West Africa for slaves. The sciences found increasing support from a bourgeoisie imbued with the values of hard work, thrift and the desire to make useful and profitable discoveries.

Service industries such as banking were also beginning to hold their own. Following the upheavals of the Reformation and its aftermath, the eighteenth century was a time of peace and prosperity for Switzerland; the incomes of Swiss mercenaries serving in foreign armies filled the coffers of Swiss banks, thereby making Geneva a financial centre to which the French king, despite Geneva's Calvinism, turned for loans. Genevan banks supported the ancien régime in France until its downfall; Jacques Necker (1732-1804), a Genevan banker, was Louis XVI's last hope to save the country's ailing finances. Bankers figured prominently in the Franco-Swiss Boy de la Tour and Delessert families, who played such an important role in Rousseau's botany.

However, despite their success, these industries did not supply sufficient opportunities for the expanding population; many young Swiss men sought their fortunes outside their homeland, which in turn played a key role in promoting developments that favoured scientific practice in eighteenth-century Switzerland. The Swiss were active overseas as merchants, bankers, pastors, physicians, scholars and tutors. Swiss commercial families placed their members at key locations – London, Paris, Geneva – helping to form what J. F. Bosher called 'The Protestant International'.[31]

Nationalbewusstseins und Erziehung eines neuen Bürgers, ed. Michael Böhrer, *et al*, (Geneva, 2000), p.447-57 (456), quoting figures provided by François de Capitani. This population increase, as Blum notes, contradicts Rousseau's theory in *Projet de constitution pour la Corse* that agriculture, not industry and commerce, promotes population growth.

30. Alfred Dufour, *Histoire de Genève*, 3rd edn (Paris, 2004), p.89.

31. Bosher, 'Huguenot merchants and the Protestant International in the seventeenth century'.

The Swiss made their particular mark outside their homeland as mercenaries, an enterprise that proved to be extremely lucrative for both the soldiers and their home cantons; the latter received payment in return for sending soldiers into foreign armies; a famous artefact of this tradition is the Pope's Swiss guard.[32] In 1748 77,000 Swiss served in foreign regiments, of whom 23,000 served in France; it was said that '[l]e Suisse est le meilleur soldat du monde'.[33] This way of earning foreign hard currency was controversial, having been decried by Béat Louis de Muralt (1665-1749) and Haller for corrupting Swiss values of thrift, hard work and honesty.

Rousseau held that in order for a country to remain happy and prosperous such trends must be checked, or better yet, never allowed to take root. According to his reasoning, material riches relate inversely to true prosperity, which lies outside the cash nexus in a subsistence economy.[34] Trade – both internal and external – manufacturing and luxury are all deleterious to the common good, causing inequality to increase and population to diminish. Political life suffers in turn; oligarchy replaces aristocratic government, the rich have their way and the rest suffer not only poverty, but exclusion. Citizens become subjects, they no longer fight their own wars and factions take over the *res publica.* Instead of employing rhetoric, those in power whisper at divans.[35]

In Rousseau's view, Switzerland's traffic in mercenaries favoured such dangerous developments. Like other Swiss moralists such as Muralt and Haller, Rousseau condemned cultural contamination by returned mercenaries, but even more, he foresaw the consequences of the wealth they brought into the country:

> Quand les Suisses [...] ont commencé à communiquer avec d'autres nations ils ont pris goût à leur manière de vivre et ont voulu l'imiter; il se sont aperçus que l'argent était une bonne chose et ils ont voulu en avoir; sans productions et sans industrie pour l'attirer, ils se sont mis en commerce eux-mêmes, ils se sont vendus en détail aux puissances, ils ont acquis par là précisément assez d'argent pour sentir qu'ils étaient pauvres; les moyens de le faire circuler étant presque impossibles dans un pays qui ne produit rien et qui n'est pas maritime, cet argent leur a porté de nouveaux besoins sans augmenter leurs ressources. Ainsi leurs premières aliénations de Troupes les ont forcés d'en

32. 'Es gibt zu viele Männer mit zu wenig Arbeit und zu wenig Raum.' Furthermore, the oligarchic cantons such as Bern preferred to export any 'Gewaltpotenzial', preferably with a profit. Under the direction of their home cantons, the Swiss formed their own regiments, had their own officers and special status in the armies of foreign powers. Joëlle Kunz, *Schweizer Geschichte – einmal anders* (Geneva, 2006), p.54; see also p.36-37.

33. Jean-Jacques Rousseau, *Deux lettres à M. le Mareschal Duc de Luxembourg (20 et 28 janvier 1763)*, ed. Frédéric S. Eigeldinger (Neuchâtel, 1977), p.70; see also Capitaine [Paul Emanuel] de Vallière, *Honneur et fidelité: histoire des Suisses au service étranger* (Neuchâtel, n.d.).

34. Rousseau, *Corse*, *OC*, iii.914, 922.

35. Rousseau, *Essai sur l'origine des langues*, *OC*, v.428.

faire de plus grandes et de continuer toujours. La vie étant devenue plus dévorante, le même pays n'a plus pu nourrir la même quantité d'habitants.[36]

The Swiss were therefore becoming impoverished as they became more embedded in the international cash nexus.[37] Subsistence household autarchy is the only economy which assures liberty by assuring equality: 'C'est ainsi qu'au milieu de leurs précipices et de leur vallons[,] chacun vivant sur son sol parvint à en tirer tout son nécessaire [...] à ne désirer rien au délà'.[38]

The significance of foreign service for Switzerland is disputed; according to some, it actually promoted the Swiss sense of national identity.[39] Rousseau observed that the Swiss, who invented the illness of 'Heimweh' (which he called 'Hemé' and 'Hemvé'), longed for Switzerland when they left it.[40] The sound of the famed ranz des vaches could bring them to tears or worse – flight from their regiments. However, their longing for their homeland diminished with the deterioration in their mores: 'ayant perdu le goût de leur première simplicité, ils ne la regrettent plus quand on la leur rappelle'.[41]

Rousseau could not or would not accept that Switzerland's lack was also her gain; foreign service opened up a variety of career paths in armies, diplomacy, churches, learned academies, banks, and trading companies; it offered exposure to new ideas, stimulated the cultivation of the sciences in the home country and was instrumental in positioning Switzerland at the forefront of scientific and technological fields such as geology, cartography, engineering, mechanical drawing, mathematics and physics.[42] Medicine and her traditional handmaiden, botany, also

36. Rousseau to the maréchal de Luxembourg, 20 January 1763; *CC*, xv.50; see also Rousseau, *Projet de constitution pour la Corse*; *OC*, iii.915. The translator of Rousseau's botanical writings into English, the Reverend Thomas Martyn (1735-1825), likewise commented on the high cost of living in Switzerland, but notes that one lives better in Switzerland. See Thomas Martyn, *Sketch of a tour through Swisserland: with an accurate map. a new edition. to which is added a short account of an expedition to the summit of Mont Blanc, by M. De Saussure, of Geneva* (London, G. Kearsley, 1788), p.16.
37. Rousseau, *Projet de constitution pour la Corse*, *OC*, iii.924; Rousseau to the maréchal de Luxembourg, 20 January 1763; *CC*, xv.50.
38. Rousseau, *Projet de constitution pour la Corse*, *OC*, iii.914. See also Céline Spector, 'Rousseau et la critique de l'économie politique', in *Rousseau et les sciences*, ed. Bernadette Bensaude-Vincent and Bruno Bernardi (Paris, 2003), p.237-56 (249).
39. '[T]hat foreign residence of the Swiss played a constitutive role in the formation of Swiss national consciousness', Simone Zurbuchen, 'Berliner "Exil" und Schweizer "Heimat": Johann Georg Zimmermanns Reflexionen über die Rolle des Schweizer Gelehrten', in *Schweizer im Berlin des 18. Jahrhunderts*, ed. Martin Fontius and Helmut Holzhey (Berlin, 1996), p.57-68 (60-61).
40. A Swiss doctor described 'Heimweh' or 'nostalgie' in a treatise of 1688. Leigh, *CC*, xv.67, note (g).
41. Rousseau, *Dictionnaire de la musique*; *CC*, v.924.
42. See Fueter, *Geschichte der exakten Wissenschaften*.

benefited significantly from Swiss physicians serving abroad as military surgeons, or academics.

Most eighteenth-century botanists had a medical background and earned their living as physicians; Swiss botanists were no exception. However, of the Swiss universities, only Basel offered medical education, and aspiring physicians often chose to study abroad – in Protestant Holland at Leiden, in Germany, at Montpellier in France, which had a tradition of toleration. As we have seen, this turning toward the outside had extremely important consequences for Swiss botany, bringing its practitioners into early contact with the cutting-edge classification and nomenclature of Linnaeus.

A noteworthy example of an expatriate savant was the physician, Albrecht von Haller (discussed below, p.78-86), who was lured from Bern to the German kingdom of Hannover to take up a professorship at its flagship university in Göttingen in 1736; he returned to his native Bern in 1753. Another notable case was Johann Georg Zimmermann of Brugg (Aargau) (1728-1795), physician ('Leibarzt') in Hannover to George III of England; two of the Neuchâtel physicians connected with Rousseau's introduction to botany – Laurent Garcin and Abraham Gagnebin (1707-1800) – served in military units in Holland, Southeast Asia and France, respectively.

These expatriate Swiss savants were competing with the best in their fields, and many attained membership in foreign academies; moreover, they enjoyed greater impact and recognition abroad than they would have had at home.[43] The names of famous eighteenth-century Swiss savants with an international profile included the mathematicians: Leonhard Euler (1707-1783), Daniel, Johann and Jakob Bernoulli (1700-1782; 1710-1790 and 1759-1789 respectively) of Basel; the biologists: Albrecht von Haller, Charles Bonnet and Abraham Trembley; and the geologists: Jean-André Deluc and Horace-Bénédict de Saussure (1740-1799), the last four all of Geneva.[44] This career pattern enabled even tiny Neuchâtel to boast botanists of the first rank.

Some Swiss savants departed, never to return. Daniel Bernoulli and Leonhard Euler served on the staff of foreign academies. Their careers illustrate the influence of Swiss expatriates in the sciences of the period. Euler divided his career between Russia and Germany, having been

43. Fueter, *Geschichte der exakten Wissenschaften*, p.123.

44. Another important source of scientific impetus was the clergy: those who resided in the countryside and found themselves with time on their hands and the ready means to employ it in the study of natural history, in particular. Fueter, *Geschichte der exakten Wissenschaften*, p.100ff. Their theology supported these enterprises, as crucial to finding evidence of God's beneficence in the Creation. Hence, in Switzerland, as elsewhere, the clergy made important contributions to science during the Enlightenment.

invited to the Imperial Russian Academy of Sciences at St Petersburg in 1727, where he joined his compatriot, Daniel Bernoulli. In 1741 Euler went to Berlin at the invitation of Frederick the Great to direct the mathematics division of the Akademie; after twenty-five years in Berlin, he returned to Russia at the invitation of Catherine II, and died there.

The botanist, Johann Amman (1707-1741), born in Schaffhausen, was another Swiss who, like Euler, spent much of his career in Russia and made a major contribution not only to Russian botany, but also to the field in general. An indication of his importance is that Rousseau had a copy of Amman's 1739 work, *Stirpium rariorum in imperium Rutheno sponte provenientium,* in his last botanical library at Ermenonville.[45] Like Garcin, Amman studied medicine at Leiden and then worked as an assistant to Sir Hans Sloane (1660-1753) in London; through the latter he developed important international botanical connections. He corresponded with Linnaeus about the latter's *Flora lapponica* (which contained plants related to or the same as those around St Petersburg), and like his colleague, Johann Siegesbeck (1686-1755), critiqued Linnaeus's artificial sexual system of classification, but in a more collegial and less adversarial way.[46] Amann apparently sent specimens to Sloane, while specimens from Sloane and other correspondents form part of the important botanical collections that Amman developed for the Kunstkammer of the St Petersburg Academy of Sciences. Amman was closely involved in the creation of the Academy's Botanic Garden (1736-1738), was professor of botany in St Petersburg and a member of the St Petersburg Academy of Sciences.[47] Like Euler, Amann never returned to Switzerland; he died in Russia.

Rousseau's friend, Jean André Deluc (1727-1817) of Geneva epitomised the eighteenth-century expatriate Swiss savant. In 1773 he was named 'reader' to Queen Charlotte of Great Britain; he was elected a Fellow of the Royal Society, London, a corresponding member of the Paris Académie royale des sciences and honorary professor of geology and philosophy at the University of Göttingen in 1798. He was the first to use the term 'géologie', and published leading works in geology and meteorology.

45. Johann Amman, *Stirpium rariorum in imperio Rutheno sponte provenientium* (St Petersburg, Imperial Academy of Sciences, 1739). Concerning Rousseau's copy of this work, see Girardin to Du Peyrou, 4 October 1778; *CC*, xlii.11. Why Rousseau was interested in the St Petersburg flora is not known.

46. See Margery Rowell, 'Linnaeus and botanists in eighteenth-century Russia', *Taxon* 29:1 (February 1980), p.15-26. The controversy surrounding the artificial sexual system is discussed in Chapter 5.

47. W. J. Bryce, *A Botanist's paradise: the establishment of scientific botany in Russia in the eighteenth century* (Ystalyfera, 2008), p.72-73, 77; Dimitry D. Sokoloff, *et al.*, 'The history of botany in Moscow and Russia in the 18th and early 19th centuries in the context of the Linnaean collection at Moscow University (MW)', *Huntia* 11:2 (2002), p.129-91 (150-51).

v. Genevan science

Deluc's achievements reflected the favourable conditions for science in eighteenth-century Geneva, a centre of enlightened thought. Deluc's longevity and long residence overseas spread the reputation of Geneva – the 'Protestant Rome' – as an enlightened city that favoured practical science. Together with Haller, Amman, Euler, Bernoulli and many others, Deluc helped place Switzerland, still thought in many quarters (especially France) to be backward, on the map of enlightened Europe.[48]

Geology was, above all, a Genevan endeavour in the eighteenth century. Deluc coined the term 'geology', while his patrician compatriot, H. B. de Saussure, author of *Voyage dans les Alpes* and another important founder of geology, gave a famous account of his ascent of Mont-Blanc in 1787. Geneva's emergence as a centre of the republic of science influenced the reception of modern ideas about the natural world in all of francophone Switzerland.[49]

Geneva was the city of Rousseau's birth and exercised an immense influence on his life and work. Yet the influence of Genevan science on Rousseau's botanical studies has, to my knowledge, hardly, if ever, been mooted. Rousseau's scientific proclivities can in fact be traced to his Genevan origins and his continuing contact with Genevan science into the mid-1760s. These proclivities included his inclination toward observation, empiricism and experimentation and his rejection of the Philosophes as 'ardents missionnaires d'athéisme et très impérieux dogmatiques'.[50] Like other participants in the Swiss Enlightenment, Rousseau accepted empirical methods of investigating nature, while rejecting the atheism that often accompanied it. Genevan savants happily employed cutting-edge scientific methods drawn from Dutch and English science; they were inspired by Newtonian physics while preserving a conservative Protestant theology, resisting the deistic and atheistic tendencies of the French philosophes.[51]

Genevan science, like science generally in Switzerland, developed somewhat more slowly than in England, France, Prussia and Russia, due in large part to the lack of official support; for example, the Société de physique et d'histoire naturelle of Geneva was not founded until 1790. However, appearances are somewhat deceiving. While it might be easy to attribute the slow development of science in Geneva to Jean Calvin (1509-1564), he in fact held the sciences in high esteem as 'ruisseaux qui

48. Concerning eighteenth-century Switzerland's reputation for primitivism, see Roger Francillon, 'The Enlightenment in Switzerland', in Patrick Coleman *et al* (eds), *Reconceptualizing nature*, p.13-27 (13), quoting Marie-Jean Hérault de Séchelles.

49. Sigrist, *L'Essor*, p.14.

50. Rousseau, *Rêveries*, *OC*, i.1016.

51. Dawson, *Nature's enigma*, p.65; Francillon, 'The Enlightenment in Switzerland', p.23-24.

découlent de la fontaine de l'esprit de Dieu'.[52] Calvin was not prepared, of course, to give unlimited latitude to human curiosity, for he regarded 'les secrets de la nature' as 'secrets de Dieu', that human knowledge should not penetrate. His esteem for the sciences was compatible with his vision for the Académie (theological seminary) that he founded in Geneva. Calvin's international influence, Geneva's role as a centre of the Reformation and her reception of Protestant refugees exposed the city to many external intellectual influences.

Calvin was particularly interested in medicine, an interest stimulated by the repeated return of the plague.[53] In the sixteenth century there were attempts, often undermined by financial considerations, to establish medical instruction at Geneva. Dissections of cadavers (usually of executed criminals) formed part of medical instruction as early as 1564, the year of Calvin's death, and the botanist Jean Bauhin (1541-1613; brother of the more famous Caspar Bauhin, author of the *Pinax*) was appointed in 1568 to lecture twice a week on medicine. Geneva thus offered a 'terrain nourricier' for the sciences.[54]

Calvin's positions have to be distinguished from those of his dogmatic, Aristotelian successors such as Theodore Beza (1519-1605), who arrested the development of Genevan science.[55] However, by the late seventeenth century the Académie had abandoned dogmatic Calvinism for a more latitudinarian theology and tolerance for scientific investigation that, Perrot argues, reflected the thinking of Calvin himself.[56] This trend had begun with the appointment of the Cartesian, Jean-Robert Chouet (1642-1731), as professor of philosophy at the Académie in 1669. Chouet, who had taught at the more liberal (or heretical) theological academy at Saumur, in France, was determined 'to make sure that Genevan science did not develop in isolation from the main European intellectual currents'.[57] Geneva thus dates its acceptance of modern science (for example heliocentrism, Newton's theories) from Chouet's tenure at the Academy. In the late seventeenth century the Académie was reformed by the natural theologian, Jean-Alphonse Turrettini (1671-1737), who espoused the argument from design and a critical reading of Scripture. This had important consequences for the teaching of science: 'with Turrettini's reform [...] came the recognition that natural philosophy could reinforce orthodox religion'.[58]

52. Quoted from Calvin's commentary on Job in Perrot, *Jean Calvin*, p.5, p.17, n.5.
53. Perrot, *Jean Calvin*, p.6, 35.
54. Perrot, *Jean Calvin*, p.49.
55. Perrot, *Jean Calvin*, p.31.
56. Perrot, *Jean Calvin*, p.5-12.
57. Dawson, *Nature's enigma*, p.64-65.
58. Victoria P. Dawson, 'Foundations of natural history in eighteenth-century Geneva', in Coleman *et al.*, (eds), *Reconceptualizing nature*, p.31-42 (33).

In the early eighteenth century Gabriel Cramer (1704-1752) and Jean-Louis Calandrini (1703-1758) brought mathematical and scientific instruction to Calvin's Académie. They trained many important scientists, including Abraham Trembley (1710-1784) and Charles Bonnet (1720-1793), who today would be called biologists, and Jean Jallabert (1712-1768), an expert on electricity. Together, Bonnet and Trembley transformed the understanding of reproduction, most notably through Trembley's famous experiments on the polyp (hydra) and Bonnet's demonstration of parthenogenesis in aphids. Their dramatic findings prompted metaphysical questions that they themselves – as conservative members of the Calvinist oligarchy – found disquieting.[59]

It might seem that Rousseau, the self-educated son of a watchmaker, would have had little, if anything, to do with elite citizens who pursued science and socialised together (such as Trembley, Horace Bénédict de Saussure and Bonnet, who were related to each other). The teaching of the sciences in the Académie of Geneva should likewise be expected to have played no role in his scientific awareness. Yet by the time of his visit to Geneva in 1754, Rousseau was quite aware of Geneva's distinguished record in the investigation of nature, and gained access to the most eminent members of the Genevan intelligentsia. One of these contacts was Jallabert, who was the Académie's professor of mathematics and experimental physics.[60]

Botany, the handmaiden of medicine, had had significant practitioners in Geneva from the sixteenth century onwards, becoming renowned through the work of the French Huguenot brothers, Jean and Caspar [Gaspard] Bauhin, referred to above.[61] Jean Bauhin, who had been called to teach medicine in Calvin's Académie in 1568, authored a *Historia plantarum universalis* that was prepared for publication in 1650-1651 by another Genevan physician, Dominique Chabrey (1610-1669).[62] Chabrey also published a truncated version of this work as *Omnium stirpium sciagraphia et icones* (Geneva, 1678) that Rousseau acquired in his late botanical phase.[63]

59. Dawson, *Nature's enigma*, p.53.

60. Rousseau read the *Discours sur l'origine d'inégalité* (but not the Epistle Dedicatory) to Jallabert, 'qui en parut transporté'. Rousseau, *Confessions*, *OC*, i.394.

61. Caspar Bauhin authored the famous *Pinax*, which Rousseau considered an indispensable resource in synonymy. Rousseau to Malesherbes, 17 April 1772; *CC*, xxxix.36-41.

62. Even though he considered Chabrey's works 'assez médiocres', Augustin Pyramus de Candolle (1778-1841) chose a bust of Chabrey to represent seventeenth-century Genevan botany in the newly established Jardin botanique of Geneva. *Mémoires et souvenirs*, ed. Jean-Marc Drouin and Jean-Daniel Candaux *et al.* (Geneva, 2004), p.354.

63. Girardin to Du Peyrou, 4 October 1778; *CC*, xlii.11. Rousseau's annotated copy of Chabrey was discovered in the Lloyd Library, Cincinnati, OH, in 2007.

Genevan science undoubtedly influenced Rousseau's thought in other ways that have scarcely been acknowledged. For example, Rousseau's social origins connected him closely to artisans, who were adept in instrument-making and sometimes practised – empirical sciences. Thus, even though Rousseau did not come from the elite milieu that produced the majority of eighteenth-century Genevan scientists, he came from the same literate, well-educated artisanal milieu that produced the geologist, Jean-André Deluc. Both Rousseau, the unschooled polymath, and Deluc, the founder of modern geology, were sons of master watchmakers. As Rousseau recounts in the *Confessions*, reading at home with his father formed an important part of his childhood education. Deluc's father, Jacques-François (1698-1780), known as the 'bonhomme' Deluc, turned author, and wrote *Observations sur les savans incrédules*. This work excited the patrician ire of Charles Bonnet, who asserted that Deluc, 'un simple horloger', took liberties in pronouncing on religion and should stay out of politics.[64]

Rousseau was close to the Deluc family, with whom he had much in common, including an interest in science and shared political views. In September 1754 Rousseau, Thérèse, Jean-André, his brother, Guillaume-Antoine Deluc (1729-1812) and other members of the Deluc family, made a seven-day tour of Lake Geneva: 'Nous mîmes sept jours à cette tournée par le plus beau temps du monde.'[65] This journey also had a scientific dimension: 'Tous ensemble ils étaient en quête de témoins géologiques, en particulier de fossiles sur lesquels dissertaient alors à perte de vue les physiciens, professionnels ou amateurs.'[66] This voyage made an indelible impression on Rousseau, who later called upon the recollections from this journey when writing *Julie*:

> Plus j'approchais la Suisse, plus je me sentais ému. L'instant où, des hauteurs du Jura, je découvris le lac de Genève fut un instant d'extase et de ravissement. La vue de mon pays, de ce pays si chéri où des torrents de plaisirs avaient inondé mon cœur; l'air des Alpes si salutaire et si pur; le doux air de la patrie, plus suave que les parfums de l'Orient; cette terre riche et fertile, ce paysage unique, le plus beau dont l'œil humain fut jamais frappé; ce séjour charmant auquel je n'avais rien trouvé d'égal dans le tour du monde; l'aspect d'un peuple heureux et libre; la douceur de la saison, la sérénité du Climat; mille souvenirs délicieux qui réveillaient tous les sentiments que j'avais goûtés; tout cela me jetait dans des transports que je ne puis décrire, et semblait me rendre à la fois la jouissance de ma vie entière.[67]

64. Bonnet to Haller, 15 June 1762; *CC*, xi.85-88.
65. Rousseau, *Confessions*, *OC*, i.393.
66. Alexis François, *Lyrisme et géologie: le séjour de J.-J. Rousseau à Genève en 1754* (Geneva, 1941), p.9.
67. St Preux to Milord Edouard, part 4, letter 6; *Julie*, *OC*, ii.419.

The detailed attention to the earth and its productions recommended in *Emile* likewise attests to Rousseau's association with the Delucs:[68] 'Rousseau a certainement su gré à ses mentors genevois de leurs leçons de physique.'[69]

Another testimony to the Delucs' influence on Rousseau's interest in the emerging science of geology is found in a letter to the maréchal de Luxembourg from Môtiers in January 1763:

> Plus j'examine en détail l'état et la position de ce Vallon, plus je me persuade qu'il a jadis été sous l'eau, que ce qu'on appelle aujourd'hui le Val de Travers fut autrefois un lac formé par l'Areuse[,] la Cascade et d'autres ruisseaux, et contenu par les montagnes qui l'environnent, de sorte que je ne doute point que je n'habite l'ancienne demeure des poissons. En effet le sol du vallon est si parfaitement uni qu'il n'y a qu'un dépôt formé par les eaux qui puisse l'avoir ainsi nivelé. Le prolongement du Vallon, loin de descendre[,] monte le long du cours de l'Areuse de sorte qu'il a fallu des temps infinis à cette rivière pour se caver dans les abymes qu'elle forme un cours en sens contraire à l'inclinaison du terrain. Avant ces temps, contenue de ce côté de même que de tous les autres, et forcée de refluer sur elle-même, elle dut enfin remplir le vallon jusqu'à la hauteur de la première grotte que j'ai décrite, par laquelle elle trouva ou s'ouvrit un écoulement dans la galerie souterraine qui lui servait d'aqueduc.[70]

Where geology met religion, the Delucs may have differed from Rousseau, however; for Rousseau, an adherent of the ideas of Buffon on the history of the earth, may not have accepted the biblically based views of geology and natural history espoused by the Delucs. A biblical interpretation did not necessarily mean literal interpretation, since by mid-century critical reading of Scripture had been accepted. Deluc could therefore transform the six biblical days of the Creation into '*gahambars*, ou périodes composées peut-être chacune de plusieurs milliers d'années, qui ont dû précéder l'état actuel du globe'.[71] As might be expected, the Delucs opposed Buffon's theory of the formation of the earth on religious grounds. In the burning geological controversy of the time, Jean-André Deluc sided with the Neptunists, such as the German geologist Abraham Gottlob Werner (1749-1817), who held that the history of the earth is one of gradual change, in contrast with vulcanism, a cataclysmic version of the history of the earth.

Deluc also facilitated Rousseau's access to botanical instruments and

68. Rousseau, *Emile*, *OC*, iv.772. The Delucs may have also influenced Rousseau's interest in writing a 'Histoire du Valais' that remained unfinished and unpublished. François, *Lyrisme et géologie*, p.14-15; Rousseau, *Histoire du Valais*, *OC*, v.493-95.
69. François, *Lyrisme et géologie*, p.12.
70. Rousseau to the Maréchal de Luxembourg, 28 January 1763; *CC*, xv.119.
71. Original emphasis; Isaac Benguigui *Genève et ses savants: physiciens, mathématiciens et chimistes au XVIIIe et XIXe siècles* (Geneva, 2006), p.37.

painters' supplies for an optique (discussed in Chapter 7, p.228-31).[72] As the son of a master artisan, Deluc was well-connected in the higher 'artisanat', and gave Rousseau privileged access to precision optical devices such as magnifying glasses and dissecting instruments. Rousseau assured him: 'Le microscope est comme vous l'avez deviné pour la botanique, ainsi je désire qu'il ait un champ suffisant pour embrasser le pistil et les étamines d'une petite fleur. Du reste je m'en rapporte absolument à vous.'[73] Deluc even offered to make these instruments himself. Later Rousseau added specialised botanical instruments to his requests:

> Tout bien pensé, j'abandonne l'optique pour la botanique, et si *notre ami* [Jean-André Deluc] était à portée de me *faire faire les petits outils* nécessaires pour la *dissection des fleurs*, je serais sûr que son *intelligence* suppléerait avantageusement à celle des ouvriers. Ces outils consistent en trois ou quatre *microscopes*[74] de différents foyers, de petites pinces délicates et minces pour tenir les fleurs, des ciseaux très fins, canifs et lancettes pour les découper. Je serais bien aise d'avoir le tout à double, excepté les microscopes; parce qu'il y a ici quelqu'un [Du Peyrou] qui a le même goût que moi, et qui a été mal servi.[75]

Geneva, with its already highly developed technical capabilities, offered Rousseau unparalleled access to the best instruments available for pursuing his artistic and scientific interests.[76]

The Delucs remained loyal friends of Rousseau; Rousseau nursed 'le bonhomme' Jacques Francois Deluc when he fell ill at Môtiers in 1762, sending for Dr Jean-Antoine d'Ivernois (1703-1765; later Rousseau's botany teacher), 'un médecin fort habile', to tend him.[77] Deluc père was a leader of the Genevan bourgeoisie during the representations unleashed by Rousseau's persecution by the Petit Conseil of Geneva that began in June 1762. In October 1763 Deluc persuaded Rousseau to respond to the *Lettres écrites de la campagne*, and the result was the *Lettres écrites de la montagne* (1764). The Delucs sent Rousseau regular bulletins about the evolving political situation in Geneva.[78] Declaring that he

72. Rousseau to Jean-André Deluc, 20 December 1764; *CC*, xxii.256-58.
73. Rousseau to Jean-André Deluc, 20 December 1764; *CC*, xxii.256.
74. Rousseau seems to be asking for a 'simple' microscope, also known as a 'loupe'. Microscopes could be 'simples, ou loupes' or 'composés', i.e. 'microscopes proprement dits, ceux qui renversent l'image'. See Littré, 'microscope'.
75. My emphasis; Rousseau to François Henri d'Ivernois, 29 July 1765; *CC*, xxvi.106-107; see also Rousseau to Jean André Deluc, 15 December 1764; *CC*, xxii.240-41.
76. On Rousseau's interest in the instruments for doing science, see Bernadette Bensaude-Vincent, 'La nature laboratoire', in *Rousseau et les sciences*, p.155-74 (161ff.).
77. Rousseau to Jean-André and Guillaume Antoine Deluc, 16 October 1762; *CC*, xiii.215.
78. Rousseau shrewdly urged the Delucs to evade detection by using a secure means to communicate with him, rather than the postal service. Rousseau to Jacques-François Deluc, 25 October 1763; *CC*, xviii.69-72.

would love them forever, Rousseau ended his correspondence with the brothers Deluc on 24 February 1765 because he decided to abandon all contact with Geneva and all concern for its problems.[79]

vi. Helvetism meets the Alps

Helvetism, or the patriotic feeling for the whole of Switzerland (the 'Gesamtschweiz'), came to the fore in the eighteenth century; this movement was reinforced by the unprecedented scientific, literary and artistic celebration of Switzerland's most unique natural endowment – the Alps.[80] Regarded in the past as dangerous and demonic, mountain landscapes were taking on a completely new significance; Caspar Wolff's famous painting of the glacier at Lauteraar (fig. 3.1, p.72) depicts the novel scene of tourists viewing this natural wonder from under the protection of an umbrella![81]

An early Alpine explorer was Johann Jacob Scheuchzer (1672-1733): physician, mathematician, Fellow of the Royal Society, London, and the first Swiss writer to popularise the beauty of the Alps. He recorded his Alpine journeys in *Ouresphcitës Helveticus, sive itinera per Helvetiae alpines regiones* (1702, 1708 and 1723).[82] Since Scheuchzer wrote in Latin, his work was readily accessible to European elites, including English tourists, who visited this exotic European hinterland in increasing numbers throughout the century.

As one of those who popularised the Alpine landscape, Rousseau wrote about the strong, even bizarre, contrasts they presented:

> Ce n'était pas seulement le travail des hommes qui rendait ces pays étranges si bizarrement contrastés: la nature semblait encore prendre plaisir à s'y mettre en opposition avec elle-même, tant on la trouvait différente en un même lieu sous divers aspects! Au levant les fleurs du printemps, au midi les fruits de l'automne, au nord les glaces de l'hiver: elle réunissait toutes les saisons dans le même instant, tous les climats dans le même lieu, des terrains contraires sur le même sol, et formait l'accord inconnu partout ailleurs des productions des plaines et de celles des Alpes. Ajoutez à tout cela les illusions de l'optique, les pointes des monts différemment éclairées, le clair-obscur du soleil et des ombres, et tous les accidents de lumière qui en résultaient le matin et le soir; vous aurez quelque idée des scènes continuelles qui ne cessèrent d'attirer mon admiration, et qui semblaient m'être offertes en un vrai théâtre; car la perspective des monts, étant verticale, frappe les yeux

79. Rousseau to Jean-André and Guillaume Antoine Deluc, 24 February 1765; *CC*, xxiv.87-88. This episode is discussed in Chapter 4.

80. Im Hof, *Mythos Schweiz*, p.86, 92, 109.

81. Camilla Murgia, 'Le mythe de Rousseau chez Zurlauben et Laborde', *Annales de la société Jean-Jacques Rousseau* 45 (2003), p.583-617 (589).

82. Johann Jacob Scheuchzer, *Ouresphoitës helveticus, sive itinera per Helvetiae alpines regiones*, 4 vols. in 2 (Leiden, 1702; London, Henry Clements, 1708; Leiden, Pieter van der Aa, 1723).

Figure 3.1: Caspar Wolf, *Der Lauteraargletscher mit Blick auf den Lauteraarsattel* [The Lauteraar glacier with a view of the Lauteraar saddle], 1776, oil on canvas. Photo: akg-images.

tout à la fois et bien plus puissamment que celle des plaines, qui ne se voit qu'obliquement, en fuyant, et dont chaque objet vous en cache un autre.[83]

Rousseau saw the mountains as a source of physical well-being, comparable to any known medicine: 'Je doute qu'aucune agitation violente, aucune maladie des vapeurs pût tenir contre un pareil séjour prolongé, et je suis surpris que des bains de l'air salutaire et bienfaisant des montagnes ne soient pas un des grands remèdes de la médecine et de la morale'.[84] Mountain herbs already had a venerable reputation for their curative powers, exemplified by the famous 'Faltranck' (discussed below, p.76) and other tisanes. This same pure mountain environment made the peoples who inhabited these ethereal regions moral, and uncorrupted:

> En effet, c'est une impression générale qu'éprouvent tous les hommes, quoiqu'ils ne l'observent pas tous, que sur les hautes montagnes où l'air est pur et subtil [...] plus de sérénité dans l'esprit, les plaisirs y sont moins ardents, les passions plus modérés [...] Il semble qu'en s'élevant au-dessus du séjour des hommes on y laisse tous les sentiments bas et terrestres, et qu'à mesure qu'on approche des régions éthérées l'âme contracte quelque chose de leur inaltérable pureté.[85]

Rousseau was not alone in holding up mountain people as a model to more corrupt peoples.[86] Saussure, no friend of Rousseau, offered moral reflections in his *Voyage dans les Alpes* that map closely onto those of Rousseau: the people of the mountains are 'assez civilisés pour n'être pas féroces, et assez naturels pour n'être pas corrompus'.[87]

At the same time, the mountains offered immense scientific interest. From the slopes of the Alps, it was a short step to the emergent studies of geology and paleontology, in which the Swiss, especially Jean-André Deluc and Saussure, took a leading role; an early, important study of fossils, *Traité des pétrifications* (Paris, Briasson, 1742), was undertaken by the 'Pliny of Neuchâtel', Louis Bourguet (1678-1742), in collaboration with the pastor of La Chaux-du-Milieu, Pierre Cartier and Abraham Gagnebin.[88]

Rousseau never visited the Swiss Alps, but nonetheless contributed

83. Rousseau, *Julie*, *OC*, ii.77.
84. Rousseau, *Julie*, *OC*, ii.78-79.
85. Rousseau, *Julie*, *OC*, ii.78.
86. Rousseau, *Julie*, *OC*, ii.79ff.
87. Horace-Bénédict de Saussure, *Voyages dans les Alpes* (Neuchâtel, Samuel Fauche, 1779), vol.1, p.ix; quoted in Murgia, 'Le mythe de Rousseau', p.590, n.17.
88. Bourguet was a versatile member of the republic of letters, and in contact with many of its leading figures; he held the first professorship in philosophy and mathematics at Neuchâtel when the principality was just beginning to develop its cultural resources. Bourguet influenced the Neuchâtel savants discussed in Chapter 5. See Jean-Paul Schaer, 'Louis Bourguet: philosophe et naturaliste (1678-1742)', in Schlup (ed.) *Biographies neuchâteloises*, vol.1, p.17-22; Nathalie Guillod, 'Une esquisse de l'élite culturelle neuchâteloise dans la seconde moitié du XVIIIe siècle', *Revue historique neuchâteloise* 2 (2007), p.107-24.

significantly to alpine tourism with his famous descriptions of them in *Julie*:

> Supposez les impressions réunies de ce que je viens de vous décrire, et vous aurez quelque idée de la situation délicieuse où je me trouvais. Imaginez la variété, la grandeur, la beauté de mille étonnants spectacles; le plaisir de ne voir autour de soi que des objets tout nouveaux, des oiseaux étranges, des plantes bizarres et inconnues, d'observer en quelque sorte une autre nature, et de se trouver dans un nouveau monde. Tout cela fait aux yeux un mélange inexprimable, dont le charme augmente encore par la subtilité de l'air qui rend les couleurs plus vives, les traits plus marqués, rapproche tous les points de vue; les distances paraissant moindres que dans les plaines, où l'épaisseur de l'air couvre la terre d'un voile, l'horizon présente aux yeux plus d'objets qu'il semble n'en pouvoir contenir: enfin le spectacle a je ne sais quoi de magique, de surnaturel, qui ravit l'esprit et les sens; on oublie tout, on s'oublie soi-même, on ne sait plus où l'on est.[89]

The generality of these descriptions is striking; indeed, everything would have been new and surprising for Rousseau had he actually visited this region. His descriptions were in fact based on written works rather than direct observation. It appears that Scheuchzer's work, which Rousseau requested in 1767 from his Amsterdam publisher under the title *Iter alpinum*, was the basis for the descriptions on which the passages in the novel draw.[90]

However, Rousseau only obtained Scheuchzer after he wrote his famous passages on the Alps in *Julie*. Thus, it was probably another work – albeit heavily influenced by Scheuchzer – that actually shaped Rousseau's view of the Alps, especially the Valais and the St Gotthard pass: *Les Délices de la Suisse* by Abraham Ruchat (1678-1750), first published in 1714, and republished with a related work in 1730 under the title *L'Etat et les délices de la Suisse*.[91] This 'guide de voyage' was 'indispensable lorsqu'on s'intéressait aux Alpes'.[92] Rousseau's own depiction of the alpine landscape in *Julie* was 'composite', since it derived from Ruchat or Altmann and excursions in the Savoyard Alps, rather than the Swiss Alps.[93]

89. Rousseau, *Julie*, *OC* ii.79.
90. He received it from Rey via Jean Dandiran and Francois Coindet. Rousseau to Rey, 28 December 1767; *CC*, xxxiv.268; Coindet to Rousseau, 16 May 1768; *CC*, xxxv.298. See Marguerite Richebourg, *Essai sur les lectures de Rousseau* (Geneva, 1934), *passim*, which does not include it.
91. The works of Abraham Ruchat, *Les Délices de la Suisse* (Leiden, Pieter van der Aa, 1714) and Abraham Stanyan, *L'Etat de la Suisse* (Amsterdam, chez Jean Garrel/chez les frères Wetsten, 1714) were later published by the Bernese pastor Johann Georg Altmann in a combined and expanded work as *L'Etat et les délices de la Suisse en forme de relation critique* (Amsterdam, Wetsteins and Smith, 1730).
92. Claude Reichler, *La Découverte des Alpes et la question du paysage* (Geneva, 2002), p.172.
93. Reichler, *La Découverte des Alpes*, p.172. Later, in Môtiers, Rousseau became one of the Swiss exotics to be visited by European travellers, including the intrepid James Boswell.

Another important source of the burgeoning interest in the Alps and other botanically rich regions of the country were works that were both poetic and scientific, such as Albrecht von Haller's poem, *Die Alpen* (1729) and his botanical journey, *Iter Helveticum anni 1739* (Göttingen, 1740).[94] The latter recounted Haller's voyage through Switzerland, including his visit to the Creux du Van in the Jura, accompanied by Jean-Antoine d'Ivernois and Abraham Gagnebin, later Rousseau's botany teachers. Rousseau knew both of these works; he read *Die Alpen* in the translation by Vincenz Bernhard Tscharner (1728-1778), with whom he corresponded about Swiss politics.[95]

A new genre of painting complemented the literature of Swiss mountain life and exploration; Caspar Wolf depicted the most remote and exotic elements of the most extreme mountain environments for the first time (fig.3.1). Artists ventured into the Alps to depict waterfalls, glaciers, and terrifying ravines.

vii. The flowering of Swiss botany

In 1741-1742 a Genevan apothecary and a Neuchâtel physician debated the state of botany in Switzerland in the first pan-Helvetic publication, the *Journal helvétique*. This journal, published in Neuchâtel, facilitated contacts across linguistic and religious divides, in the era in which Swiss patriotism began to crystallise. The apothecary, Jean-Baptiste Tollot (1698-1773),[96] lamented the neglect of botany in Switzerland in general, as well as in his native city: 'Il est surprenant qu'en Suisse, et à Genève en particulier, ou l'on cultive les Sciences avec succès, on ait tout à fait négligé la Botanique, qui est une partie si essentielle à la Médicine.'[97] Jean-Antoine d'Ivernois (1703-1765), the Neuchâtel physician, is known to posterity as Rousseau's botany teacher; he defended Swiss botany,

94. Haller was then professor at Göttingen, in the German kingdom of Hannover. Albrecht von Haller, 'Extrait de deux pièces que Monsieur Haller, docteur en médecine, médecin du roi de la Grande-Bretagne, professeur en anatomie, en botanique et en chirurgie dans l'université de Gottingue et membre de la Société Roïale des sciences d'Angleterre et de Suède, a publiées depuis peu, sur la botanique', *Journal helvétique* (August 1740), p.172-210 (193). Haller's account was considered to be of sufficient interest for extracts of it to be published in French translation in the pan-Helvetic *Journal helvétique*.

95. In 1767 Rousseau was seeking a copy of *Iter helveticum*, which shows his sustained interest in Haller and Swiss flora. Rousseau first requested it from Duchesne by a letter of 30 June 1765; *CC*, xlix.300, and again in a letter to Rey of 28 December 1767, *CC*, xxxiv.268.

96. Jean-Baptiste Tollot was a master pharmacist of Geneva, who contributed to the *Journal helvétique*. See Jacques-Augustin Galiffe, *Notices généalogiques sur les familles genevoises*, vol.7, ed. Louis Dufour-Vernes and Eugène Ritter (Geneva, 1895), p.509.

97. 'Réflexions sur l'étude des plantes et sur leur usage', *Journal helvétique* (December 1741), p.1142-57 (1154). Even though the article is unsigned, its authorship by Tollot was confirmed by d'Ivernois (cited below).

extolling the accomplishments of generations of Swiss botanists[98] and asserting that the Swiss were the 'pères de la botanique'; for that matter, Swiss women knew the 'Plantes usuelles'.[99]

D'Ivernois had a point. Botanical explorers had long traversed the difficult terrain of Switzerland in search of plants, and the country enjoyed a rich botanical tradition; herbalists, apothecaries and physicians were well-acquainted with a variety of mountain plants with curative properties. Swiss herbal remedies were renowned well beyond the borders of the Confederation: the 'Faltranck', or 'boisson contre les chutes',[100] was adopted by French physicians, and described in the major encyclopedias.[101] Rousseau commented that '[l]es bons herboristes *suisses* ont soin d'en bien garnir leur Falltranck moins pour sa prétendue qualité vulnéraire, qu'à cause de l'excellent parfum qu'elle y répand'.[102]

The long botanical tradition in Switzerland can be traced back at least to the sixteenth-century naturalists, Benedikt Aretius (1522-1574) of Bern and Konrad Gessner (1516-1565) of Zürich; the latter was known as the 'German Pliny'. Aretius, a classicist and theologian, also made contributions to botany, in commemoration of which the genus *Aretia* bears his name. In addition to his linguistic erudition and medical

98. Jean Antoine d'Ivernois, 'Lettre à Monsieur Cartier ministre du St Evangile et pasteur de l'église de la Chaux du Milieu, dans cette souveraineté: servant d'apologie aux médecins-botanistes Suisses, contre ce qu'on a avancé dans le *Journal helvétique* de décembre 1741 p.1155 sur l'état de la botanique, en Suisse', *Journal helvétique* (January 1742), p.3-38 (31).

99. D'Ivernois, 'Lettre à Monsieur Cartier ministre', p.12, 14; see also Fritz Störi, 'Der Helvetismus des "Mercure suisse" ("Journal helvétique"), 1732-1784', doctoral dissertation, University of Zürich (Bern, 1953), p.57.

100. There were different recipes for this famed drink. One contained pyrola, sanicula, virga aurea, vinca pervinca, pes leonis and scabiosa. Jean Benjamin d'Apples, 'Mémoires sur le Faltranck', *Nouvelles de la république des lettres* (July 1709), p.17-25 (17). Another contained leaves of periwinkle, sanicle, veronica, Bugle, pes leonis, St John's Wort, Hart's tongue fern, maidenhair fern, lungwort, artemisia, betony, verbena, figwort, agrimony, small centaury, hawkweed and peppermint. Nicolas Lémery, 'Faltranck', in *Dictionnaire universel des drogues simples* (Paris, L.-Ch. d'Houry, 1759), p.355.

101. 'Mot allemand que nous avons adopté, et qui signifie boisson contre les chutes. C'est ce que nous appelons *Vulnéraires suisses*'; original emphasis; Mahon, 'Faltranck', *Encyclopédie méthodique: médecine* (Paris, Panckoucke, 1793), p.261. The author may have been John Mac Mahon, who taught at the Ecole militaire in Paris from 1766 to 1786. Several members of the Irish Jacobite Mac Mahon family were physicians in eighteenth-century France. Richard Hayes, 'Biographical dictionary of Irishmen in France: part XI', *Studies: an Irish quarterly review* 33:131 (September 1944), p.367-77 (375).

102. My emphasis; Rousseau's annotations of Nicolas François and Geneviève de Nangis Regnault, *La Botanique mise à la portée de tout le monde ou Collection des plantes d'usage dans la médecine, dans les alimens et dans les arts: avec des notices instructives puisées dans les auteurs les plus célèbres, contenant la description, le climat, la culture, les propriétés et les vertus propres à chaque plante, précedé d'une introduction à la botanique, ou dictionnaire abrégé des principaux termes employés dans cette science*, 2 vols (Paris, chez l'auteur, 1774) in BdAN [Munch] MS, Art. 26. Rousseau uses a different spelling than the more standard 'Faltranck'. See Lémery, 'Faltranck', p.355.

practice, Gessner became a professor of natural history in Zürich. His interests in natural history were wide-ranging, comprising zoology (*Historia animalium*), botany (especially plant geography) and geology.

In eighteenth-century Switzerland, botanical gardens in the strict sense were still rare due to Switzerland's fragmentation and lack of official support for the sciences. The sole academic garden in the Confederation – that of the University of Basel – served medical instruction and pharmacology. D'Ivernois argued that Switzerland's lack of botanical gardens was not a disadvantage, however: 'Il vaut toujours mieux [...] examiner [les plantes...] dans les endroits que la Nature leur a assignés.' The botany professors at Basel 'ont soin de mener souvent les étudiants à la campagne, et quelques fois jusques sur des montagnes fort éloignés'. Furthermore, gardens were not needed because 'en Suisse [...] chacun en sortant de chez soi, se trouve presque partout, dans un Jardin botanique formé par la Nature' in front of his house.[103]

At the time Tollot and d'Ivernois were debating the status of Swiss botany, a nascent botanical community was forming around figures such as Haller in Bern, and Garcin, d'Ivernois and Gagnebin in Neuchâtel, stimulated by Swiss botanists' contacts within and beyond the Swiss borders in the republic of science. In a related development, patriotic reformers, especially the Schweitzerische Gesellschaft in Bern,[104] fostered physiocratic thinking and practical reforms in agriculture.[105]

The *Journal helvétique* took a particular interest in the emergence of a specifically Swiss botany,[106] as demonstrated in an unsigned review of Haller's *Enumeratio methodica stirpium helvetiae indigenarum* (Göttingen, 1742):

> Les Suisses ont cultivé, avec succès, toutes les Sciences, depuis qu'elles ont recommencé à fleurir en *Europe*; mais on peut dire qu'ils se sont surtout distingués par leurs Travaux et leurs Progrès dans la Botanique. Les grands Hommes que la *Suisse* a vu naître, et qu'elle produit encore en ce genre, sont un des Ornements de la Patrie et le sujet de l'admiration des Curieux étrangers.[107]

D'Ivernois and other authors writing in this journal depicted botany as a *Swiss*[108] science legitimated by its very own Swiss expert:

103. D'Ivernois, 'Lettre à Monsieur Cartier, Ministre', p.36.
104. For example the Schweitzerische Gesellschaft in Bern (Swiss Society in Bern), founded in 1759.
105. Johnstone, 'The rural Socrates'.
106. Störi, 'Der Helvetismus des "Mercure suisse"', p.50. My own perusal of articles published in this journal in the 1740s confirms this finding.
107. Original emphasis; 'Idée générale du grand ouvrage de botanique', p.174.
108. They referred to 'Suisses' and 'la Suisse' irrespective of cantonal or linguistic boundaries. This journal, as noted above, was committed to superseding linguistic, cultural and religious differences.

[L]'incomparable Mr *Albert Haller*, de *Berne*, *Médecin* du Roi de la *Grande Bretagne*, *Professeur* de Botanique, d'Anatomie et de Chirurgie dans la nouvelle Université de *Göttingen*, Membre de la Société Royale *d'Angleterre* et de *Suède*. Quelle partie du monde savant n'a déjà pas retenti du bruit de la juste réputation, que la botanique lui a acquise, et qui ne s'accorde à le faire aller du pair avec les plus habiles botanistes, anciens et modernes?[109]

viii. Swiss botany in flower: Albrecht von Haller of Bern

Albrecht von Haller (fig.3.2, p.79), poet, physician and physiologist, was the midwife of Swiss botany, who almost single-handedly created alpine botany. In his *Enumeratio Methodica stirpium Helvetiae indigenarum* (1742) and *Historia stirpium indigenarum Helvetiae inchoata* (1768), he delimited the flora of Switzerland for the first time, giving a bio-geographical definition to the country, including 'allies' such as Neuchâtel (fig. 3.3, p.79). He consolidated their plant species with those of the central Swiss cantons, thereby laying a natural-historical claim to these regions for the Confederation. In Francophone Neuchâtel, where the *Journal helvétique* was published, Haller's *Enumeratio* was seen, above all, as a *Swiss* work: 'Quoique cet ouvrage soit imprimé en Allemagne, il appartient particulièrement à la *Suisse*, puisqu'il renferme la description des plantes qui s'y trouvent, avec leurs propriétés médicinales, et que son auteur, connu si avantageusement dans la république des lettres, est de la Ville de Berne'.[110] Haller's botanical work can therefore be viewed not only as scientific, but also as political, for it provided the bio-geographical contours of a greater Switzerland at the very moment that modern Swiss identity was being forged by the pan-Helvetic patriotic movement.[111]

Haller's Swiss flora was the work of a staunch patriot, scion of one of the ruling families of Bern, where he was entitled by descent to hold public office. Beset by 'Heimweh',[112] Haller gave up a prestigious post at the up-and-coming University of Göttingen as professor of anatomy, botany and surgery (1736-1753) to serve his native Bern, the largest and most powerful of the Swiss cantons. He stuck with this decision even though his talents were never fully appreciated there.[113]

109. Original emphasis; d'Ivernois, 'Lettre à Monsieur Cartier ministre', p.31.

110. My emphasis; 'Livres nouveaux: Göttingue', *Journal helvétique* (October 1742), p.83. The full title of the work was *Enumeratio methodica stirpium helvetiae indigenarum* (Göttingen, Abram Vandenhoek, 1742).

111. This development is discussed in Chapter 4.

112. This is the German word for homesickness. Rousseau, who called it Hemé, considered this a peculiarly Swiss ailment; see Rousseau to the maréchal de Luxembourg, 20 January 1763; *CC*, xv.52; *Dictionnaire de musique*, *OC*, v.924.

113. Martin Stuber and Regula Wyss, 'Der Magistrat und ökonomische Patriot', in *Albrecht von Haller Leben, Werk, Epoche*, ed. Hubert Steinke, *et al.* (Göttingen, 2008), p.347-80.

Figure 3.2: E. J. Handmann and G. L. Crusius, *Albrecht Von Haller*, engraving, by permission of the Linnean Society of London.

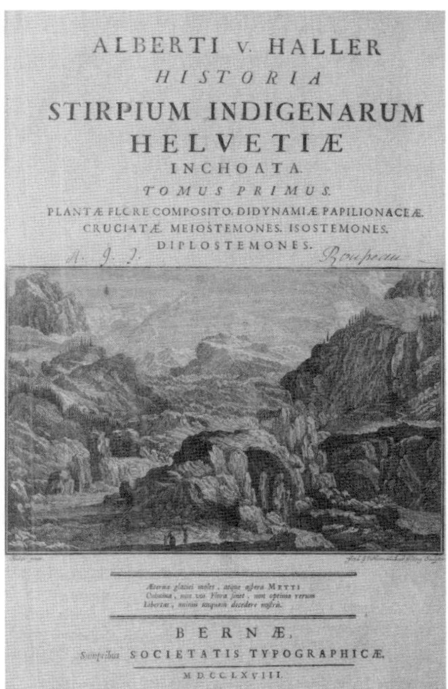

ALBERTI V. HALLER
HISTORIA
STIRPIUM INDIGENARUM
HELVETIÆ
INCHOATA.
TOMUS PRIMUS.
PLANTÆ FLORE COMPOSITO. DIDYNAMIÆ. PAPILIONACEÆ.
CRUCIATÆ. MEIOSTEMONES. ISOSTEMONES.
DIPLOSTEMONES.

BERNÆ,
Sumptibus SOCIETATIS TYPOGRAPHICÆ.
MDCCLXVIII.

Figure 3.3: Title page from Rousseau's copy of Albrecht von Haller, *Historia stirpium* (Bern, Societas Typographica, 1768), by permission of the Royal Horticultural Society, London.

Haller took up botany in 1728 as a way to improve his poor health. As a student in Basel, where he was studying mathematics, he was privileged to examine the herbarium of the great seventeenth-century Swiss botanist, Caspar Bauhin, who had attempted to group plants by families.[114] At this relatively early date, Haller already conceived of a patriotic botany: 'le Dessein qu'il a encore, de donner une "Histoire des Plantes de sa Patrie"'.[115] In 1728 Haller undertook an extensive journey on foot through Switzerland with his friend, Johannes Gessner (1709-1790), professor of physic at Zürich.[116] Haller and Gessner divided up Switzerland between them, intending to collaborate on a complete flora of the country. Gessner was unable to see the project to completion, but Haller continued it, engaging virtually every naturalist in Switzerland to help him find new plants, often paying their expenses.[117]

The collections Haller and Gessner made on this journey formed the basis for Haller's private herbarium, now in Paris.[118] In Göttingen he began a collection of alpine plants under the patronage of A. J. Hugo, royal physician in Hannover, and continued to collect plants in the Alps. Haller's renown was to a large extent based on the quality and range of his collections – his personal herbarium, now in Paris, comprises 59 folio volumes packed with 10,000 plant species.[119] His herbaria at Paris and Göttingen contain many type specimens, and plants collected by Caspar [Gaspard] Bauhin[120] which provided the raw material for Haller's first Swiss flora, *Enumeratio*,[121] later superseded by extensive field research by Haller and his collaborators. Haller opened a new chapter in the investigation of alpine flora while he was director of the Bern salt works in Roche (canton of Vaud/Waadt) (1758-1764). The culmination of these

114. Heinrich Zoller, 'Albrecht von Hallers Pflanzensammlungen in Göttingen, sein botanisches Werk und sein Verhältnis zu Carl von Linné', *Nachrichten der Akademie der Wissenschaften in Göttingen: mathematisch-physikalische Klasse* 10 (1958), p.217-51 (218). Zoller's is still considered the leading work on the subject. See Jean-Marc Drouin and Luc Lienhard, 'Botanik', in *Albrecht von Haller: Leben, Werk, Epoche*, ed. Hubert Steinke *et al.* (Göttingen, 2008), p.292-314 (311, n.23).
115. Haller, 'Extrait de deux pièces', p.210.
116. Gessner, a native of Zürich, and a descendent of the sixteenth-century naturalist, Konrad Gessner, was an early adherent of Linnaeus and at the same time an important collaborator and friend of Haller. 'Physic' denoted medicine, excluding surgery.
117. Gavin de Beer, 'Haller's *Historia stirpium*', *Annals of science* 9:1 (1953), p.1-46 (3).
118. This remarkable collection was looted from Pavia by Napoleon's troops, whence it was brought to Paris in 1796. I discuss this herbarium in Chapter 8.
119. The herbarium originally comprised 60 volumes, but for unknown reasons, now only has 59. See Marlène Margez, Cécile Aupic and Denis Lamy, 'La restauration de l'herbier Haller du Muséum national d'histoire naturelle', *Support/Tracé* 5 (2005), p.58-68 (59).
120. Zoller, 'Albrecht von Hallers Pflanzensammlungen', p.219, 223.
121. Gavin De Beer and Bernard Gagnebin, 'Abraham Gagnebin de la Ferrière d'après sa correspondance', *Bulletin de la société neuchâteloise des sciences naturelles* 80 (1957), p.45-78.

efforts was the *Historia stirpium* of 1768, published exactly forty years after Haller and Gessner's journey through Switzerland.

The *Historia stirpium* was 'made possible by the system of collaboration that he developed [...] he also invited and employed most of the naturalists of Switzerland to scour the various nooks and crannies of the Alps to find new plants, often at his personal expense'.[122] Haller enjoyed cordial relations with colleagues throughout the country and beyond; his network included 1200 correspondents, 25 per cent of whom exchanged letters with Haller about botany, especially seed and plant exchanges.[123] He wrote a total of 13,000 letters. This network, like that of other important eighteenth-century savants, was crucial to his success. Even though Haller himself undertook arduous collecting trips through the length and breadth of Switzerland, as well as in the Harz mountains in Germany, he depended on erudite collaborators such as Gagnebin and Horace-Bénédict de Saussure, not to mention local people. Haller gave credit where it was due; his *Historia stirpium* names 110 contributors, of whom forty were his contemporaries. Thirty-two of Haller's correspondents were also correspondents of Rousseau; this probably caused Haller, the staunch conservative, more discomfort than it did Rousseau, whose assessment of Haller seems to have been generally positive.[124]

In an era when studies of indigenous flora were all the rage, Haller's unprecedented, three-volume *Historia stirpium* stood out as a masterpiece in which he described 2,486 species of alpine plants, and established the study of mosses and lichens in Switzerland. The work culminated Haller's forty-year project to document the Swiss flora and put Switzerland on the botanical map; its significance for Rousseau's botanical studies and political sensibilities is therefore considerable.

Haller was an important botanist for Rousseau from the outset of his botanical studies in the Swiss Jura. Rousseau is sure to have known about Haller and his work from his guide and teacher,[125] Abraham Gagnebin, a close collaborator of Haller and his most important botanical correspondent.[126] Indeed, Rousseau maintained a high regard for Haller's contri-

122. De Beer, 'Haller's *Historia stirpium*', p.3.

123. The definitive work on Haller's correspondence is Martin Stuber, Stefan Hächler and Luc Lienhard (eds), *Hallers Netz: ein europäischer Gelehrtenbriefwechsel zur Zeit der Aufklärung* (Basel, 2005), p.371-410 (373).

124. In response to the *Contrat social* Haller wrote, 'Je crois que tout Prince chrétien doit, sur les ordres précis de Dieu même, marquer son indignation contre un blasphémateur tel que l'est dans le plus exact sens du terme le S. J. Jacques. L'exil d'un étranger me paraît bien doux et bien modéré', Albrecht von Haller to Samuel Auguste André David Tissot, 19 July 1762; *CC*, xii.55.

125. Rousseau to Du Peyrou 16 June 1765; *CC*, xxvi.44-45.

126. Urs Boschung, *et al.* (eds), *Repertorium zu Albrecht von Hallers Korrespondenz 1724-1777*, vol.1 (Basel, 2002), p.157.

butions to Swiss botany. In 1765 Rousseau told Nicolas Bonaventure Duchesne (1711-1765) that he especially desired the *Enumeratio* 's'il n'est pas d'un prix exorbitant'.[127] The appearance of *Historia stirpium* in 1768 was an important event in the Swiss botanical circles that Rousseau had frequented during his stay in Môtiers, and with which he remained in contact after his departure from the region in 1765. Rousseau's continuing enthusiasm for Haller's work was manifested by his desire to acquire the *Historia stirpium* as soon as possible after its publication in 1768:[128]

> Je me suis ruiné en livres de Botanique, et j'avais bien résolu de n'en plus acheter; cependant je sens que m'affectionnant aux *Plantes des Alpes*, je ne puis me passer de *celui de Haller*. Vous m'obligerez de vouloir bien me marquer exactement son titre, son prix, et le lieu où vous l'avez trouvé car la France est si barbare encore en botanique, qu'on n'y trouve presque aucun Livre de cette science; et j'ai été obligé de faire venir à grands frais de Hollande et d'Angleterre le peu que j'en ai [...].[129]

Pierre Alexandre Du Peyrou (1729-1794) apparently gave him a copy, for in his subsequent letter to Du Peyrou of 21 July 1769 Rousseau thanked him for 'le livre de M. Haller'.[130]

In his short history of botanical nomenclature, Rousseau called the *Historia stirpium* a 'grand et excellent traité des plantes alpines'.[131] Rousseau sold his copy, painstakingly cross-referenced with Linnaean binomial names,[132] circa 1775 (with the rest of his botanical library) to Daniel Malthus (1730-1800), who bequeathed it to his cousin, Jane Dalton (c.1742-1812), an avid botanist.[133] Rousseau used Haller's work to desig-

127. Rousseau to Duchesne, 30 June 1765; *CC*, xlix.300.

128. There has been considerable confusion about Rousseau's access to Haller's works. Leigh infers that Rousseau's letter of 19 December 1768 refers to Haller's *Iter Helveticum anni 1739* (1740). *CC*, xxxvi.212, n.(d); *CC*, xxxiv.270, n.(k). Yet on 28 December 1767 Rousseau requested Haller's *Iter Helveticum* from his Amsterdam publisher, and therefore would not have needed to ask Du Peyrou for the exact title of Haller's work on 'plantes des Alpes'. Leigh compounds the confusion, claiming that Rousseau acknowledged receipt of Haller's *Nomenclator ex historia stirpium indigenarum Helvetiae*. Rousseau to Du Peyrou, 21 July 1769; *CC*, xxxvii.114 n. (a).

129. My emphasis; Rousseau to Du Peyrou, 19 December 1768; *CC*, xxxvi.211.

130. De Beer and Bernard Gagnebin (the latter also an editor of Rousseau's collected works in French) assert that by 21 July 1769 Rousseau owned *Historia stirpium*, although they advance no evidence for this claim, simply inserting '*Historia stirpium*' within brackets after the phrase 'non plus que du Haller' in Rousseau's letter to Du Peyrou, 21 July 1769 (also in *CC*, xxxvii.114). De Beer and Gagnebin, 'Abraham Gagnebin de la Ferrière', p.60.

131. Rousseau, 'Introduction', *OC*, iv.1208.

132. A. Cook, 'Jean-Jacques Rousseau's copy of Albrecht von Haller's *Historia stirpium indigenarum Helvetiae inchoata (1768)*', *Archives of natural history* 30:1 (April 2003), p.149-56.

133. See for example Daniel Malthus to Rousseau, 24 January 1768; *CC*, xxxv.53-56. After passing through various hands, Rousseau's copy of *Historia stirpium* is in the Lindley Library of the Royal Horticultural Society, London. Malthus was the father of Thomas 'Population' Malthus. Two of Rousseau's botany books are found in the Malthus library at Jesus College, Cambridge.

nate as *Seseli halleri* (number 762 in the *Historia stirpium*) a specimen he could not identify from Linnaeus's *Species plantarum*:

> Vous trouverez, Madame, une ombellifère à laquelle j'ai pris la liberté de donner le nom de *Seseli Halleri* faute de Savoir la trouver dans le *Species* au lieu qu'elle est bien décrite dans la dernière édition des *plantes de Suisse de M. Haller nº 762*. C'est une très belle plante qui est plus belle encore en ce pays que dans les contrées plus méridionales parce que les premières atteintes du froid lavent son verdi foncé d'un beau pourpre et surtout à la couronne des graines, car elle ne fleurit que dans l'arrière-saison, ce qui fait aussi que les graines ont peine à murir et qu'il est difficile d'en recueillir.[134]

Haller, who was very religious, believed that only a natural classification system could reveal the hidden plan of Creation; toward this end he formulated a complex system based on the systems of John Ray (1627-1705) and Tournefort.[135] At the same time, his knowledge of the great variety of alpine habitats and species made him an ecologist *avant la lettre*, with a keen eye for environmental influences. He made detailed studies of the ecology of the plants he described, and required his correspondents to do the same. Such information was often lacking in floras and herbaria of the time.[136] Haller's attention to this topic was therefore of great significance in the development of botanical record-keeping.[137] His Swedish rival, Linnaeus, spent less time on such matters, devoting himself to cataloguing and naming the global flora using an artificial classification system based on a discrete set of 'characters', that is, the number and position of the stamens (male reproductive parts) and pistils (female reproductive parts).

Haller and Linnaeus exchanged plants and correspondence for a time, but they had very different views concerning not only nomenclature and taxonomy, but also variability and the fixity of species.[138] Although he was as deeply religious as Linnaeus, Haller's awareness of variation seems to have made him less wedded to Creationism. Haller's preference for a natural system led him to reject the Linnaean sexual system (see p.158-67) and binomial nomenclature (see p.234-37) as imprecise and destructive of an important tradition of plant

134. My emphasis; Rousseau to the Duchess of Portland, 21 December 1769; *CC*, xxxvii.190. Umbellifers present many difficulties, so Rousseau's problem is not surprising. See T. G. Tutin, *et al.* (eds), *Flora europaea*, 5 vols (Cambridge, 1964-1983), under 'seseli'.

135. Luc Lienhard, '"La machine botanique": zur Entstehung von Hallers Flora der Schweiz', in *Hallers Netz*, ed. Martin Stuber, Stefan Hächler and Luc Lienhard, p.371-410 (381). Haller turned to his Calvinist faith during a deep personal crisis following the deaths in succession of his first two wives and his young son in the 1730s.

136. Antoine Magnin, *Claret de la Tourrette: sa vie, ses travaux, ses recherches sur les lichens du Lyonnais d'après ses ouvrages et les notes inédites de son herbier* (Paris, 1885), p.49.

137. Lienhard, '"La machine botanique"', p.379-80.

138. Zoller, 'Albrecht von Hallers Pflanzensammlungen', p.219.

nomenclature.[139] Haller did provide Linnaean generic synonymies in his botanical works, however, citing Linnaeus's *Genera plantarum,*[140] a practice Rousseau applauded against the complete exclusion of Linnaean names by Michel Adanson (1727-1806):[141]

> M. Haller, dans son grand et excellent traité des plantes alpines [...] cite toujours les genres et quelquefois les phrases des espèces de M. Linnaeus, mais M. Adanson n'en cite jamais ni genre ni phrases. M. Haller s'attache à une synonymie *exacte*, par laquelle, quand il n'y joint pas la phrase de M. Linnaeus, on peut du moins la trouver indirectement par le rapport des synonymes.[142]

Despite this concession to the new nomenclature, Haller's sceptical attitude cost him the attention of the botanical community; his nomenclature and system of classification lost out to those of Linnaeus who, for all of his traditionalism created a nomenclature and an artificial taxonomic system that provided simplicity, efficiency and ease of access, well-suited to eighteenth-century botanists intent on cataloguing the earth's plant wealth for efficient exploitation.

Thus, despite his very modern biological perspectives, Haller was overlooked and excluded from mainstream botany.[143] Yet, ironically, his rejection of binomial nomenclature and the sexual system caused his herbaria to be neglected – a blessing in disguise. Having been so little exposed to environmental hazards, Haller's vibrant and colourful specimens can be seen today virtually as he mounted them (fig.3.4, p.85). Haller's scepticism about the Linnaean artificial sexual system was later vindicated by the triumph of the natural method (discussed in Chapter 6, p.173-99).

Haller and Rousseau never met, but were well aware of each other. Haller, scion of a ruling family in the most powerful Swiss oligarchy, corresponded with Charles Bonnet (a member of Geneva's Council of Two Hundred) about the crisis unleashed in Geneva by the condemnation of *Emile* and *Du contrat social.* Both men decried Rousseau's religious views and his doctrine of popular sovereignty, which threatened the ruling classes of Geneva, Bern and Zürich.[144] Bonnet

139. Drouin and Lienhard, 'Botanik', p.303.

140. Haller cites a purported edition of 1765, perhaps meaning the sixth edition of 1764. Haller, *Historia stirpium*, vol.1 (1768), p.xlvii; see also Linnaeus, *Species plantarum*, 2 vols (London, 1957-1959), vol.2, p.22.

141. Rousseau, 'Introduction', *OC*, iv.1208.

142. My emphasis; Rousseau, 'Introduction', *OC*, iv.1208.

143. Frans A. Stafleu, *Linnaeus and the Linnaeans: the spreading of their ideas in systematic botany, 1735-1789* (Utrecht, 1971), p.246-48.

144. Haller would have been horrified by Rousseau's criticisms of Bern in *Considérations*, *OC*, iii.1006. This was Rousseau's revenge for his expulsion from Yverdon and the Ile St Pierre. However, even in the Swiss oligarchies, Rousseau had many admirers, as attested by Bonnet himself. Bonnet to Haller, 13 August 1762; *CC*, xii.185-86.

Figure 3.4: Albrecht von Haller, *Herbarium of Albrecht von Haller*, vol.11, by permission of the Laboratoire de phanerogamie, Muséum national d'histoire naturelle, Paris.

seemed to regret that Rousseau would not burn at the stake, while Haller advocated a milder punishment, such as exile (cited above, p.81n).[145]

Haller and Rousseau were not that far apart, however, in their moral thought. In *Die Alpen*, for example, Haller expressed the same view of mountain people and their simple, uncorrupted mores that Rousseau later popularised in *Julie*.[146] Haller likewise accepted the thesis of the second *Discours* concerning the origin of evil in social life, characterising this work as an 'überaus scharfsinnige Urkunde'.[147]

In botany Haller and Rousseau were both patriotic, loving the very soil in which the Swiss plants were rooted. Rousseau shared Haller's devotion to Swiss plants,[148] having 'apporté de Suisse quelques plantes sèches qui se sont pourries en chemin; c'est un herbier à recommencer'.[149] He later told Du Peyrou: 'j'ai déja beaucoup de plantes *alpines* et des *plus rares*; cependant, comme il y en a encore un très grand nombre qui me manquent, je ne doute pas qu'il ne s'en trouve dans votre envoi qui me feront grand plaisir par elles-mêmes'.[150] In general, Haller and Rousseau both preferred indigenous to exotic flora,[151] although they at most only hinted at the latter's connections to empire, the cash nexus and slavery.[152]

Yet Haller does not seem to have understood that he and Rousseau shared a patriotic vision for Switzerland that both expressed through botany. It is to Rousseau's patriotic-botanical experience of his homeland – bio-geographically defined by Haller – that we turn in the next chapter.

145. 'Il n'y a guère que deux cents ans que nous eussions fait rôtir Rousseau: nous nous sommes bornés à faire rôtir ses Livres', Bonnet to Haller, 17 July 1762; *CC*, xii.50; see also Albrecht von Haller to Samuel Auguste André David Tissot, 19 July 1762; *CC*, xii.55.

146. Haller, *Die Alpen*, stanzas 1-31; Rousseau, *Julie*; *OC*, ii.79ff.

147. An 'extremely astute document'. Quoted in Florian Gelzer and Béla Kapossy, 'Roman, Staat und Gesellschaft', in *Albrecht von Haller: Leben, Werk, Epoche*, ed. Steinke *et al.*, p.156-81 (171).

148. '[L]e plus grand agrément de la botanique est de pouvoir étudier et connaître la nature autour de soi plutôt qu'aux Indes', Rousseau to Malesherbes, 19 December 1771, *CC*, xxxviii.302.

149. Rousseau to the Duchess of Portland, 10 October 1766; *CC*, xxxi.41.

150. My emphasis; Rousseau to Du Peyrou, 19 December 1768; *CC*, xxxiv.211.

151. A. Cook, 'Jean-Jacques Rousseau and exotic botany'. Interest in local plants was widespread in early-modern Europe: 'as they grappled with the issues of geography, identity and natural origins, many Europeans began to look *inwards* as well as outwards', Alix Cooper, *Inventing the indigenous: local knowledge and natural history in early modern Europe* (Cambridge, 2007), p.3. See also Staffan Müller-Wille, *Botanik und weltweiter Handel: zur Begründung eines natürlichen Systems der Pflanzen durch Carl von Linné (1707-1778)* (Berlin, 1999); Lisbet Koerner, *Linnaeus: nature and nation* (Cambridge, MA, 1999), p.129-30.

152. For the eighteenth-century Swiss trade in exotic goods and slaves, see Thomas David, *et al.*, *La Suisse et l'esclavage des noirs* (Lausanne, 2005).

4. Patriotism in a new key: Rousseau encounters Neuchâtel and botany

> Me voilà donc à mon foin pour toute ma nourriture, et à la botanique pour toute occupation. Déjà vieux j'en avais pris la première teinture en *Suisse* auprès du Docteur [d]'Ivernois.[1]

i. Rousseau returns to Switzerland

Switzerland was an important source of both Rousseau's political and botanical projects. To take up botany in Neuchâtel was therefore not only to learn a science and identify various plant species; it was to encounter his 'pays' in its myriad aspects – its natural history, people, economy, political forms and traditions.[2]

Yet Rousseau's return to his 'pays' was precipitous and fraught with difficulties. On 11 June 1762, the parlement of Paris condemned *Emile*, Rousseau's treatise on education, and Jean-Jacques had to leave French territory or face arrest. Fleeing the Montmorency home of the maréchal and maréchale de Luxembourg, he crossed the border of the Swiss Confederation, and descended from his carriage: 'En entrant sur le territoire de Berne je fis arrêter; je descendis, je me prosternai, j'embrassai, je baisai la terre, et m'écriai dans mon transport [...] je touche une terre de liberté'.[3]

Upon his arrival in Switzerland, Rousseau first sought refuge in Yverdon with an old friend from Paris, Daniel Roguin (1691-1771). After a brief sojourn from 14 June to 10 July 1762 as a guest of Roguin, Rousseau was ordered out of Yverdon by the canton of Bern, which controlled the pays de Vaud from 1536 to 1798. The once proud 'citoyen de Genève' did not go to Geneva, where, on 19 June 1762, *Du contrat social* and *Emile* had been burned by the public executioner, and their author faced arrest.[4]

1. My emphasis; Rousseau, *Rêveries*, *OC*, i.1060. Rousseau refers to receiving botany instruction from 'Docteur d'Ivernois', in the *Confessions* and the *Rêveries*, *OC*, i.631, 1042.
2. Rousseau seems to distinguish his 'patrie' (Geneva) from his 'pays' (Switzerland), as discussed below. This hypothesis requires further study.
3. Rousseau, *Confessions*, *OC* i.587.
4. However, as Rousseau notes, *Emile* drew the most opprobrium, while the prohibitions against the *Contrat social* were less widespread; it could even be obtained in Geneva. *Lettres écrites de la montagne*, *OC*, iii.810.

Daniel Roguin's niece, Julie-Anne-Marie Boy de la Tour, née Roguin (1715-1780), offered Rousseau a house in the village of Môtiers, located in the Val de Travers, in the Prussian principality of Neuchâtel. Mme Boy de la Tour was the widow of a Lyon merchant and banker, Pierre Boy de la Tour (1706-1758), a native of Neuchâtel, who had been a prime mover in the political life of the principality.[5] A 'CEO' in her own right, Mme Boy de la Tour carried on the family banking business after her husband's death, and became Rousseau's private banker.[6] Rousseau referred to her affectionately as 'la Maman'.[7] Her daughters, Julie Emélie Willading Boy de la Tour (1751-1826), and Madeleine-Catherine Delessert, née Boy de la Tour (1747-1816), botanised with him, and Madeleine-Catherine became Rousseau's most famous botany student. Hence, Rousseau's botanical project is intimately bound up with the Franco-Swiss Boy de la Tour and Delessert families.[8]

At the time, Neuchâtel was under Prussian rule and, like Geneva, was only an 'allié associé' of the Swiss Confederation. The hostility in Geneva and Bern that drove Rousseau to the Prussian enclave was serendipitous, for in Neuchâtel he made the acquaintance of botanists of catholic tastes, versed in local and exotic flora and classification.

In Neuchâtel Rousseau came under the suzerainty of that friend of the Philosophes, Frederick the Great of Prussia. On 16 August 1762 Frederick granted Rousseau asylum; he was naturalised in the principality on 16 August 1763, and granted the 'bourgeoisie' of the commune of Couvet on 1 January 1765, an important precaution against statelessness.[9] The king's representative in the principality, 'milord maréchal' George Keith (1686?-1778), befriended Rousseau and offered

5. See Bandelier, *Des Suisses dans la république des lettres*. The Boy de la Tours looked after Rousseau's savings and assisted him in various ways.
6. On 27 July 1762 Rousseau consigned the sum of 1000 écus (3131 livres) to the Boy de la Tour bank; this money was later transferred to an account in the name of Thérèse Levasseur, who became Rousseau's wife on 30 August 1768. See *OC*, i.1565, n.2.
7. Rousseau to Mme Delessert, 24 May 1773; *CC*, xxxix.162-67.
8. In 1771-1773 Rousseau penned his famous eight letters on botany to Madeleine-Catherine Delessert and in 1772 gave a herbarium ('Zürich') to her younger sister, Julie Emélie Boy de la Tour, who married Charles Emmanuel Willading of Bern in 1778; the herbarium was left by her daughter, Julie, and her son-in-law, Henry Faesy, to the Zentralbibliothek Zürich in 1833. See Ruth Schneebeli-Graf, *Das Zürcher Herbar von Jean-Jacques Rousseau* (Zürich, 1980). One of Mme Delessert's sons, Jules Paul Benjamin Delessert (1773-1847), used his immense wealth to advance the science of botany, as discussed in Chapter 9.
9. To the end of his life he remained a communier of Couvet, and therefore a subject of the King of Prussia. On Rousseau's nationality, see Jost, *Jean-Jacques Rousseau suisse*, p.19-39 and Jean-Pierre Jelmini, 'Rousseau neuchâtelois', *Bulletin de l'Association Jean-Jacques Rousseau* 72 (2012), forthcoming.

him protection. Milord's departure from Neuchâtel at the end of April 1763 left Rousseau exposed to the full ire of the pastorate.[10]

Rousseau initially got on well with his pastor, Frédéric-Guillaume de Montmollin (1709-1783), making his declaration of faith to the pastor on 24 August and partaking in communion with great emotion on 29 August 1762.[11] The two men engaged in lengthy theological discussions, and Montmollin even lent his carriage to Thérèse 'avec grand plaisir' so she might attend mass in a Roman Catholic town.[12] As late as January 1765 Montmollin was prepared to vouch for Rousseau's theological conformity.

This state of affairs was not to last, however. Despite the legal steps taken to ensure Jean-Jacques's tranquillity in the Prussian principality, he experienced growing difficulties with Neuchâtel's Vénérable Classe of pastors over his heterodox views on religion, including those advanced in the 'Profession du foi du vicaire savoyard', and the last chapter of *Du contrat social*.[13] The re-printing of his *Lettre à Christophe de Beaumont*, in which he rebutted the condemnation of *Emile* by the Archbishop of Paris, was likewise forbidden in Geneva even though it refuted Catholic teaching and upheld Protestantism.

In Geneva the public prosecutor, Jean Robert Tronchin (1710-1793), attacked Rousseau's political and religious ideas in the anonymous *Lettres écrites de la campagne* (1763). As a result of these attacks and condemnations, Rousseau surrendered his Genevan citizenship by a letter dated 12 May 1763; this act of renunciation was duly recorded by the Petit Conseil.[14] The surrender of citizenship was a big disappointment to Genevans such as J. F. ('le bonhomme') Deluc who hoped to recruit Rousseau to their cause; the letter was widely distributed and published in English in the *London chronicle* on 4 June 1763.

Despite significant reservations, Rousseau gave in to the entreaties of le bonhomme Deluc and replied to Tronchin in his *Lettres écrites de la montagne* (1764). In this work he presented the Reformation as a tolerant religion of free conscience in the interpretation of Scripture, and argued that the essence of Protestantism was being distorted by Genevan dog-

10. On Rousseau's admiration for Keith see Rousseau, *Lettres écrites de la montagne*, *OC*, iii.797; *Confessions*, *OC*, i.596-600.

11. *Confessions*, *OC*, i.596-605.

12. Rousseau to Dumoulin, 16 January 1763; *CC*, xv.42. This was possibly Vierrières-de-Joux in the Franche-Comté. Frédéric S. Eigeldinger, *Des Pierres dans mon jardin: les années neuchâteloises de J.-J. Rousseau et la crise de 1765* (Geneva, 1992), p.151.

13. The 'Classe' required absolute obedience of its members. Yet, '[d]e point de vue théologique, la Compagnie des pasteurs suit une ligne plus pratique que dogmatique'. Eigeldinger, *Des Pierres dans mon jardin*, p.42.

14. Rousseau to Jacob Favre, 12 May 1763; *CC*, xvi.164. Having already obtained naturalisation in Neuchâtel Rousseau had ensured he would not become stateless.

matism. Reinterpreting the *Contrat social* specifically for Geneva, he showed how the city-state could return to true republican principles.[15]

Rousseau's assertions in the *Lettres* only further inflamed the opposition. In Switzerland, 'une terre de liberté', the *Lettres* were prohibited at Bern on 23 January 1765, and at Geneva on 12 February, where they were deemed unworthy to be burned by the hangman. Rousseau's attack on the Genevan pastors annoyed the Reformed ministers in Neuchâtel, who followed suit on 18 February. The *Lettres* were burned at The Hague on 21 January and at Paris on 20 March 1765.

In early September 1765 Rousseau's much-discussed departure from Môtiers was finally precipitated by the 'lapidation' or stoning of his house by an angry mob; again he took flight, finding shelter, after various inquires in the region, on the Ile St Pierre in the Lake of Bienne (Biel). We are familiar with his botanical activities there from the cinquième promenade of the *Rêveries*, in which he promised a '*Flora Petrinsularis*'.[16] This episode in the Suisse romande ended with Rousseau's expulsion once again by Bern in October 1765, whereupon Rousseau accepted David Hume's invitation to relocate to Great Britain.

ii. The soil-politics nexus: Rousseau and the 'modèle suisse'[17]

During his Neuchâtel sojourn Rousseau analysed the Swiss socio-economic and political situation in two letters written in January 1763 to his French protector, the maréchal de Luxembourg; these letters deserve to be better known, and to join Rousseau's *Projet de constitution pour la Corse* (c.1765), and *Considérations sur le government de Pologne* (1771) as applications of his political philosophy to actual circumstances. In all of these works he stressed the political salience of a people's way of life. Neuchâtel, he argued, shared a common way of life with the rest of Switzerland: 'Pour connaître Môtiers, il faut avoir quelque idée du Comté de Neuchâtel, et pour connaitre le comté de Neuchâtel il faut en savoir de la Suisse entière.'[18] This observation demonstrates (1) Rousseau's belief that despite being a Prussian principality allied with the Confederation, Neuchâtel nevertheless belonged to Switzerland; and (2) his belief that despite its many divisions, Switzerland possessed sufficient integrity and distinctness to merit analysis as an entity apart.

15. Rousseau, *Lettres de la montagne*, *OC*, iii.685-897.
16. Rousseau, *Confessions*, *OC*, i.642; *Rêveries*, *OC*, i.1042-43. Rousseau's lyrical description of his few weeks on the island is justifiably renowned.
17. This term is adopted from Frédéric S. Eigeldinger, 'Les Montagnons: un archétype social', in *La Ville s'étend sur tout le pays: actes du colloque Rousseau, Neuchâtel et l'Europe*, ed. Frédéric S. Eigeldinger and Roland Kaehr (Neuchâtel, 1993), p.75-94 (88).
18. Rousseau to the maréchal de Luxembourg, 20 January 1763; *CC*, xv.49.

This distinctness rests on bio-geographical, cultural and political criteria rather than linguistic or religious ones.

Rousseau's modèle Suisse comprises three key features:

(1) A shared way of life determined by a harsh, unforgiving soil: pastoralism (animal husbandry); the theoretical basis is climatic determinism;
(2) Autarchy and self-sufficiency, with their corollaries: absence of a division of labour or participation in the cash nexus;
(3) These conditions promote democratic republics, in contrast with oligarchies such as Bern, i.e. the harder the life, the more democratic the regime.

The character of the soil, or more broadly, the climate, determines the Swiss character: 'les Suisses ont aussi tous à peu près les mêmes *mœurs* [...] Ils ont des *manières de vivre* qui ne changent point, parce qu'elles tiennent, pour ainsi dire, au *sol*[,] au *climat*, aux besoins divers, et qu'en cela les habitants seront toujours forcés de se conformer à ce que *la nature des lieux* leur prescrit.'[19] Thus: 'c'est de la nature du *sol* que naît le *caractère primitif* des habitants';[20] 'Des montagnes, des bois, des rivières, des pâturages. Ne croirait-on lire la description de la Suisse?'[21] The 'rudesse du pays', suited to pastoralism, keeps the people hard-working and poor; they make strong soldiers and have no time for the activities of the city – the theatre, card-playing, gossip – not to mention crime or other trouble-making.

This way of life is inhospitable to the division of labour; it therefore keeps the citizens independent of each other, of trade and of the cash nexus:

> Aussi retrouvait-on jadis dans les *Suisses* le même caractère que Diodore donne aux Corses; l'équité, l'humanité, la bonne foi [...] chacun dans sa maison exerçait *tous les arts* nécessaires; tous étaient maçons, charpentiers, menuisiers, charrons [...]. C'est ainsi qu'au milieu de leurs précipices et de leur vallons chacun vivant sur son sol parvint à en tirer tout son nécessaire [...] à ne désirer rien au délà.[22]

The famous 'Montagnons' of Neuchâtel therefore failed to adhere to this paradigm, since they sold their watches in Paris.[23]

A harsh climate and unyielding soil are inhospitable to the growth of luxury and the circulation of money, with its tendency to corrupt public officials and promote inequality. Writing to the Poles, Rousseau cites as

19. My emphasis; Rousseau to the maréchal de Luxembourg, 28 January 1763; *CC*, xv.49.
20. My emphasis; Rousseau, *Projet de constitution pour la Corse*, *OC*, iii.913.
21. Rousseau, *Projet de constitution pour la Corse*, *OC*, iii.914.
22. My emphasis; Rousseau, *Projet de constitution pour la Corse*, *OC*, iii.914.
23. Rousseau, *Lettre à d'Alembert sur les spectacles*, *OC*, v.55.

positive examples the eighteenth-century Swiss cantons where taxes were still levied in kind: 'De nos jours encore on voit en Suisse les officiers, magistrats et autres stipendiaires publics payés avec des denrées.'[24] For the same reason, Rousseau approved of the labour tax, the corvée:[25] 'Je sais qu'il est un abomination en France, mais l'est-il en Suisse? Les chemins se font aussi par corvées et personne ne se plaint.'[26] Rousseau considers money to lie at the root of the behaviour of the largest and most powerful of the cantons: 'Il se manie dix fois plus d'argent dans le seul Canton de Berne que dans tout le reste de la Suisse; aussi l'administration en est-elle inique en proportion.'[27]

While the Swiss lived in this way, they could resist foreign domination: 'Une troupe de pauvres montagnons dont toute l'avidité se bornait à quelques peaux de moutons, après avoir dompté la fierté autrichienne écrasa cette opulente et redoutable maison de Bourgogne qui faisait trembler les potentats d'Europe'.[28] The Swiss only felt poor when money began to circulate.[29]

Neuchâtel shared the 'rude' environment of the rest of Switzerland: 'Le pays, Si l'on excepte la Ville et les bords du lac, est aussi rude que le reste de la Suisse, la vie y est aussi rustique [...].' The implication is that, other factors being equal, Neuchâtel would have been republican like the other Swiss states, for 'la rudesse du pays', its perishable productions, the dispersion of both men and their livestock were common to all the Swiss cantons, and were 'le plus grand avantage de la Suisse'.[30] As long as the Swiss remained poor, they would remain relatively equal and free of corruption.

However, an important factor differentiated the Neuchâtelois from the other Swiss: 'les habitants, accoutumés à vivre sous des Princes, s'y sont encore plus affectionnés aux grandes manières; de sorte qu'on trouve ici du jargon, des airs, dans tous les états, de beaux parleurs labourant les champs, et des Courtisans en Souquenille'.[31] Rousseau recounts their love of titles, the hunt and other noble occupations. Unfortunately, the Neuchâtelois had not adhered to these ancient, aristocratic ways either, allowing themselves to become integrated in the cash nexus, and therefore poor:

24. Rousseau, *Considérations sur le gouvernement de Pologne*, *OC*, iii.1006.
25. Rousseau, *Considérations sur le gouvernement de Pologne*, *OC*, iii.1009-10.
26. Rousseau, *Projet de constitution pour la Corse*, *OC*, iii.932.
27. Rousseau, *Considérations sur le gouvernement de Pologne*, *OC*, iii.1006.
28. Rousseau, *Discours sur les sciences et les arts*, *OC*, iii.20.
29. Rousseau, *Projet de constitution pour la Corse*, *OC*, iii.916; see also Rousseau to the maréchal de Luxembourg; 20 January 1763; *CC*, xv.50.
30. Rousseau to the maréchal de Luxembourg, 20 January 1763; *CC*, xv.50.
31. Rousseau to the maréchal de Luxembourg, 20 January 1763; *CC*, xv.54.

Les trois seules marchandises qui leur en aient fourni jusqu'ici sont les fromages, les chevaux et les hommes; mais depuis l'introduction du luxe ce commerce ne leur suffit plus, et ils y ont ajouté celui des manufactures dont ils sont redevables aux réfugiés français; ressource qui cependant a plus d'apparence que de réalité; car comme la cherté des denrées augmente avec les espèces, et que la culture de la terre se néglige quand on gagne davantage à d'autres travaux, avec *plus d'argent* ils n'en sont pas plus riches.[32]

Rousseau later concedes, however, that the soil-politics nexus is not absolute:

[T]out varie en nous selon les climats; la couleur, la forme et la nature des différents peuples semblent dépendre de l'air qu'ils respirent, de la nourriture qu'ils prennent, et de la température du pays qu'ils habitent. L'immortel Montesquieu avait puisé dans Hippocrate[33] et dans Bodin le beau système de l'influence des climats, mais peut-être a-t-il un peu trop généralisé la conséquence morale qu'il fait découler de ce principe physique. On ne saurait en effet l'adopter sans beaucoup de modification, car des causes étrangères et des institutions politiques ont pu souvent aider ou détruire, augmenter ou affaiblir l'influence du climat sur les hommes.[34]

The subsistence way of life in the Swiss mountains maintains a rude equality; more than any other way of life, it favours *democratic* republics:

La Suisse en général est un pays pauvre et stérile. Son gouvernement est partout Républicain. Mais dans les cantons plus fertiles que les autres tels que ceux de Berne, de Soleure et de Fribourg le Gouvernement est Aristocratique. Dans les plus pauvres, dans ceux ou la culture est plus *ingrate* et demande un plus grand *travail* le Gouvernement est *Démocratique*. L'état n'a que ce qu'il faut pour subsister sous la plus *simple* administration.[35]

Rousseau distinguished, therefore, between two types of Swiss republics – democratic and aristocratic, or more accurately, oligarchic. He argued that the cantons that remained democratic were those least favoured by nature for human life – the rudeness of the climate and limited agricultural possibilities kept people relatively equal in their poverty, self-sufficient and outside the cash nexus; this could also be the

32. My emphasis; Rousseau to the maréchal de Luxembourg, 20 January 1763; *CC*, xv.52.
33. 'Airs, waters, places', in *Hippocratic writings*, ed. G. E. R. Lloyd, trans. J. Chadwick and W. N. Mann (London, 1983), p.148-69.
34. This little-known comment appears in Rousseau's annotations to the Regnaults', *La Botanique mise à la portée de tout le monde*, BdAN [Munch MS], art. 208. In the Regnaults' unpaginated work each plant is numbered, illustrated and given a description to which Rousseau added his notes. See also 'Notes inédites de J.-J. Rousseau sur la botanique', in *Œuvres inédites de J.-J. Rousseau*, ed. V. D. Musset-Pathay, vol.1 (Paris, 1825), which contains many lapses, and the partial 'Notes botaniques', A666; *CC*, xl.304-13.
35. My emphasis; Rousseau, *Projet de constitution pour la Corse*, *OC*, iii.906. Like the ancient Greek cities, some cantons came to dominate the others – for example the oligarchies of the 'Plateau' such as Bern and Zürich.

case for Corsica. In the Swiss lowlands, by comparison, cultivation was easier, harvests more ample and the scope for territorial conquest greater due to the gentler topography. These characteristics enabled a few families to form oligarchies that held power for centuries and kept most of the people disenfranchised. The largest and most powerful of these oligarchies of the 'plateau' was Bern.

Rousseau's hostility toward Bern was understandable; he was victimised by the oligarchy of Bern in 1762 when he was expelled from Yverdon, and again in 1765 upon his expulsion from the Ile St Pierre.[36] Bern was a hidebound oligarchy of long standing, a closed corporation that annexed its neighbours, including the pays de Vaud. Bern accepted no new citizens after 1651.[37] The oppressive political environment in the oligarchic republics drove some into voluntary or involuntary exile; for example, the Zürich physician and man of letters, Johann Georg Zimmermann (1728-1795), emigrated to Germany, remarking that, 'Obwohl republikanische Städte ihrer Verfassung nach frei sind, sind sie es ihrer Mentalität nach nicht, vielmehr durch Unterwürfigkeit gegenüber den politisch Mächtigen auf der einen, durch eingebildeten Stolz auf den Adel andereseits charakterisiert'.[38] Zimmermann believed that '[m]ehr Freiheit sei in der Monarchie zu finden.' [39]

iii. Jean-Jacques and the 'corps helvétique'[40]

Rousseau's reflections on Neuchâtel and the modèle Suisse were not merely those of a disinterested observer, for he took an active role in discussions with the progenitors of Switzerland's indigenous brand of patriotism – Helvetism – at the very moment of its creation in the early 1760s. Helvetism was grounded in neither linguistic nor confessional identities, but rather in the defensive alliance of communities and shared way of life that Rousseau had so cogently analysed in his two *Lettres* to the maréchal.

36. Rousseau, *Lettres de la montagne*, *OC*, iv.930. He condemns the Republic of Bern, which expelled him from Yverdon and the Ile St Pierre, while lavishing praise on the poorer, more democratic Swiss republics. *Considérations sur le gouvernement de Pologne*, *OC*, iii.1006.
37. François Robert, Louis XIV's geographer, criticised this policy. Kuntz, *Schweizer Geschichte*, p.162.
38. 'Although republican cities are free according to their constitution[s], according to their mentality they are not, rather, they are characterised by obsequiousness in regard to the politically powerful on the one hand, [and] by conceited pride in the nobility on the other', Simone Zurbuchen, 'Berliner "Exil" und Schweizer "Heimat"', p.65.
39. '[M]ore freedom was to be found under a monarchy', Zurbuchen, 'Berliner "Exil" und Schweizer "Heimat"', p.63.
40. The most definitive treatment of Rousseau's connection to Switzerland is by Jost, *Jean-Jacques Rousseau suisse*; see also Walter Stutzer, 'Jean-Jacques Rousseau und die Schweiz: zur Geschichte des Helvetismus', doctoral dissertation, University of Zürich (Zürich, 1950), and Gaspard Vallette, *Jean-Jacques Rousseau, genevois*, 2nd edn (Geneva, 1911).

Rousseau's reflections on Switzerland resulted from long observation and thought. The way was therefore prepared for his patriotic encounter with Switzerland well ahead of his hasty departure for Yverdon in June 1762; this encounter culminated his long-standing interest in various aspects of the 'corps helvétique': moral and cultural as well as political. Switzerland was his 'pays', not merely a Weberian ideal type.[41] This engagement with Switzerland as his 'pays' has, however, taken a back seat to Rousseau's relationship to Geneva, and has consequently received little attention in the literature.[42]

Rousseau used the term 'Suisse' before he styled himself 'citoyen de Genève'. In 1741, he identified himself to the Académie royale des sciences (Paris) as 'Suisse de Nation'.[43] Rousseau had used Swiss pastoral authors such as Muralt, Solomon Gessner (1730-1788) and Haller as sources for the rapturous descriptions of alpine landscapes and their virtuous inhabitants in *Julie*.[44] He situated *Julie* in the Alps and on Lake Geneva in order to contrast idyllic, virtuous Switzerland with France. Thus, St Preux is proud that for the pastoralists of the high Valais, the 'seul compliment qu'ils [lui] firent' was to call him 'Suisse'. Rousseau's identification of himself and St Preux as 'Suisses' shows his loyalty toward Switzerland, understood in the broad sense invoked by the Helvetic movement, and seconded by Haller's bio-geographical definition of Switzerland in his botanical works.[45]

Thus, Rousseau's surrender of his Genevan citizenship in 1763 may have left him without a patrie, but it did not leave him without a pays –

41. 'Cet idéal n'etait pas une abstraction, puisque Rousseau prenait pour modèle les campagnards et les montagnards suisses', Sven Stelling-Michaud, 'Projet de constitution pour la Corse', *OC*, iii.ccx.

42. But see Jost, *Jean-Jacques Rousseau suisse*; Stutzer, 'Jean-Jacques Rousseau und die Schweiz'; Leonore Speerli, 'Rousseau und Zürich: vom Erscheinen des ersten Discours bis zum Ausbruch der Revolution in Frankreich', doctoral dissertation, University of Zürich (Zürich, 1941), the principal source for Francis Cheneval, 'The reception of Rousseau's political thought by Zurich's "Patriots"', in M. Bohler, *et al.* (eds), *Republikanische Tugend: Aussbildung [sic] eines schweizer Nationalbewusstseins und Erziehung eines neuen Bürgers* (Geneva, 2000), p.425-46.

43. [Dortous de Mairan], 'Extraits des registres de l'Académie royale de sciences' 61 (1742), *CC*, i.317, A45. Leigh claimed it would be 'a gross anachronism' to consider Rousseau 'Swiss': R. A. Leigh, 'Jean-Jacques Rousseau and the myth of antiquity', in *Classical influences on western thought A.D. 1650-1870: proceedings of an international conference held at King's College, Cambridge March 1977*, ed. R. R Bolgar (Cambridge, 1979), p.155-68 (160).

44. Rousseau, *Julie*, *OC*, ii.77-81. On Rousseau's reception of Muralt, see Arthur Ferrazzini, *Béat de Muralt et Jean-Jacques Rousseau* (La Neuveville, 1951); see also Stutzer, 'Jean-Jacques Rousseau und die Schweiz', p.50.

45. 'War Genf ihm Vaterstadt und Vorbild in manchem, so war die Schweiz ihm ein Ideal, und sie war ihm zweifelos wichtiger [If Geneva was his native city and model in many respects, Switzerland was for him an ideal, and was for him doubtless more important]', Störi, 'Der Helvetismus des "Mercure suisse"', p.35.

the pays of which he had given a utopian vision in *Du contrat social*: 'Quand on voit chez le plus heureux peuple du monde des troupes de paysans régler les affaires de l'Etat sous un chêne et se conduire toujours sagement, peut-on s'empêcher de mépriser les raffinements des autres nations [...]?'[46] While he praised the alpine republics of his pays, he perhaps never completely dissociated himself from Geneva, the 'patrie' of his birth. In this regard, the letter in which he surrendered his Genevan citizenship contains this telling phrase: 'ma *Patrie* en me devenant étrangère ne peut me devenir *indifférente*: je l[ui] reste *attaché* par un tendre Souvenir, et je n'oubli[e] d'elle que ses outrages'.[47]

Rousseau took considerable interest in the reform movement then gaining momentum in the major cities of the Swiss Confederation: Basel, Bern and Zürich – all largely German-speaking.[48] He was already corresponding with Swiss physiocrats and reformers well before his flight to Yverdon in June 1762. In Bern these contacts included the Landvogt, Vincenz Bernhard Tscharner, translator of Haller's *Die Alpen* and Johann Rudolf Tschiffeli (1716-1780), an agricultural reformer; both were founders of the Schweitzerische Gesellschaft in Bern.[49] Rousseau also had correspondents from the circle around Johann Jakob Bodmer (1698-1783), professor of patriotic history at the Collegium Carolinum in Zürich: the pastor, Leonhard Usteri (1741-1789), a notable correspondent of Rousseau,[50] the city doctor, Johann Kaspar Hirzel, author of *Le Socrate rustique* and Professor Caspar Hess (1727-1800).

46. Rousseau, *Du contrat social*, Book 4, ch. 1, *OC*, iii.437. See also *OC*, iii.114, 118, 907 regarding the election of small groups of citizens to govern. *Pologne*, one of the works in which he praises the poor, alpine republics, was composed in 1770-1771. The 'inventory' that ten-year old Poles should prepare would therefore be of Switzerland, not Geneva. See Patrick Coleman, '"Aimer la patrie et les lois", l'objet de l'éducation républicaine chez Rousseau', in M. Böhler, *et al.* (eds), *Republikanische Tugend*, p.459-70 (461). For the dating of this work, see Jean Fabre, 'Considérations sur le gouvernement de Pologne', *OC*, iii.ccxxxvii.
47. My emphasis; Rousseau to Jacob Favre, 12 May 1763; *CC*, xvi.164.
48. Rousseau served as a model for this movement for political, social and economic transformation. Usteri refers to the patriots, Johann Jakob Breitinger (1701-1776) and Johann Jakob Bodmer, as 'nos Socrates ou Rousseaus, et vos amis. Je souhaite que vous les embrassiez bien tôt.' Usteri to Rousseau, 20 January 1763; *CC*, xv.82. See Rousseau's persecution in Switzerland in *Lettres écrites de la montagne*, *OC*, iii.688, 796-97.
49. The 'Schweitzerische Gesellschaft in Bern' (Swiss Society in Bern) was founded in 1759 by seven 'Suisses', including Tschiffeli and Daniel von Fellenberg (1736-1801); the first was the principal instigator for the society, while the second was its first president. Note that this group described itself as 'Schweitzerische', i.e. as 'Swiss', rather than as Bernese (after the canton). It was later called Ökonomische Gesellschaft zu Bern (= Société Oeconomique de Berne).
50. This correspondence comprised 26 letters from Usteri, and 18 from Rousseau. *Correspondance de J.-J. Rousseau avec Leonard Usteri*, ed. Paul Usteri and Eugène Ritter (Zürich and Geneva, 1910).

In 1762, the year of Rousseau's return to his homeland, these Swiss 'patriots' founded the Helvetic Society (Helvetische Gesellschaft, Société helvétique), the world's first patriotic association, in Schinznach, near Brugg. The reformers had both local (cantonal) and national goals: (1) the extension of political rights beyond the small number of patrician families that had long controlled their cantons, and (2) encouraging Swiss patriotism across linguistic and religious divides, following the pioneering work of the *Journal helvétique* of Neuchâtel, later known as the *Mercure suisse.*

Many in the patriotic movement admired Rousseau, as is evident in a letter from Tscharner about the founding of the Schweitzerische Gesellschaft in Bern:

> [U]ne Société de Citoyens animés de ces sentiments d'humanité, de désintéressement et de liberté, que *respirent vos ouvrages*, et, ce qui est bien plus persuasif, vos actions et toute votre vie. Je ne doute pas un instant que Vous n'applaudissez aux vues expliquées dans l'imprimé que je suis chargé de Vous communiquer. Vous pouvais regarder tout cela comme *votre ouvrage; vos écrits* [...] nous ont fait présumer, que ce serait l'époque pour réunir les efforts de toute part, et se flatter de réussir à détrôner tout à fait quelques préjugés qui tyranniseent ou tout le genre humain ou au moins des nations et des Sociétés particulières, aux dépends de leur félicité.

In closing, Tscharner writes: 'Nous attendons, Monsieur, des leçons de sagesse du Philosophe de Montmorency, et des leçons de patriotisme du Citoyen de Genève; mais avant toute chose votre amitié en retour de la nôtre.'

Appended to Tscharner's letter were four prize questions posed by the Schweitzerische Gesellschaft:

1. Quels sont les moyens de tirer un peuple de la corruption? Et quel est le plan le plus parfait qu'un législateur puisse suivre à cet égard?
2. Est-il des préjugés respectables qu'un bon Citoyen doive se faire un scrupule de combattre publiquement?
3. Quel peuple a jamais été le plus heureux?
4. Par quels moyens pourrait-on resserrer les liaisons et l'amitié entre les Citoyens des diverses Républiques, qui composent la confédération helvétique?[51]

To the last question, Rousseau responded: 'il n'y a qu'un *Suisse,* ou quelqu'un qui connaisse à fond la constitution *Physique,* Politique et Morale du *corps helvétique* qui puisse la traiter avec Succès'.[52] Rousseau's

51. My emphasis; Tscharner to Rousseau, 5 April 1762; *CC*, x.184-86.
52. My emphasis; Rousseau to Tscharner, 29 April 1762; *CC*, x.227.

response demonstrates the special character that he attributed to Switzerland, a character that only a Swiss could understand.

Swiss patriotism was closely allied to economic and agricultural concerns, and the movement offered a Swiss peasant as the paragon of thrift and sound practice. In 1761 Rousseau received a long letter from Usteri detailing the intensive farming techniques and prudent household and farm management practised by that most famous of Swiss peasants, Jacob Gujer ('Kleinjogg' or little Jacob; 1716-1785), immortalised in Hirzel's *Socrate rustique*.[53] He was impressed that in Switzerland men of science praised such a man: 'Quelque étonnant que soit le héros de votre livre, l'Auteur ne l'est pas moins à mes yeux. Il y a plus de respectables paysans que de savants qui les respectent et qui l'osent dire. Heureux le *pays* où les Kliyoggs cultivent la terre, et ou des Hirzels cultivent les lettres! L'abondance y règne, et les vertus sont en honneur'.[54]

Gujer became the object of a cult;[55] he was visited by European notables such as Johann Caspar Lavater (1741-1801), Johann Wolfgang von Goethe (1749-1832), Mirabeau the elder and Prince Ludwig-Eugen of Württemberg (1731-1795).[56] The Prince of Württemberg discoursed with the prince of peasants at the annual meeting of Helvetische Gesellschaft in Schinznach in 1765.

During his time in Môtiers Rousseau received visitors from the major cities of the Confederation, and corresponded with them. In September 1762 he recounted to Usteri a visit to Môtiers by Professor Caspar Hess and his bride; following his discussions with Hess, Rousseau vowed to undertake 'un pélerinage au séjour de la raison, des moeurs, du zèle patriotique, de tout ce qui peut intéresser un ami de l'humanité'.[57] Indeed, Rousseau started a journey on foot to Zürich in June 1763, but turned back due to bad weather. He became rather disillusioned with his prospective hosts in Zürich,[58] but in 1770 he nonetheless asked Jakob-Heinrich Meister (1744-1826) of Zürich for 'beaucoup de nouvelles' of 'Messrs Bodmer, Breitinguer, Hirzel et Weguelin', leaders of the Zürich patriotic movement.[59]

53. Usteri to Rousseau, 20 November 1761; *CC*, ix.254-64.
54. My emphasis; Rousseau to Hirzel, 12 November 1764; *CC*, xxii.46.
55. For a discussion of the cult of the noble peasant, see Amy Wyngaard, *From savage to citizen: the invention of the peasant in the French Enlightenment* (Newark, DE, 2004), p.139, 209, n.6.
56. The Prince raised his daughter according to the precepts of *Emile*: 'sa fille qu'il élève par rapport au physique aussi bien que par rapport au moral tout à fait par vos règles est l'enfant le plus robuste, le plus agile, le mieux portant que j'ai vu'. Usteri to Rousseau, 2 June 1764; *CC*, xx.133.
57. Rousseau to Usteri, 2 September 1762; *CC*, xiii.7.
58. Jost, *Jean-Jacques Rousseau suisse*, vol.1, p.444.
59. Jakob Heinrich Meister to Johann Heinrich Meister, 22 July 1770; *CC*, xxxviii.71. Jakob Heinrich Meister (1744-1826), son of Johann Heinrich, was consecrated pastor in Zürich in 1763, but defrocked for his *De l'origine des principes religieux* (1768). He moved to Paris in

Given Rousseau's Swiss patriotism, the botany of the pays took on a peculiar, lifelong resonance that is evident in his last botanical library, inventoried after his death by the marquis de Girardin. This small library of fourteen works contains two works on alpine flora: a folio edition of Albrecht von Haller's *Enumeratio methodica stirpium helvetiae* (Göttingen, 1742) and an octavo edition of Giovanni-Antonio Scopoli's flora of the Slovenian Alps, *Flora carniolica* (Vienna, 1760), which was dedicated to Haller. Another Swiss work was Chabrey's *Omnium stirpium sciagraphia et icones* (Geneva, 1678), referred to in Chapter 3.[60] Rousseau likewise regarded the *Pinax* (Basel, 1623) of Caspar [Gaspard] Bauhin as a Swiss work: 'les phrases de Tournefort étant presque toutes tirées de Gaspard Bauhin, tout l'honneur qu'en cela les Français veulent faire à leur compatriote remonte à *un Suisse* en toute équité'. [61] Rousseau's annotations to the Regnaults' *La Botanique mise à la portée de tout le monde*, made in 1777-1778, refer to Swiss medicines, e.g. the Falltranck, and cuisine prepared in 'mon pays'.[62]

iv. Neuchâtel 1762

With the exception of the Napoleonic interregnum between 1806 and 1814, Neuchâtel was under Prussian rule from 1707 until 1848.[63] During Rousseau's residence in the principality it was an allié associé, and hence not yet a full member, of the Swiss Confederation. However, there was no doubt at the time that 'l'Etat de Neuchâtel fut partie de la Suisse' and that it was 'un véritable membre du Corps helvétique [...].Il n'y a donc aucune incompatibilité entre l'appartenance prussienne et l'attirance helvétique.'[64] Helvetism had reached the Suisse romande.[65]

1769, becoming secretary to Baron Grimm and editor of the *Correspondance littéraire, philosophique et critique (1773-1813). He took refuge in England in 1792, returning to Zürich in 1794; Napoleon Bonaparte appointed him president of the commission in charge of introducing the Act of Mediation to Zürich. He was a member of the Grand Council of Zürich (1803-1826).*

60. Chabrey continued the work of the Swiss botanist, Jean Bauhin (1541-1613), who was the brother of Caspar Bauhin, author of the famous *Pinax*.

61. My emphasis; BdAN [Munch MS], article 190.

62. See BdAN [Munch MS], art. 26, and 240. Rousseau served Bernardin de St Pierre '*un mets suisse; un pot-pourri de lard, de mouton, de légumes et de châtaignes*'. Original emphasis; Jacques-Henri Bernardin de St Pierre, 'Essai sur J.-J. Rousseau', *Œuvres complètes de Jacques-Henri-Bernardin de St Pierre: mélanges*, ed. L. Aimé-Martin (Brussels, 1820), p.28.

63. In 1707 Neuchâtel passed from the French royal house of d'Orléans-Longueville to the house of Prussia upon the death of its last heir, Marie de Nemours (1625-1707). During the Napoleonic interregnum, Neuchâtel was governed by Napoleon's chief of staff, Louis Alexandre Berthier.

64. Philippe Henry, 'Les relations politiques entre Neuchâtel et Berlin au XVIIIe siècle', in *Schweizer in Berlin des 18. Jahrhunderts*, ed. Martin Fontius and Helmut Holzhey (Berlin, 1996), p.33-44 (41).

65. Jost alleges Neuchâtel was a 'terre helvétique' for centuries; see Jost, *Jean-Jacques Rousseau*

Helvetism, 'le courant de l'indigénat helvetique'[66] inspired Colonel Abraham Pury (1724-1807), a local notable (colonel in the militia of the Val de Travers), and erstwhile botanising friend of Rousseau, to advocate the 'cantonnement' of Neuchâtel, that is, its adhesion to the Swiss Confederation: 'Par sa position géographique, par ses traités d'alliance avec les villes du Plateau [Zürich et Berne], par son histoire dès le Moyen Age, par le volonté de son destin, Neuchâtel d'allié protégé se doit de devenir membre souverain à part entière du louable Corps helvétique'.[67] Pury wrote a three-part history of Neuchâtel in which he claimed the principality had belonged to the original Helvetic Confederation. It was not published until the nineteenth century and that it was a falsification was not known at the time.[68]

By the mid-eighteenth century, the city of Neuchâtel had approximately 3000 inhabitants, while the canton as a whole had about 32,000.[69] Rousseau says the canton was densely populated by French standards.[70] This population density resulted largely from the revocation of the Edict of Nantes in 1685, which had unleashed the second Réfuge, sending French Protestants fleeing to, among other places, the Suisse romande.

In comparison with Geneva, Neuchâtel society remained under the sway of the deeply conservative Vénérable Classe of pastors well into the eighteenth century, and was slow to join the Republic of Letters.

---

suisse, vol.1, p.64ff. By the 1780s twenty-five Neuchâtelois were involved with the Helvetische Gesellschaft; of these, twenty were members of the elite intellectual circle, the Société du jardin. Louis-Edouard Roulet, 'La Société helvétique et les Neuchâtelois', *Musée neuchâtelois* (1991), p.211-23 (213).

66. Roulet, 'La Société helvétique', p.212.

67. Louis-Edouard Roulet, 'Abraham Pury,' ed. Michael Schlup, in *Biographies neuchâteloises*, vol.1, p.225-28 (227).

68. [Abraham Pury], *Mémoires sur le comté de Neuchâtel en Suisse par le chancelier de Montmollin* (Neuchâtel, 1831) and *Extraits des chroniques ou Annales autrefois successivement par les chanoines du chapitre de Notre-Dame de Neuchâtel* (Neuchâtel, 1839, 1884). The deception is documented in Jules Jeanjaquet, 'L'auteur de la chronique des chanoines et de mémoires du chancelier de Montmollin', *Musée neuchâtelois* 38 (1951), p.3-14, 43-54. Discussions of Pury's relationship with Rousseau rarely mention Pury's Helvetism. But see Eigeldinger, *Des Pierres dans mon jardin*, p.124-27.

69. Jaucourt, 'Neuchâtel', in *Encyclopédie*, vol.11, p.109 gives the figure for the canton, while an anonymous article, also entitled 'Neuchâtel', gives the population figure for the city. This latter article may have been authored by Frédéric-Samuel Ostervald. See *Encyclopédie*, vol.11, p.113. Historians give a figure of 32,300 for 1752. See Philippe Henry, 'L'évolution démographique', in *Histoire de pays de Neuchâtel 2: de la Réforme à 1815* (Hauterive, 1991), p.140-57 (146).

70. Rousseau to the maréchal de Luxembourg, 28 January 1763; *CC*, xv.120. Rousseau asserted, however, that the export of mercenaries was depopulating Switzerland and raising the cost of living: 'La vie étant devenue plus dévorante, le même pays n'a plus pu nourrir la même quantité d'habitants. C'est la raison de la dépopulation que l'on commence à Sentir dans toute la Suisse.' Rousseau to the maréchal de Luxembourg, 20 January 1763; *CC*, xv.50.

Neuchâtel had no academy (seminary to train pastors) or other institutions of higher education, and few individuals were positioned to form a vanguard of the Enlightenment.

Nonetheless, the authorities realised the importance of cultivating local talent, appointing the eminent polymath and member of the republic of letters, Louis Bourguet (1678-1742), to a newly created chair of philosophy in 1731. Bourguet's work ranged from antiquities to paleontology; the latter interest culminated in his highly regarded *Traité des pétrifications*, co-authored with Pierre Cartier, pastor of La Chaux-du-Milieu, and published in Paris.[71] The death notice for Bourguet in the *Mercure suisse* stresses his participation in the 'République des Lettres':

> Le 31. de ce Mois Mr Louis Bourguet, Professeur en Philosophie et en Mathématiques [...] mourut en cette Ville [Neuchâtel] d'un accès d'Asthme, dans la 64. Année de son âge. La bonté de son Caractère, la pureté et la douceur de ses Mœurs, sa Piété et son Amour pour la Religion [...] le font regretter de tous ceux qui le connaissent. Il était en Correspondance particulière avec les Savants les plus Illustres de l'Europe; ce qui joint aux Ouvrages dont il a enrichi la République des Lettres, prouve que le Monde savant fait une perte considérable en sa Personne.[72]

By 1732 the Neuchâtel Société typographique was participating in the Swiss republic of letters with the publication of the *Journal helvétique ou Recueil de pièces fugitives de littérature choisie, de poésies, de traits d'histoire, ancienne et moderne, de découvertes des sciences et des arts, de nouvelles de la république des lettres, et de diverses autres particularités intéressantes et curieuses, tant de Suisse, que des païs étrangers*. This 'revue littéraire et d'information générale [...] agissait comme un ferment dans la principauté'.[73] It was cited, for example, in the *Encyclopédie*, 'Neuchâtel'.[74] Later it was known as

71. This work included the famous 'étoile marine petrifiée', said to be 'absolument unique', from the natural history cabinet of the Gagnebin brothers of La Ferrière in the Jura, discussed below. This unique fossil, which achieved renown through Bourguet's *Traité*, was depicted in fig.438 of plate 9. See Samuel Frédéric Ostervald, *Description des montagnes et vallées du pays de Neuchâtel en 1764*, ed. Victor Benoit (Neuchâtel, 1861), p.106.

72. Anon., *Mercure Suisse/Journal helvetique* (December 1742), p.30.

73. Michel Schlup, 'Neuchâtel aux alentours de 1769', in *L'Edition neuchâteloise au siècle des Lumières: la société typographique de Neuchâtel (1769 à 1789)*, ed. Michel Schlup (Hauterive, 2002), p.15-28 (25).

74. While Jaucourt refers to the *Journal helvétique* only once in the article, he may have used it in researching the topic. Remarkably, his reference to Neuchâtel concerns Rousseau's first botany teacher, Dr Jean-Antoine d'Ivernois: 'Les montagnes sont couvertes de simples dont on fait le thé Suisse et l'eau vulnéraire, il y en a des espèces très-rares. M. le docteur d'Yvernois, médecin du roi [de Prusse] dans cette souveraineté, botaniste célèbre, en a donné une savante description dans le journal helvétique, qui s'imprime à *Neuchâtel*'. Original emphasis; 'D. J.' [Louis, chevalier de Jaucourt], 'Neuchâtel', in *Encyclopédie, ou Dictionnaire raisonné des sciences, des arts, et des métiers, par une société de gens de lettres*, ed. Denis Diderot and Jean Le Rond d'Alembert CD-ROM (Paris, 2000).

the *Mercure suisse*. As its two names suggest, this was a literary gazette and all-round information source for and about the Swiss republic of letters, including the Suisse romande and the German-speaking cantons. For example, the editors of the *Journal helvétique* stressed the importance of Haller's work for all of Switzerland, including the principality of Neuchâtel.

The success of the *Journal helvétique* was a harbinger of both Helvetism and Enlightenment in Neuchâtel society: 'Un monde nouveau se dessine dès les années 1760. La principauté est touchée aussi par l'effervescence philosophique et littéraire qui gagnent l'ensemble de l'Europe.'[75] In the 1750s an elite 'secte philosophique' began to emerge, including Rousseau's protector, George Keith and Pierre-Alexandre Du Peyrou, Rousseau's close friend for many years. In 1759 twenty retired military officers, businessmen and magistrates formed a men's-only reading club known as the 'Société' or 'Cercle du Jardin'; this club accepted members of the most powerful families in the principality: de Luze, Pourtalès, Montmollin, Boyve, Chaillet, Pury and Sandoz.[76]

Like many Swiss cantons, Neuchâtel exported mercenaries, who served foreign masters from the king of France to the Dutch East India Company. The majority of patrician families in the eighteenth century had one or more officers in foreign military service; a 1749 estimate found forty-eight in French service, twenty-six in Holland and four in Sardinia; for the century as a whole there were at least 500 officers and 3500 soldiers in the service of these three nations.[77] As noted earlier, Rousseau considered the Swiss 'commerce' in men deleterious for the country.[78] The chevalier de Jaucourt, by contrast, considered it beneficial for the principality:

> Il y aurait évidemment plus à perdre qu'à gagner pour S. M. le roi de Prusse, si les Neuchâtelois abandonnaient ou suspendaient l'exercice d'un droit qui dans des circonstances telles que celles qui affligent aujourd'hui l'Europe, est la sauvegarde de leur pays. Quoique le goût pour le commerce ait affaibli chez eux celui que les portait généralement autrefois à prendre le parti de armes ils ont cependant encore un nombre considérable d'officiers qui servent avec distinction.[79]

75. Michel Schlup, 'Neuchâtel aux alentours de 1769', p.26.
76. Eddy Bauer, 'Le cercle du jardin – esquisse historique d'une société', in *La Société du jardin de Neuchâtel* (Neuchâtel, 1963), p.38ff.; see also Nathalie Guillod, 'Une esquisse de l'élite culturelle neuchâteloise'.
77. Henry, 'L'évolution démographique', p.157.
78. Rousseau to the maréchal de Luxembourg, 20 January 1763; *CC*, xv.50. The translator of Rousseau's botanical writings into English, Thomas Martyn, noted the high cost of living in Switzerland, but observed that one lives better in Switzerland. Thomas Martyn, *Sketch of a tour through Swisserland*, p.16.
79. 'Jaucourt, 'Neuchâtel', in *Encyclopédie*. While much of the information concerning

As elsewhere in Switzerland, retired military officers played an important role; Rousseau's friend and fellow botanist, Colonel Abraham Pury, was one of these.

Frédéric Samuel Ostervald (1713-1795), banneret (militia leader) of Neuchâtel, played a leading role in the intellectual life of the canton during the latter part of the eighteenth century, authoring the indispensable *Description des montagnes et vallées du pays de Neuchâtel en 1764*, organising courses of instruction and helping create the Société du jardin.[80] In 1769 he and his son-in-law, Jean-Elie Bertrand (1713-1797), organised the Société typographique de Neuchâtel (STN), which was 'one of the most important publishers of French books in the eighteenth century'.[81] This venture approached 'the most powerful publisher in France, Charles-Joseph Panckoucke' to enter into an agreement to publish the *Encyclopédie* of Diderot and d'Alembert, which had fallen afoul of French censors following the attempted regicide of Louis XV in 1757.[82]

The STN invoked as one of its advantages the freedom of the press under the rule of the philosopher-king, Frederick the Great of Prussia: 'On sait que le comté de Neuchâtel est un des pays les plus libres de la Suisse, en sorte qu'il n'y aurait aucun obstacle à redouter de la part du gouvernement et du magistrat.'[83] Despite the difficulties Rousseau was to experience there, Neuchâtel offered a greater degree of freedom for controversial ideas than many European states. There was no official censor, as in France, and the titular overlord, who tried to shield Rousseau from theological extremists, had protected Voltaire and La Mettrie.[84]

The presence of the STN on the international publishing scene showed that Neuchâtel's size posed no obstacle to its playing a major

Switzerland in the *Encyclopédie* is defective, the article 'Neuchâtel' is '[s]olidement documenté', based, perhaps, on information from Ostervald's *Description*. See Charly Guyot, *Le Rayonnement de l'Encyclopédie en Suisse française* (Neuchâtel, 1955), p.74; James Doolittle, 'Jaucourt's use of source material in the *Encyclopédie*', *Modern language notes* 65:6 (June 1950), p.387-92 (387).

80. Jacques Rychner and Michel Schlup, 'Frédéric Samuel Ostervald: homme politique et éditeur (1713-1795)', *Biographies neuchâteloises*, vol.1, p.197-201.

81. Robert Darnton, *The Business of Enlightenment: a publishing history of the Encyclopédie* (Cambridge, MA, 1979), p.3.

82. Darnton, *The Business of Enlightenment*, p.35.

83. Letter of 25 July 1769 from the STN to Panckoucke, quoted in Darnton, *The Business of Enlightenment*, p.38.

84. Self-censorship and some official censorship arose from concerns about the political and religious sensitivities of France and Bern. Frederick doubted he could protect controversial people in the principality: 'J'ai voulu, dans ce pays, protéger Jean-Jacques, on l'a chassé; j'ai demandé qu'on ne persécutât point un certain Petitpierre, je ne l'ai pu obtenir', Frederick II to Voltaire, 16 September 1771; *Œuvres complètes de Voltaire*, vol.122, p.79-81.

role in the Enlightenment and the republic of letters; a small number of extremely active individuals were sufficiently well-educated, well-connected and affluent to stage a major publishing sensation well away from major publishing centres such as Paris and Amsterdam.

The Neuchâtel publishing concern, which split into two companies in 1773, produced a number of important works, including Jean-Frederic Ostervald's famous *Bible* (1744), the 'Vulgate protestante', Emer de Vattel's great work of international law, *Droit des gens* (1758),[85] a new edition of Holbach's *Système de la nature* (which cost Ostervald the post of banneret in 1771), the *Essai sur le despotisme* by the marquis de Mirabeau (1775-1776) and of course, the *Encyclopédie*. Samuel Fauche, the partner who left the STN, went on to publish such successful works as Louis-Sébastien Mercier's *Tableau de Paris* (1781), Charles Bonnet's *Œuvres d'histoire naturelle et de philosophie*, and the first volume of Saussure's *Voyages dans les Alpes* as well as the latter's *Essai sur l'hygrométrie* (1783).

Another Neuchâtelois who contributed to the diffusion of Enlightenment ideas in the region was Jean-Rodolphe de Vautravers (originally Johann Rodolph von Valltravers of the canton of Bern), (1723-1814 or 1815). Vautravers, a naturalised Englishman, Fellow of the Royal Society of London and man of many contacts, made the acquaintance of Linnaeus in 1760-1761 while visiting Sweden, and was elected a member of the Royal Academy of Uppsala. He became acquainted with Rousseau in 1763, provided him with the English translation of *Emile* and attempted to persuade him to stay at Rockhall, his estate near Bienne, before and after the 'lapidation' of Môtiers; he even made inquiries of one of the 'Magistrats les plus difficiles' as well as the mayor of Bienne.[86] While Vautravers was no great man of science, his contacts in England, Sweden, Germany and Switzerland, his well-connected English wife (niece of the Bishop of London) and enterprising character made him the kind of person to promote the circulation of correspondence that was the lifeblood of the republic of science; this he did, for example, by relaying messages to Linnaeus from Swiss naturalists such as Haller and Gessner.[87]

Such powerful international linkages were typical for Neuchâtel, which enjoyed an international presence that far exceeded its small

85. This has been called 'la plus belle impression typographique neuchâteloise du XVIIIe siècle. Mais il s'agit aussi d'un des grands textes des Lumières qui devait faire autorité dans la pratique diplomatique et le règlement de différends internationaux', Michel Schlup, 'Aperçu de l'imprimerie et de l'édition neuchâteloises avant 1769', in Schlup (ed.), *L'Edition neuchâteloise au siècle des Lumières*, p.29-59 (48).

86. Vautravers to Rousseau, 4 February 1765; *CC*, xxiii.286-88.

87. See Gavin R. de Beer, 'Rodolph Valltravers, F. R. S.', *Notes and records of the Royal Society of London* 4:2 (October 1946), p.216-26.

size.[88] This presence was being daily increased by the enormous wealth its citizens generated directly or indirectly from slavery, a venture that Rousseau deplored. Commercial activity inevitably involved the triangular trade and the Caribbean plantation system. For example, Rousseau's friend and literary executor, Pierre-Alexandre Du Peyrou, inherited a fortune in Surinam, including plantations worked by slaves; he gave funds to build a quay along the lake.[89] He also built himself a magnificent mansion in the French style, devoted himself to the Enlightenment and served as a standard bearer for Rousseau by overseeing, at much personal cost, the posthumous publication of Rousseau's writings.

David de Pury (1707-1786), another benefactor of Neuchâtel, became extremely rich in Lisbon through the diamond and slave trades; he endowed his native city with a magnificent library and city hall.[90] Jacques-Louis de Pourtalès (1722-1814), nicknamed the 'nabab neuchâtelois',[91] had the most wide-reaching success, amassing an immense fortune through global networks of industry, trade and finance. His company, Pourtalès et Compagnie became a global mercantile powerhouse.[92] Pourtalès 'donne une extension mondiale, non seulement à l'entreprise qu'il venait de fonder, mais à l'industrie de l'indiennage dans son ensemble'[93] because his Neuchâtel firm produced the calico that was shipped from Nantes to West Africa, where discriminating African buyers exchanged it for slaves, who were shipped to the Caribbean.[94] On the final leg of the journey, the sugar and chocolate the slaves produced returned to Europe.[95] Pourtalès donated a hospital to Neuchâtel. Whether Rousseau was aware of the true nature of the

88. Kuntz, *Schweizer Geschichte*, p.209.
89. In his will, Du Peyrou settled a sum of money on each of his slaves. Charly Guyot, *Un Ami et défenseur de Rousseau. Pierre-Alexandre Du Peyrou* (Neuchâtel, 1958), p.218.
90. Portugal was at the time the greatest slave-trading nation; Pury traded 42,000 slaves from Angola to Brazil. David, *et al.*, *La Suisse et l'esclavage*, p.24.
91. Paul and Jacques-Louis de Pourtalès came from a noble family of Languedoc that established itself in Neuchâtel at the beginning of the eighteenth century, and was ennobled by King Frederick II of Prussia in 1750. Their father, Jérémie, co-founded a banking house in Neuchâtel. Paul de Pourtalès was mayor of Verrières from 1776 to 1781.
92. Jacques-Louis de Pourtalès inherited his father's banking house and established calico production in Neuchâtel. See David, *et al.*, *La Suisse et l'esclavage*, p.25. A Pourtalès bank remains in business in the Bahamas.
93. Bauer, 'Le Cercle du jardin', p.44.
94. So many Swiss participated in this trade in Nantes that it was said that an entire Swiss canton inhabited the right side of the Loire. David, *et al.*, *La Suisse et l'esclavage*, p.20.
95. This relatively unknown aspect of Swiss economic and social history is documented in David, *et al.*, *La Suisse et l'esclavage*; see also Kuntz, *Schweizer Geschichte*, p.216-17. The Swiss participated financially in many aspects of the infamous triangular trade by furnishing merchandise or funding it as an investment. David, *et al.* found 100 expeditions in which Swiss investment was involved. Some investors did not realise they had invested in the slave trade, such as the Zürich theologian, Leonhard Meister.

Pourtalès business interests is unknown, but during his later years in Paris Rousseau received visits from both Jacques-Louis and Paul de Pourtalès (1735-1821).[96]

v. Neuchâtel science

> [C]'est toute la Suisse romande qui s'ouvre aux sciences, en particulier, Neuchâtel (Bourguet, Garcin, puis Agassiz) et le Pays de Vaud.[97]

It may seem curious that Rousseau started his study of botany in Neuchâtel, where there was little formal educational or academic infrastructure – no botanical garden, learned academy or university. Indeed, one might well ask: why botany, of all things, in Neuchâtel? Yet even though Neuchâtel could boast none of the academic advantages of Paris, Leiden or Basel, it offered natural riches beyond the wildest dreams of Haller, who botanised there in the summer of 1739 with Jean-Antoine d'Ivernois and Abraham Gagnebin, local botanists who would later instruct Rousseau.[98] Haller called the Creux du Van in the Val de Travers 'ce fameux Jardin de Botanique formé par la Nature', and gushed about the 'Plantes de ce Pais qu'il n'avait point encore trouvées, ou qu'il ne croïoit même pas naître en Suisse'.[99] D'Ivernois added: 'La satisfaction que nous eûmes alors, de voir un botaniste du premier ordre, dans l'endroit du monde peut être le plus curieux par ses simples, nous ravit encore'.[100] Haller recorded this voyage in his *Iter helveticum anni 1739*, published in 1740 and extracted in the *Journal helvétique*.[101]

The countryside around Môtiers endeared itself to Rousseau. In Môtiers he found a varied landscape filled with wildflowers in summer, rocky crags, ibex and, above all, just the right amount of solitude[102] in the midst of a 'mélange bizarre' of asymmetrically disposed houses, workshops in ravines, and sheep on rocks:

> [L]es maisons, au lieu d'être alignées, Sont dispersées Sans Symétrie et Sans ordre, comme on dit qu'étaient celles de l'ancienne Rome. On ne croit plus

96. 'Je voudrais, Madame, que Monsieur Pourtalès eut été aussi content de mon accueil que je l'ai été de son passeport. J'ai tâché au moins de lui témoigner à lui et à sa compagnie le cas que je ferai toujours de votre souvenir et de vos amis', Rousseau to Marianne Françoise de Luze-Warney, 23 June 1772; *CC*, xxxix.75.
97. Sigrist, *L'Essor*, p.14.
98. Haller, 'Extrait de deux pièces', p.183.
99. Haller, 'Extrait de deux pièces', p.181.
100. Jean Antoine d'Ivernois, 'Lettre à Monsieur Cartier', p.32.
101. Rousseau requested this work from Duchesne by a letter of 30 June 1765, *CC*, xlix.300, and again in a letter to Rey of 28 December 1767; *CC*, xxxiv.268.
102. Rousseau to the maréchal de Luxembourg, 20 January 1763; *CC*, xv.112; Rousseau, *Lettres à Malesherbes*, *OC*, i.1031.

parcourir des déserts quand on trouve des clochers parmi les Sapins, des troupeaux Sur des rochers, des manufactures dans des précipices, des ateliers Sur des torrents. Ce mélange bizarre a je ne Sais quoi d'animé de vivant qui respire la liberté le bien être, et qui fera toujours du pays où il Se trouve un Spectacle unique en Son genre, mais fait Seulement pour des yeux qui Sachent voir.[103]

As a Swiss, Rousseau could appreciate this unique spectacle.

The appearance of this region differed dramatically from that of today. In that wood-hungry era, contemporary illustrations (fig.4.1 and 4.2, p.108)[104] show it as deforested: 'Au premier coup d'œil, le Spectacle, tout grand qu'il est, semble un peu nud, on voit très peu d'arbres dans la vallée; ils y viennent mal et ne donnent presque aucun fruit'.[105] Referring to the Swiss case, Rousseau warns against deforestation in *Projet de constitution pour la Corse*:

La Suisse était jadis couverte de bois en telle abondance qu'elle était incommodée. Mais tant pour la multiplication des pâturages que pour l'établissement des manufactures on les a coupés sans mesure et sans règle; maintenant ces forêts immenses ne montrent que les rochers presque nus. Heureusement, avertis par l'exemple de la France, les Suisses ont vu le danger et y ont mis l'ordre autant qu'il a dépendu d'eux. Il reste à voir si leurs précautions ne sont pas trop tardives; car si malgré ces précautions leurs bois diminuent journellement il est clair qu'ils doivent enfin se détruire.[106]

In his famous account of the region, written, like *Corse*, in the mid-1760s, Ostervald took a dimmer view of the Swiss response to deforestation:

On ne s'attache pas avec assez de soin à la conservation et à l'économie des forêts dont le dépérissement total rendra quelque jour ces contrées inhabitables. Croirait-on que dans un pays qui en était autrefois entièrement couvert, la cherté actuelle et toujours progressive du bois ait réduit ces peuples à se servir de tourbe et à tirer de leurs marais des racines de sapins à demi-pourries?[107]

Both accounts show that the re-forested Swiss Jura of today presents a very different aspect than it did to Rousseau.

Nonetheless, in his second letter from Switzerland to the maréchal de

103. The 'treize quartiers' are the thirteen cantons of the eighteenth-century Swiss Confederation. Rousseau to the maréchal de Luxembourg; 20 January 1763; *CC*, xv.112.

104. Jean-Benjamin de Laborde and Beat Fidel Baron de Zurlauben, *Tableaux de la Suisse ou Voyage pittoresque fait dans les XIII cantons du corps helvétique, représentant les divers phénomènes que la nature y rassemble, et les beautés dont l'art les a enrichis*, 13 vols (Paris, Clousier, 1780; Lamy, 1784-1786). This work served, as did many others mentioned here, to promote Helvetism. See Camilla Murgia, 'Le mythe de Rousseau chez Zurlauben et Laborde'.

105. Rousseau to the maréchal de Luxembourg; 20 January 1763; *CC*, xv.113.

106. Rousseau, *Projet de constitution pour la Corse*, *OC*, iii.926-27.

107. Ostervald, *Description des montagnes et vallées du pays de Neuchâtel*, p.49-50.

Figure 4.1: *IIe vue de Môtiers-Travers et de ses environs*, engraving from Jean-Benjamin de Laborde and Beat Fidel Baron de Zurlauben, *Tableaux de la Suisse ou Voyage pittoresque fait dans les XIII cantons du corps helvétique, représentant les divers phénomènes que la nature y rassemble, & les beautés dont l'art les a enrichis* (Paris, Lamy Libraire, 1780), courtesy of the Library of Congress, Washington, DC.

Figure 4.2: *Torrent du Val-Travers, vue à peu de distance de la Maison du Philosophe de Genève, dans le Comté de Neuchâtel*, engraving from Laborde and Zurlauben, *Tableaux de la Suisse*, courtesy of the Library of Congress, Washington, DC.

Luxembourg, dated 28 January 1763, Rousseau extolled the ample natural endowments of the country, even in winter:

> Si la première impression de tout cela n'est pas agréable, elle change insensiblement par un examen plus détaillé, et dans un pays où l'on croyait avoir tout vu du premier coup d'œil on se trouve avec surprise environné d'objets chaque jour plus intéressants. Tout le pays est plein de curiosités naturelles qu'on ne découvre que peu à peu, et qui par ces découvertes successives lui donnent chaque jour l'attrait de la nouveauté. La *botanique* offre ici ses *trésors* à qui saurait les connaitre, et souvent en voyant autour de moi cette profusion de plantes rares, je les foule à regret sous le pied d'un ignorant.[108]

Here Rousseau first botanised, filling his head – metaphorically – with straw: 'Mes courses dans *nos montagnes* si riches en plantes m'ont donné du goût pour la botanique, cette occupation convient fort à une machine ambulante à laquelle il est interdit de penser';[109] 'Il ne me vient jamais une idée vertueuse et utile que je ne voie à côté de moi la *potence* ou *l'échafaud*: avec un Linnaeus dans la poche et du *foin* dans la tête j'espère qu'on ne me pendra pas.'[110]

Rousseau later recalled the Swiss landscape with nostalgia. Writing from England to Mme de Luze-Warney, Rousseau recounted his fond memories of the region: 'Ce n'est qu'auprès de vous, Madame, que je pourrais trouver une Société préférable à la Solitude. Pour la former dans cette province il y faudrait transporter votre famille entière, une partie de Neuchâtel et presque tout Yverdon. Encore après cela, comme l'homme est insatiable, me faudrait-il vos *bois*[,] vos *monts*[,] vos *vignes*, enfin tout, jusqu'au *lac* et ses poissons.'[111] In 1769, after an unsuccessful herborisation on Mont Pilat (Dauphiné), he lamented that 'la montagne est fort triste, inculte, déserte, et n'a rien de l'admirable variété des montagnes de Suisse'.[112]

The principality was not only rich in rare and medicinal plants, but also home to the 'Montagnons', the ingenious mountain-dwellers whom Rousseau glorifies in the *Lettre à d'Alembert*. They were sufficiently sophisticated in their skill sets to impress the Parisians, as did the famed watchmakers, Abraham-Louis Breguet (1747-1823) and Ferdinand

108. My emphasis; Rousseau to the Maréchal de Luxembourg, 28 January 1763; *CC*, xv.114.

109. My emphasis; Rousseau to Duclos, 2 December 1764; *CC*, xvxii.148.

110. My emphasis; Rousseau to Malesherbes, 11 November 1764; *CC*, xxii.44.

111. My emphasis; Rousseau to Mme de Luze-Warney, 10 May 1766; *CC*, xxix.199-200; see also Rousseau to the maréchal de Luxembourg; 20 January 1763; *CC*, xv.121; '[J]e passerais ici sans regret le reste de ma vie, si j'y pouvais voir quelquefois ceux qui me la font encore aimer.'

112. Rousseau to Du Peyrou, 16 September 1769; *CC*, xxxvii.142.

Berthoud (1727-1807).[113] Rousseau wrote in the *Lettre à d'Alembert* that '[d]ans une petite ville, on trouve [...] plus d'esprits originaux, plus d'industrie inventive, plus de choses vraiment neuves parce ce qu'on est moins imitateur [...], chacun tire plus de lui-même, et met plus du sien dont tout ce qu'il fait; parce que l'esprit humain, moins étendu [...] s'élabore et fermente mieux dans la tranquille solitude'.[114] These original spirits put the long winters to good use:

> Il leur reste encore du *loisir* pour inventer mille instruments divers [...] qu'ils *vendent aux étrangers*, dont plusieurs parviennent même *à Paris* [...] et, ce qui parait incroyable, chacun réunit à *lui seul* toutes les professions diverses dans lesquelles se subdivise l'horlogerie, et fait tous ses outils *lui-même*.
> Ce n'est pas tout; ils ont des livres utiles et sont passablement instruits; ils raisonnent sensément de toutes choses, et plusieurs avec esprit [...] vous prendriez le poêle d'un Paysan pour un atelier de mécanique et pour un cabinet de physique expérimentale. Tous savent un peu dessiner, peindre, chiffrer; la plupart jouent de la flûte, plusieurs ont un peu de musique et chantent juste. Ces arts ne leur sont point enseignés par des maitres, mais leur passent, pour ainsi dire, par la tradition.[115]

They were also literate:

> Aussi appelle-t-on les Neuchâtelois les Gascons de la Suisse. Ils ont de l'esprit et ils se piquent de vivacité; ils lisent, et la lecture leur profite; les paysans mêmes sont instruits; ils ont presque tous un petit recueil de livres choisis qu'ils appellent leur bibliothèque; ils sont même assez au courant pour les nouveautés; ils font valoir tout cela dans la conversation d'une manière qui n'est point gauche [...].[116]

Rousseau's account was corroborated by Ostervald, who stressed the Montagnons' inventiveness, citing, for example, the watch-making industry they created from scratch. The first watchmaker in the region, Daniel Jean-Richard (1665-1741) was self-taught from a watch brought from London, made all his own tools and taught others the craft.[117]

In addition to watchmaking, Neuchâtel manufactured calico, which played a crucial role in the triangular trade, while merchant enterprises such as Pourtalès and Compagnie (discussed above) adapted to complex international conditions in which they were able to operate competi-

113. Breguet is a noted luxury watch brand. Berthoud moved to Paris, worked on the problem of longitude, and was elected a Fellow of the Royal Society of London.

114. My emphasis; Rousseau, *Lettre à d'Alembert*; *OC* v.55. The 'petite ville' could be one of the watch-making towns of the region such as Le Locle or La Chaux-des-Fonds.

115. My emphasis; Rousseau *Lettre à d'Alembert sur les spectacles*, *OC*, v.56. Ostervald concurs: *Description des montagnes et vallées du pays de Neuchâtel en 1764*, p.23-24.

116. Rousseau to the maréchal de Luxembourg; 20 January 1763; *CC*, xv.54. This quote supports Jost's thesis that Neuchâtel had long been a 'terre helvétique'.

117. Ostervald, *Description des montagnes et vallées du pays de Neuchâtel en 1764*, p.61-62.

tively. The Neuchâtel publishing industry played a leading part in the later Enlightenment, prospecting for publications in cities such as Paris. Likewise, in the emerging sciences, Neuchâtel savants showed a readiness to take up new ways of doing science and handling information, as exemplified by their adoption of Linnaean binomial nomenclature at a relatively early date, and their acquaintance with the artificial sexual system by the 1740s, if not earlier.

In mid-eighteenth-century Neuchâtel botany three individuals stand out: Dr Abraham Gagnebin, who worked closely with Haller and ranks among one of the great experts on plants of western Switzerland; Dr Jean Antoine d'Ivernois, another Haller correspondent who authored the first flora of the region and Dr Laurent Garcin, who travelled widely and had a genus named for him by Linnaeus; all three enjoyed international reputations and helped to put Neuchâtel on the botanical map. Together with Haller's cousin, Friedrich-Salomon Scholl of Bienne, they constituted a 'comité botanique' that collaborated with Haller, and decisively influenced Rousseau's introduction to botany.[118]

vi. Laurent Garcin (1683-1752)

> Mr. *Garcin*, notre Ami commun, Membre de la Société Roïale des Sciences, d'Angleterre et Correspondant de celle de *Paris*, a donné un nouveau lustre aux Botanistes *Suisses*, en fixant son séjour parmi eux.[119]

Of the three savants referred to above, Garcin[120] was probably the most famous in his own day, but most neglected by posterity; despite his contributions to Neuchâtel botany, his name is completely absent from the accounts of Rousseau's introduction to botany in the principality. This absence is unfortunate, given the importance of his services to medicine, but also to botany, meteorology and palaeontology.

Born in France in 1683, Garcin became a refugee at two years of age

118. Jules Thurmann, *Abraham Gagnebin de la Ferrière* (Porrentruy, 1851), p.16.

119. Original emphasis; d'Ivernois, 'Lettre à Monsieur Cartier'. p.31. While d'Ivernois states that Garcin was a 'Membre' of the Royal Society of London, he was in fact 'F. R. S.', i.e. a Fellow. The Royal Society, founded in 1660, followed the precepts of Sir Francis Bacon.

120. For biographies of Garcin, and his son, Jean Laurent, see Philippe Sirice Bridel, *Le Conservateur suisse, ou Recueil complet des étrennes helvétiennes* (Lausanne, 1831), vol.13, p.98-127; Bridel (1757-1845), known as 'le Doyen', fostered Swiss literary culture and was a friend of Garcin fils. Frédéric Alexandre Marie Jeanneret and James-Henri Bonhôte, *Biographie neuchâteloise*, vol.1 (Le Locle, 1863), 'Laurent Garcin' and 'Jean Laurent Garcin', p.373-88, follow Bridel virtually verbatim. Marcel Jacquat, 'Laurent Garcin: médecin-chirurgien, naturaliste (1683-1752)', *Biographies neuchâteloises*, vol.1, p.103-109.

with his Huguenot father, who obtained naturalisation in Neuchâtel. Garcin studied medicine with Herman Boerhaave at Leiden, served as a physician and surgeon in a Dutch regiment for sixteen years and for nine years in the VOC (Dutch East India Company). He recounts his career in a letter to the editors of the *Journal helvétique*:

> J'ai eu occasion de voyager commodément pendant 16 Années en Espagne, en Portugal et en Flandres, étant au Service des Etats Généraux des Province Unies; et j'y ai continuellement exercé dans leurs Troupes, avant et après la Paix d'*Utrecht*, la Médecine et la Chirurgie avec agrément, en y occupant les principaux Postes, de ma Profession. J'ai aussi été pendant 9 autres Années, c'est-à-dire, dès l'Année 1720. Jusqu'en 1729 sur les vaisseaux de leur incomparable Campagne dans les Indes Orientales. C'est ce qui m'a fourni la facilité de faire plusieurs Observations sur l'Histoire naturelle[,] la Phisique et la Médecine, en m'attachant spécialement à tout ce qui pouvoit me perfectionner dans la Pratique de cette dernière. Et comme tous les Hommes sont appelés à contribuer autant qu'il dépend d'eux au bien de la Société, je me ferai un devoir de lui être utile, tant par mes Observations, que par les Remèdes que j'ai découvert.[121]

The VOC was a major driving force in the European appropriation of the world's plant resources; indeed, the principal premise of eighteenth-century botany was that trading companies, states or individuals should identify, acquire and transplant useful, so-called 'exotic' plants to Europe or other locations for profitable exploitation.[122] The list of transplants is very long indeed, including such familiar daily commodities as cotton, tea, coffee and sugar, but also the tulip, the potato, the breadfruit, the Cedar of Lebanon and, of course, a whole range of spices and condiments, without which European food was bland and even unhealthy. Botanical entrepreneurs such as Sir Joseph Banks (1743-1820), who organised the transplantation of tea production from China to India, understood that political power and enormous profits stood to be gained from the successful practice of 'exotic botany'.[123] Efforts at acclimatisation could become obsessive; the Swedish classifier, Carolus Linnaeus, claimed to have naturalised tea in the Uppsala Botanical Garden after many failed attempts over a number of years.[124]

121. Original emphasis; Laurent Garcin, 'Lettre aux éditeurs du *Journal helvétique*, à l'occasion de quelques remèdes nouveaux et expérimentés, que M. Garcin a découverts dans ses voyages dans les Indes', *Journal helvétique, ou Annales littéraires et politiques de l'Europe et principalement de la Suisse* (September 1744), p.257-79 (269-70).

122. See A. Cook, 'Jean-Jacques Rousseau and exotic botany'.

123. This term comes from the title of a work by Sir James Edward Smith (1759-1823), founder of the Linnean Society of London, *Exotic botany*, 2 vols (London, 1804-1805).

124. Carolus Linnaeus, 'Potus Theae', defended 7 December 1765 by Petrus C. Tillaeus, *Amoenitates academicae* (Stockholm, Laurentius Salvius, 1769), vol.7, p.239; Carolus

Garcin's career illustrates the global character of the botanical work of this era. For example, he was an adept of exotic botany, professing to know 'toutes les plantes' depicted by the German-Dutch naturalist and illustrator, Maria Sibylla Merian (1647-1717), in her *Metamorphosis insectorum surinamensium* (Amsterdam, 1705); Garcin lent or gave this work to Gagnebin.[125] He was conversant with leading works of exotic botany such as the first flora of southern India, *Hortus malabaricus* by Hendrik Adriaan van Reede tot Drakenstein (1636-1691).[126]

During his voyages to the Cape, India, the Dutch East Indies and Ceylon, Garcin was guided by his teacher, Boerhaave; he made plant collections and learned about Indian and Chinese medicine during his travels.[127] Boerhaave gave him letters of introduction to VOC governors and directors in south and Southeast Asia and instructed him to observe the clinical method of Asian physicians; this instruction follows on the work of other investigators of Asian flora such as Reede, who focused on local knowledge.[128]

Garcin not merely collected valuable information from other cultures, he also participated in the translation and transformation of knowledge in the contact zone and beyond. He lauded the Brahmin doctors he encountered: 'à l'égard [des maladies] du corps, ils sont certainement bien versés dans l'Art de les guérir. Ils possèdent des secrets depuis une grande Antiquité.'[129] During his travels Garcin likewise consulted Arabs, Persians and many others. He found the Brahmins and the Chinese to be the best, although he did not consider their work beyond reproach:

> J'ai observé que les Brachmanes de *Malabar* et de *Bengale* [...] et les médecins Chinois du Malacca, de Sumatra, et de Java y excellaient le plus. Les Européens qui y sont en grand nombre s'en servent souvent, et sont témoins

Linnaeus, 'Usus historiae naturalis', defended 17 May 1766 by Matthew Aphonin, in *Amoenitates academicae*, vol.7, p.422-23.

125. Laurent Garcin to Abraham Gagnebin, 27 June 1743, reproduced in H. Brandt (ed.), 'Lettres inédites adressées à Abraham Gagnebin', *Actes de la société jurassienne d'émulation* (1931), p.215-23 (218).

126. Laurent Garcin, 'A letter from Dr Laurence Garcin, of Neuchâtel, F. R. S. to Sir Hans Sloane Bart, late P. R. S. concerning the cyprus of the ancients: done from the French by W. Watson, F. R. S.', *Philosophical transactions* 45 (1748), p.564-78 (571).

127. Laurent Garcin, 'Lettre aux éditeurs', *Journal helévetique*.

128. While there are studies of Reede's *Hortus indicus malabaricus*, 12 vols (Amsterdam, Johan van Someren and Jan van Dyck, 1678-1703), there has been '[l]ittle attention [...] focused on the [...] Indian Ocean'. On the Indian subcontinent, 'Europeans discovered a world that was already familiar to them, one already dominated by trade and the presence of Muslims, their perennial, yet well-known, rivals'. Kapil Raj, *Relocating modern science* (Basingstoke/New York, 2007), p.30. Thus, no established canons of 'European' or 'non-European' knowledge can be applied to the knowledge exchanged and transformed in the Indian Ocean contact zone. Raj, *Relocating modern science*, p.31.

129. Garcin, 'Lettre aux éditeurs', *Journal helvétique*, p.276.

des Cures surprenantes qu'ils font tant parmi les Naturels de leur Pays, que parmi les étrangers qui y résident.
[...]
Ces médecins Indiens, seraient les meilleurs Médecins du Monde, si avec leur Méthode et leurs bons remèdes, ils avaient autant de connaissance dans l'Anatomie et dans la Physique qu'en ont ceux de l'Europe; mais c'est ce qui leur manque presque entièrement.[130]

Garcin sent some of his plant specimens to Nicolaas Laurens Burman (1733-1793) for inclusion in the *Flora Indica* (1768),[131] but his herbarium, if he had one, has not survived. Garcin also communicated his findings to learned societies – the Royal Society of London and the Académie royale des sciences, Paris. He was a corresponding member of the Académie royale des sciences, Paris,[132] and a fellow of the Royal Society of London. His papers on the mangosteen[133] and other botanical discoveries were published by these academies.[134]

Consequently, Linnaeus became aware of Garcin: 'Le chevalier Linné a été si content des services que Garcin a rendus, que pour en immortaliser le souvenir, il a donné le nom de *Garcinia* à un bel arbre de la famille des Mangostans, originaire des Moluques et acclimaté à Java.'[135] Naming this genus for Garcin complied with Linnaeus's dictum: 'Nomini generica, ad Botanici optime meriti memoriam conservandum constructa, sancte servanda sunt'.[136]

130. Original emphasis; Garcin, 'Lettre aux éditeurs', *Journal helvétique*, p.278-79. Other Europeans were also in Bengal to find useful plants. Raj cites the long-forgotten, unpublished *Jardin de Lorixa* of Nicolas L'Empereur (c.1660-1742), containing information about 722 plant species. Raj, *Relocating modern science*, p.31-59.
131. Nicolaas Burman was the son of the more famous Johannes Burman (1707-1779).
132. 'Especially for French-speaking foreigners, correspondence with the Paris Academy was a means of establishing scientific priority and reputation'; on the special role of the 'correspondant' in the Paris Académie royale des sciences, see Dawson, *Nature's enigma*, p.18-19.
133. See for example Laurent Garcin, 'The settling of a new genus of plants, called after the Malayans, mangostans; by Laurentius Garcin, M. D. and F. R. S. Translated from the French by Mr Zollman, F. R. S.' *Philosophical transactions* 38 (1733-1734), p.232-42. Mangosteen was first reported in François de Vitré Martin, *Description du premier voyage faict aux Indes orientales par les Français: contenant les moeurs, loix, etc. des Indiens, une description et remarque des animaux, épiceries, et fruicts que se trouvent aux Indes* (Paris, Laurens Sonnius, 1609), p.21. A tincture of the shell is used to treat bowel 'looseness'.
134. See for example Laurent Garcin, 'A letter from Dr Laurence Garcin, of Neuchâtel, F. R. S. to Sir Hans Sloane Bart', p.564-78.
135. Bridel, *Conservateur suisse*, p.101. Linnaeus's description was based on 'unseen Clifford material as a suitable type [specimen], but none is in existence'. Charles Jarvis, *Order out of chaos: Linnaean plant names and their types* (London, 2007), p.532. The type specimen of *Garcinia mangostana* L. is not extant in the Burman herbarium at the Conservatoire et jardin botaniques of Geneva. Personal communication, Nicolas Fumeaux, Conservatoire et jardin botaniques, Geneva.
136. Carolus Linnaeus, *Philosophia botanica* (New York [Coldicote], 1996 [1751]), p.171, section

Garcin showed his cosmopolitanism in preferring the Swede to the Swiss; Haller, he suggested, was able, but offered no significant breakthroughs:

> Je ne doute pas que Mr Haller ne trouva mauvais de lui faire apercevoir qu'il s'est souvent trompé dans la fin de la Botanique et sur son système même. C'est pourquoi vous ferez bien de ne lui en rien dire et que ce que nous en savons reste entre nous deux. Cela n'empêche pas que nous ne le regardions pour l'un des plus habiles Botanistes de l'Europe et que sa méthode composée en partie de celle des autres, ne donne de grandes ouvertures à ceux qui suivront.

He reserved his highest praise for the Swedish innovator, Linnaeus:

> Je crois cependant que Mr Linnaeus qui doit être considéré comme le plus habile du temps ne doit pas être fort content du procédé de Mr Haller à son égard, car celui-ci parait affecter de le corriger et d'être meilleur observateur, c'est de quoi je doute. D'ailleurs il est bien certain que Mr Haller n'aurait pas été si loin sans Mr Linnaeus qui lui a ouvert bien des Sentiers et frayé chemin'.[137]

After the conclusion of his peripatetic phase, Garcin returned to Europe to see his teacher, Boerhaave, in Leiden, to receive his degree in Reims and to marry in Geneva. He settled in Neuchâtel, becoming a member of its scientific scene, where he made many important contributions. In 1739 he changed botany in Neuchâtel forever by introducing Linnaeus's new system of sexual classification: 'Monsieur [d]'Ivernois vous aura appris sans doute l'arrivée de Monsieur le Docteur Garcin qui a rapporté de ces voyages un *nouveau Système de Botanique* très différent de celui dont on se sert aujourd'hui. L'auteur est un certain *Linnaeus Suédois*.'[138] Garcin probably knew the sexual system from his time in the Dutch Republic, where Linnaeus's earliest works were published. The sexual system was first explicated in the *Systema naturae* (Leiden, 1735). The Neuchâtel botanists thus became aware of Linnaeus and his artificial system of sexual classification, while continuing to collaborate with

238. 'Generic names that have been formed to perpetuate the memory of a botanist who has done excellent service should be religiously preserved', Linnaeus, *Philosophia botanica* (2003), p.185, section 238. This has been called 'a naming system abstract in relation to the properties of plants, but concrete in relation to the history of botany in Europe'. Londa Schiebinger, *Plants and empire: colonial biosprospecting in the Atlantic world* (Cambridge, MA, 2004), p.201. Yet this naming system did not operate strictly in accordance with Linnaeus's early prescriptions. See Alexandra Cook, 'Linnaeus and Chinese plants' in *Notes and records of the Royal Society*, 64 (June 2010), p.121-38.

137. Laurent Garcin to Abraham Gagnebin, 27 June 1743, reproduced in H. Brandt, ed., 'Lettres inédites adressées à Abraham Gagnebin', p.219.

138. My emphasis; Abraham Gagnebin to Haller, 25 November 1739, quoted in Marcel S. Jacquat, 'Abraham Gagnebin: médecin (1707-1800)', p.101. I discuss this system at length in Chapter 5.

Haller and use his works. They occupied a middle ground in the controversy between Linnaeus and Haller concerning both classification and nomenclature.

In Neuchâtel Garcin practised medicine, integrating medical knowledge acquired in India and the Dutch East Indies into his medical practice. His botanical and medical cosmopolitanism brought Asian botany and medicine to the doorstep of ordinary Neuchâtelois; he used remedies obtained in Asia: 'quelques remèdes nouveaux ou à l'indienne'.[139] Garcin reported great success in using a Brahmin remedy that he called 'pilules madurines' to treat digestive problems and 'les maladies du sexe'. He asserted that the 'pilules' work by 'rectifiant le ton des fibres nerveuses et la digestion alterée ou dépravée', thereby returning the patient to a condition of 'gaieté' and relieving 'pesanteur du corps'. These 'pilules' were on offer in apothecary shops in Basel, Yverdon and Vevey; they were probably available in Neuchâtel from Garcin himself.[140] He also promoted hydrotherapy.[141]

Most importantly, Garcin taught botany, the basis for pharmaceutical preparations.[142] Garcin may have been the first to do so in Neuchâtel, where no medical training was available.[143] Garcin's course followed that at the Jardin du roi and elsewhere:

> Mr Garcin [...] fera en français, comme on le pratique à Paris, un cours complet de cette Science [...] Il leur enseignera et expliquera, par les véritables Principes, et en suivant *une Méthode* claire, distincte, et aisée, les *meilleurs Systèmes* des Auteurs modernes. Il les *combinera en un seul*, autant qu'il lui sera possible, pour la rendre plus commode et plus intelligible.[144]

Like Garcin, Rousseau hybridised a variety of approaches into one, rather than simply following the course set out by a particular author, such as Tournefort, Haller or Linnaeus. As Rousseau never refers to Garcin, it is impossible to say whether he derived inspiration from this Neuchâtel precursor or not. However, it is entirely possible that

139. Laurent Garcin to Abraham Gagnebin, 27 June 1743 reproduced in H. Brandt, ed., 'Lettres inédites adressées à Abraham Gagnebin', p.218-19.

140. Laurent Garcin, 'Lettre de M. Garcin sur la nature et l'usage des pilules madurines, avec des remarques intéressantes sur les effets du thé', *Journal helvétique*, (May 1747), p.473-90 (481). See also 'Avis', *Journal helvétique* (April 1745), p.389.

141. According to Jeanneret and Bonhôte, Garcin preferred ordinary water to mineral water, and tied its use to a particular 'regime' or diet. Jeanneret and Bonhôte, *Biographie neuchâteloise*, vol.1, p.376.

142. At least 40% of modern pharmaceuticals are synthetic versions of substances derived from plants, moulds and other naturally occurring substances. E. O. Wilson, *The Diversity of life* (Cambridge, 1992), p.285.

143. Anon., 'Cours de botanique à ceux qui veulent apprendre cette science', *Journal helvétique* (February 1747), p.143-46 (145).

144. My emphasis; Anon., 'Cours de botanique', p.145.

Rousseau's teachers, Gagnebin and d'Ivernois, both of whom knew Garcin, may have been inspired by his teaching methods, even if they were already able botanists.

The course was to be taught in the field, in the manner of Linnaeus's *herbationes Upsaliensis* or the herborisations of the Paris botanists:

> à la Campagne, en parcourant les Vallées et les Montagnes de la Souveraineté de Neuchâtel, pour en exercer la pratique par les Démonstrations. Il fera connaître toutes les Plantes du Pais, en même temps que les usuelles dans la médecine, avec leurs vertus, en donnant, tant en Latin, qu'en français, la définition de leurs classes, de leurs genres, et de leurs Espèces.

Furthermore, '[i]l mettra, par sa méthode les disciples en état de connaître par eux-mêmes, sur de bons principes les plantes étrangères, et les ranger à leur propre place dans le Système le plus naturel'.[145] Finally, he intended to proceed 'jusques fort avant dans l'Automne, ainsi que le pratiquent les professeurs en botanique des principales universités d'Europe'.[146] The significance of Garcin's teaching for Rousseau's study of botany has hitherto gone completely unnoticed.

Garcin contributed to botany in other ways. He collected rare plants for major botanical gardens, suffering paralysis of the fingers as a result:

> Cette Incommodité irrémédiable est une suite d'un mauvais rhumatisme que je gagnai dans vos Montagnes, en y allant chercher des Plantes rares, tant pour l'histoire naturelle que pour la médecine qui m'étoit demandées pour les jardins de botanique de Paris, de Londres, et d'Amsterdam. J'aurais bien voulu continuer plus longtemps des Recherches si utiles et si satisfaisantes, auxquelles M. votre Frère [Abraham Gagnebin] s'est appliqué d'une manière si distingué, comme Mr. Haller dans son grand ouvrage de botanique, lui en rend un témoignage glorieux; mais ma santé ne m'a pas permis la continuation de ces courses pénibles.[147]

Garcin was consulted as an expert on the East Indies, exotic flora and medicines, and contributed to a revised edition of the *Dictionnaire universel de commerce* published in Geneva in 1742. As a testimonial to Garcin's work, the editor wrote:

> Nous sommes de moins certains que l'on goûtera principalement les nouveaux Articles manuscrits, qui nous ont été fournis par Mr le Docteur *Garcin*, membre de la Société Royale de Londres, et correspondant de celle de Paris; ayant voyagé longtemps (1720-1729) dans les Indes Orientales; en

145. Anon., 'Cours de botanique', p.146.
146. Anon., 'Cours de botanique', p.145-46.
147. Laurent Garcin, 'Lettre de M Garcin sur les œufs philosophiques, servant de réponse à celle de M. [Daniel] Gagnebin, insérée dans le journal de mars dernier', *Journal helvétique* (June 1745), p.517-29 (529). This 'ouvrage' was probably Haller's *Enumeratio* (1742). Garcin's lack of feeling in his fingers allowed him to burn his hand to the bone during a chemistry experiment, causing a fatal gangrene infection.

Botaniste, et Naturaliste, il a recueilli avec soin et exactitude des Mémoires curieux et fort instructifs sur les drogues, les épiceries, les bois, et autres marchandises que l'on tire de ces pays. Personne ne pouvait être mieux en état que lui d'y travailler, et de corriger exactement et avec précision les fautes de tant de voyageurs qui ont publié mille fables, manque d'un examen assez attentif, et qui n'ont pas été assez expérimentés pour écrire avec justesse, n'ayant ramassé que confusément tout ce qu'on leur débitait. Au contraire, Mr *Garcin*, non moins amateur de la vérité, qu'éclairé et curieux, n'a pu qu'être bien instruit par les autres et par lui-même de tout ce dont il parle [...].[148]

Among Garcin's contributions to this substantial compendium was a new article on the much-vaunted ginseng (presumably the root, the part nomally used), in which he refused to credit reports of the miraculous powers of this plant. Instead, Garcin was sceptical of ginseng's purported effects, especially on the 'parties de la génération, et pour lesquelles les Chinois le recherchent si fort pour se satisfaire'. Quinine produces 'effets réels' whereas those of ginseng 'paraissent qu'imaginaires'. Garcin believed ginseng 'fortifie l'estomac, et en cela elle fera commune avec bien d'autres', but that it tires the body by over-stimulation: 'il n'y a point de doute qu'au lieu de fortifier le corps véritablement, elle ne l'énerve en peu de temps: car il est certain que tout ce qui excite les forces outre mesure, les abrège d'autant'.[149] He also doubted the health claims for tea, opposing conventional wisdom of the time, which saw tea, like ginseng, as a panacea; *Panax* (panacea) is the generic name for ginseng.[150]

Garcin's contributions to exotic botany show that by the 1740s Neuchâtel was aware of the wider world beyond Europe in which Neuchâtel calico could be exchanged for slaves, and Asian species could be discussed in an encyclopedia published in Geneva. Hence, in Neuchâtel Rousseau may have become more informed about the exotic botany he had critiqued in *Julie*.[151] The penetration of Europe by the wider botanical world was unavoidable, for it was now possible for ginseng root, imported from China or Canada, to be exchanged in his humble Paris apartment for coffee from the Ile de Bourbon, a far-flung French colony in the Indian Ocean.[152] He himself transplanted an exotic

148. Original emphasis; Jacques Savary de Bruslons, *Dictionnaire universel de commerce*, ed. Philémon-Louis Savary, 2 vols (Geneva, chez les Héritiers Cramer & Frères Philibert, 1742), p.xxxvii.

149. Laurent Garcin, 'Gin-seng', *Dictionnaire universel de commerce*, vol.2, p.628.

150. Laurent Garcin, 'Lettre de M. Garcin sur la nature et l'usage des pilules madurines', p.483ff. Another tea sceptic, John Coakley Lettsom, wrote *The Natural history of the tea tree: with observations on the medical qualities of tea and effects of tea-drinking* (London, Edward and Charles Dilly, 1772). On the naming of ginseng, see Cook, 'Linnaeus and Chinese plants', p.131.

151. Cook, 'Jean-Jacques Rousseau and exotic botany'.

152. In June 1772 Rousseau gave Bernardin de St Pierre (1737-1814) ginseng root in exchange

species, *Hibiscus esculentus* L. (= *Abelmoschus esculentus* Moench) in France, even though he also criticised this practice.[153]

Garcin's significance for our story extends well beyond his own accomplishments to those of his son, Jean-Laurent (1733-1781). Garcin fils, also an accomplished botanist, was recruited to prepare some 'chiffons' of Rousseau into *Fragmens pour un dictionnaire des termes d'usage en botanique* for the first posthumous edition of Rousseau's works, the *Collection complète* published by Du Peyrou and Moultou in Geneva in [1780-]1782. While Garcin's role in this endeavour has never been acknowledged in print, we know of it from correspondence exchanged by Du Peyrou and Moultou.[154] Despite his current obscurity, Garcin père played a key role in the development of Neuchâtel botany, while his son participated in the posthumous dissemination of Rousseau's botanical writings; these contributions deserve recognition long overdue in the story of Rousseau and botany.

vii. Jean Antoine d'Ivernois (1703-1765)

> Me voilà donc à mon foin pour toute ma nourriture, et à la botanique pour toute occupation. Déjà vieux j'en avais pris la première teinture en *Suisse* auprès du Docteur [D]'Ivernois.[155]

According to Rousseau's own testimony, Dr Jean-Antoine d'Ivernois administered the first tincture of botany to Rousseau; this statement contains a double meaning, since a tincture can be either a dye or a medicine.[156] One meaning is consistent with the study of plants being the only 'pharmacie' that his soul requires.[157] The second meaning is that botany stained him; it could not be washed away. Both of these meanings

---

for coffee from the Ile de Bourbon (now Réunion, a département of metropolitan France); Bernardin was posted to the nearby colony of Ile de France (now Mauritius) as an engineer, and wrote *Voyage à l'Ile de France* (1773). Bernardin de St Pierre, 'Essai sur J.-J. Rousseau', p.28.

153. Rousseau to Latourrette, 22 February 1770; *CC*, xxxvii.246.

154. Du Peyrou to Moultou, 10 May 1780; *CC*, xliv.231. I discuss the role of Garcin fils in the preparation of the *Fragmens pour un dictionnaire* in Chapter 9.

155. Rousseau, *Rêveries*, *OC*, i.1060. My emphasis.

156. Thurmann contradicts Rousseau's testimony, claiming Abraham Gagnebin, not d'Ivernois, was Rousseau's first botany teacher. Jules Thurmann, *Abraham Gagnebin de la Ferrière*, p.13-14, 104-105, n.6. See also Frédéric-Alexandre-Marie Jeanneret and James-Henri Bonhôte, *Biographie neuchâteloise* (Le Locle, 1863), vol.1, 'Jean-Antoine d'Ivernois', p.524-27. In their article on Gagnebin they state that d'Ivernois was Rousseau's first botany teacher. Jeanneret and Bonhôte, 'Abraham Gagnebin', *Biographie neuchâteloise*, vol.1, p.357-61 (360). See also de Beer and Gagnebin, 'Abraham Gagnebin de la Ferrière', p.46-47, 49, 51.

157. Rousseau, *Fragments*, *OC*, iv.1251.

are highly suggestive. There is certainly irony in Rousseau's suggestion that d'Ivernois, a medical doctor, administered medicine for the soul, not the body. Rousseau usually kept his distance from physicians, whom he regarded with justifiable scepticism, but d'Ivernois was an exception, whom Rousseau described as 'fort habile' – high praise, coming from Rousseau.[158]

D'Ivernois had studied philosophy at Basel and medicine at Montpellier, a medical school that educated Protestants even after the revocation of the Edict of Nantes.[159] The medical faculty of Monpellier, founded in the Middle Ages, was one of the oldest and most illustrious in Europe, next to Leiden; from the sixteenth century, Montpellier physicians attended the king of France.

Montpellier was no less renowned for botany. The botanical garden of Montpellier, founded in the sixteenth century by Pierre Richer de Belleval (1564-1632), is the oldest in France; the Jardin royal des plantes médicinales (the Jardin du roi, discussed in Chapter 2) was founded in Paris in 1635 by the king's Montpellier-trained physician, Guy de la Brosse. Until the appointment of Buffon as intendant of the Jardin du roi in 1739, Montpellier graduates maintained a secure foothold in the Paris medical and botanical establishments. Montpellier botanists such as François Boissier de Sauvages de la Croix[160] and his student, Antoine Gouan (a correspondent of Rousseau), were early adherents of Linnaean botany and made Linnaeus a foreign member of the Société royale des sciences of Montpellier in 1743. The Société's recruitment was international, including Swiss savants such as Bonnet, Jallabert, Cramer (discussed in Chapter 3) and Georges Louis le Sage, all of Geneva.[161] While d'Ivernois finished his medical studies in the late 1720s, before the Linnaean turn in Montpellier botany, he may have remained in contact with teachers or peers, especially Swiss compatriots. Linnaean botany's ascendency in Montpellier may have influenced d'Ivernois' acceptance of Linnaeus as one of three important botanists, alongside Tournefort and Haller.

D'Ivernois played an important role in Neuchâtel botany, where he

158. Rousseau called d'Ivernois to attend le bonhomme Deluc, who fell ill on a visit to Môtiers. Rousseau to Jean-André and Guillaume Antoine Deluc, 16 October 1762; *CC*, xxiv.215.

159. James Livesey, 'London by the light of Montpellier: scientific networks between northern Europe, Britain and the Languedoc 1680-1789', *SVEC* 2010:04, p.85-102 (90).

160. Sauvages authored *Methodus foliorum* (The Hague, n.p., 1751), which Rousseau used, and an unprecedented classification of diseases, *Nosologia methodica sistens morborum classes, genera et species, juxta Sydenhami mentem et botanicorum ordinem* (Amsterdam, Frères De Tournes, 1763).

161. Livesey, 'London by the light of Montpellier', p.95-96, 99. Livesey speculates that the presence of Genevans can be explained not by long-standing contact between Montpellier and Geneva, but rather, by their Newtonian orientation, which tied them to the same scientific culture that inspired science in Montpellier.

participated in the comité botanique.[162] He botanised and corresponded with Haller in 1739; this linked him to Haller's immense scientific network. At the age of thirty, d'Ivernois was appointed physician to the King of Prussia for his services to medicine. He was also elected to the Academy of Sciences of St Petersburg, whose members included eminent Swiss savants such as the mathematician, Leonhard Euler and the botanist, Johann Amman.

D'Ivernois was justly famous in the region and beyond for his two-volume catalogue of the plants of Neuchâtel and Valangin, which Haller consulted in preparing the *Historia stirpium* of 1768.[163] D'Ivernois' catalogue, which remains unpublished to this day, 'fut le point de départ de tout ce qui a été publié plus tard sur ce sujet. D'Ivernois jouissait d'une grande considération comme médecin et naturaliste.'[164] This unique work was organised 'par ordre alphabétique, en suivant littéralement l'*Enumeratio stirpium Helvetiae*, de Haller, qui fut son ami'.[165] D'Ivernois enumerated 1227 species with extensive synonymies to Bauhin, Haller, Linnaeus (his pre-*Species* nomenclature) and other important botanists.[166] This accounted for about fifty per cent of the species enumerated in Haller's *Historia stirpium*. D'Ivernois' knowledge of the botanical literature was, to judge by the catalogue, vast.

D'Ivernois prefaced his enumeration with a review of the classification systems of Tournefort, Haller and Linnaeus, in that order; concerning the last, he notes: 'Ce système est fondé uniquement sur le Sexe des Plantes.' Similarly, the Genevan apothecary (and d'Ivernois' interlocutor), Tollot,[167] matter-of-factly discussed the sexual idea of the flower.[168] Hence, it seems that in the Calvinist Suisse romande plant sexualism was

162. Charles H. Godet, 'Enumération des végétaux vasculaires qui croissent dans le canton de Neuchâtel', *Mémoires de la société des sciences naturelles de Neuchâtel*, vol.2 (Neuchâtel, 1839), p.iv-v.

163. Jean-Antoine d'Ivernois, *Catalogue méthodique des plantes qui croissent naturellement dans la Souveraineté Neufchâtel et Valangin: fait l'Hiver de l'an 1745 à 1746*, 2 vols, p.14, BPUN MsA 147. D'Ivernois provided synonymies from many authors.

164. Thurmann, *Abraham Gagnebin*, p.13.

165. Godet, 'Enumération des végétaux vasculaires qui croissent dans le canton de Neuchâtel'.

166. It has been claimed that d'Ivernois used Linnaean binomial names in his catalogue. Marie-Marguerite Duckert-Henriod, 'Jean-Antoine d'Ivernois, médecin et botaniste (1703-1765)', *Biographies neuchâteloises*, vol.1, p.143-47 (144). He could only have used them if he had seen Linnaeus, *Oländska och Gothländska Resa* (Stockholm & Uppsala, Gottfried Kiesewetter, 1745), but this is unlikely. See Marie Åsberg and William Stearn, 'Linnaeus's Öland and Gotland journey 1741', *Biological journal of the Linnean Society* 5:1 (March 1973), p.1-107 (6).

167. Tollot is discussed in Chapter 3, p.75,77.

168. 'Réflexions sur l'étude des plantes et sur leur usage', p.1142-57. Even though the article is unsigned, its authorship by Tollot is confirmed by [Jean-Antoine] d'Ivernois, 'Lettre à Monsieur Cartier Ministre' p.3.

adopted without much, if any, of the moral concern about the sexual system articulated by the St Petersburg botanist, Johann Siegesbeck (discussed in Chapter 5, p.121-64).

Rousseau said so little about the botanical instruction he received that we can only make educated guesses about it. To judge by d'Ivernois' Neuchâtel flora, Rousseau would have learned the classification systems of Tournefort, Haller (1742) and Linnaeus (sexual system). In his flora d'Ivernois followed the alphabetic order of Haller's *Enumeratio*.[169] We know that he was well-acquainted with the work of both Tournefort and Haller, and approved of their contributions to botany. D'Ivernois died in 1765, the same year as Rousseau's departure from the region, but his influence was lasting, and his memory revered.

Another influence on Rousseau may have been Du Peyrou, who suffered from gout, and no doubt received the standard medical advice to take exercise. Botany could be a pleasant way to pass the time while walking. Rousseau advised him to take a dose of botany with his exercise: 'si vous pouvez vous soutenir jusqu'au temps de notre caravane elle vous garantira d'être arrêté durant le reste de l'année, vu que la goute n'a point de plus grand ennemi que l'exercice pédestre. Vous devriez prendre la botanique par *remède* quand vous ne la prendriez pas par goût.'[170] The correspondence between Du Peyrou and Rousseau lends credence to the theory that Du Peyrou, who promoted Linnaeus's 'Méthode',[171] encouraged Rousseau to study botany:

> A bon compte nous en profitons et nous Herborisons copieusement du moins, si ce n'est utilement, et mon Herbier s'en ressent [...] Mais pour ne plus être dans cette vilaine dépendance d'un second, ignare, j'ay pris des informations sur les Auteurs élémentaires. C'est *Linnaeus* qu'il faut avoir comme le plus *moderne*, et ce qui vaut mieux, comme celui dont la Méthode bien saisie conduit à la connaissance des plantes les plus inconnues [...] Moquez-vous hardiment, Monsieur, de moi, de ma vanité et de mon savoir, ce doit aussi entrer dans nos conventions, mais corrigez mes fautes et mes Erreurs [...].[172]

Rousseau's response suggests that in 1764 Du Peyrou may have been the more accomplished in botany: 'J'en étais ici, Monsieur, quand j'ai reçu vôtre *docte* lettre. Je suis charmé de vos progrès; je vous exhorte à continuer; vous serez nôtre *maître*, et vous aurez tout l'honneur de notre

169. Godet, 'Enumération'.

170. My emphasis; Rousseau to Du Peyrou, 29 April 1765; *CC*, xxv.204.

171. The reference to 'Méthode' probably refers to the artificial sexual system of classification since Du Peyrou says '[il] conduit à la connaissance des plantes les plus inconnues'. However, loosely used, he might also mean Linnaeus's binomial nomenclature which he used systematically in *Species plantarum* (1753).

172. My emphasis; Du Peyrou to Rousseau, 7 October 1764; *CC*, xxi.227.

futur *savoir*.'[173] By June 1765 Rousseau seems to have shared Du Peyrou's liking for Linnaeus:

> Si je reste un jour de plus, je suis pris: je pars donc, mon cher Hôte, pour la Ferrière où je vous attendrai avec le plus grand empressement mais sans m'impatienter. Ce qui achève de me déterminer est qu'on m'apprend que vous avez commencé à sortir. Je vous recommande de ne pas oublier parmi nos provisions, café, sucre, cafetière, briquet, et tout l'attirail pour faire quand on veut du café dans les bois. Prenez Linnaeus[174] et Sauvages, quelque Livre amusant, et quelque jeu pour s'amuser plusieurs [...].[175]

Du Peyrou was in turn taking botanical instruction from the town physician, Frédéric-Samuel Neuhaus, who had studied under Haller at Göttingen, and was, according to Du Peyrou, author of a *Catalogue raisonné des plantes usuelles en Suisse*.[176] Dr Neuhaus could have also stimulated Rousseau's interest in botany.[177] Yet d'Ivernois' knowledge of the local flora can be demonstrated beyond doubt, and this fact, coupled with Rousseau's own testimony, establishes his primacy as Rousseau's first instructor in botany.

viii. Abraham Gagnebin (1707-1800)

> Gagnebin, de la Ferrière, le plus intrépide nomenclateur de plantes qui ait peut-être jamais existé sur ce globe [...] avait dans la tête et présent à la mémoire, *douze à quinze mille noms de végétaux* avec les diverses phrases caractéristiques employées par les plus célèbres botanistes.[178]

173. My emphasis; Rousseau to Du Peyrou, 10 October 1764; *CC*, xxi.238.
174. This is probably the *Species plantarum* (1753 or 1762-1763), referred to above.
175. Rousseau to Du Peyrou, 11 June 1765; *CC*, xxvi.32-33. 'Sauvages' probably refers to Sauvages's *Methodus foliorum*. Rousseau requested this work from Duchesne, 19 May 1765; *CC*, xxv.301-302. He sold his copy, heavily annotated with Linnaean binomial names, to Daniel Malthus c.1775, and it was willed with Thomas Malthus's collection to Jesus College, Cambridge. See Cook, 'Jean-Jacques Rousseau's copy of Albrecht von Haller's *Historia stirpium*'; Henry Cheyron, 'Ray et Sauvages annotés par Jean-Jacques Rousseau'.
176. Du Peyrou to Rousseau, 20 November 1764; *CC*, xxii.86, 87, n.(h). Du Peyrou states that Neuhaus's catalogue included the uses and names of Swiss plants. However, this work has proven impossible to locate; no manuscript or publication by Neuhaus is in the BPUN. Personal communication, Anne-Lise Grobéty, BPUN, December 2008.
177. Personal communication, Michel Schlup, 16 October 2009.
178. My emphasis; François Louis d'Escherny, *Mélanges de littérature, d'histoire, de morale et de philosophie*, vol.3 (Paris, 1811), p.42-43. D'Escherny (1733-1815), a Neuchâtel noble, shared an interest in music with Rousseau, but the two fell out when d'Escherny brought Ostervald (who had opposed the printing of Rousseau's works while a magistrate in Neuchâtel) to see Rousseau in Paris. D'Escherny was well-connected in Vienna, Württemburg, Prussia and Paris. He supported the French Revolution, but was nearly executed for having the 'air d'un aristocrate et d'un traître'. Jeanneret and Bonhôte, 'D'Escherny', *Biographie neuchâteloise*, vol.1, p.314-20 (317).

Figure 4.3: Anonymous, *Abraham Gagnebin*, oil on canvas, Société jurassienne d'émulation © Marcel S. Jacquat.

The third important member of Neuchâtel's comité botanique was Abraham Gagnebin (1707-1800) of the village of La Ferrière in the Swiss Jura (fig.4.3, p.124). Like Garcin and d'Ivernois, he was a physician, served abroad as a doctor in French military units and was one of Haller's most important botanical collaborators. Gagnebin was legendary for knowing thousands of plant species by sight. According to the comte d'Escherny (1733-1815),

> [S]on extrême modestie, une certaine simplicité de caractère, donnaient un nouveau prix à son immense érudition. Il dominait le règne végétal, je dirais presque, il régnait sur ce règne, mais il s'ignorait lui-même, et pendant que tant de pygmées se croyant des géants, lui se rapetissait au point d'être extrêmement flatté que des hommes célèbres eussent seulement, en passant, articulé son nom. C'est en baissant les yeux et rougissant, qu'il nous disait avec une naïveté piquante: on trouve mon nom dans tels livres de M. le baron de Haller, de M. de Réaumur.[179]

Gagnebin collected plants for Haller, is acknowledged in Haller's works and was one of Haller's most important botanical correspondents, sending Haller 118 letters from 1739 to 1773.[180] Gagnebin's 'véritable titre scientifique, c'est sa longue et fructueuse collaboration avec Haller'.[181] D'Ivernois is supposed to have introduced Gagnebin to Haller, who made excursions with them in 1739 in the Neuchâtel Jura and the Creux du Van.[182]

Rousseau and Gagnebin may have become acquainted through their mutual friend, Pierre Alexandre Du Peyrou.[183] Gagnebin served as botanical guide to Rousseau and his 'collège de botanique', which included Du Peyrou, and Abraham Pury[184] as recounted by Rousseau in a letter to Pury of June 1765:

> Quant à l'âne aux provisions, je l'approuve fort; c'est un cortège dont je

179. D'Escherny, *Mélanges de littérature*, p.42-43.

180. Gagnebin is considered 'vielleicht der wichtigste botanische Mitarbeiter AvHs [perhaps Haller's most important botanical collaborator]', Boschung *et al.* (eds), *Repertorium zu Albrecht von Hallers Korrespondenz 1724-1777*, vol.1, p.157.

181. Thurmann, *Abraham Gagnebin*, p.47.

182. Jeanneret and Bonhôte, *Biographie neuchâteloise*, vol.1, p.358; d'Ivernois, 'Lettre à Monsieur Cartier ministre', p.31.

183. Thurmann, *Abraham Gagnebin*, p.42, 47; Jeanneret and Bonhôte, 'Abraham Gagnebin', *Biographie neuchâteloise*, p.359-60. Du Peyrou later requested plant specimens from Gagnebin for Rousseau. Gagnebin to Haller, 1 March 1769; *CC*, xxxvii.63; concerning the dating of this letter, see de Beer and Gagnebin, 'Abraham Gagnebin de la Ferrière', p.70.

184. Relations between the two men were complicated because Du Peyrou had been infatuated with Julie-Régine Le Chambrier (1729-1791), whom Pury married in 1750. Frustrated in his suit with the mother, Du Peyrou married her daughter, Henriette-Dorothée, when he was forty and she was eighteen. The marriage was unsuccessful; Du Peyrou suffered from various ailments and was deaf at a relatively young age.

compte tirer parti plus que personne: il faut aussi que nous élisions entre nous un trésorier ou intendant qui se charge de toutes les fournitures et de la bourse. Comme vous êtes celui des *quatre* qui conçoit le mieux le pays et le seul qui en parle la langue, j'opine qu'on vous prie de vous charger de ce soin, espérant que vos bons offices auront plus de succès dans ce voyage qu'il n'en eurent à l'abord de Chasseron.[185]

The comte d'Escherny recalled that the expeditions were well-organised:

M. Leclerc était le pourvoyeur. M. Du Peyrou avait soin des herbiers. Le colonel De Pury était notre éclaireur, il portait la boussole, car dans la sombre épaisseur des forets, on peut se guider qu'en connaissant le nord [...] j'avais de plus la garde du café et l'emploi de la faire: muni d'un briquet que je garde précieusement, c'était moi qui dans le bois allumais le feu [...] et à donner au café sa juste cuisson. Rousseau, comme le plus âgé, était le capitaine de la petite troupe, chargé de la discipline du corps, et d'y maintenir l'ordre et la subordination.[186]

According to d'Escherny, the Linnaean artificial system of sexual classification provided their guide:

Ce fut [...] en 1764, que nous nous trouvâmes tous réunis au val de Travers, M. du Peyrou, le colonel de Pury, Rousseau et moi. Il fut question alors d'herboriser en grand sur le Jura, d'étudier le *système sexuel* de Linnaeus, d'y puiser les principes généraux de la science, d'observer sur chaque fleur à l'aide d'une loupe, les étamines et les pistils, d'étudier dans chaque plante les caractères assignés à chacune des vingt-quatre classes de l'ingénieuse nomenclature de ce grand naturaliste. Rousseau nous recommanda de nous munir, chacun de nous, d'un *Systema naturae* de Linnaeus [...] j'appris à reconnaître et à retenir les noms de plus de trois cents plantes avec leurs synonymes, d'après Linnaeus, Tournefort, Gaspard Bauhin, etc.[187]

Gagnebin was well-positioned to serve as a guide both to living plants and to dried ones in herbaria since he and his brother Daniel (1709-1781), a medical doctor and physicist, possessed enormous natural history collections, including thousands of botanical specimens, the majority of which seem to have been lost. The brothers also collected minerals, fossils and other curiosities. They contributed to Bourguet and Cartier's famous work on paleontology, *Traité des pétrifications* (1740), discussed above, p.73.

Gagnebin gave Rousseau botanical instruction, although its extent is unknown. In June 1765 Rousseau visited Gagnebin for ten days in La Ferrière, but suffered with quinsy (i.e. tonsillitis) during the entire

185. My emphasis; Rousseau to Pury, 10 June 1765; *CC*, xxvi.31.

186. D'Escherny, *Mélanges de littérature*, p.66. Note Rousseau's role as disciplinarian and leader of the group, which carried supplies for collecting and drying specimens.

187. My emphasis; d'Escherny, *Mélanges de littérature*, p.41-42.

visit.[188] He therefore probably saw more of Gagnebin's famous collections than he did of plants in the countryside. Thurmann asserts that Gagnebin was in fact Rousseau's 'véritable maître' in botany, an assertion that probably also applies to Rousseau's apprenticeship in making herbaria, since Gagnebin possessed such rich collections.[189] At Du Peyrou's request, Gagnebin supplied Rousseau with hundreds of plants after the latter's departure from Switzerland in 1765,[190] and even requested plant samples from Haller for Rousseau.[191] However, Gagnebin assured the politically conservative Haller,[192] 'je ne le fais qu'à la sollicitation de Mr Du Peyrou'.[193]

A considerable number of Gagnebin's herbarium specimens survive in the Swiss herbarium of the Institut de botanique of the Université de Neuchâtel.[194] Other specimens are in Haller's herbarium at the Muséum national d'histoire naturelle, Paris, but they were attached by Haller, who, unlike Linnaeus or Rousseau, placed many specimens on the page.[195] Thanks to recent efforts, some of these have been remounted with the original labels included.[196] In some cases these labels constitute veritable essays on synonymy and the medical usages of the plants in question – one label comprises sixteen tiny pages written in a minute, yet

188. Rousseau to Du Peyrou, 11 June 1765; *CC*, xxvi.32.

189. Thurmann, *Abraham Gagnebin*, p.43. See also Rousseau, *Deux lettres à M. le mareschal duc de Luxembourg*, ed. Eigeldinger, p.96. D'Ivernois had a herbarium which, it has been suggested, was dispersed in the herbarium of Swiss plants in the Institut de botanique, Université de Neuchâtel. However, none of d'Ivernois' specimens have been located to date. Personal communication, Claire Jaquier Kaempfer, 15 March 2012.

190. Gagnebin to Haller, 3 March 1769; *CC*, xxxvii.63. Leigh gives the date of this letter as 3 March, but de Beer and Gagnebin date it 1 March 1769 in 'Abraham Gagnebin de la Ferrière', p.70.

191. Rousseau to Du Peyrou, 19 December 1768; *CC*, xxxvi.210; Gagnebin to Haller, 12 April 1769; *CC*, xxxvii.81.

192. Haller's reception of Rousseau's thought was more nuanced than generally thought, recognising affinities between his own ideas and those of Rousseau – he characterised the second *Discours* as an 'überaus scharfsinnige Urkunde [extremely astute document]'. Gelzer and Kapossy, 'Roman, Staat und Gesellschaft', p.171. See also Otto Sonntag, 'Introduction', *The Correspondence between Albrecht von Haller and Horace-Bénédict de Saussure* (Bern, 1990), p.44.

193. Gagnebin to Haller, 1 March 1769; *CC*, xxxvii.63 and 12 April 1769; *CC*, xxxvii.81.

194. Consulted by this author, 26 May 2008; with Marcel S. Jacquat, retired director of the Musée d'histoire naturelle of La Chaux-de-Fonds, Switzerland, and François Felber, former director of the botanical garden of Neuchâtel. Some specimens have been remounted, with the original labels, while many others remain in their original form, ordered chronologically in Gagnebin's working herbarium. See also de Beer and Gagnebin, 'Abraham Gagnebin de la Ferrière', p.78.

195. Haller used wax to attach his specimens, which effectively damages them. Consultation of Haller's herbarium, 25 August 2004. The author is grateful to Mme Cécile Aupic and Professeur Gérard-G. Aymonin for their kind assistance on this occasion.

196. These efforts have been spearheaded by Marcel S. Jacquat, who has greatly increased our knowledge of Gagnebin's contributions to science.

legible, hand. Gagnebin's annotations on his specimens attest to his detailed knowledge of contemporary pharmacology, medicine and botany.[197]

These labels likewise attest to Gagnebin's devotion to Linnaean binomial nomenclature and his knowledge of the works of Haller and other pre-Linnaean authors. His technique for preparing the final product is difficult, if not impossible, to observe, since the specimens have in some cases been remounted, while in others we simply have his 'herbier de travail', a working collection assembled in some haste and in chronological order, not according to a classification system.[198]

That Gagnebin was a careful student of Linnaeus is confirmed by his minutely annotated copy of the first edition of *Species plantarum* (1753), given to him by Haller in 1755. This has been described as 'l'ouvrage le plus remarquable' donated to the Bibliothèque publique et universitaire de Neuchâtel by the Neuchâtel botanist and bibliophile, Jean-Frédéric de Chaillet (1747-1839):

> Les pages sont couvertes de notes manuscrites de Gagnebin, écrites à l'encre dans les marges et même entre les lignes du texte; le savant a ajouté mille petits renseignements: il indique notamment dans quels autres répertoires l'espèce est signalée et donnant parfois l'appellation française ou le lieu où il peut l'observer. Cet exemplaire ne vaut pas seulement par ces notes manuscrites. Il est encore truffé de gravures tirées d'autres ouvrages et de dessins des plantes dont l'une - la reprise de la Sibérie à fleur jaune - est signée par Frédéric-Guillaume Gagnebin (1744-1810), le cinquième des six enfants d'Abraham, chronométrier de son état.[199]

Gagnebin's collections in botany, geology and other fields were so extensive that he was compelled to abandon his own home to them and take up residence in the local inn; he gave an important ornithological collection to the French entomologist, Réaumur.

In 1764 Ostervald asserted that Gagnebin 'a formé un herbier vivant très ample'; in addition the brothers 'cultivent sur le sommet de ces montagnes des plantes de la Chine et du Canada'.[200] Godet claims that Gagnebin also naturalised high alpine plants in the Jura, which were erroneously taken to be indigenous.[201]

The Gagnebins' renowned collections became a tourist attraction on

197. I discuss herbaria in Chapter 8.
198. The Gagnebins' cabinet of plants, animals, minerals, fossils, coins and shells is discussed in Marcel S. Jacquat, 'Une page régionale d'histoire des sciences relue récemment: 3. Abraham Gagnebin (1707-1800) et son cabinet de curiosités à la Ferrière', *Bulletin de la société neuchâteloise des sciences naturelles* 123 (2000), p.23-26.
199. Michel Schlup, *et al.* (eds), *Bibliothèques et mécènes: deux siècles de donations à la bibliothèque de Neuchâtel* (Neuchâtel, 2006), p.190; see also p.188.
200. Ostervald, *Description*, p.104.
201. Godet, *Enumération*, p.vi.

the Grand Tour.[202] Visitors included Jean André Deluc, the Scottish naturalist Thomas Blaikie (1750-1838) and the English diplomat and geologist, John Strange (1732-1799), who formed a low opinion of Gagnebin's herbarium.[203] Du Peyrou gives this account of the Gagnebins' natural history cabinet:

> Depuis ma précédente lettre, je me suis informé auprès des Messrs Gagnebin, du prix de leur Cabinet d'histoire naturelle. Ils m'ont répondu qu'ils l'estimaient mille Louis, et m'en ont fourni un catalogue très détaillé, mais trop volumineux pour vous être envoyé, avant une nécessité indispensable. Je comptais aller voir moi-même cette Collection que l'on dit *précieuse et abondante*, mais l accident de ma Mère[204] m'arrête chez moi, et vraisemblablement quand je me trouverai plus libre, la Saison Sera trop mauvaise pour aller dans ces Montagnes de la Ferrière[,] lieu de domicile des frères Gagnebin.[205]

Gagnebin's herbarium is believed to have been sold c.1795 to Captain Benoît des Ponts, from whom it was acquired by Louis Chapuis, pharmacist at Boudry, Switzerland (canton of Neuchâtel).[206] This collection undoubtedly fostered Rousseau's interest in making herbaria, and Gagnebin's meticulous note-taking was an example to follow.

ix. Conclusion

The unique character of the Neuchâtel landscape and its flora, Neuchâtel's status as an ally of the Swiss Confederation, and its connections with the republic of letters, all contributed to a milieu favourable for doing botany in a flexible and progressive way. Physician-botanists such as d'Ivernois, Gagnebin and Neuhaus, shared their knowledge with Rousseau, a Genevan exiled from both his 'patrie' and from France, his home of many years. They played a key role in encouraging his interest in plants and the science of botany through their spirit of open-minded, yet critical, inquiry, exemplified in the first instance by the botanist who influenced them all: Laurent Garcin.

I argue that Rousseau's teachers in Helvetia mediatrix were open-minded, taxonomic pluralists, who taught the classification systems of

202. De Beer and Gagnebin, 'Abraham Gagnebin', p.71ff.; Ostervald, *Description*, p.103-104.

203. De Beer and Gagnebin, 'Abraham Gagnebin', p.71.

204. Du Peyrou's mother, who lived with him in Neuchâtel, had suffered 'l'affreux malheur de se casser l'os de la cuisse aux 3/4 de sa partie supérieure. Il est plus aisé, Monsieur, de Sentir que d'exprimer ma consternation, ma douleur et mes angoisses', Du Peyrou to Coindet, 6 October 1768, *CC* xxxvi.131.

205. My emphasis; Du Peyrou to Coindet, 6 October 1768, *CC*, xxxvi.131. The collection had been for sale for some time. In 1765 the asking price was 660 Louis. Jacquat, 'Abraham Gagnebin, médecin', p.97; Jacquat, 'Une page régionale d'histoire des sciences'.

206. 'M. le capitaine Benoît des Ponts [...] avait formé un herbier considérable, qui appartient maintenant à M. Chapuis, pharmacien à Boudry', Godet, *Enumération*, p.vi.

Tournefort, Haller and Linnaeus.[207] Prepared by this solid grounding, Rousseau was ready to take up the natural method pioneered by Adanson and the Jussieus in Paris, the subject of Chapter 6. In order to appreciate Rousseau's botanical project, it is therefore to its Helvetic origins that we must look.

207. For a taxonomic pluralist or a 'promiscuous realist' there is 'no unique way of carving nature at its joints; different systems of classification are equally legitimate'. Alexander Bird and Emma Tobin, 'Natural kinds', *Stanford encyclopedia of philosophy* (http://plato.stanford.edu/entries/natural-kinds/).

5. Sex, plants and classification

> Ce grand observateur [Linnaeus] est à mon gré le seul avec Ludwig[1] qui ait vu jusqu'ici la botanique en naturaliste et en philosophe.[2]
>
> [C]e n'est pas une nomenclature des perroquets qu'il s'agit d'acquérir, mais une *science réelle*, et l'une des sciences les plus aimables qu'il soit possible de cultiver.[3]

i. Introduction

Rousseau claimed that studying nature, which cannot lie, fosters wisdom and virtue, raises our thoughts to the Author of nature, calms the passions and purges the soul of bitterness and vengeful thoughts.[4] Employed as part of a 'pratique de soi', botany is 'une des sciences les plus aimables qu'il soit possible de cultiver'.[5] While botany may seem alien to the usual techniques of the self, such as meditation and reflection, Rousseau's botany is a meditation – a meditation on plant form, on the visible similarities and differences among plants – in short, classification, as it was understood in eighteenth-century Europe.[6]

1. Christian Gottlieb Ludwig (1709-1773), botanist, Hofrat, professor of medicine at Leipzig and author of *Institutiones historico physicae regni vegetabilis* (Leipzig, Gleditsch, 1742; revised edn 1757). In 1765-1766 Goethe took his meals with Ludwig in Leipzig, where he listened to discussions about Buffon, Linnaeus and Haller. Johann Wolfgang von Goethe, *Dichtung und wahrheit*, bk 6, Lieselotte Blumenthal and Erich Trunz (eds) in *Goethes Werke*, ed. Erich Trunz, et al (Hamburg, 1955), vol.9, p.257. Linnaeus considered Ludwig an 'Institutor [Legislator]', *Philosophia botanica* (1751), section 23, although Ludwig questioned some of Linnaeus's ideas. See Frans A. Stafleu, *Linnaeus and the Linnaeans*, p.242-44.
2. Rousseau, *Confessions*, *OC*, i.643; see also Rousseau to Linnaeus, 21 September 1771; Jean-Marc Drouin, *L'Herbier des philosophes* (Paris, 2008), p.15; *CC*, xxxviii.267; Jean-Marc Drouin, 'Rousseau: lecteur de Linné', *passim*.
3. My emphasis; Rousseau to Mme Delessert, 2 May 1773; *CC*, xxxix.151; see Rousseau, *Emile*, *OC*, iv.350.
4. Rousseau to Mme Delessert, 22 August 1771; *CC*, xxxviii.251; Rousseau to the Duchess of Portland, 3 September 1766; *CC*, xxx.313-14.
5. Rousseau to Mme Delessert, 2 May 1773; *CC*, xxxix.151.
6. Much controversy surrounds classification in philosophy of science. Modern philosophers distinguish between 'naturalists' ('scientific realists'), who hold that natural classification is possible, and different kinds of 'conventionalists' ('constructivists') who consider natural classification either impossible to achieve (the weak variety espoused for example by Locke) or driven by interests and external factors (culture, economy, etc.), and not based in anything essential. For epistemic conventionalists 'kinds [species, genera, however defined] are merely an artifact of the particular cultural or sociological framework within which we develop our concepts'. Alexander Bird and Emma Tobin, 'Natural kinds', *Stanford encyclopedia of philosophy:* http://plato.stanford.edu/entries/natural-kinds/

Eighteenth-century botany sought to establish a rigorous and universal system of order, a 'mathesis' of similarities and differences among plants; this mathesis could be avowedly artificial, or strive to be natural.[7] Famously, Linnaeus wrote: 'Filium Ariadneum Botanices est Systema sine quo Chaos est Res herbaria'.[8] The various taxonomic systems of the era were based on a selection of external characteristics: 'la méthode rapproche les objets semblables, et séparent ceux qui diffèrent; de là naissent les divisions, des sous-divisions que l'esprit saisit bientôt, et qui se gravent facilement dans le souvenir'.[9]

The engraving of similarities and differences on the memory was seen as analogous to learning the grammar or words of a language, and indeed botanists employed analogies to grammar, dictionaries and language in their writings.[10] According to Linnaeus, father of the artificial sexual system, the most well-known and most controversial, classification system of the era, difference and resemblance forms the core of the science: 'Botanices paucissimis debetur nitor et certitude Scientiae, idque praecipue Auctoribus systematicis, quorum exempla sequendo debemus continuare excolere, ac perficere Divisionem Plantarum Systematicam.'[11] The system – that is to say, the selection of plant characteristics that should be considered salient in classification, and how to weight them – was indispensable to the science as it was understood and practised in the eighteenth century.

Eighteenth-century botany did not concern itself with internal plant

7. Why this was a major imperative is taken up, albeit briefly, in later sections of this chapter, p.148 and following. For discussions of classification as historically situated, rather than as a purely abstract activity, see Barry Barnes, *et al.*, *Scientific knowledge: a sociological analysis* (Chicago, IL, 1996), Geoffrey Bowker and Susan Leigh Star, *Sorting things out: classification and its consequences* (Cambridge, MA, 1999); Klein and Lefèvre, *Materials*, p.1, 9, 49ff., 66ff. Locke was sceptical about the possibility of a taxonomy true to nature. John Locke, *An Essay concerning human understanding*, ed. R. Woolhouse (London, 1997), part 3.
8. 'The Ariadne's thread of botany is system without which botany is chaos': Linnaeus, *Philosophia botanica* (2003) [1751], p.113, section 156.
9. Marc Antoine Flerieu de Claret de Latourrette, and Jean-Baptiste François Rozier, *Démonstrations élémentaires de botanique pour l'usage de l'école vétérinaire* (Lyon, Jean-Marie Bruyset, 1766), vol.1, p.5. Rozier was co-author of this work, although this has generally not been acknowledged.
10. On one account, eighteenth-century natural history 'est une science, c'est-à-dire une langue, mais fondée et bien faite'. Michel Foucault, *Les Mots et les choses* (Paris, 1967), p.148. This view is supported by the observation in 1766 that Tournefort's method operates like 'une espèce de *Dictionnaire*'. Original emphasis; Latourrette and Rozier, *Démonstrations élémentaires*, p.95.
11. 'The lustre and achievement of the science is due to a *very few* botanists and among those especially to authors of *systematics*, whose example we should follow by continuing, improving and perfecting the systematic division of plants.' My emphasis; Carolus Linnaeus, *Systema naturae 1735: facsimile of the first edition*, introduction and translation by M. S. J. Engel-Ledeboer and H. Engel (Nieuwkoop, 1964), part 4, p.22, section 3.

systems such as reproduction, respiration and nutrition; this was the preserve of 'physiologie végétale'. Contemporary textbooks made this distinction explicit: 'l'objet de la Botanique étant de fournir les moyens de *reconnaître* et de *distinguer* les plantes, les recherches des Botanistes ne doivent essentiellement porter que sur leurs *parties extérieures*. L'examen des *organes internes* appartient au Physicien qui cherche à découvrir les lois de la végétation.'[12] Botany at this time 'n'est rien d'autre que la nomination du visible'.[13] Natural history gave a 'privilège presque exclusif de la vue'.[14] It was not felt that the interior workings of the plant needed to be understood in order for identifying marks to be delineated and clear distinctions made. With the emergence of the science of evolutionary biology, such a clear separation between inside and outside is no longer possible.

Yet this was an era in which many physiological breakthroughs occurred. The study of plant nutrition was pioneered by Marcello Malpighi and Nehemiah Grew, who separately discovered the vascular systems of plants in the seventeenth century.[15] Particularly important was eighteenth-century research into respiration or gas exchange by Stephen Hales, whose *Vegetable staticks* (1727) was translated into French by Buffon, and by Jan Ingen-Housz, F. R. S., (1730-1799), whose investigation of the effect of light on plants led to the discovery of photosynthesis. The physiology of reproduction nevertheless intersected with classification in systems based on the reproductive organs, such as Linnaeus's artificial sexual system and the natural method.

Rousseau elected to study botany, which was synonymous with taxonomy, but he claimed to dislike the long-running 'dispute' about systems and methods, which is to say contemporary disagreements about plant characteristics salient to classification, and what weight to give them: 'se bornant à la botanique de cabinet et de jardin tout au plus, au lieu d'observer les végétaux dans la nature, on ne s'occupe que de *systèmes* et de *méthodes*; matière éternelle de dispute qui ne fait pas connaître une plante et ne jette aucune véritable lumière sur l'histoire naturelle et le règne végétal'.[16] In 1763 there were reported to be fifty-five 'systèmes universels' and fourteen 'systèmes particuliers', that is to say, sixty-nine in all.[17] Most of these had proliferated within the previous sixty to seventy years.

12. My emphasis; Latourrette and Rozier, *Démonstrations élémentaires*, p.24.
13. Foucault, *Les Mots*, p.144.
14. Foucault, *Les Mots*, p.144-45.
15. Drouin, *L'Herbier*, p.53ff.
16. My emphasis; Rousseau, *Rêveries*, *OC*, i.1069-70.
17. Michel Adanson, *Familles des plantes*, repr. edn with an introduction by Frans A. Stafleu (New York, 1966), p.lxxxviii. The work was originally published in 2 volumes in Paris, 1763-1764.

Rousseau suggested that by studying plants outdoors, and outside the garden, we move beyond this 'matière éternelle de dispute' to discover new plants.[18] It is tempting to see in this comment the critical stance Rousseau adopted as an outsider toward mainstream thinking in any number of fields. However, we should not allow this polemic to mislead us into believing that Rousseau was not interested in taxonomy. We shall see that Rousseau was in fact a careful student of systems and methods, who made well-reasoned taxonomic choices reflecting the cutting-edge botany of the day; this is why Goethe acclaimed Rousseau's botanical writings.

The focus on taxonomic relationships enabled eighteenth-century botany to define itself as a distinct field, separated from medicine and pharmacology, to which it had long been subservient: 'L'art d'étudier les plantes par la combinaison de leurs mixtes qui composent leur substance, l'art d'en déterminer les *vertus médicinales* vraies ou fausses par l'expérience et l'observation toujours imparfaite et trompeuse soit par l'analyse chimique encore plus fautive n'a rien de commun avec la botanique.'[19] In this Rousseau concurred with eighteenth-century botanists such as Linnaeus who were keen to emancipate their science from medicine.[20]

It may seem easy to agree with Rousseau's critical comments about taxonomy, since today it 'undoubtedly suffers from [...] an image problem'.[21] Yet while evolutionary biology has been generally deemed to have displaced the eighteenth-century concern with classification, this is a misleading impression; Darwin's insights into evolution were inspired by his own taxonomic research, a fact that is little known today. In *On the origin of species* Darwin writes: 'It is truly a wonderful fact [...] that all animals and all plants throughout all time and space should be related to each other in group subordinate to group.'[22] Naturalists of Darwin's time still considered classification the key to unlocking the relationships among organisms.

My claim is that Rousseau's technique of the self entailed studying taxonomy in a rigorous way; this claim may appear to contradict Rousseau's criticism of systematisers, and the 'de-naturing' that occurs,

18. See Du Peyrou: 'C'est Linnaeus qu'il faut avoir comme le plus *moderne*, et ce qui vaut mieux, comme celui dont la méthode bien saisie conduit à la connaissance des plantes les *plus inconnues*.' My emphasis; Du Peyrou to Rousseau, 7 October 1764; *CC*, xxi.227.
19. My emphasis; Rousseau, *Fragments de botanique*, *OC*, iv.1249.
20. Linnaeus advocated the separation of botany from medicine even though he was a physician by training.
21. Mary P. Winsor, 'Taxonomy was the foundation of Darwin's evolution', *Taxon* 58:1 (February 2009), p.43-49 (43).
22. Charles Darwin, *On the origin of species*, facsimile edn, introduced by Ernst Mayr (Cambridge, MA, 1964), p.128.

for example, in the typical garden. It also contradicts his allegedly lazy wanderings in the *Rêveries*: 'La botanique est l'étude d'un *oisif* et *paresseux* solitaire [...]. Il se promène, il *erre* librement d'un objet à l'autre, il fait la revue de chaque fleur avec intérêt et curiosité.'[23] From statements such as this, Jean Starobinski concludes that 'Jean-Jacques herborise en collectionneur, et non pas en naturaliste. C'est pour lui une occupation, un amusement, plutôt qu'une véritable action.'[24] It has likewise been suggested that Rousseau rejects a scientific 'statut' for botany, engaging in a 'rejet global des sciences'.[25]

Yet Rousseau's purportedly nonchalant wanderings from plant to plant were guided by close study of the literature, theory and practice of botany; without such a command of the fundamentals of the discipline, he could not have made the collections and observations to which his letters and herbaria attest. Without a good knowledge of 'systemes et méthodes' the *Lettres élémentaires sur la botanique* to Mme Delessert could not have been written. Rousseau alluded to botanical rigour even in the most autobiographical of his works.[26] In Book XII of the *Confessions* he explains that to realise a real benefit from botany, he must observe the minute details of plant organisation and structure: 'Cette constante analogie et pourtant cette variété prodigieuse ne transporte que ceux qui ont déjà *quelque idée* du *système végétal*'. [27] The mere spectator of nature, by contrast, can muster only 'une admiration stupide et monotone'. Rousseau is no mere spectator; he proceeds by comparative and analogical study to familiarise himself with plant morphology, thereby equipping himself for a rigorous examination of the similarities and differences among a wide range of plants.[28]

ii Ethnobotany and classification

Foucault argued that the creation of a mathesis, a universal science of order, characterises the episteme of the 'classical age', that is, the era in which classification based on the visible was a system of signs or a

23. My emphasis; Rousseau, *Rêveries*, *OC*, i.1069.
24. Starobinski, *Rousseau: la transparence et l'obstacle*, p.279. It is unclear how Starobinski distinguishes the collector from the naturalist, since eighteenth-century naturalists usually collected specimens. Modern botany still relies upon specimen collection, which is indispensable to botanical study, although Linnaeus distinguishes between collectors and systematists. Linnaeus, *Philosophia botanica* (2003), p.15, sections 7, 8.
25. Pierre Saint-Amand, 'Rousseau contre la science: l'exemple de la botanique dans les textes autobiographiques', *SVEC* 219 (1983), p.159-67 (159).
26. Cook, 'The septième promenade of the *Rêveries*', p.12; 'Idées et pratiques scientifiques dans la correspondance botanique', p.279-82.
27. My emphasis; Rousseau, *Confessions*, *OC*, i.641.
28. The method of analogy permeates the *Lettres*. See for example Rousseau to Mme Delessert, 22 August 1771; *CC*, xxxviii.254.

language. Yet classifications – information retrieval systems based on similarities and differences among things – are as old as human knowledge itself, and human life as we know it depends intimately upon them. In order to think about things in the world, we need a way to discriminate, divide and distinguish them from one another; one way to do this is by giving different names to things we take to be different and give the same name to things we consider similar: for example, dogs and cats differ in many ways, which we acknowledge by giving them different generic and specific names (such as *Canis familiaris* and *Felinus domesticus*), but they are nonetheless mammals and quadrupeds, and hence belong to together in higher taxa, or groups, such as the order Carnivora and the class Mammalia.

Creating order for ourselves in this way is a foundational epistemological exercise, and shows that 'science is dependent on the operations of common sense'; only then should it be analysed within the framework of philosophy of science.[29] Ethnobotanists such as Scott Atran and Claude Lévi-Strauss argue that cataloguing and naming the natural world is not in the first instance preoccupied with meeting immediate physical needs: rather, 'son premier objet n'est pas d'ordre pratique. Elle répond à des exigences intellectuelles, avant, ou au lieu, de satisfaire à des besoins.'[30] The intellectual needs served by such exercises are dictated by an 'exigence d'ordre [...] à la base de la pensée que nous appelons primitive, mais seulement pour autant qu'elle est à la base de toute pensée: car c'est sous l'angle des propriétés communes que nous accédons plus facilement aux formes de pensée qui nous semblent très étrangères'.[31] Hence, in indigenous cultures of Mexico and Peru 'more than a third of the named plants in both societies have no known social use, nor are they poisonous or pestiferous'.[32]

Atran rejects the time-honoured view that utility accounts for why people from time immemorial have created fine-grained classifications and nomenclature for the plant world. Anthropologists find highly nuanced botanical vocabulary among so-called primitive peoples, vocabularies that easily compete with those of modern languages in nuance, accounting for subtle differences among plants, for example not only by use, but also by size, woodiness and other characteristics;[33] 'It suffices here to signal [...] that tribal peoples *simultaneously* recognise all of

29. Scott Atran, 'Covert fragmenta and the origins of the botanical family', *Man* 18 (1983), p.51-71 (67).
30. Claude Lévi-Strauss, *La Pensée sauvage* (Paris, 1962), p.16.
31. Lévi-Strauss, *La Pensée sauvage*, p.17.
32. Scott Atran, *Cognitive foundations of natural history: towards an anthropology of science* (Cambridge, 1990), p.20.
33. Atran, *Cognitive foundations*, p.31.

these divisions of the plant world for what they are, and do not confuse their merits'.[34] In many parts of the world knowledge of local flora is still taken for granted: 'Among the Ifugao of northern Luzon (Philippines) [...] major sub-categories of rice are based on overall plant morphology and then on the glutinous or waxy quality of the endosperm. The former distinction divides all rice into either *tinawon* (or *tiniyalgo*) "yearly", "winter", or "dry-season" rice; or *pinidwa* (or *linawang*) "second crop", "summer", or "wet-season' varieties'.[35]

Lévi-Strauss recounts the story of an anthropologist who was perplexed when handed a large number of botanical specimens as language learning aids by the African tribesman who was acting as her instructor. He presumably chose plants for the important role they play – so he thought – in everyone's life, and assumed that she would recognise the plants and begin learning his language by attaching names to them. Like many modern westerners, however, the anthropologist knew little, if anything, about plants. According to Lévi-Strauss, 'elle ne s'était jamais intéressée aux richesses et à la diversité du monde végétal, alors que les indigènes tenaient une telle curiosité pour acquise'.[36] As she herself states, 'these people are farmers: to them plants are as important and familiar as people[...] every man, woman and child knew literally hundreds of species. None of them could ever believe that I could not if only I would.'[37]

Detailed knowledge of plants among autochthonous peoples should not surprise us, according to Lévi-Strauss, since 'l'agriculture ne se confond-elle pas avec le savoir du botaniste'.[38] The kind of detailed knowledge of plants possessed by supposedly primitive people was considered worthy of comment even in eighteenth-century Europe, when a large percentage of the European population still lived on or near the land. In 1800 the librarian of the Museum d'histoire naturelle (Paris), Georges Toscan, cites the testimony of the botanist Jacques Labillardière (1755-1834),[39] a member of the voyage (1791-1793) led by

34. Original emphasis; Atran, *Cognitive foundations*, p.24.

35. Harold C. Conklin, *Ethnographic atlas of Ifugao: a study of environment, culture, and society in northern Luzon* (New Haven, CT, 1980) p.20. The elderly women who supervise the selection of rice for planting distinguish among over 500 varieties, each of which is grown under unique conditions found on the terraced mountainsides. Personal communication, Gerald W. Sullivan.

36. Lévi-Strauss, *La Pensée sauvage*, p.9.

37. Elenore Smith Bowen [Laura Bohannan], *Return to laughter: anthropological novel* (Garden City, NY, 1964), p.16, quoted in French translation in Lévi-Strauss, *La Pensée sauvage*, p.10.

38. Lévi-Strauss, *La Pensée sauvage*, p.6.

39. Toscan (1756-1826), founding librarian of the Muséum d'histoire naturelle (Paris), wrote *Histoire du lion: de la ménagerie du muséum national d'histoire naturelle et de son chien* (Paris, Cuchet, 1795), concerning the Museum's zoo, which came from Versailles during the Revolution.

the French naval officer Antoine Raymond Joseph de Bruni d'Entrecasteaux (1739-1793) to locate the lost ships of the French explorer Jean-François de Galaup, comte de La Pérouse (1741-1788?): 'Labillardière, dans la relation de son voyage à la recherche de Lapérouse, observe que les sauvages de la baie des Roches, dans la Nouvelle-Hollande, ont donné des noms particuliers à chaque végétal; et il assure que leurs connaissances en botanique sont invariables'.[40] Labillardière's testimony points not only to the unerring accuracy of indigenous botanical knowledge, but also to an ability among his European contemporaries that he believed was being lost.

This is not to deny that plants would have been named and classified from the earliest times as an essential part of the human and non-human animal food supply; early human vocabularies most certainly featured the names of useful plants. During the greater part of human history, most people lived directly on and from the land, so they had to know something about plants – this applied not only to farmers, but also to physicians, pharmacists and herbalists, all of whom needed a first-hand knowledge of medicinal plants; hence most noted botanists from the eighteenth century and before were medical doctors. For example, 'une attitude d'esprit véritablement scientifique'[41] facilitated the rapid acclimatisation of New World plants in the Philippines, 'adoptées et nommées par les indigènes, qui, dans bien des cas, semblent même avoir redécouvert leurs usages médicinaux, rigoureusement parallèles à ceux qui étaient traditionnels au Mexique'.[42] As Toscan writes:

> Les peuples sauvages nous offrent à cet égard un art qui nous étonne. Le parti qu'ils tirent de la plupart des productions végétales des leurs pays pour divers ouvrages appropriés à leur manière de vivre, est un objet de curiosité qui nous amuse, mais qui prouve, en même temps, combien ces peuples nous *surpassent* dans la connaissance de la véritables propriétés des plantes, et combien la Nature a préparé de ressources à l'homme qui ne s'est pas écarté de ses lois.[43]

iii. Sex, medicine and the history of botany

Rousseau gave a history of botany in the essay customarily presented as the 'Introduction' either to the *Lettres élémentaires* or to the *Fragmens pour*

40. Georges Toscan, *L'Ami de la nature, ou Choix d'observations sur divers objets de la nature et de l'art* (Paris, 1800), p.54, n.1. Labillardière was responsible for identifying over 100 plant species, and the genus *Billardiera* is named for him.
41. Lévi-Strauss, *La Pensée sauvage*, p.23.
42. Lévi-Strauss, *La Pensée sauvage*, p.22.
43. My emphasis; Toscan, *L'Ami de la nature*, p.54.

un dictionnaire des termes d'usage en botanique.[44] In this text he tells his version of the story of botany, especially nomenclature.[45] Rousseau agrees that civilised man had lost his ability to order the plant world by means of fine-grained distinctions. He depicts an alienated relation to plants in which human ills take priority over a direct study of natural beings:

> Le premier malheur de la Botanique est d'avoir été regardé dès sa naissance, comme une partie de la Médecine. Cela fit qu'on ne s'attacha qu'à trouver ou supposer des vertus aux plantes, et qu'on négligea la connaissance des plantes mêmes [...] On ne cherchait des plantes que pour trouver des remèdes, on ne cherchait pas des plantes mais des simples [...] on connaissait fort bien les remèdes, on ne laissait pas de connaitre fort mal les plantes; et c'est tout ce que j'avance ici.[46]

In this depiction the human relationship with plants was for centuries purely medicinal and pharmacological, leading to a very much reduced understanding of the plant realm than that exhibited by so-called savage peoples: 'Chacun connaissait de vue cinq ou six plantes de son canton auxquelles il donnait des noms au hasard enrichis de vertus merveilleuses qu'il lui plaisait de leur supposer.'[47]

Rousseau neglected to mention that sexual difference in plants was recognised long before the early-modern European era, although this understanding was admittedly imperfect.[48] Looking back on pre-modern botany, the eighteenth-century botanist and systematist, Carolus Linnaeus (of whom more will be said later), acknowledged that traditional botanical knowledge included a limited awareness of sexual difference in plants; he cites the Arabs' practice of manual fertilisation of the date palm by 'flores masculi supra foemineos suspendendi semper

44. This was first done in 1780-1782. As Cheyron has convincingly argued, there is no evidence that Rousseau intended this text as an introduction to the *Fragmens pour un dictionnaire*. Cheyron, 'Ray et Sauvages annotés par Jean-Jacques Rousseau', p.86. It should therefore be considered a stand-alone essay on the history of botanical nomenclature, which is why Deville placed it *after* the other botanical writings in *La Botanique de J.-J. Rousseau, contenant tout ce qu'il a écrit sur cette science, augmentée de l'exposition de la méthode de Tournefort, de celle du système de Linné, d'un nouveau dictionnaire de botanique et de notes historiques*, ed. Albéric Deville, 2nd edn (Paris, 1823).

45. The history of botany that I relate here compresses a long and complex history into a brief account that terminates with the work of Linnaeus and the Jussieus in the late eighteenth century.

46. Rousseau, 'Introduction', *OC*, iv.1201.

47. Rousseau, 'Introduction', *OC*, iv.1202.

48. I rely on the discussion of ancient views of plant sexuality in Mark Elvin, 'Curiosity: some examples from Xie Zhaozhe 謝肇淛 (1567-1624) and Gilbert White (1720-1793), and others, of the observation of nature in late-imperial China and early-modern Europe', unpublished MS.

errant, si quis fructose obtineretur'.[49] Hence, '[h]is omnibus sexuii plantarum inhaere re necesse fuit, mares in foeminam actionem eadem methodo promoturis, qua hodieque vtuntur; rem itaque adeo apertam ignorare prorsus nequinere, in his certe arboribus'.[50] Yet the experience of the Arabs and Phoenicians since ancient times did not settle the matter of plant sexuality: 'Nec minus vetustissimos scriptores sexus plantarum diserte mentionem facere constat. Sed quam parum tamen solidi cognition illorum habuerit, quamque lubrico innixa fuerit fundamento, inde patet, quod mares et foeminas separatas saepe docuerint, vbi non errant.'[51]

Linnaeus believed one of these shaky ancient authorities was the fourth-century BCE Greek philosopher, Theophrastus, who was a student (and/or colleague) of Aristotle, as well as Aristotle's successor at the Lyceum. Theophrastus reported the belief that wild plants exhibit male and female characteristics,[52] but he warned that 'we must not assume that in all respects there is complete correspondence between plants and animals'.[53] This warning was prescient, given eighteenth-century attempts to analogise plant sexuality from animal sexuality, an effort fraught with pitfalls, given phenomena such as rampant hermaphrodism, and asexual reproduction by shoots.[54]

Rousseau, on the other hand, called Theophrastus 'le seul botaniste de l'antiquité' because he took a systematic, rather than solely pharmacological, approach to plant study.[55] Theophrastus's plant treatises relied on a classification of plant habit into 'tree, shrub, under-shrub, [and]

49. Carolus Linnaeus, *Disquisitio de quaestione ab Academia Imperiali Scientiarum Petropoli in annum MDCCLIX pro praemio proposita: sexum plantarum argumentis experimentis nouis, praeter adhuc iam cognita, vel corroborare, vel impugnare, praemissa expositione historica et physica omnium plantae partium, quae aliquid ad foecundationem et perfectionem seminis et fructus conferre creduntur: ab eadem Academia die VI. Septembris MDCCLX in conuentu publico praemio ornata* (St Petersburg, Academy of Sciences, 1760), p.5; 'suspending the male flowers over the female, in order to obtain the fruit'; Linnaeus, *A Dissertation on the sexes of plants*, trans. James Edward Smith (Dublin, L. White, 1786), p.5.

50. Linnaeus, *Disquisitio*, p.3; 'It is altogether impossible that they should have been ignorant of a circumstance, which, in these trees at least, is so apparent'; Linnaeus, *Sexes of plants*, p.2.

51. Linnaeus, *Disquisitio*, p.5. 'Nor is it less certain, that the oldest writers have expressly mentioned the sexes of plants. But how little real knowledge of the matter they possessed, and on what slight grounds they held it, appears from their having frequently described plants as being severally male and female, which were not so.' Linnaeus, *Sexes of plants*, p.5-6.

52. Theophrastus, *Enquiry into plants*, trans. Sir Arthur Hort (London and Cambridge, 1916), I.xiv.3-5, p.101.

53. Theophrastus, *Enquiry into plants*, I.I.3-4, p.7.

54. The ways in which naturalists tried to understand and account for plant sexuality by various analogies are documented in François Delaporte, *Le Second Règne de la nature: essai sur les questions de la végétalité au XVIIIe siècle* (Paris, 1979), p.101-60.

55. Rousseau, *Rêveries*, *OC*, i.1063.

herb'.[56] His distinction between woody and herbaceous plants was used for nearly two thousand years, including by one of the great seventeenth-century botanists, Tournefort.[57] Arber remarks that by 'grouping together different species of oaks, willows, etc. [...] [Theophrastus] had some conception of genus'.[58]

However, Theophrastus was not available in Latin until Theodore of Gaza's (1398-1478) translation in the mid-fifteenth century, so Albertus Magnus (c.1123-1280) instead used a spurious work called *De plantis*, translated by Alfredus Anglicus from its Arabic version, which came from the now long-lost Greek of Nikolaos Damaskenos (first century CE), usually agreed to be the 'Pseudo-Aristotle' to whom it was traditionally attributed.[59] *De plantis* combined Aristotle, Theophrastus and other elements, and it was read in the Middle Ages very differently than it had been in antiquity. Albertus Magnus made some empirical observations regarding leaf anatomy and propagation, and refined the classification system which he found in the versions of ancient texts available to him. In the main, however, there was little systematic plant classification, not to mention any attention to morphology or natural affinities, for almost two thousand years.

On the other hand, the preoccupation with medicinal plant properties was long-standing. Eagerly sought were any works that could help identify medicinally useful plants – hence the attraction of works by ancient writers such as Dioscorides, Galen and Pliny the Elder (Gaius Plinius Secundus; 23-79 CE).[60] For centuries, pharmacological botany was based on *De materia medica*, the Latin title of a work by Dioscorides (c.40-90 CE), a Greek physician in the service of the Roman army (fig.5.1 and 5.2, p.142). He provided a guide to 520 species of plants considered to have medicinal effects, of which fifty-four have been placed on the World Health Organisation's list of essential medicinal plants.[61]

The focus on pharmaceutical 'virtues' of plants led to their alphabetical or medical arrangement in herbals and textbooks. Dioscorides presented his *materia medica* (pharmacopia) in alphabetical order, ac-

56. Theophrastus, *Enquiry into plants*, I.II.6-iii.1, p.23.

57. Edward Lee Greene, *Landmarks of botanical history*, ed. Frank N. Egerton, with Robert P. McIntosh and Rogers McVaugh, part 2 (Stanford, CA, 1983), p.948.

58. Agnes Arber, *Herbals: their origin and evolution, a chapter in the history of botany 1470-1670*, 3rd edn (Cambridge, 1986), p.166.

59. Atran, *Cognitive foundations*, p.128.

60. Rousseau, 'Introduction', *OC*, iv.1202. He used the famous edition of Pliny, edited by Jean Hardouin: *Caii Plinii secundi historiae naturalis libri XXXVII: quos interpretatione et notis illustravit Joannes Harduinus [...] in usum serenissimi Delphini*, 3 vols in 2 vols, folio (Paris, A.-U. Coustelier, 1723).

61. Loïc Girre, *Traditions et propriétés des plantes médicinales: histoire de la pharmacopée* (Paris, 1997), p.34.

Figure 5.1: Artist unknown, *Dioscorides arboriste*, engraving, courtesy of the National Library of Medicine, Bethesda, MD.

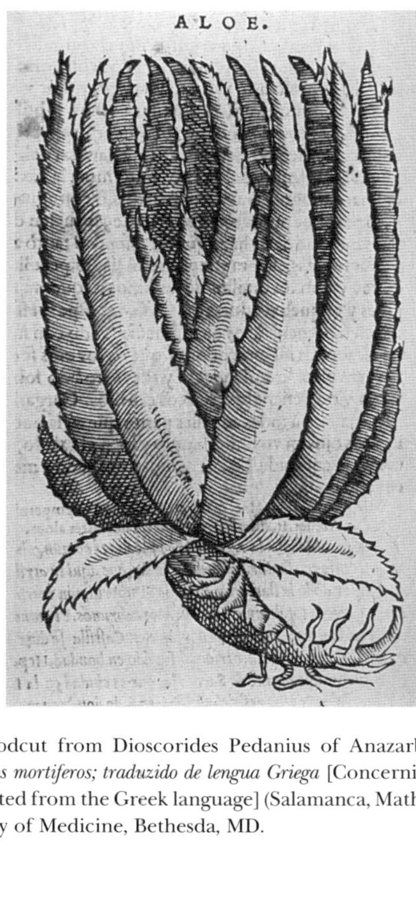

Figure 5.2: 'Aloe', woodcut from Dioscorides Pedanius of Anazarbos, *Acerca de la materia medicinal, y de los venenos mortiferos; traduzido de lengua Griega* [Concerning materia medica, and deadly poisons; translated from the Greek language] (Salamanca, Mathias Gast, 1566), courtesy of the National Library of Medicine, Bethesda, MD.

cording to the properties of the individual drugs.[62] Rousseau criticised Pliny and Dioscorides[63] for appropriating plants to medicine.[64] He also took the trouble to procure an important edition of the *Materia medica* of Dioscorides, found in the Basel edition of the works of Pietro Andrea Matthioli (1500-1577).[65]

However, Rousseau mainly referred to these two ancient authors for polemical effect: 'grâce à un certain Dioscoride, grand compilateur de recettes et à ses commentateurs, la médecine s'est tellement emparée des plantes transformées en simples qu'on n'y voit que ce qu'on n'y voit point, savoir les prétendues vertus qu'il plait au tiers et au quart de leur attribuer'.[66] Michel Adanson (1727-1806), the famous anti-Linnaean, noted that faulty divisions made on the basis of plant properties can even be dangerous.[67] For example, different parts of the same plant may have different and even opposed properties, while plants with the same properties may possess that property to different degrees. These criticisms are not surprising in the Enlightenment, which viewed the Middle Ages as a time of superstition and ignorance. Subsequent investigation has shown that organising plants according to their medicinal virtues can have classificatory value, since plants of the same family often possess similar medicinal properties,[68] as Augustin Pyramus de Candolle demonstrated for twenty-one families of flowering plants in 1804.[69]

Rousseau thought medieval botany strayed from its true task – the

62. http://penelope.uchicago.edu~grout/encyclopaedia_romana/aconite/materiamedica.html.
63. He used the dictionary of Ambroise Calepino (1453–1511), first published in 1502: 'Au lieu du Dictionnaire in-4° Grec et Latin [...] j'ai pensé qu'il vaudroit mieux qu'on y mit mon vieux Dictionnaire de Calepin in folio, qui, étant plus gros, doit tomber plus aisément sous la main, et qui contient encore plus surement les noms Grecs de plantes tirés de Dioscoride et de Théophraste pour lesquels seuls ce Dictionnaire me fait besoin', Rousseau to Richard Davenport, 31 January 1767; *CC*, xxxii.77. He had it shipped to England. Rousseau to Du Peyrou, 1 January 1766; *CC*, xxviii.146.
64. Rousseau referred to Dioscorides in his commentary on the Regnaults', *La Botanique mise à la portée de tout le monde* no.230: L'Herbe aux cuillers (*Cochlearia officinalis* L.), ou l'Herbe au Charpentier, 295: L Oeil de Christ (*Aster amellus* L.), and 305, La Mille-feuille. Regarding the last, Rousseau wrote: 'Ceci me paraît dit bien affirmativement: rien n'est plus douteux, ce me semble, que le temps où Dioscoride a vécu, et il n'est pas non plus très certain que l'aster dont parle cet auteur soit celui qu'on décrit ici', Munch MS (see Bibliography/ Manuscripts/France
65. Pietro Andrea Matthioli, *Opera quae extant omnia: hoc est, commentarii in vi. libros pedacii Dioscoridis Anazarbei de medica materia* (Basel, Johann Koenig, 1674). Rousseau to Guy, 20 January 1767, *CC*, xxxii.56.
66. Rousseau, 'Introduction', *OC*, iv.1202.
67. Adanson, *Familles des plantes*, p.lxxviii, quoted in Latourrette and Rozier, *Démonstrations élémentaires*, p.7.
68. Arber, *Herbals*, p.171.
69. Augustin Pyramus de Candolle, *Essai sur les propriétés médicales des plantes, comparées avec leurs formes extérieures et leur classification naturelle* (Paris, 1804).

study of plants themselves. He gave a pejorative account of medieval botany that bears a more than passing resemblance to that given by Linnaeus in his *Philosophia botanica* (1751). Rousseau named some of the culprits in the endeavour to put plants to medical use, including Myrepsus,[70] Hildegard of Bingen,[71] Paulus Suardus[72] and Arnald of Villanova or Villa Nova.[73]

Rousseau's list of misguided medieval botanists follows the same a-chronological order as that of Linnaeus, who names '*Myrepsus*', '*Hildegardis* Abedrissa', and '*Suardus*' as 'Obscuri [obscure authors]' and 'Barbari'. The Greek '*Myrepsus*' is permitted to make a second appearance as 'Graece', possibly a more eminent category, since the Greeks had been famous in antiquity for their medical knowledge.[74] Despite his status as a physician to Popes and professor at the famous medical faculty of Montpellier, '*Villa nova*', is mentioned in Linnaeus's list of 'Obscuri' only as a man of medicine; he does not rate a mention among the 'Patres' [fathers] of botany.[75]

Whatever their capabilities in the medical field, Rousseau saw these 'Docteurs' diminishing plant study by (1) reducing the plant to its use for the material body, a vain enterprise that results from the created needs

70. Nicolas Myrepsos (13th c.) was a Greek physician of the Byzantine court and at Alexandria. See Rousseau, 'Introduction', *OC*, iv.1202.
71. Hildegard of Bingen (1098-1179), a German abbess, mystic and herbalist, accepted 'not only marvellous and occult virtues of natural substances such as herbs and gems, but also the power of words and incantations', although her 'discussion of vegetation is for the most part limited to medicinal properties of herbs, which are effective without the addition of fantastic ceremonial'. See Lynn Thorndike, *A History of magic and experimental science during the first thirteen centuries of our era* (New York, 1923), vol.2, p.141-42. See Rousseau, 'Introduction', *OC*, iv.1202.
72. Paulus Suardus (fl. c.1500), author of *Antidotarium* (Venice, 1551), was an apothecary from Bergamo (now in Switzerland), although he is identified as 'Mediolanensis' (of Milan) in Carolus Linnaeus, *Bibliotheca botanica* (Amsterdam, Salomon Schouten, 1736), p.11. John Lewis Heller, 'Linnaeus's *Bibliotheca botanica*', *Taxon* 19:3 (June 1970), p.363-411 (387, 407), notes that Linnaeus's *Bibliotheca botanica* contains many bibliographical errors and 'sins of commission'. Following Tournefort, 'Isagoge in rem herbarium', in *Institutiones rei herbariae*, 3 vols (Paris, de l'Imprimerie royale, 1700), p.20, Linnaeus refers pejoratively to Suardus's work as 'quid absurdius?' See Rousseau, 'Introduction', *OC*, iv.1202.
73. Arnald (1240-1311) was 'a leading figure' in the scholastic medicine that was developing at the famous medical school of Montpellier c.1300; he authored about a dozen works of medical theory that advocated compounded and mineral medicines. He treated Pope Clement V and other important persons, and likewise authored treatises of 'experiments' that 'depend[ed] principally on vegetable simples and their products'. See Michael McVaugh, 'The *Experimenta* of Arnald of Villanova', *The Journal of medieval and renaissance studies* 1:1 (1971), p.107-18 (109-10). See Rousseau, 'Introduction', *OC*, iv.1202.
74. The italics denote less important authors; the names of those deemed to be 'heroes or leading figures' by Linnaeus such as Dioscorides, Pliny, and Theophrastus appear in capital letters. See Heller, 'Linnaeus's Bibliotheca botanica', p.383.
75. Linnaeus, *Philosophia botanica* (1751), section 9.

and diseases of civilised humanity (for example too rich a diet); (2) ignoring plant 'structure' and 'économie végétale' and (3) perpetrating such a nomenclatural, taxonomic and epistemological confusion that no one could possibly know what they were talking about:

> Ces plantes avaient des noms vulgaires différents dans chaque canton [...] et quand leur récipés [recettes] couraient dans d'autres pays, on ne savait plus de quelle plante il y était parlé; chacun en substituait une à sa fantaisie, sans autre soin que de lui donner le même nom. Voilà tout l'art que les Myrepsus, les Hildegardes, les Suardus, les Villanova et les autres Docteurs de ces temps-là mettaient à l'étude des plantes, dont ils ont parlé dans leurs livres, et il ferait difficile peut-être au peuple d'en reconnaître une seule sur leurs noms ou sur leurs descriptions.[76]

The results could be fatal for patients, and were almost fatal for botany; the confusion of names was linked to a lack of ordering devices, schemes or technologies to organise both the known plants and those coming into Europe from Islamic and Greek pharmacology. These were to be invented and refined in the early-modern era; as Rousseau was aware, the problem that plagued early-modern botany already afflicted medieval botany: the Babel of names – multiple names for the same plant, or names that did not actually belong to the plants to which they were being attributed.

> Any potential practical usefulness [of the new pharmacopeia] in medicine was, however, offset by the increased possibilities of confusion in substance identification and nomenclature provided by the multiplication of translated or transliterated plant names. Plant naming was of course unsystematic in all the languages involved [...] The rendering of plant names from Greek and Arabic into Latin was accomplished in some instances by translation, in some by transliteration; different authors might choose different ways of rendering the same name. As a result the same plant might be known by several different names, or a completely different plant might be identified with one mentioned by an ancient or Islamic author.[77]

Hence, presumably reputable commentaries on ancient authors could lead to unfortunate, even fatal mistakes due to the application of the wrong name to a plant, and the wrong plant to the patient's ailment. Rousseau notes the potential for fatal nomenclatural error: the true pimpernel ('le Mouron', *Anagallis arvensis* L.) can be confused with chickweed ('le Mouron des bonnes femmes', also known as 'margeline', *Alsine media* L.), which in common parlance was also known as pimpernel. Chickweed is good for birds but true pimpernel kills them.[78]

76. Rousseau, 'Introduction', *OC*, iv.1202.

77. Nancy G. Siraisi, *Medieval and early renaissance medicine: an introduction to knowledge and practice* (Chicago, IL, 1990), p.143.

78. Rousseau to Madeleine-Catherine Delessert, 9 August 1773; *CC*, xxxix.181.

Unfortunately, 'many scholastic "naturalists" assumed that the local flora of the temperate regions of Europe could be exhaustively apportioned among the Mediterranean plants depicted in ancient sources'.[79] The works of famous medieval physicians, Averroes, Maimonides, Avicenna and the famous *Regimen* of the medical school of Salerno were therefore all but useless for northern Europeans.[80]

The sixteenth-century German physician, Leonhard Rauwolf (1535-1596), was so frustrated by the nomenclatural confusion and lack of access to medicinal plants described by ancient authors that he undertook a dangerous trip to the Near East in 1573-1576 in order to identify plants referred to by Dioscorides that did not grow in northern Europe.[81] A work based on Rauwolf's famous herbarium at Leiden[82] was published by Johan Fredrik Gronovius (1686-1762), under the title *Flora orientalis* (1755); the plants were presented according to the artificial sexual system of Linnaeus (discussed below, p.151 and following), but the names were indexed in both Latin and indigenous tongues.[83] Despite being an avowed critic of medicinal botany, Rousseau owned a copy of the *Flora orientalis* at his death.[84]

A variation on the theme of plant as medicine was the so-called doctrine of signatures, which provided a seemingly more sophisticated way in which 'simples' or plant remedies could be researched by seeking a resemblance between a plant and the body part requiring treatment (for example, hepatica resembles the liver and was therefore used in liver treatments). The doctrine of signatures was (re-)introduced in the six-

79. Atran, *Cognitive foundations*, p.127.

80. The famous medical school at Salerno enjoyed a long history, finally being closed by Napoleon in 1811.

81. Rauwolf was the first European to describe the preparation of coffee and the public baths of the Near East in his *Aigentliche Beschreibung der Raiss inn die Morgenländer* (Lauingen, Leonhart Reinmichel, 1582). Three editions appeared in 1582-1583, so this may have been a popular work. An English translation was published in 1738. He is memorialised by the generic name, *Rauwolfia*. See Karl H. Dannenfeldt, *Leonhard Rauwolf: sixteenth-century physician, botanist, and traveler* (Cambridge, MA, 1968).

82. Rauwolf's herbarium comprised four volumes, the first three containing plants from southern France, northern Italy and Switzerland; each volume contained 106 leaves, with plants glued on each side, framed by cardboard that prevented the specimens from touching, while in the fourth, containing 364 Levantine specimens, plants are glued to one side of the page. Swedish troops took the herbarium from the Munich Kunstkammer to Stockholm during the Thirty Years' War; Queen Christina gave it to her Dutch librarian, Isaac Vossius, at whose death in 1688 it was purchased by the University of Leiden.

83. Rauwolf's work was used by Jacques Daléchamps (1513-1588) in his *Historia generalis plantarum*, 2 vols (Lyon, G. Rovillium, 1586-1587) and Jean Bauhin, in his *Historia plantarum universalis*, 3 vols (Yverdon, n.p., 1650-1651). Specimens from Rauwolf's herbarium were consulted by Joachim Camerarius (1534-1598), Jacob Breyne (1637-1697), Robert Morison and Leonard Plukenet. Dannenfeldt, *Rauwolf*, p.229-30.

84. A copy of *Flora orientalis* (Leiden, Wilhelm de Groot, 1755) was found with Rousseau's botany books at his death, Girardin to Du Peyrou, 4 October 1778; *CC*, xlii.11, 22, n.ff.

teenth century by Paracelsus and other 'savants [...] un peu charlatans [...] prétendaient établir une correspondance rigoureuse entre telle partie du corps humain et telle plante dont un caractère remarquable rappelait en quelque manière l'organe concerné et qui se trouvait ainsi désigné de guérir de ses maux'. The physician and alchemist, Paracelsus (1493-1541), together with Giambattista Della Porta (1535?-1615), may have popularised an already ancient idea as a guide to the medicinal properties of plants.[85]

In fact, many of the plants thought to show in their appearance a correspondence or sympathy with a particular part or disease of the human body do exhibit medicinal properties, although these may not exactly conform to the doctrine of signatures.[86] Brosse argues that those attempting to read the signs in nature did so not in a spirit of omniscience, but rather, in a spirit of humility before the power of nature or of God, who had provided everything that human beings require; they simply had to decipher the code that unlocks this secret knowledge.[87] Their approach called up another ancient doctrine, that like cures like: *similia similibus curantur*, the basis for modern homeopathy.

We first encountered the doctrine of signatures and like curing like in Chapter 1. I argue there that Rousseau applied this concept in a metaphorical way to botany, as the cure that is drawn from the disease (the sciences), much like a vaccine.[88] I do not mean to suggest, however, that Rousseau gave the doctrine of signatures a literal value; he did not advocate the search for 'simples' drawn from plants. Indeed, his stance was quite the opposite. According to the reasoning offered in the second *Discours*, *Emile*, the botanical writings and the *Rêveries*, Rousseau agreed with the assumption of the doctrine of signatures that God has provided us with everything we require, but he denied that we need medicine. Since the doctrine of signatures is wedded to medicinal and instrumental needs of the material, socialised body, the doctrine cannot point the way toward health, but rather, is just another symptom of the diseases of civilisation.[89]

85. Suzanne Amigues, *Etudes de botanique antique* (Paris, 2002), p.149ff. See the laudatory discussion of Paracelsus and the doctrine of signatures in Jacques Brosse, *La Magie des plantes* (Paris, 1990), p.100ff. Della Porta, a Neopolitan nobleman, authored *Magiae naturalis* in 1558. This highly popular work was translated into English in 1658. An academy he founded to investigate the secrets of nature was closed down by the Inquisition for suspected dealing in occult matters, and his works were banned by the Roman Catholic Church from 1592 to 1598.

86. Brosse, *La Magie des plantes*, p.105ff.

87. Brosse, *La Magie des plantes*, p.109.

88. Rousseau, *Discours sur les sciences et les arts*, *OC*, iii.26 ; see also Starobinski, *Le Remède dans le mal*, p.166ff.

89. See for example Rousseau, *Rêveries*, *OC*, i.1063-66.

The body of social man is afflicted by unnatural diseases of his own making, resulting from 'monstrueux mélanges des aliments [...] leur pernicieux artificiel assaisonnements, aux denrées corrompues, aux drogues falsifiées, aux friponneries de ceux qui les vendent, aux erreurs de ceux qui les administrent, au poison des Vaisseaux dans lesquels on les prépare [...] les maladies épidémiques engendrées par le mauvais air parmi les multitudes d'hommes rassemblées'.[90] Beyond these all-too-common dangers of civilised life, there are 'métiers mal-sains qui abrégent les jours ou détruisent le temperament' such as the occupational hazards of working with heavy metals and posions like lead, copper, arsenic and mercury.[91] God did not mean for us to handle these substances, so why should he have provided us with plants to cure the diseases occasioned by them?

Hence the doctrine of signatures was, like any other medicinal use of plants, to be rejected: 'Il ne manque à la botanique pour être une étude délicieuse que d'être otée au médecin et rendue au naturaliste.'[92] As examples of the benefits of avoiding medicines, he offers savages and peasants, who are robust without physicians or medicines.[93] Herbal medicine was implicated, therefore, as were heavy metals such as mercury, used to treat venereal disease.

Rousseau took his medicine from the *study* of plants themselves rather than from plant substances: 'Si l'étude des plantes me purge l'âme[,] c'est assez pour moi[;] je ne veux point d'autre pharmacie.'[94] Pharmaceutical substances therefore do not form part of the 'magasin d'aliments donnés par la nature à l'homme et aux animaux [...] jamais il ne m'est venu à l'esprit d'y chercher des drogues et des remèdes'.[95]

iv. Botany in crisis – the search for order

For Rousseau, the Renaissance botanists likewise followed a faulty procedure by taking the ancients as their guide: 'Au lieu d'étudier les plantes sur la terre, on ne les étudiait plus que dans Pline et Dioscoride, et il n'y a rien si fréquent dans les Auteurs de ce temps-là, que d'y voir nier l'existence d'une plante par l'unique raison que Dioscoride n'en a pas parlé.'[96] Some Renaissance botanical authors continued to use the

90. Rousseau, *Discours sur l'origine de l'inégalité*, *OC*, iii.204, n.9. He is referring, for example, to eating and drinking vessels made from lead.
91. Heavy metals were known hazards in the eighteenth century. See *OC*, iii.1365, n.3-5.
92. Rousseau, *Fragments de botanique*, *OC* iv.1250.
93. See for example Rousseau, *Discours sur l'origine de l'inégalité*, *OC* iii.138-39.
94. Rousseau, *Fragments de botanique*, *OC* iv.1251.
95. Rousseau, *Rêveries*, *OC*, i.1064-65.
96. Rousseau, 'Introduction', *OC*, iv.1202.

classification of plants outlined by Theophrastus; a typical herbal of the time, *New Kreütter Buch* (Strasbourg, 1539) by Hieronymus Tragus, grouped plants as 'herbs, shrubs and trees, according to the traditional classification scheme'.[97] This tripartite division died hard; in late seventeenth-century France, Tournefort still used these categories. The first to reject the traditional triad of plant groupings outlined by Theophrastus was the German botanist, August Quirinus Rivinus (Bachmann) (1652-1723), in the late seventeenth century.[98]

Sixteenth-century herbalists such as Hieronymus Tragus (Jerome Bock) (1498-1554), Carolus Clusius (Charles de l'Ecluse) (1526-1609) and Matthias de Lobel (L'Obel or Lobelius) (1538-1616), who was botanist to King James I of England, attempted realistic depiction of plants, but their frame of reference remained largely medical and pharmaceutical. Their ordering of plants was often alphabetical by plant name. Rousseau nevertheless valued the work of Renaissance botanists, and went to great trouble and expense to acquire their works to supplement the purely textual plant descriptions offered by Linnaeus and other eighteenth-century authorities.[99]

Due to nomenclatural confusion, lack of consistent classificatory criteria and information overload, all was not well in Renaissance botanical gardens and apothecaries' shops, but help was on the way in the persons of Caspar and Jean (1541-1613) Bauhin: 'ces *deux illustres frères*, qui ont plus fait eux seuls pour le progrès de la botanique, que tous les autres ensemble qui les ont précédés et même suivis jusqu'à Tournefort. *Hommes rares*, dont le savoir immense et les solides travaux consacrés à la botanique, les rendent dignes de *l'immortalité* qui leur ont acquise.'[100] Over forty years Caspar Bauhin compiled his masterpiece, *Pinax theatri botanici*, comprising 6000 species, their descriptions and synonymies to earlier authors. Caspar [Gaspard] Bauhin was 'l'auteur commun', that is, the common point of reference, for both Tournefort and Linnaeus,[101] but 'quoique presque toute la nomenclature de Tournefort soit tirée mot à mot du *Pinax* on croiroit à lire les botanistes françois qu'il n'a jamais existé ni Bauhin ni *Pinax*, au monde'.[102] Caspar Bauhin's elder brother, Jean, co-authored with Heinrich Cherler the less well-known, but nonetheless important, *Historia plantarum universalis*, which described 5226 plants and was published posthumously

97. Arber, *Herbals*, p.166.
98. Stafleu, *Linnaeus and the Linnaeans*, p.241-42.
99. I discuss Rousseau's reception of Renaissance botanical works in Chapter 7, p.217-25.
100. My emphasis; Rousseau, 'Introduction', *OC*, iv.1203.
101. Rousseau to Malesherbes, 17 April 1772; *CC*, xxxix.37.
102. Rousseau to Malesherbes, 17 April 1772; *CC*, xxxix.38.

in 1560-1561.[103] The work was completed by Dominique Chabrey of Geneva.[104]

For Linnaeus the Bauhins were 'DESCRIPTORES [describers]', and Caspar Bauhin is named among the 'SYNONYMISTAE [synonymists]' of botany.[105] Indeed, following Linnaeus, it is commonly thought that the *Pinax* was nothing more than a register (the literal meaning of the term 'pinax') of plant names, a mere list that did not substantively advance plant classification: 'Le *Pinax* de Bauhin [...] n'est encore qu'un registre des formes et des synonymes des plantes connues à son époque, ouvrage pratique mais limité dans son ambition; il ne fournit en rien la clef pour une compréhension du monde vivant.'[106] The works of both Bauhins allegedly suffer from 'un ordre incompréhensible'.[107]

Yet even one of their critics acknowledges that the Bauhins grouped related species and 'établissent une meilleure discrimination des espèces entre elles, qui conduira plus tard à l'établissement des genres et des espèces comme unités de base de la taxinomie'.[108] Bauhin's *Pinax* had 'a real power of organisation'[109] because the logical division employed by Caspar Bauhin yielded a definition in which: 'le nom est l'essence de la chose et constitue un polynôme de structure variable'.[110] Bauhin based his definitions on specimens sent to him by his large correspondence network, so he worked in an empirical way consistent with the Baconian norms of science, and did not simply rely upon received authority. Moreover, *Pinax* broke new ground by discarding medical and pharmaceutical concerns 'in favour of the concepts of the specialised botanical community'.[111]

Yet even the *Pinax* could not address disorder and information overload because the botanical world of early-modern Europeans was becoming ever larger, bursting the confines of ancient schemes of classification – new species from Asia and the New World were being

103. The Bauhins' portraits graced the French neo-Linnaean fêtes with those of Lamarck and Candolle. Pascal Duris, 'Le Néolinnéisme en France au XIXe siècle', ed. Thierry Hoquet, *Les Fondements de la botanique: Linné et la classification des plantes* (Paris, 2005), p.131-54 (138).

104. Chabrey's work, which Rousseau owned at his death, is discussed in Chapters 3 and 4 p.67, 99.

105. Linnaeus, *Philosophia botanica* (1751), p.7, section 12, p.14, section 39.

106. Thierry Hoquet, 'Linné dans le texte', in Hoquet (ed.), *Les Fondements de la botanique*, p.155-75 (156).

107. Magnin-Gonze, *Histoire de la botanique*, p.86.

108. Magnin-Gonze, *Histoire de la botanique*, p.87.

109. Philippe Selosse, 'The underlying pattern of the renaissance botanical genre pinax', in *Opening windows on texts and discourses of the past*, ed. Janne Skaffari, *et al.*, (Amsterdam & Philadelphia, PA, 2005), p.161-78 (176).

110. Philippe Selosse, 'Peut-on classer le vivant?', *Linné et la systématique aujourd'hui*, ed. D. Prat, *et al.* (Paris, 2008), p.77.

111. Selosse, 'The underlying pattern', p.169.

imported in ever-increasing numbers, the result of exploration, trade and outright bio-piracy.[112] The important botanical gardens at Padua, Montpellier, Uppsala, Amsterdam, Paris, Chelsea, Kew and Leiden became home to exotic transplants such as the tulip, the potato, corn, tobacco, the tomato, coffee, chocolate and tea.

The massive influx of previously unknown plants into European botanical gardens, herbaria and seed collections was part of the driving force behind the impetus to develop new schemes of classification. Trading companies such as the VOC (Dutch East India Company) not only traded plant products – that is, spices – but also entered the plant business to ensure a supply of local, and presumably appropriate, medications to their overseas personnel.[113] All of these plants had to be clearly identified and named.

By the eighteenth century, information overload forced Linnaeus to ignore traditional Greek appellations of plants in *Critica botanica*; he noted the failure of the 'Commentatores' to assign the correct names to plants described in the works of ancient authorities such as Theophrastus and Dioscorides.[114]

v. Systematics transformed

In the Renaissance the flower began to have a sexual character; this insight was to have great consequences for botany, preparing the way for the artificial sexual system of Linnaeus in the eighteenth century, as well as the classifications that succeeded it.[115] Andrea Cesalpino (1519-1603), professor of medicine and botany at Pisa, pursued a systematic classification based on the fruit and seeds, rather than the leaf, as had been the practice of the ancients: 'In morphology he was the first botanist to recognise and state that the various plant organs [...] do not arise haphazardly [...] but in definite and regular positions relative to one another.'[116] He advanced the radical proposition that '[c]lassification must be based on [...] morphological characters of plant organs and parts, not on the properties of the plants in relation to man or any other

112. Cook, 'Jean-Jacques Rousseau and exotic botany'.

113. H. Cook, *Matters of exchange*, p.305. See for example van Reede tot Drakenstein, *Hortus malabaricus*, on which Linnaeus relied. Carolus Linnaeus, 'Ratio operis', *Genera plantarum*, 2nd edn (Leiden, C. Wishoff, 1742), p.xiii, section 24.

114. Carolus Linnaeus, *Critica botanica*, p.117, section 244; Hort trans., p.92; see also John Considine, 'Ancient Greek among the eighteenth-century languages of science: Linnaeus, Dillenius, and the lexicographical record', *International journal of the classical tradition* 16:3-4 (September/December 2009), p.330-43 (341).

115. Arber, *Herbals*, p.168-71.

116. A. G. Morton, 'Marginalia to Andrea Cesalpino's work on botany', *Archives of natural history* 10:1 (1983), p.31-36 (31).

accidental superficial feature'.[117] For Linnaeus, Cesalpino was 'primus verus systematicus'.[118] Cesalpino's view of the flower constituted a significant departure from medieval pharmacological botany; he therefore helped liberate botany from medicine, and set it on a morphological footing: 'from the first days of the revival of natural history in the Renaissance, the idea was commonplace that what the naturalist was doing when characterizing a species by its morphology, that is, its outward form, was seeking marks by which blood relatives could be recognized'.[119] Cesalpino remained important, for Malesherbes had *De plantis libri XVI* in his library and Rousseau searched for it there in 1772.[120]

Rousseau owed a great debt to the English botanist, John Ray (1627-1705).[121] Unlike most botanists of his time, Ray was not a physician, which undoubtedly gave him credit in Rousseau's eyes; furthermore, he laid the foundation for modern classification: 'Il se fonda dans leur arrangement, sur l'ensemble de toutes les parties, la durée et la grandeur, la perfection, le lieu de la naissance, le nombre des pétales, les capsules des graines, les fleurs, les calices et les feuilles; sous ce point de vue il forma trente-trois classes.'[122] In his *Methodus plantarum nova* (1682) Ray divided plants into monocotyledons (one seed leaf) and cotyledons (two seed leaves), providing a fundamental point of departure for the natural method of classification that succeeded the sexual system (discussed below) in the late eighteenth century.[123] Rousseau owned this work and can therefore be presumed to have known about Ray's important division of the plant kingdom;[124] Rousseau suggested, moreover, that Ray had come closest to the elementary method for learning botany: 'Ray me paraît avoir approché plus qu'aucun autre de cette méthode élémentaire.'[125] Rousseau also owned and annotated another landmark work of Ray (fig.5.3 and 5.4, p.154), the third edition of his *Synopsis methodica stirpium britannicarum* (1724), edited by the Oxford botany

117. Andrea Cesalpino, *De plantis libri XVI* (Florence, Apud Georgium Marescottum, 1583), quoted in Morton, 'Marginalia', p.32, but Cesalpino's page is not supplied.
118. Linnaeus, *Philosophia botanica* (1751), p.18, section 54; 'a fructist and the first true systematist'; Linnaeus, *Philosophia botanica* (2003), p.31, section 54.
119. Mary P. Winsor, 'Setting up milestones: Sneath on Adanson and Mayr on Darwin', in *Milestones in systematics*, ed. David M. Williams and Peter L. Forey (Boca Raton, FL, London, New York, Washington, DC, 2001), p.10.
120. Rousseau to Malesherbes, 19 January 1772; *CC*, xxxix.17.
121. An important study of Ray is Charles E. Raven, *John Ray, naturalist: his life and works* (Cambridge, 1950); see also C. E. Raven, *English naturalists from Neckam to Ray: a study of the making of the modern world* (Cambridge, 1947).
122. Latourrette and Rozier, *Démonstrations élémentaires*, p.19.
123. See Agnes Arber, 'A seventeenth-century naturalist: John Ray', *Isis* 34:4 (Spring 1943), p.319-24 (320).
124. Malthus to Rousseau, 24 January 1768; *CC*, xxxv.53.
125. Rousseau to Gouan, 26 December 1769; *CC*, xxxvii.195.

professor and cryptogam expert, J. J. Dillenius; this 'was for many years the most consulted book on British plants, the pocket companion of British apothecaries, physicians, clergymen and country gentlemen [...] and from its 1724 edition, the one edited by Linnaeus's contemporary J. J. Dillenius [...] Linnaeus gained most of his knowledge of the British flora'.[126] It was given to him by the Duchess of Portland. Clearly Ray was an important author for Rousseau, as both a theorist and a describer of plants.[127] It is my contention that Rousseau's acquaintance with Ray's ideas laid the basis for Rousseau's later adoption of Antoine Laurent de Jussieu's natural method, which used Ray's division according to the number of seed leaves as a primary plant characteristic.

While Cesalpino and Ray had greatly expanded the morphological basis for botany, Rudolf Jakob Camerer (Camerarius) (1665-1721), professor of medicine and director of the botanical gardens at Tübingen, added considerable understanding to the sexual character of plants with his pollination experiments, reported in *De sexu plantarum epistola* of 1694. This work was fundamental for the sexual turn in botany in the eighteenth century, most notably in the artificial sexual system of Linnaeus.[128] Camerarius thus gave empirical support to Cesalpino's work. By 1701 Leibniz was writing about Camerarius's findings and later referred to it in his *Nouveaux essais sur l'entendement humain* (completed in 1704, published posthumously in 1765).[129]

However, the work of Camerarius, like many scientific breakthroughs, was slow to catch on;[130] a number of important authorities, including Joseph Pitton de Tournefort (1656-1708) of the Jardin du roi, refused to accept it.[131] Tournefort, 'ignotâ sexuum actione[,] stamina

126. William T. Stearn, 'Preface', in John Ray, *Synopsis methodica stirpium britannicarum editio tertia 1724*, and Carl Linnaeus, *Flora Anglica 1754 & 1759*, facsimile edns (London, 1973), p.ix.

127. The Duchess of Portland to Rousseau, 10 September 1766; *CC*, xxx.348. His annotated copy of the *Synopsis* is in the Old Library, Jesus College, Cambridge in the collection bequeathed to the college by Thomas Malthus, the economist, and son of Daniel Malthus, who purchased Rousseau's botany library in 1775. See Appendix 1.

128. This sexual turn is documented in Delaporte, *Le Second règne*, p.101ff. Plant sexualism was pushed to the limit by sometimes strained analogies to the animal kingdom.

129. Drouin, *L'Herbier*, p.54.

130. The role of the nectary was contested up to the twentieth century, with many botanists ascribing to it an exclusively excretory role. See Jacob Lorch, 'The discovery of nectar and nectaries and its relation to views on flowers and insects', *Isis* 69:4 (December 1978), p.514-33 (529-32). See also Rousseau to Mme Delessert, 24 May 1773; *CC*, xxxix.163: 'Un caractère particulier aux Renoncules, c'est qu'elles ont toutes à l'onglet de leurs pétales une espèce de petite écaille ou vessie ou *nectaire* comme disent généralement les botanistes, et ce caractère suffit seul pour les distinguer [...] de toutes les autres rosacées.'

131. Jacques Rousseau, 'Sébastien Vaillant: an outstanding 18th century botanist', *Essays in biohistory and other contributions presented by friends and colleagues to Frans Verdoorn*, ed. P. Smit and R. J. Ch. V. ter Laage (Utrecht, 1970), p.195-228 (204).

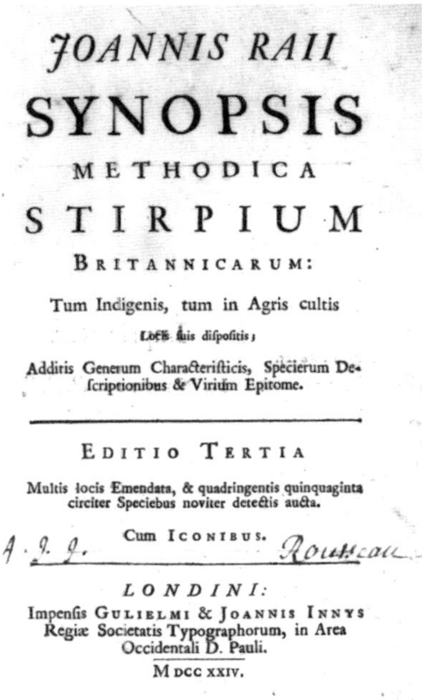

JOANNIS RAII

SYNOPSIS

METHODICA

STIRPIUM

BRITANNICARUM:

Tum Indigenis, tum in Agris cultis

Locis suis dispositis;

Additis Generum Characteristicis, Specierum Descriptionibus & Virium Epitome.

EDITIO TERTIA

Multis locis Emendata, & quadringentis quinquaginta circiter Speciebus noviter detectis aucta.

Cum ICONIBUS.

J. J. Rousseau

LONDINI:

Impensis GULIELMI & JOANNIS INNYS Regiæ Societatis Typographorum, in Area Occidentali D. Pauli.

M DCC XXIV.

Figure 5.3: Title page from Rousseau's copy of John Ray, *Synopsis methodica stirpium britannicarum*, 3rd edn, ed. J. J. Dillenius (London, William & John Innys, 1724), by permission of the Old Library, Jesus College, Cambridge.

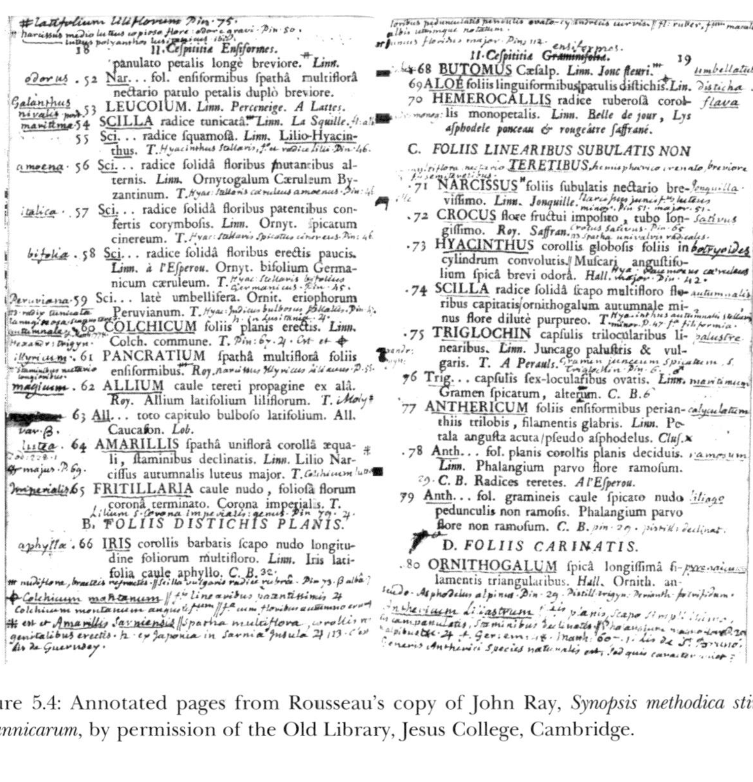

18 II. *Cespititiæ Ensiformes.*

panulato petalis longè breviore. *Linn.*

52 Nar... fol. ensiformibus spathâ multiflorâ nectario patulo petalis duplò breviore.

53 LEUCOIUM. *Linn. Perceneige. A Lattes.*

54 SCILLA radice tunicatâ. *Linn. La Squille.*

55 Sci... radice squamosâ. *Linn.* Lilio-Hyacinthus.

56 Sci... radice solidâ floribus nutantibus alternis. *Linn.* Ornytogalum Cæruleum Byzantinum.

57 Sci... radice solidâ floribus patentibus confertis corymbosis. *Linn.* Ornyt. spicatum cinereum.

58 Sci... radice solidâ floribus erectis paucis. *Linn. à l'Esperou.* Ornyt. bifolium Germanicum cæruleum.

59 Sci... latè umbellifera. Ornit. eriophorum Peruvianum.

60 COLCHICUM foliis planis erectis. *Linn.* Colch. commune.

61 PANCRATIUM spathâ multiflorâ foliis ensiformibus. *Roy.*

62 ALLIUM caule tereti propagine ex alâ. *Roy.* Allium latifolium liliflorum.

63 All... toto capitulo bulboso latifolium. All. Caucafon. *Lob.*

64 AMARILLIS spathâ uniflorâ corollâ æquali, staminibus declinatis. *Linn.* Lilio Narcissus autumnalis luteus major.

65 FRITILLARIA caule nudo, foliosâ florum coronâ terminato. Corona imperialis.

B. *FOLIIS DISTICHIS PLANIS.*

66 IRIS corollis barbatis scapo nudo longitudine foliorum multifloro. *Linn.* Iris latifolia caule aphyllo. *C. B.*

II. *Cespititiæ Graminifoliæ.* 19

68 BUTOMUS Cæsalp. *Linn. Jonc fleuri.*

69 ALOE foliis linguiformibus patulis distichis. *Lin.*

70 HEMEROCALLIS radice tuberosâ corollis monopetalis. *Linn. Belle de jour, Lys asphodele ponceau & rougeâtre saffrané.*

C. *FOLIIS LINEARIBUS SUBULATIS NON TERETIBUS.*

71 NARCISSUS foliis subulatis nectario brevissimo. *Linn. Jonquille.*

72 CROCUS flore fructui imposito, tubo longissimo. *Roy. Saffran.*

73 HYACINTHUS corollis globosis foliis in cylindrum convolutis. Muscari angustifolium spicâ brevi odorâ. *Hall.*

74 SCILLA radice solidâ scapo multifloro floribus capitatis ornithogalum autumnale minus flore dilutè purpureo. *T.*

75 TRIGLOCHIN capsulis triloculoribus linearibus. *Linn.* Juncago palustris & vulgaris. *T. A Perauls.*

76 Trig... capsulis sex-locularibus ovatis. *Linn.* Gramen spicatum, alterum. *C. B.*

77 ANTHERICUM foliis ensiformibus perianthiis trilobis, filamentis glabris. *Linn.* Petala angusta acuta/pseudo asphodelus. *Clus.*

78 Anth... fol. planis corollis planis deciduis. *Linn.* Phalangium parvo flore ramosum.

C. B. Radices teretes. *A l'Esperou.*

79 Anth... fol. gramineis caule spicato nudo pedunculis non ramosis. Phalangium parvo flore non ramosum. *C. B.*

D. *FOLIIS CARINATIS.*

80 ORNITHOGALUM spicâ longissimâ filamentis triangularibus. *Hall.* Ornith. an-

Figure 5.4: Annotated pages from Rousseau's copy of John Ray, *Synopsis methodica stirpium britannicarum*, by permission of the Old Library, Jesus College, Cambridge.

respuerat'.[132] He regarded the reproductive organs as having purely excretory functions. Despite his agamism, or anti-sexualism, Tournefort remained a towering figure in botany well into the eighteenth century; he was known principally for his *Elémens de botanique* (1684), published in an augmented Latin edition as *Institutiones rei herbariae* (1700), which arranged 8000 species in 22 classes. His artificial system of classification, based in the first instance on the corolla (mono-, poly- or apetalous), was relatively easy to learn and remained popular for some time, especially in France (fig. 5.5, p.156). 'La clarté de cette méthode, sa précision, sa généralité, lui méritèrent dès son origine, la préférence sur toutes celles qui avaient paru.'[133] Over twenty other authors adopted it (that is, about one-third of the systems in existence in 1763), including Charles Plumier, Jacques Barrelier, J. J. Dillenius, Pier Antonio Micheli, Herman Boerhaave and Bernard de Jussieu. Admiration for Tournefort's work was virtually universal, even by sexualists such as Linnaeus and Rousseau. In 1766 Claret de Latourrette and Jean-Baptiste François Rozier (1734-1793) based their textbook on the method of Tournefort, merely 'l'enrichissant des découvertes' of Linnaeus.[134] The durability of Tournefort's system was impaired, however, by his neglect of the division of the plant kingdom based on the cotyledon.[135]

Rousseau accorded Tournefort great respect despite that botanist's denial of the sexualist thesis. Rousseau refers to him reverentially as one of the 'grands Botanistes très bien oculés'.[136] Each of Rousseau's *Lettres élémentaires* begins with a Tournefortian analysis of the corolla, a facet of Rousseau's teaching of botany has yet to receive the attention it deserves. Tournefort likewise excercised considerable influence on the texts that were posthumously gathered together and published as Rousseau's *Fragmens pour un dictionnaire des termes d'usage en botanique*.[137]

Cesalpino and Camerarius paved the way for the work of eighteenth-century sexualists such as Sébastien Vaillant and Carolus Linnaeus. On

132. Tournefort, 'ignorant of the functions of the sex organs, disdained the stamens'. Jussieu, 'Introductio', in *Genera plantarum secundum ordines naturales disposita, juxta methodum in horto regio parisiensi exartam, anno M.DCC.LXXIV* (Paris, Herissant and Barrois, 1789), p.xxxi, translated in Stevens, *The Development of biological systematics*, p.350.

133. Latourrette and Rozier, *Démonstrations élémentaires*, p.20. The orthography of Latourrette's name varies: 'La Tourrette', 'La Tourette', or 'Latourrette'. The last is attested in Latourrette's own hand. F. A. Stafleu and Richard S. Cowan, *Taxonomic literature*, 2nd edn, vol.2 (Utrecht, 1979), p.764 under 'Latourrette', but Magnin prefers 'La Tourrette'. *Claret de la Tourrette*, p.119.

134. In his list of key botanical works, Tournefort's name is one of a very few that is printed in capital letters. Linnaeus, *Philosophia botanica* (2003), p.14, section 6.

135. Edward Lee Greene, *Landmarks of botanical history*, part 2, p.948.

136. Rousseau to Mme Delessert, 16 July 1772; *CC*, xxxix.81.

137. Takuya Kobayashi, '*Fragments pour un dictionnaire des termes d'usage en botanique*: processus de rédaction', *Bulletin de l'association J.-J. Rousseau* 62 (2003), p.3-24.

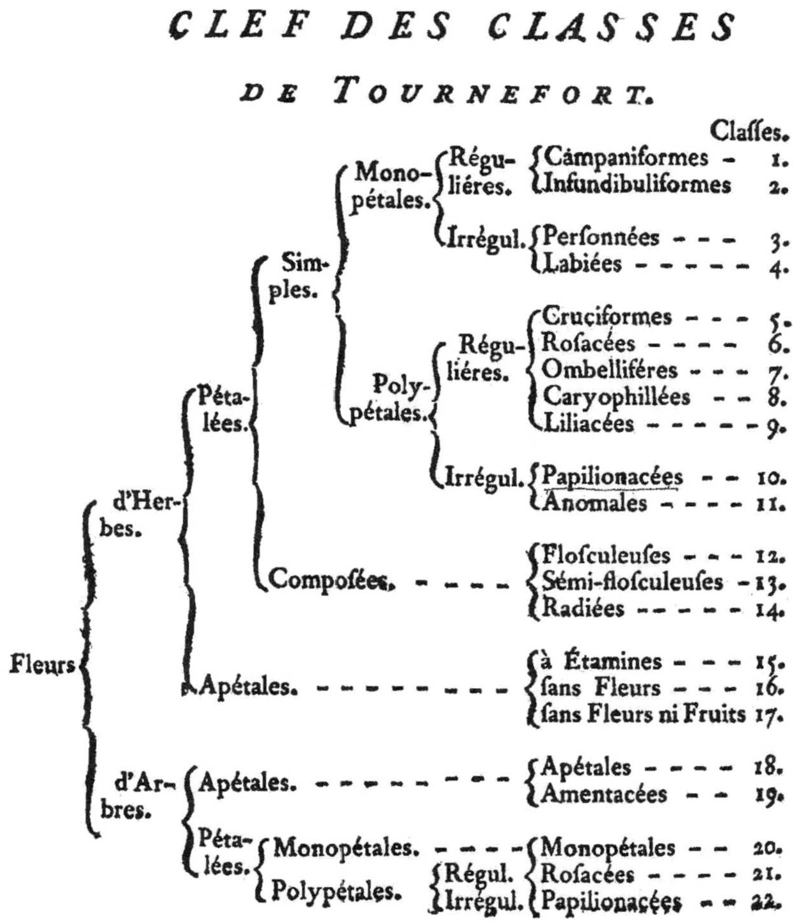

CLEF DES CLASSES

DE TOURNEFORT.

Claſſes.

- Fleurs
  - d'Herbes.
    - Pétalées.
      - Simples.
        - Monopétales.
          - Régulières.
            - Campaniformes - 1.
            - Infundibuliformes 2.
          - Irrégul.
            - Perſonnées - - - 3.
            - Labiées - - - - - 4.
        - Polypétales.
          - Régulières.
            - Cruciformes - - - 5.
            - Roſacées - - - - 6.
            - Ombelliféres - - - 7.
            - Caryophillées - - 8.
            - Liliacées - - - - - 9.
          - Irrégul.
            - Papilionacées - - 10.
            - Anomales - - - - 11.
      - Compoſées. - - - -
        - Floſculeuſes - - - 12.
        - Sémi-floſculeuſes - 13.
        - Radiées - - - - - 14.
    - Apétales. - - - - - - - - -
      - à Étamines - - - 15.
      - ſans Fleurs - - - 16.
      - ſans Fleurs ni Fruits 17.
  - d'Arbres.
    - Apétales. - - - - - - - - - -
      - Apétales - - - - 18.
      - Amentacées - - 19.
    - Pétalées.
      - Monopétales. - - - -
        - Monopétales - - 20.
      - Polypétales.
        - Régul.
          - Roſacées - - - - 21.
        - Irrégul.
          - Papilionacées - - 22.

Figure 5.5: *Clef des classes de Tournefort*, from [Marc Antoine Louis Claret de Fleurieu de Latourette and Jean Baptiste François Rozier], *Démonstrations élémentaires de botanique*, (Lyon, Bruyset, 1766), vol.1, p.82.

10 June 1717, Vaillant, a demonstrator of plants and director of cultivation at the Jardin du roi, Paris, who was standing in for the 'ordinary' professor of botany, Antoine de Jussieu (1686-1758),[138] delivered his inaugural lecture, which was to change botany dramatically. Vaillant brought the sexualist thesis not only to botanists, but also to medical students, and the interested public.[139] While the lecture caused a commotion, Vaillant proved so popular that the students asked the intendant of the garden, Guillaume-Crescent Fagon (1638-1718), to permit Vaillant to continue lecturing after Jussieu's return.[140]

The ground-breaking lecture was published in 1718 and again in 1727 in a bilingual French and Latin edition in the Low Countries (to avoid censorship) as *Discours sur la structure des fleurs/Sermo de structura florum*; this work formed a cornerstone of the sexual system of classification.[141] Vaillant stated his position clearly: '*Les organes qui constituent les différents sexes des plantes, sont deux principaux. Savoir les* Etamines et *les* Ovaires'. He did not hesitate to criticise Tournefort's resistance to the sexualist thesis, noting that the plant parts Tournefort considered to be 'les plus viles et les plus abjectes [...] soient véritablement des plus nobles'.[142] These criticisms put Vaillant at odds with the establishment of the Jardin du roi, whose doyen, Antoine de Jussieu, an eminent botanist, opposed the new ideas about plant sexuality. For this reason Vaillant did not receive an éloge from the Académie royale des sciences at his death, even though he had been a botaniste associé of that institution,[143] and had made other important contributions to his field, for example with his Paris flora, *Botanicon parisiense*, published posthumously in 1727.[144]

138. Antoine de Jussieu was the older brother of the botanists Bernard and Joseph de Jussieu, and an uncle of Antoine Laurent de Jussieu. Antoine de Jussieu's success as a physician enabled him to sponsor his younger brothers' botanical studies. He thereby launched the botanical dynasty that was to play a significant role in modern taxonomy and to dominate the Jardin du roi/des plantes into the nineteenth century. Bernard and Antoine Laurent de Jussieu's contributions to the natural method influenced Rousseau, as discussed below.

139. Rousseau, 'Sébastien Vaillant', p.208-209.

140. Rousseau, 'Sébastien Vaillant', p.199, 206.

141. Sébastien Vaillant, *Sermo de structura florum: discours sur la structure des fleurs* (Leiden, Pieter van der Aa, 1718). For details of the publication of this work, including Vaillant's pretence that it was published without his permission, see Michel Guédès, 'Notules de bibliographie botanique XVI-XXI', *Archives of natural history* 11:2 (1983), p.195-97. Guédès notes that Linnaeus owned a copy of yet another edition that appeared in 1728.

142. Original emphasis; Vaillant, *Discours*, p.10.

143. Rousseau, 'Sébastien Vaillant', p.208-209.

144. Rousseau's copy of the 1743 edition, annotated with Linnaean binomials, is in Houghton Rare Book Library, Harvard University. It contains the following note: 'This book belonged to J.-J. Rousseau, and the handwriting is his. It was given to me by Mrs [Jane] Dalton, who knew him well.' 'Mrs Dalton' (c.1742-1817) was a cousin of Daniel Malthus, who left her 'all [his] Botanical books in which the name of Rousseau is written'. See

vi. The artificial sexual system of classification

Despite the hostility of his contemporaries, Vaillant's work was not consigned to the dustbin of history. Starting in the 1730s it received a new lease on life through the work of Carolus Linnaeus, a Swedish professor of medicine and botany who set out to transform taxonomy on a global scale (fig.5.6, p.159).[145] He also advocated the independence of botany from medicine, which Rousseau applauded. Linnaeus not only praised Vaillant's work in his prize-winning dissertation on the sexes of plants, *Disquisitio de quaestione ab Academia Imperiali Scientiarum Petropoli*[146] of 1760, but also employed Vaillant's insights into plant sexuality to create a new system of classification, the artificial sexual system.

Linnaeus was a versatile botanist and horticulturalist, but he is best known for two innovations, both of which he presented as self-appointed 'legislator' for the 'commonwealth' of botany[147]: (1) his artificial sexual system of classification (fig. 5.7, p.159), published in *Systema naturae* (1735) and (2) his binomial names, published in *Species plantarum* (1753).[148]

Over time, the artificial sexual system and binomial nomenclature were separated and had very different fates. Under pressure from more natural classifications, the sexual system was falling into disuse by the early nineteenth century; in 1836 the British botanist, John Lindley (1799-1865), declared it dead.[149] Darwin's theory of evolution put taxonomy on a new footing; species have a genealogy, and are spatiotemporally continuous.

Linnaean binomial nomenclature, on the other hand, has been codified, and continues to be used; its persistence presents significant problems, given that the 'Linnaean categories [genera and species] themselves correspond to little if anything in nature; and their number is insufficient for representing life's diversity [...] Furthermore, they saddle biologists with the task of assigning ontologically vacuous and misleading

---

Cook, 'Jean-Jacques Rousseau's copy of Albrecht von Haller's *Historia stirpium*', p.149; see also Richmond Laurin Hawkins, 'Manuscripts of Jean-Jacques Rousseau at Harvard University', *The Romantic review* 20 (July-September 1929), p.209-21 (212-17).

145. For an introduction to Linnaeus's life and thought, see Wilfrid Blunt, *The Compleat naturalist: a life of Linnaeus*, with the assistance of William T. Stearn (London, 1971), especially Stearn's Appendix, 'Linnaean classification, nomenclature, and method', p.242ff.

146. Linnaeus, *Disquisitio, passim.*

147. Linnaeus, *The 'Critica botanica' of Linnaeus*, trans. A. F. Hort, p.xxiii-iv.

148. By common consensus, *Species plantarum* (1753) is taken as the starting point of modern botanical nomenclature.

149. John Lindley, *A Natural system of botany, or, A systematic view of the organization, natural affinities, and geographical distribution, of the whole vegetable kingdom*, 2nd edn (London, 1836), p.vii. See the claim that 'Linnaeus's system has been in use for over 200 years'. Patricia Fara, *Sex, botany and empire* (Duxford, 2003), p.20.

Figure 5.6: Jean Haagen and A. Roslin, *Carolus Linnaeus*, oil on canvas, 1775, by permission of the Linnean Society of London.

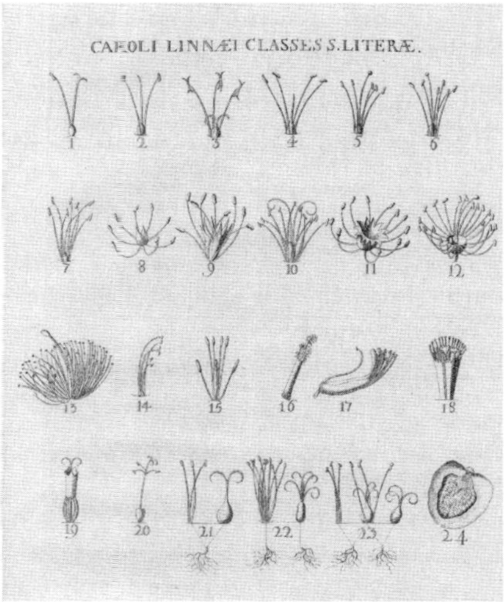

Figure 5.7: Georg Dionysius Ehret, *Classes literæ*, engraving from Linnaeus, *Systema naturae*, facsimile edn (Stockholm, 1907 [Leiden, 1735]), by permission of the Linnean Society of London.

ranks to taxa'.[150] Many biologists believe Linnaeus's nomenclature should be abandoned in favour of an identification system that better reflects the character and diversity of organisms.

Linnaeus presented his innovative system of sexual classification in *Systema naturae* of 1735, and elaborated it in *Fundamenta botanica* (1736), and *Philosophia botanica* (1751); in the last he states: 'Ego Sexuale Systema secundum numerum, proportionem et situm *staminum* cum pistillis elaboravi'.[151] The system was artificial because it selected only certain plant features to be used as the basis for classification across the plant kingdom. The artificial sexual system 'created 24 classes of plants based primarily on the number [...] and arrangement [...] of [the male] stamens, culminating in *Cryptogamia* (plants without proper flowers). Each class, was in turn, subdivided into orders, usually (but not always) based on the number and arrangement of female parts [pistils] of the flower'.[152]

While Linnaeus never claimed that the sexual system was anything but artificial, he nevertheless advocated the foundational character of the sex organs, claiming that all living things are descended 'ex omni specie viventium unicum sexus';[153] plants are just as much alive as animals, and every living thing comes from an egg, as evidenced in plants by the cotyledon(s).[154] He defended his genera, the cornerstone of the system, as natural on the same grounds: 'indestructibile fundamento fructificationis'.[155] He claimed that '[s]i vel mille detegerentur nova Genera, non unicam Notam eam ob causam Generi Naturali proximo vel addere vel auferre necesse est, uti in aliis omnibus inevitabile fuit'.[156]

150. Marc Ereshefsky, *The Poverty of the Linnaean hierarchy* (Cambridge & New York, 2001), p.238. Linnaeus thought there might be about 10,000 phanerogams (flowering plant) species, whereas we now know at least 200,000.

151. Original emphasis; Linnaeus, *Philosophia botanica* (1751), p. 24, section 68. 'I have worked out a sexual system according to the number, relative size, and position of the *stamens*, together with the pistils.' Linnaeus, *Philosophia botanica* (2003), p.37, section 68.

152. Charles Jarvis and Phillip Cribb, 'Review: Linnaean sources and concepts of orchids', *Annals of botany* (2009), p.1-12 (1).

153. Linnaeus, *Philosophia botanica* (1751), p.86, section 132; from 'a single sexual pair of every species'. Linnaeus, *Philosophia botanica* (2003), p.99, section 132.

154. Linnaeus, *Philosophia botanica* (2003), p.101, sections 134-35. On the basis of such claims, Linnaeus has been called an 'essentialist', although recent research discussed in Chapter 6 calls this assessment into question. See for example Ereshefsky, *The Poverty of the Linnaean hierarchy*, *passim*.

155. Linnaeus, 'Ratio operis', *Genera plantarum*, p.xl, section 18α; '[T]he indestructible foundation of fructification.' Staffan Müller-Wille and Karen Reeds, 'A translation of Carolus Linnaeus's introduction to *Genera plantarum* (1737)', *Studies in the history and philosophy of biological and biomedical sciences* 38 (2007), p.563-72 (569), section18α.

156. Linnaeus, 'Ratio operis', *Genera plantarum*, p.xl, section 18*β*; 'Even if a thousand new genera were discovered, it would not be necessary to add or remove a single feature because of a neighbouring natural genus, whereas this was inevitable in all other systems'.

By an ingenious calculation, based on sub-dividing the fructification into thirty-eight parts, he arrived at the conclusion that the maximum number of genera would not exceed 5736.[157]

Much has been written about this system that is not correct: 'The customary reference to the Linnaean system as a basically simple but ingenious arithmetical system founded upon the number of stamens and pistils is, in fact, misleading. The principal divisions originate in the definitive marks for genera and species, namely, number, shape, proportion and situation. Absolute number is only one among four of Linné's considerations.'[158] Hence, Linnaeus's approach is more subtle and perceptive than many reductionist accounts would have it. For example, shape and situation have a geometric as well as a numerical character; they were used to formulate specific definitions, i.e. the polynomial and the generic name formed the plant name. Shapes that could enter into the definition included those of leaves, the various parts of the fruit-body, roots and branches.

The sexual system followed contemporary epistemological norms by maintaining the distinction between primary and secondary qualities.[159] Primary qualities refer to physical properties such as size and shape 'caused by the interaction of our particular perceptual apparatus with the primary qualities of the object', while secondary qualities refer to properties such as scent and colour that 'do not resemble the powers that cause them'.[160]

Linnaeus based his sexual system on *primary* qualities alone. These are quantifiable or measurable and apprehensible by the sense of sight: 'NOTA CHARACTERISTICA omnis erui debet a Numero, Figura, Proportione, et Situ omnium Partium Fructificationis

---

Staffan Müller-Wille and Karen Reeds, 'A translation of Carolus Linnaeus's introduction to *Genera plantarum* (1737)', p.563-72 (569).

157. Linnaeus, *Philosophia botanica* (2003), p.130, section 167. See Ereshefsky, who states that Linnaeus arrived at 3884 combinations based on thirty-one parts. *The Poverty of the Linnaean hierarchy*, p.202.

158. James L. Larson, *Reason and experience: the representation of natural order in the work of Carl von Linné* (Berkeley, CA, 1971), p.58.

159. This distinction was first broached by the ancient atomists, and elaborated by Descartes, Locke, Boyle and Berkeley. See for example Locke, *An Essay*, p.135ff. Rousseau is certain to have been conversant with these theories of perception. See Foucault, *Les Mots*, p.144-45.

160. William Uzgalis, 'John Locke', *The Stanford encyclopedia of philosophy* (Winter 2010 edn), Edward N. Zalta (ed.) (URL http://plato.stanford.edu/archives/win2010/entries/locke/). Considerable controversy surrounds Locke's account of the distinction between primary and secondary qualities. Berkeley went further than Locke, holding the secondary qualities to exist only in the mind of the perceiver.

differentium'.[161] Merely accidental (place, time, use) or variable features (colour, taste, size, season of flowering), otherwise known as secondary qualities, were ineligible to count as differentiae.[162]

In *Genera plantarum* (1789) the French botanist, Antoine Laurent de Jussieu (discussed further in Chapter 6, p.178-88), praised the sexual system's capacity to determine membership in a class 'ac sine praeceptore agnoscendis conducant'.[163] On the other hand, he noted that the system requires careful examination of very delicate parts (stamens and pistils) using a lens or dissecting needle with a risk of error. Moreover, the selection of a select set of parts on which to base the system was arbitrary, with some classes being quite natural – obviously showing a family resemblance – and others not. This was evident well before being challenged by the theory of evolution. Jussieu offered a long list of '[i]ndè primarias ex staminum numero et proportione notas minùs stabiles deducens, discordes in iisdem sectionibus congerit plantas'.[164]

vii. The provocative legacy of Linnaeus's sexual system

> This was a period of keen debate about generation [...]. Botanists, committed to finding uniform laws of nature, increasingly drew analogies between plant and animal life and thus transferred their fascination with animal sexuality onto plants.[165]

Despite its demise, the artificial sexual system of classification awakened great attention upon its first publication, and this interest has continued to the present, from the puritanical criticisms of the eighteenth-century German naturalist, Johann Georg Siegesbeck (1686-1755), to more recent feminist history of science. Siegesbeck (1686-1755), director of the St Petersburg Apothecary's Garden and professor of botany at the St Petersburg Academy of Sciences, condemned the Linnaean sexual system as prurient, immoral and against God's law concerning monogamy.[166] Linnaeus's friends, Johann Friedrich Gleditsch (1714-1786)

161. Linnaeus, *Philosophia botanica* (1751), p.116, section 167. 'Every CHARACTERISTIC FEATURE ought to be elicited from number, shape, relative size and position of all the different parts of the fruit-body', Linnaeus, *Philosophia botanica* (2003), p.129, section 167.
162. Linnaeus, *Philosophia botanica* (2003), p.130, section 167; p.220-36, section 258-74.
163. '[W]ith little effort, and [...] without an instructor', Jussieu, *Genera plantarum*, p.xxxii, trans. in Stevens, *Development of biological systematics*, p.352.
164. '[D]isparate plants grouped together [by the Linnaean sexual system] in the same sections', Jussieu, *Genera plantarum*, p.xxxiii, trans. in Stevens, *Development of biological systematics*, p.352.
165. Schiebinger, *Nature's body: gender in the making* (Boston, MA, 1994), p.37.
166. Despite his notoriety as a vocal, but ineffective, critic of the artificial system, Siegesbeck

and Johannes Browallius (1707-1755), a Lutheran minister and later bishop of Åbo (now Turku), in Finland, defended him; their replies were matched by vituperative replies from Siegesbeck.[167] Linnaeus refrained from replying directly to his critic, but secretly took revenge on Siegesbeck.[168] Linnaeus received vindication from the prize awarded for his *Disquisitio de quaestione ab Academia Imperiali Scientiarum Petropoli* [*Dissertation on the sexes of plants*] by the St Petersburg Academy of Sciences in 1760.

The artificial sexual system enjoyed a dramatic success in spite of the sixty-nine systems from which it was possible to choose.[169] Eighteenth-century botanical explorers both in Europe and overseas adopted the artificial sexual system because it offered ease of access, and could purportedly be learned in one year without a teacher.[170] By the last quarter of the eighteenth century, the sexual system had been widely accepted across Europe (although it was less popular in France),[171] because it provided a ready ordering device for the European appropriation of exotic plants.[172] Botanists needed such a system because of 'information overload.'[173]

---

was 'apparently a good botanist'. Dimitry D. Sokoloff, *et al.*, 'The history of botany in Moscow and Russia', p.150; 158-61. His fifteen-volume herbarium is in the Herzog August Bibliothek, Wolftenbüttel, Germany.

167. Browallius and Linnaeus fell out, however, after the former tried to lure Linnaeus's fiancée, Sara Lisa Morea, away from her betrothed during his absence from Sweden in 1735-1738. The trivial names assigned to plants of the genus *Browallia* reflected the rise and fall of their friendship.

168. Linnaeus's response – to transmit seeds of *Sigesbeckia* as *Cuculus ingratus* ('ungrateful cuckoo') – to Siegesbeck, while damaging to Russian-Swedish botanical exchange, did not end the traffic in botanical specimens (including Asian plants) to Uppsala via Russia; Siegesbeck himself gave or sold over 200 hundred Siberian plants to Linnaeus's collaborator, Baron Sten Carl Bjelke (1709-1753), while other specimens were sent to Linnaeus by Grigorii Akinfeevich Demidov (1715-1761). Furthermore, a number of Russian students studied with Linnaeus. See Rowell, 'Linnaeus and botanists in eighteenth-century Russia', p.16, 19-20.

169. Michel Adanson, *Familles des plantes* (1966), p.lxxxviii.

170. Linnaeus, *Philosophia botanica* (2003), p.111, section 151.

171. '[T]o take up Linnaean taxonomy was still a controversial act in Britain during the 1760s and 1770s.' Janet Browne, 'Botany in the boudoir and garden: the Banksian context', *Visions of empire: voyages, botany, and representations of nature* ed. David Philip Miller and Peter Hans Reill (Cambridge, 1996), p.154.

172. See Browne, 'Botany in the boudoir', p.154; John Dean, 'Controversy over classification: a case study from the history of botany', in *Natural order: historical studies of scientific culture*, ed. Barry Barnes and Steven Shapin (London/Beverly Hills, CA, 1979), p.215; A. Cook, 'Politics of nature and voyages of exploration: some purposes and results', in *Voyages and exploration in the north Atlantic from the Middle Ages to the 17th century*, ed. Anna Agnarsdottir, 2nd edn (Reykjavik, 2001), p.129.

173. Linnaeus, *Disquisitio*, p.4; 'they became so overwhelmed with the multitude of species, as almost to despair of finding the way in or out of their gardens; both the [East and West]

The artificial sexual system of classification had more important critics than Siegesbeck. In France, Georges Louis Leclerc, comte de Buffon, the powerful intendant of the Cabinet du roi (which included the Jardin du roi), rejected both Linnaean binomials and the sexual system, arguing that Linnaeus grouped species contrary to their genera, and hence nature;[174] he declared that in nature there are only individuals, although he also upheld the biological species concept based on the reproductive barrier between species.[175] Buffon's research was mainly zoological, and his critique of Linnaean nomenclature did not readily apply to botany, which had to name tens of thousands of species. In fact, Buffon's critique of Linnaeus's concept of the relation of species to genus may have arisen from a misunderstanding.[176]

Rousseau admired Buffon, describing him in the second *Discours* as 'une de ces autorités respectables pour les philosophes, parce qu'elles viennent d'une raison solide et sublime'.[177] While Rousseau relied on the first three volumes of Buffon's great *Histoire naturelle*,[178] he ultimately rejected Buffon's materialism, barely concealed atheism and scepticism about final causes. He also found Buffon's account of generation inadequate:[179] 'Le Pline de notre siècle voulant développer le mystère de la

Indies daily furnishing them with so many novelties, that no memory was strong enough to retain them'. Linnaeus, *Sexes of plants*, p.3.

174. It should be noted that Buffon was an important figure in the development of the idea of evolution over a hundred years before Darwin published his theory in *On the origin of species* in 1859. Buffon extended the age of the earth from the date of Creation in 4004 BCE – estimated by Bishop Ussher in the seventeenth century – to over 70,000 years. See Georges, Louis Leclerc, comte de Buffon, *Histoire naturelle, générale et particulière*, 36 vols (Paris, Imprimerie royale, 1749-1788), vol.2.

175. There is a large literature on the Linnaeus-Buffon controversy. See for example Thierry Hoquet, *Buffon-Linné: éternels rivaux de la biologie?* (Paris, 2007); Jacques Roger, *Buffon: a life in natural history*, trans. Sarah L. Bonnefoi, ed. L. Pearce Williams (Ithaca, NY 1997), p.309ff.; Phillip R. Sloan, 'Buffon, German biology and the historical interpretation of biological species', *British journal for the history of science* 12:2 (July 1979), p.109-53.

176. Roger, *Buffon*, p.320; see also Linnaeus, *A Dissertation on the sexes of plants*, p.43-45, n.vi-viii.

177. Rousseau, *Discours sur l'origine de l'inégalité*, *OC*, iii.195; see also p.231, where Rousseau groups Buffon with such astute observers of men as Montesquieu, Diderot, d'Alembert and Condillac. For Rousseau's relationship with Buffon, see Roger, *Buffon*, p.347-49.

178. Roger, *Buffon*, p.347. Hulliung argues that Rousseau followed Buffon as a 'natural historian'. Yet Hulliung ignores Rousseau's work in botany, concluding: 'Rousseau pronounced "the generation of living and organised bodies [...] [the] abyss of the human mind" and devoted his attention to other matters', Hulliung, *The Autocritique of Enlightenment*, p.182, quoting *Emile*, p.580; see also p.172.

179. Buffon writes that it is 'très vraisemblable [...] qu'il existe [...] une infinité de petits êtres organisés [...] composé de parties organiques vivantes qui sont communes aux animaux et aux végétaux, que ces parties organiques sont des parties primitives et incorruptibles, que l'assemblage de ces parties forme à nos yeux des êtres organisés, et [...] la reproduction ou la génération n'est qu'un changement de forme [...] la destruction de l'être

génération s'est vu forcé de recourir à un principe inintelligible et inconciliable avec les lois connues de la mécanique et du mouvement.'[180] So while he respected Buffon and his works, Rousseau disagreed with the intendant of the Jardin du roi, preferring Linnaeus, the northern sage.

Rousseau had no patience with Linnaeus's other French critics either: 'Je sais avec quel mépris on affecte en France de traiter ce grand Naturaliste, mais le reste de l'Europe l'en dédommage, et la postérité l'en vengera'.[181] While sympathetic to the natural method, he also criticised one of its major French progenitors, Michel Adanson, for his anti-Linnaean stance: 'M. Linnaeus et ses livres sont tout-à-fait nuls pour M. Adanson et ses lecteurs [...]'.[182]

Unlike other defunct systems of classification, Linnaeus's artificial system of sexual classification continues to be discussed, and its manifold implications examined. It has been argued that Linnaeus, perhaps more than any other botanical author, sexualised plants and taxonomy, analogising plants' parts to human reproductive organs and using elaborate conjugal images to explain his system.[183] 'Though by the 1720s most botanists had come to recognise that plants engaged in sexual reproduction, Linnaeus's system of sexual classification popularised the idea [...] he placed the sex life of plants at the centre of botanical study and illustration [...]. Botany thus became, oddly enough, the most explicit discourse, in the public domain, on sexuality during the period.'[184] Linnaeus did not leave the matter there: he 'consistently went beyond biological analogies in order to import social analogies into his descriptions of plant life [...]'.[185]

The sexual system thus provided a way to discuss sexuality without direct reference to humans; it could inspire a double gesture of repugnance and attraction facilitated by the ambiguity of plant sexuality.[186] However, this discourse varied widely from the explicitly sexual work of

---

organisé, se fait par la division de ces mêmes parties'. Georges Louis Leclerc, comte de Buffon, *Histoire naturelle, générale et particulière*, vol. 2, p.24.

180. 'Le Pline de notre siècle' is Buffon. See Rousseau, *Lettres morales* 3, *OC*, iv.1096.

181. Anticipating objections, Rousseau defended the objectivity of his views on Linnaeus: 'Ce que je dis est assurément sans partialité et par le seul amour de la vérité et de la justice; car je ne connais ni M. Linnaeus, ni aucun de ses disciples, ni aucun de ses amis', Rousseau to Vincent-Louis Dutens, 5 February 1767; *CC*, xxxiv.99.

182. Rousseau, 'Introduction', *OC*, iv.1208. This comment is made in the context of a discussion of nomenclature rather than classification.

183. '[T]he CALYX is the *bedroom*, the COROLLA is the curtain'. Linnaeus, *Philosophia botanica* (2003), p.105, section 146.

184. Alan Bewell, '"On the banks of the South Sea": botany and sexual controversy in the late eighteenth century', in *Visions of empire: voyages, botany, and representations of nature*, ed. David Philip Miller and Peter Hans Reill (Cambridge, 1996), p.174.

185. Bewell, '"On the banks of the South Sea"', p.175.

186. For example hermaphrodism. François Delaporte, *Nature's second kingdom*, p.138.

Erasmus Darwin, *The Loves of the plants* (part 2 of *The Botanic garden*, 1791), to the de-sexualised botany of the Quaker author of children's books, Priscilla Wakefield.[187] This fascination with plant sexuality is far from obsolete, as suggested by the recent publication of a lavish book of photographs portraying the magnified reproductive organs of plants.[188]

Feminist scholars claim that Linnaeus's theories reflected important social and ideological trends of the mid-eighteenth century[189] by creating a system founded on sexual difference; on materialist grounds, the female was now recognised as distinct from the male, not simply as an imperfect version of him: 'the fundamental divisions Linnaeus devised as the basis of his new botanical taxonomy recapitulated the sexual hierarchy of Western Europe'.[190] Ann Shteir characterises Linnaeus's position thus:

> The sexuality that infuses all parts of nature's realm, so his argument goes, also shapes the conduct of male and female, assigning each a preordained part to play in continuing the cycle of life. Linnaeus presents as 'natural' what we should read as his discourse about sexuality [...]. Linnaeus's assumptions about sexual functions in plant reproduction reflect traditional gender concepts. Thus he assigned a higher ranking to the class, a unit based on the stamens (the male part) and a lower ranking to the order, a unit based on the pistil (the female part). He also represented the male part in plant reproduction as active and the female part as receptive.[191]

The leading proponents of this position argue: 'Linnaeus simply brought traditional notions of gender hierarchy whole cloth into science';[192] 'Linnaean botany expresses a contemporary obsession with sex and gender difference.'[193]

Leading historians of botany reject this interpretation, however. For instance, while it is true that 'sur le plan d'observation, le nombre des étamines est un critère plus facile à utiliser que la structure du pistil', the conjugal relations of male and female plant parts constitute 'une inversion ludique de l'ordre patriarcal' because of the large number of plants in which one female enjoys the favours of multiple males.[194] Staffan

187. George, *Botany, sexuality and women's writing*, p.48; Ann B. Shteir, '"The pleasing objects of our present researches": women in botany', in *Women and history*, ed. Valerie Frith (Toronto, 1995), p.149.
188. Rob Kesseler and Madeline Harley, *Pollen: the hidden sexuality of flowers* (London, 2004).
189. Schiebinger, *Nature's body*, p.37-38.
190. Londa Schiebinger, 'The private life of plants: sexual politics in Carl Linnaeus and Erasmus Darwin', in *Science and sensibility: gender and scientific enquiry, 1780-1945*, ed. Marina Benjamin (Oxford, 1991), p.123.
191. Shteir, *Cultivating women*, p.16. A similar argument is made in Schiebinger, *Nature's body*, p.17.
192. Schiebinger, *Nature's body*, p.17.
193. Shteir, *Cultivating women*, p.17.
194. Jean-Marc Drouin, *L'Herbier des philosophes*, p.189.

Müller-Wille goes even further, asserting that 'Linnaeus's sexual system does not possess a hierarchical structure at all, but rather is the result of a systematic combination of classes.'[195] It is therefore by no means self-evident that the artificial sexual system of classification expresses the principles of patriarchal domination and hierarchy that have been attributed to it.

viii. Was Rousseau a sexualist? Was he a Linnaean?

> Tout ce qui l'entoure ne fait que mieux le convaincre que l'objet premier de la nature est de se *reproduire* en liberté, sans l'aide ni le frein de quiconque.[196]

With Rousseau's introduction to the sexual system, readers might expect this to be the conclusion of our taxonomic journey; for most accounts of Rousseau and botany, the Linnaeus of the sexual system is the undisputed hero of the tale.[197] In the *Confessions* and the *Rêveries* Rousseau refers more often to Linnaeus than to any other botanist, reinforcing the impression of an elective affinity between his botany and the sexual system.[198] Consequently, Rousseau's botany has been freighted with the same patriarchal and sexist implications as the sexual system has.[199] His *Lettres élémentaires* to Mme Delessert allegedly provide a 'Linnaean-inspired unveiling of the "mysteries of vegetation"',[200] replete with flirtatious comments to his correspondent, conveying 'a latent meaning which conceals the onset of romance'.[201] Thus Rousseau, 'Europe's greatest advocate of sexual complementarity', is presumed to have taken

195. Staffan Müller-Wille, 'Collection and collation: theory and practice', p.548.
196. My emphasis; Odile Richard-Pauchet, '"Voici des fruits, des fleurs, des feuilles et des branches": une lecture érotique des lettres élémentaires sur la botanique de Rousseau', *Revue de l'AIRE*, 34 (2008), p.197-214 (209).
197. See for example Roger L. Williams, *Botanophilia in eighteenth-century France: the spirit of the Enlightenment* (Dordrecht, 2001), p.95. Williams claims Rousseau could not have known the natural method, as its major theorist, Antoine Laurent de Jussieu, only published the full version in 1789. However, as I show in Chapter 6, Jussieu's 1789 treatise culminates a long period of gestation during which Rousseau had direct and sustained contact with both Antoine Laurent, and his uncle, Bernard, creators of the natural method.
198. Rousseau, *Confessions*, *OC* i.643; *Rêveries*, *OC* i.1043, 1064.
199. See Jacques Berchtold, 'Le framboisier nourricier: Rousseau et l'érotisme végétal de la tradition pastorale', *Etudes Jean-Jacques Rousseau* 14-15 (2003-2004), p.23-49; Odile Richard-Pauchet, 'Voici des fruits'; Joel Schwartz, *The Sexual politics of Jean-Jacques Rousseau* (Chicago, IL, 1984), p.111-12.
200. George, *Botany, sexuality and women's writing*, p.51.
201. George, *Botany, sexuality and women's writing*, p.50; see also Richard-Pauchet, 'Voici des fruits', p.203. The suggestion of romantic feelings on Rousseau's part for Mme Delessert is far-fetched; his attitude toward the daughter of his long-time friend, Mme Boy de la Tour, would more accurately be described as avuncular.

up botany *because of* the artificial sexual system.[202] Yet the seeming consistency between Linnaeus and Rousseau's views on botany should not be regarded as over-determined or rigid; it is in fact highly contingent. I argue that Rousseau's discovery of Linnaeus should in fact be considered no more than a fortuitous accident.

In this regard two facts are crucial. First, there can be no question of the two men having known of each other's works as they were producing their major theoretical positions. Linnaeus barely knew who Rousseau was (even after naming a genus for Rousseau), while Rousseau probably did not know anything about Linnaeus until late 1764, well after writing the important works – *Emile* and *Julie* – in which sexual complementarity plays a key role. Hence, the seeming affinity of their notions of sexual difference can only be seen as part of a larger, European tendency in which 'interest in plant sexuality coincided with a keen interest in the exact differentiation of sexual character in animals and humans'.[203]

Second, Rousseau's adoption of Linnaeus was hardly over-determined but rather a late development, given that he was already interested in botany prior to the death of Claude Anet in 1734, and before the publication of Linnaeus's *Systema naturae* in 1735. Between 1735 and 1764 Rousseau had not worked in botany, although he was acquainted with chemists' plant experiments. It therefore seems unlikely that Rousseau's first concern in Môtiers was to find a *sexual* approach to plant study. It seems more likely that he became *re-attracted* to plants for the reasons detailed in the *Confessions*: the contentment he saw in Anet's eyes after a botanical expedition.[204] Indeed, when he began botanising, his starting point was uncertain: 'Je donnerais tout au monde pour savoir la botanique; c'est la véritable occupation d'un corps ambulant et d'un esprit paresseux; je ne répondrais pas que je n'eusse la folie d'essayer de l'apprendre si je savais par *où* commencer.'[205]

Linnaeus claimed his system could be learned in one year without a teacher. This, coupled with Du Peyrou's enthusiasm for Linnaean taxonomy, may have influenced Rousseau to study the sexual system.[206] In the summer of 1765, when he abandoned the optique for botany, he sought the magnifying glass he would need to examine the stamens and pistils, a key operation for using the sexual system:[207] 'Le microscope est comme vous l'avez deviné pour la botanique, ainsi je desire qu'il ait un

202. Schiebinger, *Nature's body*, p.219-20, n.66.
203. Schiebinger, *Nature's body*, p.37-38.
204. Rousseau, *Confessions*, *OC*, i.180.
205. My emphasis; Rousseau to Mme de Boufflers, 26 August 1764; *CC*, xxi.70.
206. Du Peyrou extolled this system to Rousseau as the most 'moderne' and best suited to finding the least well-known plants. Du Peyrou to Rousseau, 7 October 1764; *CC*, xxi.227-28.
207. Rousseau to François Henri d'Ivernois, 29 July 1765; xxvi.105-107.

champ suffisant pour embrasser le *pistil* et les *étamines* d'une petite fleur.'[208] This interest had not been evident a few months before when Rousseau wrote to Mme de Boufflers.

According to the well-known account in the cinquième promenade of the *Rêveries*, Rousseau employed Linnaean sexual classification on the Ile St Pierre in autumn 1765:

> J'allais une loupe à la main et mon *Systema naturae* sous le bras, visiter un canton de l'île que j'avais pour cet effet divisée en petits carrés dans l'intention de les parcourir l'un après l'autre en chaque saison. Rien n'est plus singulier que les ravissements, les extases que j'éprouvais à chaque observation que je faisais sur la structure et l'organisation végétale et sur le jeu des *parties sexuelles* dans la fructification, dont le système était alors tout à fait *nouveau* pour moi. La distinction des *caractères génériques*,[209] dont je n'avais pas auparavant la moindre idée, m'enchantait en les vérifiant sur les espèces communes en attendant qu'il s'en offrît à moi de plus rares. La fourchure des deux longues *étamines* de la brunelle, le ressort de celles de l'ortie et de la pariétaire, l'explosion du fruit de la balsamine et de la capsule du buis, mille petits jeux de la *fructification* que j'observais pour la *première fois* me comblaient de joie, et j'allais demandant si l'on avait vu les cornes de la brunelle comme La Fontaine demandait si l'on avait lu Habacuc.[210]

He was not exaggerating. Hermaphrodite plants, which are the norm, contain flowers of both sexes, but the male flowers mature after the female ones to prevent autofecundation; orchids and palms imitate insect pollinators of the opposite sex so that the pollinator tries to copulate with the plant, thereby being covered in pollen; for this purpose the palm 'distille un parfum identique à celui qu'émet la femelle dans cette espèce'.[211] Some plants offer sweet nectars fragrant enough to attract insect pollinators from afar, and this is their sole purpose – a fact that some botanists were unwilling to grasp in the eighteenth century, and that was not universally accepted until much later.[212] In some plants, the colours of the 'centre producteur de nectar' are ones that only insect eyes can detect.[213] For a botanical neophyte such as Rousseau, this was riveting stuff; indeed, these 'mille petits jeux' continue to shock and amaze to this day, and are little known, except to specialists.

208. My emphasis; Rousseau to Jean André Deluc, 20 December 1764; *CC*, xxii.256.
209. These are the parts of the 'fruit-body', i.e. the fructification. Linnaeus, *Philosophia botanica*, (1751) sections 159, 165, 167.
210. My emphasis; Rousseau, *Rêveries*, *OC*, i.1043. Davis argues that the correct reference is not the Old Testament book of Habacuc, but Baruch: 'that is, a non-apocryphal for an apocryphal book of the Bible'. This substitution has a number of possible meanings. Davis, *The Autobiography of philosophy*, p.175-76.
211. Brosse, *La Magie*, p. 65.
212. Lorch, 'The discovery of nectar'. Many botanists held that nectar existed to nourish the plant!
213. Brosse, *La Magie*, p.61.

Rousseau's preoccupation with the 'parties sexuelles' of plants continued his long-standing interest in reproduction or *generation*, as it was known in the eighteenth century. This was *the* biological topic 'du jour', and any natural philosopher worth his salt was interested in it. Rousseau joined a veritable army of savants seeking to understand generation in insects, plants, and animals: Bonnet's aphids, Trembley's polyp, and Haller's chickens.[214]

Rousseau portrayed generation as a profound mystery: 'La seule génération des corps vivants et organisés est l'abîme de l'esprit humain.'[215] In plants it still was. Circa 1770, critical questions in plant reproduction remained unresolved: (1) whether sexual reproduction occurred, (2) the mechanics of sexual reproduction, and (3) the location of the plantlet.[216] Hermaphrodism was assumed to mean auto-fecundation, but this was later proven to be only rarely the case; insect pollination was falsely deemed exceedingly unreliable,[217] and various analogies drawn between the plant and animal kingdom were being investigated. Hence, both agamist and sexualist theories of plant generation remained on the table; convincing answers to these questions were not forthcoming until the late eighteenth and early nineteenth centuries.

Rousseau had concluded that chemistry did not offer any answers or insights to the mysteries of generation, because chemistry studies what is dead, not what is living. Chemistry's failure to address the question of generation was, as we have seen, one of the reasons why Rousseau rejected it. He wanted to observe the *living* plant, and perhaps dissect it, but he refused to crush it in an apothecary's mortar.[218]

Rousseau was undoubtedly aware of other methods and systems from the beginning of his botanical studies in Switzerland, where taxonomic pluralism was the rule rather than the exception. In Neuchâtel the systems of Tournefort and Haller were taught alongside Linnaean botany by Jean Antoine d'Ivernois and Abraham Gagnebin. Moreover, Laurent Garcin, who had brought Linnaeus to Neuchâtel in his pocket, had offered in his botany course '*une Méthode* claire, distincte, et aisée, [en suivant] les *meilleurs Systèmes* des Auteurs modernes. Il les *combinera en un seul*, autant qu'il lui sera possible, pour la rendre plus commode et plus intelligible.'[219]

Not surprisingly, Rousseau was already diversifying his reading in the

214. See Delaporte, *Nature's second kingdom*, p.91ff.

215. Rousseau, *Emile*, *OC*, iv.580; see also Rousseau, *Institutions chimiques*, p.100-101.

216. Delaporte, *Nature's second kingdom*, p.103.

217. Delaporte, *Nature's second kingdom*, p.130.

218. Rousseau, *Fragments de botanique*, *OC*, iv.1249-50.

219. My emphasis; 'Cours de botanique', p.145. At this writing, we lack the details of the method Garcin taught.

first flush of his enthusiasm for the sexual system. Six months after starting to study botany (mid-1765) he was seeking works by Tournefort and Vaillant, one an agamist, one a sexualist and both pre-Linnaean.[220] Throughout the 1760s Rousseau continued to acquire books, including Renaissance herbals, the works of Haller and contemporary works that presented hybrid methods of arrangement.[221] Furthermore, he personally sought out experts on alternative methods such as Claret de Latourrette, abbé Rozier, the Jussieus, André Thouin, Claude Richard (1705-1784), Antoine Gouan, Michel Adanson and Barbeu Dubourg (all discussed in succeeding chapters).

In 1769 Rousseau testified that his passion for 'le système de Linnaeus' was tempered 'après en avoir senti le vide'[222] and he speculated in a letter to the Montpellier botanist, Antoine Gouan, that John Ray, a father of the natural method, most closely approached an elementary method for learning botany.[223] Rousseau likewise developed a fascination with plant *habit*, a key concept of the natural method that perhaps rivalled his initial fascination with reproduction. The 'disciple' of Linnaeus, therefore withheld unqualified approval of the artificial system of his 'maître'.[224]

Yet just as Rousseau seemed to be distancing himself from the sexual system of his 'maître', he was also singing Linnaeus's praises to a visiting Swede.[225] He even wrote an admiring letter to Linnaeus himself: 'je tire un profit plus réel de votre *Philosophia botanica* que de tous les livres de morale [...] continuez d'ouvrir et d'interpréter aux hommes le livre de la nature [...] je vous lis, je vous étudie, je vous médite, je vous honore et je vous aime de tout mon cœur'.[226] How could Rousseau both revere Linnaeus and pursue other systems of classification besides the sexual system?

The answer may be quite simple. There were in effect two Linnaeuses: the Linnaeus of the sexual system and the Linnaeus of the natural

220. Rousseau had 'un extrême besoin de deux livres de botanique, l'un de Tournefort, en trois volumes in-4°, intitulé *Institutiones rei herboriae*; ou [...] le livre du même auteur intitulé *Elemens de botanique*; l'autre livre est de M. Vaillant, intitulé *Botanicon parisiense* in-folio; [...] ne laissez pas, je vous en prie, de me les envoyer le plus tôt que vous pourrez en me marquant le prix'. Rousseau to Duchesne, 28 April 1765; *CC*, xxv.197. Rousseau's acquisition of these and other botanical works is discussed in detail in Chapter 7.

221. See for example Rousseau to Duchesne, 30 June 1765; *CC*, xlix.299-300. In this letter Rousseau requests not only two key works of Linnaeus, *Genera plantarum* and *Philosophia botanica*, but also Garsault's *Les Figures de plantes et animaux d'usage en médecine*, 6 vols (Paris, Didot le jeune, 1764-1765) and the two works on Swiss flora by Haller, *Iter helveticum* and *Enumeratio*. See Appendix 1.

222. Rousseau, *Confessions*, *OC* i.643. For the dating of this text, see 'Chronologie' in *OC*, i.cxv.

223. Rousseau to Gouan, 26 December 1769; *CC*, xxxvii.195

224. Björnståhl to Gjörwell, 1 September 1770; *CC*, xxxviii.93.

225. Björnståhl to Gjörwell, 1 September 1770; *CC*, xxxviii.93-94.

226. Rousseau to Linnaeus, 21 September 1771; *CC*, xxxviii.267.

method; these were of course one and the same man, who exposed his key theories in one great *summa*, the *Philosophia botanica* of 1751. In that work he gave a final formulation to the artificial sexual system, while also presenting the main principles of its successor, the natural method, the orders it would contain (section 77), the habitual characters it would use, and in what way (sections 163 and 168). This was taken further by Adanson, the Jussieus and others, whose contributions and influence on Rousseau I discuss in the next chapter. For Rousseau, I argue, Linnaeus was a genius and a philosopher precisely because his work comprehended both the artificial sexual system and the natural method.

ix. Conclusion

Contrary to the received view, therefore, an interest in generation hardly limited Rousseau or any of his contemporaries to the artificial sexual system; the reproductive organs continued to figure prominently in the natural method of Antoine Laurent de Jussieu, the subject of the next chapter.[227] By the early 1770s Rousseau was actively employing the concept of plant habit and the families of the natural method, which he applied in the *Lettres élémentaires*. Rousseau could thereby be termed a taxonomic pluralist for whom different systems of classification are equally legitimate.[228] Rousseau's early enthusiasm for the sexual system should therefore be understood as a *phase*, not as a loyalty in which he remained ideologically frozen until his death in 1778.

227. Stevens, *Development of biological systematics*, p.33; Jussieu, 'Exposition d'un nouvel ordre des plantes', p.179-80.

228. Bird and Tobin, 'Natural kinds'. Rousseau's metaphysics, on the other hand, was monist; in other words, he believed there was *one* natural order that could be carved up in different ways. The idea of 'carving nature at its joints' goes back to Plato. Ereshefsky, *The Poverty of the Linnaean hierarchy*, p.17.

6. 'Se tracer un plan à sa guise':[1] Rousseau and the natural method

> [L]a Botanique, [est] non une Science de mémoire et de nomenclature, mais une *Science nouvelle*, qui ait ses combinaisons et ses affinités comme la Chimie, ses problèmes comme la Géométrie.[2]
>
> Inspirées par le bon sens et le génie de Jussieu, [les *Lettres élémentaires sur la botanique*] sont encore aujourd'hui ce que les commençants peuvent lire de plus clair, en abordant l'étude de la Botanique.[3]

i. Toward a natural method

Since the publication of the *Philosophia botanica* in 1751, no serious botanist believed that the sexual system could be the final classification system for all time, or that it would provide the solution to information overload. The system's limitations and transitory nature were understood better than anyone by its creator; Linnaeus sounded the system's death knell and the coming of its successor, the natural method. Indeed, Linnaeus 'was the first to draw a clear terminological distinction between artificial and natural systems, which was praised by later naturalists'.[4] One of Rousseau's nineteenth-century editors distinguished system from method in the following way:

> Le *système* est un arrangement, un *ordre général* fondé partout sur les mêmes

1. Jacques Barbeu Dubourg, *Le Botaniste françois: comprenant toutes les plantes communes et usuelles, disposées suivant une nouvelle méthode, et décrites en langue vulgaire*, 2 vols (Paris, chez Lacombe, Libraire, 1787), vol.1, p.144.
2. My emphasis; Antoine Laurent de Jussieu, 'Examen de la famille des renoncules', *Histoire de l'Académie royale des sciences. Année M.DCCLXXIII. Avec les mémoires de mathématique et de physique pour la même année, tirés des registres de cette académie* (Paris, de l'Imprimerie Royale, 1777), p.216. This is an early version of Antoine Laurent de Jussieu's natural method.
3. Augustin Pyramus de Candolle, *Histoire de la botanique genevoise* (Geneva & Paris, 1830), p.20. Candolle's *Théorie élémentaire de la botanique, ou, Exposition des principes de la classification naturelle et de l'art de décrire et d'étudier les végétaux* (Paris, 1813), in which he expounds his theory of classification, is considered the foundation of modern plant morphology, and is still used today.
4. Müller-Wille, 'Collection and collation', p.550. There was considerable imprecision, however, in how authors used these terms. See Stevens, *Development of biological systematics*, p.11-13.

principes, soit que l'auteur ne fasse usage *que d'une seule partie*, soit qu'il emploie un petit nombre de parties qui aient entre elles une analogie bien marquée; tel est le système de Linné. La *méthode*, au contraire, est un arrangement fondé sur des principes moins fixes, moins déterminés, et dont l'auteur peut s'écarter toutes les fois qu'il le juge nécessaire ou avantageux pour remplir l'objet qu'il se propose; telle est la *méthode florale* de Tournefort.[5]

In contrast with the sexual system, which classified plants based on the number and position of the stamens and pistils, the natural methods were based on a range of plant parts and characteristics, although how this was exactly done varied from one author to another.[6] These methods acknowledged a number of family groupings above the level of species and genera – many of them long-recognised in folk taxonomy; these families were developed further not only by Linnaeus, but most notably, by the French botanists, Michel Adanson, and Bernard and Antoine Laurent de Jussieu.[7]

Linnaeus had in fact offered the sexual system only as a way-station on a longer journey toward a natural method. It has also been suggested that he:

probably intended [the artificial system] as a mere substitute for the Natural System, for which he found the world in his day unprepared, to be relinquished as soon as the principles of the latter could be settled: as, indeed, is obvious from his writings, in which he calls the Natural System *primum et ultimum in botanicis desideratum*. He could not have expected that his artificial system should exist when the science had made sufficient progress to enable botanists to revert to principles of natural arrangement, the temporary abandonment of which had been solely caused by the difficulty of defining its groups.[8]

By his own admission, the ultimate desideratum of botany would be a natural system: '[n]ullum Systema Plantarum Naturale, licet unum vel alterum propius accedat, adhucdum constructum est; nec ego heic Systema quoddam Natruale contendo (forte alia vice ejus Fragmenta exhibebo); necque Naturale construe potuit, antequam omnia, ad nostrum System pertinentia notissima sunt. Interim tamen Systemata artificiala, defectu Naturalis, omnino necessaria sunt'.[9] He later con-

5. My emphasis; J.-J. Rousseau, *La Botanique de J.-J. Rousseau*, ed. Deville, p.247.
6. For example, Antoine Laurent de Jussieu took the cotyledon as the primary character, drawing on the work of the seventeenth-century British botanist, John Ray.
7. The Jussieu botanical dynasty originated in Lyon. See Appendix 2.
8. Lindley, *A Natural system of botany*, p.vi-vii; emphasis in original.
9. 'No natural system of plants [...] has so far been constructed, nor do I contend that this system is really natural (perhaps some other time I may issue the fragments of one); nor can it become a natural system before all details in connection with our system become known. In the meantime, however, as long as a natural system is lacking, artificial systems

firmed this position, 'Methodus Naturalis hinc ultimus finis Botanices est et erit'.[10] Linnaeus's point was not lost on his adherents – even the staunchest adherents of the sexual system admitted that the artificial sexual system might one day reach its use-by date.[11]

However radically the sexual system might seem to differ from a natural method, there were ways to assimilate the artificial system to the natural one: '*Fructificatio*, recentiorum inventum, viam primario aperuit ad Methodum naturalem, sed Haec etiamnum ita intelligitur, ut omnes classes detegat.'[12] He expected that his artificial system of sexual classification would eventually be superseded by the natural method which would arrange plants in a completely different way than the sexual system had done: 'METHODI NATURALIS: Fragmenta studiose inquirenda sunt. Primum et ultimun hoc in Botanicis desideratum est. Natura non facit saltus. Plantae omnes utrinque affinatatem monstrant, uit Territorium in Mappa naturae geographica'.[13]

Hence, to understand Linnaeus and his legacy solely in the context of plant sexuality is to misconstrue and underestimate his contribution to botany: 'Linnaeus himself regarded the sexual system as artificial, not for accidental, but for systematic, reasons. Every new species discovered could, and in many cases actually did, prompt its failure by exhibiting characteristics that were represented inadequately by the sexual system.'[14] His was therefore not a scholastic, hermetically sealed system, as has been alleged, but rather, an open-ended, inductive and empirical project of collation, compilation and revision. Historians of science have shown that Linnaeus was not a hidebound traditionalist, propounding Aristotelian essentialism, as has often been suggested, but rather, an

will definitely be needed.' Linnaeus, *Systema naturae*, part 4, section 12, p.23 (the Latin section is unpaginated). Foucault gives a different interpretation of this passage: *Les Mots*, p.153.

10. Linnaeus, *Philosophia botanica* (1751), p.137, section 206; '[T]he natural method is the ultimate end of botany, and will continue to be so'. Linnaeus, *Philosophia botanica* (2003), p.149, section 206.

11. Martyn, *Letters*, p.15, n.(r).

12. Linnaeus, *Philosophia botanica* (1751), p.101, section 163. 'The *fruit-body* [...] first opened the way to the natural method, but even now this is not so well understood as to discover all the classes'. The emphasis is original; Linnaeus, *Philosophia botanica* (2003), p.115, section 163, a disquisition on habit, of great relevance to the natural method.

13. Linnaeus, *Philosophia botanica* (1751), p.27, section 77. The metaphor of plant groups contiguously arranged on a map dominated late eighteenth-century botanical thought: 'The fragments of the NATURAL METHOD are to be sought out studiously. This is the beginning and end of what is needed in botany. Nature does not make leaps. All plants exhibit their contiguities on either side, like territories on a geographical map.' Linnaeus, *Philosophia botanica* (2003), p.40, section 77. This metaphor denoted 'a multidimensional network of taxonomic relations'. Müller-Wille, 'Collection and collation', p.553.

14. Müller-Wille, 'Collection and collation', p.559.

innovator.[15] Understood in this light, the artificial sexual system was a provisional information retrieval system.[16] Yet, while well-known at the time, Linnaeus's commitment to the natural method is barely acknowledged in recent accounts of the history of botany that paint him as wedded to a gendered, hierarchical science; this commitment is relegated to the footnotes, if acknowledged at all.[17]

In progressing toward the natural method, Linnaeus revived the concept of habit, an important notion for Tournefort.[18] In French habit is called 'air de famille' or 'port', and in Latin facies propria[19] or habitus plantae:[20]

> 163. HABITUS est conformitas quaedam Vegetabilium affinium et congenerum, in *Placentatione, Radicatione, Ramificatione, Intorsione, Gemmatione, Foliatione, Stipulatione, Pubescentia, Glandulatione, Lactescentia, Inflorescentia, aliisque.*
> *FACIES EXTERNA* vulgo a Botanicis antea dictus est Habitus.
> *C. Bauhinus & Veteres egregie* ex Habitu plantarum divinarunt earundem affinitates, ut ipsi Systematici sæpius deflexerint, ubi Habitus recta duxerat.
> Methodus naturalis est ultimus finis Botanices (§77).
> *Fructificatio*, recentiorum inventum, viam primario aperuit ad Methodum naturalem, sed nec Haec etiamnum ita intelligitur, ut omnes classes detegat.[21]

15. Müller-Wille, 'Collection and collation', p.542. Against the view that Linnaeus was an essentialist, see Mary P. Winsor, 'Linnaeus's biology was not essentialist', *Annals of the Missouri botanical garden* 93 (2006), p.2-7.
16. Müller-Wille, 'Collection and collation', p.559. On this account, Linnaeus might now be considered a 'promiscuous realist' in respect to natural kinds, since he was committed to naturalism and realism, but recognised the possibility of carving up the plant world in different ways. Bird and Tobin, 'Natural kinds'. Stevens, who argues that for Linnaeus, Lamarck and Jussieu '[e]ven a natural classification would have seemed an oxymoron to them; there was nowhere to divide nature.' *Development of biological systematics*, p.256.
17. Schiebinger, *Nature's body*, p.17, 216, n.8; see also Schiebinger, 'Gender and natural history', in *Cultures of natural history* , ed. N. Jardine, J. A. Secord and E. C. Spary (Cambridge, 1996), p.163-77 (170-73). In this essay Linnaeus is portrayed as having developed only the sexual system.
18. Linnaeus earlier condemned botanists' recourse to habit in 'Prefatio', *Genera plantarum*, ¶11. See Müller-Wille and Reeds, 'A translation', p.567, ¶11.
19. 'Facies' is the Latin term for 'the appearance or expression of the face'. *Oxford English dictionary* online: 'facies'.
20. These are synonyms, the last three used by Tournefort, Latourrette and Rozier, *Démonstrations élémentaires*, p.27.
21. Linnaeus, *Philosophia botanica* (1751), p.101, section 163; emphasis in original. 'HABIT is a certain conformity of vegetables that are allied and related in *placentation, radication, ramification, intorsion, gemmation, foliation, stipulation, pubescence, glandulation, lactescence, inflorescence* and other things. Formerly, *the EXTERNAL APPEARANCE* was commonly called habit. *C(aspar)* Bauhin and the ancients have excellently divined the affinities of plants from their habit, so that the systematists themselves have quite often gone astray, where habit had drawn the right [conclusions]. The natural method is the ultimate purpose of

Latourrette and Rozier described habit as the 'caractère habituel':

> Il consiste dans la conformation générale d'une plante, considérée suivant le résultat et l'ensemble de toutes ses parties, dans leur position, dans leur accroissement, dans leurs grandeurs respectives, et tous autres rapports qui les rapprochent ou les différencient entre elles. On peut le comparer à la *physionomie* qui résulte de toutes les modifications des traits du visage.[22]

The habit played an important role in the natural method: 'French taxonomists, under Bernard de Jussieu's lead, were intent on following the Linnaean approach to the natural system through the use of the *habitus*, rather than using the sexual system based on a single part'.[23] In 1789 Antoine Laurent de Jussieu wrote, 'plantae habitu simillimae caractere et naturâ conformes habentur et vice versâ'.[24] The notion of habit likewise served as a unifying principle of the *Lettres élémentaires*.

It is noteworthy that Linnaeus presented the study of habit as an esoteric activity: '168. Habitus (163) *occulte* consulendus est, ne genus erroneum lævi de caussa fingatur. Experientia rerum magistra, primo intuitu ex facie externa, plantarum familias sæpius divinat. Exempla confirmant regulam: Isopyrum, Nigella, Helleborus, Caltha diversa. Sambucus et Ebulus; Trifolium et Triphylloides comungenda.'[25] That being said, the study of habit was in fact habitual for Linnaeus, as his son recalled: 'Ce n'était que l'habitude de mon père de reconnaître les plantes par leur aspect extérieur.'[26] This practice did not contradict *Philosophia botanica*, section168, for Linnaeus believed that experts should interpret nature for the unlearned – they and only they should consult the habit in private.[27]

The esoteric teaching of the habit thus pointed the way toward some version of the natural method, as conceived by a number of botanists.[28]

---

botany (section 77). The *fruit-body*, a thing come upon by comparatively modern (botanists), first opened the way to the natural method, but even now this is not so well understood as to discover all the classes.' Original emphasis; Linnaeus, *Philosophia botanica* (2003), p.115, section 163.

22. Rozier and Latourrette, *Démonstrations élémentaires*, p.27-28; emphasis in original.
23. Original emphasis; Phillip R. Sloan, 'Buffon, German biology', p.147, n.56; see also p.147-48, n.57.
24. '[P]lants most similar in *habit* will agree in character and nature and vice versa.' Jussieu, *Genera plantarum*, p.xix; trans. S. Rosa in Stevens, *Development of biological systematics*, p.339.
25. My emphasis. Linnaeus, *Philosophia botanica* (1751), p.117, section 168. 'The habit (163) should be carefully studied in *private*, to prevent the contriving of an erroneous genus for insufficient cause. Experience, the mistress of affairs, quite often guesses the families of plants at first sight from their external appearance. Examples confirm the rule: Isopyrum, Nigella, Helleborus and Caltha are different. Sambuccus and Ebulus; Trifolium and Triphylloides, should be linked.' Linnaeus, *Philosophia botanica* (2003), p.130, section 168.
26. Quoted in Hoquet, *Buffon-Linné: éternels rivaux*, p.57.
27. Lorraine Daston and Peter Galison, *Objectivity* (New York, 2007), p.59.
28. 'Un grand nombre de naturalistes hésitaient pourtant à se déterminer pour l'une ou

This contribution was acknowledged by none other than one of the main authors of that method, Antoine Laurent de Jussieu:

> Quelques Savants très estimés, ont déjà ébauché cet ouvrage; je ne citerai que les principaux. M. Linnaeus a donné ses ordines naturales; M. Bernard de Jussieu, mon oncle, a établi, dans le Jardin du Trianon, une suite de familles naturelles, qui n'ont que des rapports éloignés avec l'arrangement du botaniste suédois. M. Adanson, dans ses *Familles des plantes*, suit un plan différent de l'ordre de Trianon, mais qui en approche plus que celui de M. Linnaeus.[29]

Jussieu therefore considered Linnaeus to have contributed to the development of the natural method, even if it was not the same method as that developed by other botanists such as Jussieu's uncle, Bernard de Jussieu, father of an important natural method of plant classification. The elder Jussieu taught his method at the Jardin du roi (Paris) beginning in the 1740s and in 1759 arranged the Jardin du Trianon at Versailles according this method.[30] Bernard de Jussieu's 'natural orders' were published for the first time in 1789 by Antoine Laurent de Jussieu in *Genera plantarum*.[31] However, the younger Jussieu went beyond his uncle's method in the arrangement he put in place in 1774 at the Jardin du roi:[32] 'Haec dudum quaesita ordinatio, inter omnes longè praecipua, sola verè uniformis ac simplex, affinitatem legibus constanter obtemperans, est *methodus* dicta *naturalis*.'[33] The method of Antoine Laurent de Jussieu was elaborated further by the Genevan botany professor, Augustin Pyramus de Candolle.[34]

---

l'autre des méthodes'; 'On voit aussi paraître des ouvrages qui combinent les deux méthodes de Linnaeus et de Jussieu'; Pascal Duris, *Linné et la France* (Geneva, 1993), p.153, 154. These authors can therefore be regarded as 'promiscuous realists' in current philosophical usage.

29. Jussieu, 'Examen de la famille des renoncules', p.216; emphasis in original.
30. See Michel Guédès, 'Duchesne, Buisson, Durande, early followers of the natural method of the Jussieus', *Taxon* 22:2-3 (May 1973), p.211-19 (211). Under the patronage of Louis XV, a keen naturalist, the Trianon garden housed the largest number of plant species in France. Marie Antoinette transformed it into a landscape garden.
31. 'Ordines naturales in horto Ludovici XV Trianonensi dispositi anno 1759', in Jussieu, *Genera plantarum secundum ordines naturales disposita, juxta methodum in horto regio parisiensi exartam, anno M.DCC. LXXIV* (Paris, Herissant and Barrois, 1789), p.lxiii-xx. For a translation, see Stevens, *Development of biological systematics*, p.382-84.
32. Antoine Laurent de Jussieu, 'Exposition d'un nouvel ordre des plantes adopté dans les démonstrations du jardin royal', *Histoire de l'Académie royale des sciences. Année M.DCCLXXIII. Avec les mémoires de mathématique et de physique pour la même année, tirés des registres de cette académie* (Paris, de l'Imprimerie royale, 1778), p.175-97.
33. 'This long-desired arrangement, far superior to all others, alone truly uniform and simple, always in conformity with the laws of affinities, is the so-called *natural method*'. Original emphasis; Jussieu, *Genera plantarum*, p.xxxv; trans. S. Rosa in Stevens, *Development of biological systematics*, p.355.
34. Candolle, *Théorie élémentaire de la botanique*.

The sexual system of Linnaeus relied on a small set of characters common across a very large number of plants, for example the parts of the fruit-body – the stamens and pistils – while a natural method was built on 'une analogie fondée sur la ressemblance de *plusieurs parties*, et le caractère de la famille devient le fruit d'une spéculation qui équivaut à celles des Sciences les plus abstraites'.[35] Such a method therefore grouped those plants 'related in a very large number of characters',[36] rather than those sharing a few characters common to all plants. A natural system is therefore *holistic* rather than *selective*:

> On a donc cherché des caractères plus solides encore, plus constants, plus généraux. On les a nommés caractères naturels. Ils ont été tirés de *l'ensemble* et de la *combinaison* des parties les plus essentielles de la végétation; la fleur, le fruit, la graine, la disposition des tiges et des branches, etc. Tous les divers accidents de chacune de ces parties, rapprochés et comparés, ont conduit à des divisions naturelles et déterminées.[37]

Antoine Laurent de Jussieu chose the cotyledon as the primary character to differentiate the plant families, a choice that harked back to John Ray in the seventeenth century. Cotyledons are the seedling leaves, which are absent in ferns, singular in onions, double in beans and several in pines.[38] From this primary division Jussieu advanced by subordination of characters to the number of petals, and finally to the insertion of the stamens (the male reproductive organs), calyx, and corolla below the ovary (hypogynous), borne on top of it (perigynous) or joined together at a common pont of insertion, usually below the ovary (epigynous) (fig.6.1, p.80).[39] Jussieu therefore assigned a key role to the reproductive organs:

> Le pistil et les étamines sont donc les parties *essentielles et principales*; elles concourent à former un individu nouveau qui commence dans la germination une nouvelle vie [...] La reproduction des êtres est le complément des ouvrages de la Nature, le plus haut point où elle puisse atteindre [...] L'appareil de la fécondation, l'importance des organes qui l'opèrent immédiatement [...] tout annonce que la *graine* formée à si grands frais est la partie essentielle par excellence, le centre des divisions primitives de l'ordre naturel.[40]

'As with Linnaeus, characters of the fructification were privileged; the most essential characters were considered to be those involved in the

35. My emphasis; Jussieu, 'Examen de la famille des renoncules', p.216.
36. Stevens, *Development of biological systematics*, p.35, quoting Jussieu, *Genera plantarum*, p.xxxix.
37. My emphasis; Latourrette and Rozier, *Démonstrations élémentaires*, p.11.
38. Stevens, *Development of biological systematics*, p.33.
39. In a posthumously published, revised introduction to his *Genera plantarum*, Antoine Laurent de Jussieu refers to these characteristics as bundles, or fasces. Stevens, *Development of biological systematics*, p.388.
40. My emphasis; Jussieu, 'Exposition d'un nouvel ordre des plantes', p.179-80.

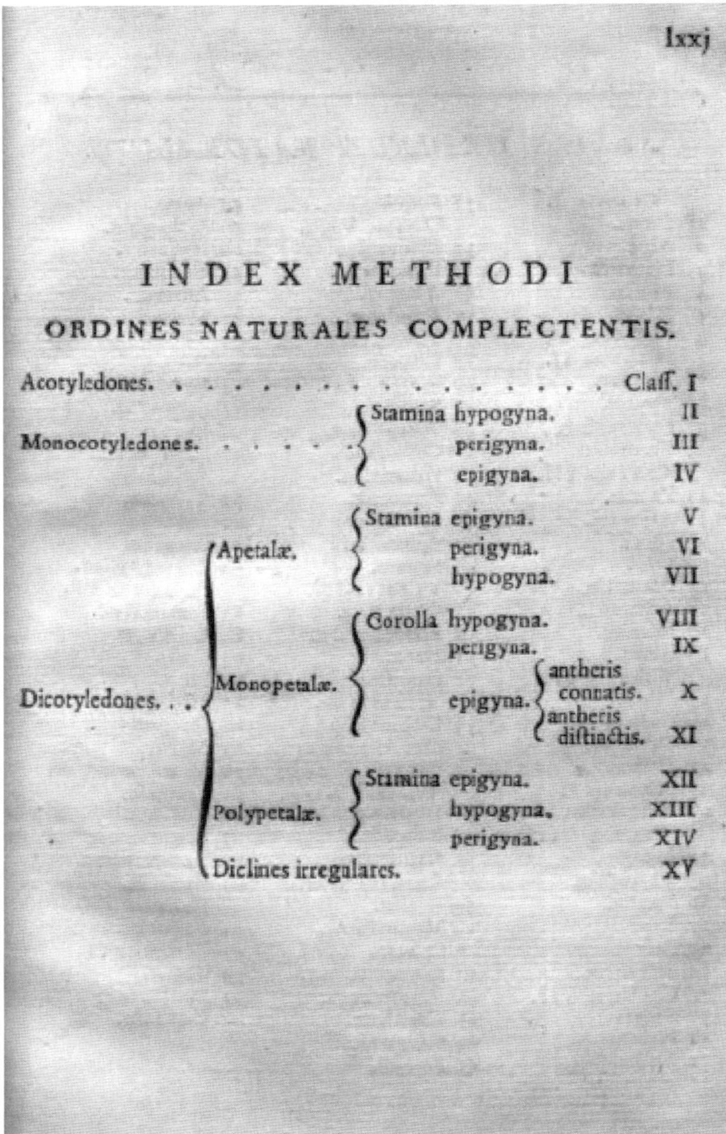
lxxj

INDEX METHODI

ORDINES NATURALES COMPLECTENTIS.

Acotyledones.				Claſſ. I
Monocotyledones.		Stamina hypogyna.		II
		perigyna.		III
		epigyna.		IV
Dicotyledones.	Apetalæ.	Stamina epigyna.		V
		perigyna.		VI
		hypogyna.		VII
	Monopetalæ.	Corolla hypogyna.		VIII
		perigyna.		IX
		epigyna.	antheris connatis.	X
			antheris diſtinctis.	XI
	Polypetalæ.	Stamina epigyna.		XII
		hypogyna.		XIII
		perigyna.		XIV
	Diclines irregulares.			XV

Figure 6.1: Antoine Laurent de Jussieu, *Genera plantarum* (Paris, Herissant and Barrois, 1798), p.lxxi, courtesy of the Biodiversity Heritage Library (online).

most important function of the plant: reproduction.'[41] Jussieu's method incorporated elements from various antecedent systems of classification:

> on y retrouve toutes les familles et les divisions primitives du Trianon; les plantes apétales, monopétales, et polypétales y sont distinguées comme dans la méthode de M. Tournefort. De plus étant fondée *en partie* sur les *étamines*, elle se rapporte en cela au *système* de M. *Linnaeus*. Les genres de cet Auteur, quoique trop minucieux [...] sont cependant les *meilleurs* [...] les espèces sont les *mieux* déterminées; sa nomenclature par les triviaux soulage la mémoire, ses phrases descriptives donnent une idée de la plante décrite. De ses divers ouvrages, celui-là est le meilleur et le plus utile; n'ayant rien de mieux à proposer, nous croyons devoir adopter ses genres, ses espèces, sa nomenclature.[42]

Jussieu, whose 'touchstone' was continuity, was also indebted to Linnaeus and others for the concept of plants arranged contiguously on the 'mappa naturae':

> Les plantes répandues sur le globe paraissent devoir former entr'elles une *chaîne continue*, dont les deux extrêmes sont l'herbe la plus petite et l'arbre le plus élevé. Par une *gradation insensible* on s'élèvera de l'une à l'autre en disposant de suite celles dont l'affinité est marquée par un plus grand nombre de rapports. Cet ordre, qui est *celui de la Nature* [...] il présente une utilité plus réelle. Le raisonnement, appuyé de l'expérience, démontre que les plantes conformes dans leurs caractères, jouissant aussi des mêmes propriétés, de sorte que l'ordre naturel une fois donné, on pourrait déterminer leurs vertus pars des signes extérieures.[43]

The seven families he considered fundamental to his work exemplified this concept.[44]

The philosophical-metaphysical standpoint of the 'gradation insensible' was rich in implications. It fitted into a larger idea, that of the Great Chain of Being, according to which all the beings in nature were thought to be continuous, as if connected by a chain.[45] This idea was certainly

41. Stevens, *Development of biological systematics*, p.33.

42. My emphasis; Jussieu, 'Exposition d'un nouvel ordre des plantes', p.193. Similarly: 'Among the various Authors who have labored successfully in the investigation of a natural method, the principal ones are Linnaeus, Bernard de Jussieu and Adanson'. Jussieu, *Genera plantarum*, p.xxxv; trans. S. Rosa in Stevens, *Development of biological systematics*, p.355.

43. My emphasis; Antoine Laurent de Jussieu, 'Exposition d'un nouvel ordre des plantes', p.175. Jussieu goes on to remark that the ancients lacked an idea of the natural order and the 'caractères sensibles' of plants because they were preoccupied with the 'vertus des plantes'. Rousseau echoed this comment in 'Introduction', *OC*, iv.1201.

44. Jussieu, 'Exposition d'un nouvel ordre des plantes', p.178. Stevens, *Development of biological systematics*, p.27-28. Jussieu's seven fundamental families are discussed below (p.196 and following) in the context of the families Rousseau discusses in the *Lettres*.

45. The 'chain' was a common way of theorising nature, often with theological and/or teleological implications. See Arthur O. Lovejoy, *The Great chain of being: a study of the history of an idea* (Cambridge, MA, 1936).

familiar to Rousseau, who spoke of a '*chaîne* de rapports' before he knew the Jussieus and their views.[46] Ultimately, there could be no gaps in the chain, and therefore no discrete groups, as was the case in the sexual system.[47]

In the 1760s, when Rousseau began to study botany, the natural method was therefore still under construction, and some of the most important eighteenth-century treatises, especially those of Antoine Laurent de Jussieu, had not yet been published. However, the natural method was already sufficiently well-known that botanists such as Latourrette and Rozier discussed it in their textbook of 1767. It is important to bear in mind that the emergence of the natural method as an important rival to the sexual system did not eliminate the reproductive organs from classification; to the contrary: the natural method also considered the apparatus of reproduction indispensable.

Late eighteenth-century botany was turning its back on the sexual system while preserving the reproductive apparatus as a key character of the newer methods: 'Linnaean binomials formed the inscriptions in the book of nature, but the Linnaean [artificial sexual] system as its table of contents was a failure [...] it was the so-called natural system that was to guide the reader around the book.'[48]

ii. Rousseau and the progenitors of the natural method

By the time he wrote Book XII of the *Confessions* in 1769, Rousseau had come to recognise the 'vide' of the sexual system, although he was not able to cure himself of his passion for it.[49] Even before taking up the natural method in the *Lettres*, Rousseau indicated the direction of his thought in response to the popular textbook by his Lyon friends, Latourrette and Rozier: *Démonstrations élémentaires pour l'usage de l'école vétérinaire de Lyon*; he found this work

> [T]rès bon, mais [...] il suit les systèmes de Tournefort et de Linnaeus, au lieu qu'il faudrait un *système élémentaire* pour commencer l'étude des plantes, avant de prendre les grands systèmes de les classer. Celui de Linnaeus apprend à mieux observer les plantes que l'on connait, mais il n'apprend pas

46. My emphasis; Rousseau, *Confessions*. *OC*, i.641. In the *Lettres* he portrayed an '*enchaînement* de merveilles qui tiennent tout esprit sain qui les observe dans une continuelle admiration.' My emphasis; Rousseau to Mme Delessert, 2 May 1773; *CC*, xxxix.157.
47. See Stevens, *Development of biological systematics*,p.256. Stevens argues that the theory of continuity precluded carving up nature at all, which disagrees with my claim that Linneaus and Rousseau were promiscuous realists, prepared to carve up nature at the joints for various purposes.
48. Stevens, *Development of biological systematics*, p.267.
49. Rousseau, *Confessions*, *OC*, i.643.

à les connaître. C'est un système pour les maîtres, il nous en faudrait un pour les écoliers.[50]

By late 1769 Rousseau was already thinking about how to do for botany what he had done for music; botany, like music, could be made more accessible and easier to learn:[51]

> L'analyse de la fructification me fatigue, me rebute, m'épuise, et je m'y perds. Je crois les méthodes des botanistes très bonnes pour classer les plantes *déjà connues*; mais je suis persuadé qu'il en faut une autre pour les *étudier*, et voilà ce que vous autres, savants, qui n'écrivent que pour vos semblables, n'avez jamais su voir jusqu'ici. *Ray* me paraît avoir approché plus qu'aucun autre de cette méthode élémentaire.[52]

He believed this approach would entail a more gradual and less abstract method than that of Tournefort or Linnaeus; as Rousseau wrote only a month later, in January 1770: 'J'avoue [...] que les difficultés que j'ai trouvées dans l'étude des plantes m'ont donné quelques idées sur les moyens de la *faciliter* et de la rendre utile aux autres, en suivant le fil du Système végétal par une *méthode plus graduelle* et *moins abstraite* que celles de Tournefort et de tous ses successeurs sans en excepter *Linnaeus* lui-même.'[53]

Thus, in 1770 the way forward was by a natural method rather than by an artificial system; we can safely assume that well before encountering the main practitioners of the natural method in Paris in the 1770s, Rousseau had already consulted the outline of the natural method given by Linnaeus in sections 77, 163 and 168 of the *Philosophia botanica*. Surveying his subject in this fashion, Rousseau felt at liberty to 'tracer un plan à sa guise', drawing upon alternatives to the sexual system and developing a botanical pedagogy aligned with cutting-edge ideas.

Following the precepts of *Emile*, Rousseau focused on things themselves, not their *names*,[54] which are purely arbitrary, and exist independently of

50. My emphasis; Rousseau to Clappier, 23 December 1768; *CC*, xxxvi.215.
51. Cook, 'The languages of music and botany', *passim*.
52. My emphasis; Rousseau to Gouan, 26 December 1769; *CC*, xxxvii.195. Rousseau means the British botanist, John Ray, and his classification according to the cotyledons. Rousseau owned a copy of Ray's important theoretical work, *Methodus plantarum nova* (London, Henry Faitborne & John Kersey, 1682), and annotated his copy of Ray's *Synopsis* exhaustively (see fig.5.4, p.154).
53. My emphasis; Rousseau to Claret de Latourrette, 26 January 1770; *CC*, xxxvii.212. Leigh considered this letter evidence for Rousseau's authorship of *Fragmens pour un dictionnaire*: 'C'est à vous qu'il faut renvoyer toutes les exhortations que vous me faites sur l'entreprise d'un Dictionnaire de botanique'. Leigh, 'Notes explicatives', *CC*, xxxvii.214, n.(c); Ducourthial, *La Botanique selon Jean-Jacques Rousseau*, p.294. However, I argue in Chapter 9 that the authorship of the *Fragmens pour un dictionnaire* is uncertain. See also Rousseau to Clappier, 23 December 1768; *CC*, xxxvi.215.
54. '[I]l n'y a point de science de mots'. Rousseau, *Emile*, *OC*, iv.350; see also p.447.

the system or method of classification.[55] The sense by which shape, size, number and position of plant parts will be apprehended is *sight*: 'Avant de leur apprendre à nommer ce qu'ils voient, commençons par leur apprendre à le voir. Cette science oubliée dans toutes les éducations doit faire la plus importante partie de la leur.'[56] This procedure complements the epistemological distinction between primary and secondary qualities (discussed above, p.161-62), which Rousseau applied in his works, following mainstream epistemology, as well as Linnaeus.

Rousseau not only studied the major systems and methods with care, but also enjoyed privileged access to some of their most important progenitors, namely the Jussieus, and possibly Michel Adanson,[57] whose *Familles des plantes* (1763-1764)[58] figures in Rousseau's annotations of *La Botanique* of Regnault, and was found in his small botanical library at Ermenonville.[59] While this facet of Rousseau's botanical formation is largely absent from the literature,[60] it was well-known in the 1770s that Rousseau botanised:[61] 'après avoir poursuivi mille chimères, finis, comme J.-J. Rousseau, par herboriser'.[62] It was also known that he counted Bernard de Jussieu and 'le neveu', Antoine Laurent de Jussieu,

55. 'Il est vrai que les noms sont arbitraires', 'Introduction', *OC*, iv.1209. I discuss Rousseau's nominalism in Chapter 7, p.237.
56. Rousseau to Mme Delessert, 16 July 1772; *CC*, xxxix.82. Rousseau emphasises the cultivation of the sense of sight throughout the *Lettres*, following the precepts in *Emile*, *OC*, iv.391, 396, 447-51. See also Martin Rueff, 'Apprendre à voir: l'optique dans la théorie de l'homme', in *Rousseau et les sciences*, ed. B. Bensaude-Vincent and B. Bernardi (Paris, 2003), p.193-214 (206-207).
57. Rousseau sought contact with Adanson in 1769. Coindet to Rousseau, 9 May 1768, *CC*, xxxv.291.
58. Adanson, *Familles des plantes*. Adanson lived in the Jussieu household for ten years, and learned Bernard de Jussieu's method. He claimed Antoine Laurent had appropriated the differential weighting of characters from his (Adanson's) work without attribution. Sloan, 'Buffon, German biology', p.147-48, n.57. Adanson rejected binomial names as well as the sexual system.
59. See for example art. 218, 246 and 251, BdAN [Munch MS]. Rousseau's comments suggest a good knowledge of Adanson's work.
60. See for example Ducourthial, *La Botanique*, p.86-88; Williams, *Botanophilia in eighteenth-century France*, p.95. Giovanni Incorvati, 'Seul avec la nature et vous: le promeneur solitaire rencontre Linné et ses disciples', *Annales de la société Jean-Jacques Rousseau* 42 (1999), p.281-326.
61. Louis P. de Bachaumont, *Mémoires secrets pour servir à l'histoire de la république des lettres en France depuis MDCCLXII jusqu'à nos jours ou journal d'un observateur* (London, John Adamson, 1777), vol.5, p.170. However, Bachaumont's account appeared considerably later than Mercier's (see n.62 below).
62. Louis Sébastien Mercier, *L'An deux mille quatre cent quarante: rêve s'il en fut jamais* (London [Paris], 1771), vol.1, p.261. In Mercier's utopia 'certains hermites [...] vivent dans les forêts [...] pour herboriser. Ils y vivent par choix, par amour: ils se rendent ici [...] afin de nous enseigner plusieurs découvertes précieuses.' Mercier, *L'An deux mille*, p.259. Botany retains its importance in the year 2440!

among his botanising companions.[63] These contacts occurred frequently during the years 1770-1774, at precisely the moment Rousseau was composing the *Lettres élémentaires sur la botanique* to Mme Delessert.

Rousseau probably knew the Jussieus' work before his return to Paris in June 1770, since students and associates of the elder Jussieu had already produced various natural methods. For example, Antoine-Nicolas Duchesne (1747-1827) published a natural method in *Manuel de botanique* (1764),[64] which Du Peyrou may have shown him soon after its publication:

> Rien de nouveau dans mes découvertes que l'annonce d'un ouvrage intitulé *Manuel de botanique* par Mr Duchesne fils à Paris. L'Eloge qu'en font les Journaux m'a Séduit, et malgré ma juste défiance pour de pareils Prôneurs, j'en veux courir les hasards, et me procurer cet ouvrage, moins dans un objet de Simple curiosité, que pour l'utilité dont il peut m'être dans mes projets de Bâtiments.[65]

Du Peyrou suggested that Rousseau may have acquired Duchesne's work: 'je vous prie, Monsieur, de me marquer le titre de l'ouvrage de Tournefort que vous me conseillez d'avoir, et celui du *Manuel de botanique* que j'ai vu chez vous'.[66]

Other natural methods were published by Michel Adanson[67] in *Familles des plantes*,[68] and by Barbeu Dubourg in *Le Botaniste françois* of 1767. Jacques Barbeu Dubourg was a physician, a friend of Benjamin Franklin (whose works he translated)[69] and possibly also an acquaintance of Rousseau.[70] *Le Botaniste françois* (1767) comprises two volumes, the first a natural method of classification, the second a guide to the Paris flora.[71] Dubourg writes: 'en confrontant les plantes dans *toutes leurs parties*, j'en

63. '[I]l herborise avec M. de Jussieu', Bjôrnståhl to Gjôrwell, 1 September 1770; *CC*, xxxviii.95. See Antoine Laurent de Jussieu, 'Sixième notice historique sur le muséum', *Annales du muséum d'histoire naturelle* 11 (1808), p.1-39 (13-14, n.1).
64. Antoine-Nicolas Duchesne, *Manuel de botanique, contenant les propriétés des plantes utiles pour la nourriture, l'usage de la médecine, employées dans les arts, d'ornemens pour les jardins et qu'on trouve à la campagne aux environs de Paris* (Paris, Didot le jeune and Panckoucke, 1764).
65. My emphasis; Du Peyrou's 'projets' were his house and gardens in Neuchâtel. Du Peyrou to Rousseau, 20 October 1764, *CC*, xxi.281.
66. My emphasis; Du Peyrou to Rousseau, 23 December 1764; *CC*, xxii.276.
67. As noted above (p.184n) Rousseau sought contact with Adanson in 1769.
68. See note 58.
69. Alfred Owen Aldridge, 'Jacques Barbeu Dubourg: a French disciple of Benjamin Franklin', *Proceedings of the American Philosophical Society* 95:4 (17 August 1951), p.331-92 (375).
70. Aldridge, 'Jacques Barbeu Dubourg', p.339-40; see also Jacques Barbeu Dubourg, 'Réflexion sur ce qu'il a plu à Jean-Jacques Rousseau', *Gazette de médecine* 49 (19 June 1762), p.385-90.
71. See also Sébastien Vaillant, *Botanicon parisiense* (Leiden, Pieter van der Aa, 1723). Rousseau's annotated copy of a later edition (Leiden and Paris, Briasson, 1743) is at Houghton Library, Harvard University; see also Thomas François Dalibard, *Florae parisiensis prodromus* (Paris, Durand and Pissot, 1749), a Linnaean version of Vaillant.

trouve des quantités qui ont tant d'affinité entr'elles à divers égards, qu'il semble que la Nature même invite à les rapprocher les unes des autres; c'est ce qu'on appelle des *familles* naturelles'.[72] He delineated 'vingt-sept familles plus distinctes, et mieux prononcées que le reste, et véritablement si naturelles, qu'elles sont presque universellement reconnues, et révérées comme telles'.[73] These account for '[p]lus de trois quarts des plantes'; the others have been 'réservées pour un plus ample examen'.[74] For Dubourg the natural method was hardly a settled matter: 'En attendant que la *méthode naturelle* dont on n'a eu jusqu'ici qu'une *idée vague*, soit démontrée et, pour ainsi dire, homologuée, chaque Ecrivain peut se tracer un plan *à sa guise*.'[75]

Rousseau knew Dubourg's work sufficiently to recommend it for botanical journeys and to refer to it in his notes. [76] He recommended the second volume of *Le Botaniste françois* to the abbé Baurin prior to their expedition to Mont Pilat in August 1769.[77] Some of Rousseau's *Fragmens pour un dictionnaire*, for example 'Fruit' and 'Plante', echo Dubourg, which is not surprising if, as I hypothesise, the *Fragmens pour un dictionnaire*, consist, as least in part, of Rousseau's reading notes. Finally, Dubourg's view of botany as an 'amusement très innocent' foreshadows a central appeal of botany for Rousseau.[78]

Like any good eighteenth-century botanist, Rousseau formed a correspondence and exchange network;[79] some members participated at a distance, while others were close at hand.[80] Rousseau's relations with both Bernard and Antoine Laurent de Jussieu fall into the latter category, and this may explain the absence of letters, since they met at the Jardin du roi:

> [L]e mauvais temps qu'il fait et la saison qui se gâte me punissent d'avoir négligé le Jardin du Roy tandis qu'il faisait beau, et me mettent hors d'état de vous rendre compte quant à présent du *plantago uniflora*[81] et des autres

72. Original emphasis; Dubourg, *Botaniste*, p.114.
73. My emphasis; Dubourg, *Botaniste*, p.115.
74. Dubourg, *Botaniste*, p.116.
75. My emphasis; Dubourg, *Botaniste*, p.144.
76. Rousseau refers to Dubourg's discussion of *Prunus* in *Botaniste*, vol.1, p.154. Rousseau, *Fragments de botanique*; *OC*, iv.1253.
77. Rousseau to Pierre Guy, 11 June 1768; *CC*, xxxv.309; Rousseau to the abbé Baurin, 8 August 1769; *CC*, xxxvii.119-20. This expedition is famous for having been 'désastreux'. Rousseau to Laliaud, 27 August, 1769; *CC*, xxxvii.129-30; see also Rousseau to Du Peyrou, 16 September 1769; *CC*, xxxvii.142.
78. Dubourg, *Botaniste*, p.vii-viii.
79. See Cook, 'Réseaux d'échange', *passim*.
80. For example the Duchess of Portland, Claret de Latourrette, Pierre Clappier and Antoine Gouan.
81. *P. uniflora* L. (= *Littorella juncea* P. J. Bergius) has a lakeside habitat: '*Habitat ad* Europae

plantes curieuses dont j'aurais pu vous parler si j'avais su mieux profiter des bontés de M. [Bernard] de Jussieu. Je ne désespère pas pourtant de profiter encore de quelque beau jour d'automne pour faire ce pèlerinage et aller recevoir pour cette année les adieux de la syngénésie.[82]

Rousseau relayed his discussions with Bernard de Jussieu concerning a papyrus specimen that Latourrette had brought back from Naples:

J'ai parlé à M. [Bernard] de Jussieu du papyrus que vous avez rapporté de Naples; il doute que ce soit le vrai papier Nilotica. Si vous pouviez lui en envoyer soit plante soit graines, soit par moi soit par d'autres j'ai vu que cela lui ferait grand plaisir, et ce serait peut-être un excellent moyen d'obtenir de lui beaucoup de choses qu'alors nous aurions bonne grâce à demander, Quoi que je sache bien par expérience qu'il est charmé d'obliger gratuitement: mais j'ai besoin de quelque chose pour m'enhardir quand il faut demander.[83]

In a letter to Malesherbes he returned to the papyrus: 'j'ai eu l'honneur de vous envoyer quelques plantes curieuses et entre autres le vrai papier qui jusqu'ici n'était point connu en France, pas même de M. de Jussieu'.[84] Rousseau clearly held Bernard de Jussieu in high regard.

Rousseau joined botanical expeditions led by Bernard and Antoine Laurent de Jussieu, who helped him search for plants such as the elusive *Plantago monanthos*:

[J]'ai suivi M. [Bernard] de Jussieu dans sa dernière herborisation, et je la trouvai si tumultueuse et si peu utile pour moi que quand il en aurait encore fait j'aurais renoncé à l'y suivre. J'ai accompagné son neveu [Antoine Laurent de Jussieu] l'année dernière moi vingtième à Montmorency, et j'en ai rapporté quelques jolies plantes, entre autres la *Lysimachia tenella* [L.] que je crois vous avoir envoyées. Mais j'ai trouvé dans cette herborisation que les indications de Tournefort et de Vaillant[85] sont très fautives ou que depuis eux bien des plantes ont changé de sol.[86]

---

littora lacuum'. Linnaeus, *Species plantarum: a facsimile of the first edition 1753*, with introductory essays by William T. Stearn, 2 vols (London, 1957), vol.1, p.115; see also Dalibard, *Florae parisiensis prodromus*, p.30.

82. Rousseau to Latourrette, 28 September 1770; *CC*, xxxviii.114-15. Syngenesia is the nineteenth class in the sexual system.

83. Rousseau to Latourrette, 26 November 1770; *CC*, xxxviii.147. '[L]e vrai papier Nilotica' may refer to Egyptian papyrus (*Cyperus papyrus* L.). Linnaeus, *Species plantarum* (1753), p.47. Rousseau wished to reciprocate the 'bontés de M. de Jussieu' with specimens or seeds, as was customary in the republic of science. See Cook, 'Les Réseaux d'échange', p.93-94; Goldgar, *Impolite learning*, p.12ff.

84. Rousseau to Malesherbes, 19 December 1771; *CC*, xxxviii.302.

85. Joseph Pitton de Tournefort, *Institutiones rei herbariae* (Paris, de l'Imprimerie royale, 1700); Sébastien Vaillant, *Botanicon parisiense* (Paris, Briasson, 1743); Rousseau to Duchesne, 28 April 1765; *CC*, xxv.197.

86. Rousseau to Malesherbes, 17 April 1772; *CC*, xxxix.38.

Another letter sheds further light on Rousseau's relations with these botanists:

> J'aurai mauvaise grâce, ayant fait une recherche vaine, de vous faire valoir une herborisation que j'ai faite à Montmorency l'été dernier avec la Caterve du Jardin du Roi;[87] Mais il est certain qu'elle ne fut entreprise de ma part que pour trouver le plantago monanthos, que j'eus le chagrin d'y chercher inutilement. M. [Antoine Laurent] de Jussieu le jeune qui vous a vu sans doute à Lyon aura pu vous dire avec quelle ardeur je priai tous ces Messieurs, sitôt que nous approchâmes de la queue de l'étang de m'aider à la recherche de cette plante; ce qu'ils firent, et entr'autres M. T[h]ouin avec une complaisance et un soin qui méritait un meilleur succès.[88]

In response to an invitation from one of the Jussieus, Rousseau wrote:

> Si car à cette heure-là je ne suis pas arrivé, c'est signe que je ne pourrai profiter pour ce jour-là de l'invitation obligeante de M. de Jussieu, à qui je vous prie de faire mes remerciements en attendant que je remplisse ce devoir moi-même. Recevez aussi les miens avec mes salutations. Je me fais une fête de cette partie; car il y a double plaisir d'herboriser en si bonne et aimable compagnie.[89]

These letters suggest cordial relations among Rousseau, the Jussieus and their entourage.

Rousseau also met Claude Richard, head gardener of the Trianon garden at Versailles, and a close collaborator of Bernard de Jussieu. Richard knew Bernard de Jussieu's natural orders, and assisted in recording them in 1759.[90] Rousseau was keen to visit the Trianon garden during the Court's absence[91]:

> J'allai hier voir M. d'Aubenton au jardin du Roy, j'y rencontrai en me promenant M. Richard[,] jardinier de Trianon[,] avec lequel je m'empressai, comme vous jugez bien, de faire connaissance. Il me promit de me faire voir son jardin qui est beaucoup plus riche que celui du Roy à Paris [...] Je prendrai pour voir Trianon plus à mon aise quelque moment où la Cour ne

87. The 'Caterve' of the Jardin du roi included André Thouin, Antoine Laurent de Jussieu, students and/or friends. Jussieu, 'Sixième notice', p.13-14, n.1. Thouin was head gardener of the Jardin du roi and its successor, the Jardin des plantes; in 1806 he became professor of cultivation. He organised a highly effective network that sent him botanical specimens from all over the world. Emma Spary, 'Scientific symmetries', *History of science* 62 (2004), p.2-46.
88. Rousseau to Latourrette, 25 January 1772; *CC*, xxxix.22.
89. Rousseau to Thouin, 16 July 1771; *CC*, xxxviii.239.
90. Jussieu, 'Sixième notice', p.11, n.1.
91. Richard's fame reached Linnaeus. See Linnaeus to Claude Richard, 24 September 1765 in A. Landrin, *Correspondance inédite de Linné avec Claude Richard et Antoine Richard* (Versailles, 1863). p.19. In the 1770s the Trianon garden was transformed into a jardin anglais-chinois at the order of Marie Antoinette; the plants were transferred to the Jardin du roi. See Jean Dominique Bourzat, *Une Dynastie de jardiniers et de botanistes: les Richard de Louis XV à Napoléon III* (Paris, 2009), p.21.

sera pas à Versailles, et je tâcherai de me fournir à double de tout ce qu'on me permettra de prendre, afin de pouvoir vous envoyer ce que vous pourriez ne pas avoir.[92]

Thus, while we cannot be certain of the amount and intensity of contact that occurred between Rousseau and these major theorists of the natural method, the available evidence suggests that direct, even privileged, contact with them significantly influenced Rousseau's approach to classification and other botanical questions.

iii. The *Lettres élémentaires sur la botanique*, 1771-1774

[J]e devins auteur presque malgré moi.[93]

Rousseau composed his famous eight *Lettres élémentaires sur la botanique* to Madeleine-Catherine Delessert, née Boy de la Tour (1747-1816), between 1771 and 1774. In these letters Rousseau acted on his disillusionment with the sexual system, announced in Book XII of the *Confessions*, developing his own way of presenting plants, informed by the natural method.

The letters were published posthumously for the first time in the *Collection complète de J.-J. Rousseau* (Geneva, [1780]-1782) as the *Lettres élémentaires sur la botanique*.[94] Given their posthumous publication, Rousseau's purpose in writing and revising them has occasioned much speculation.[95] R. A. Leigh argues that Rousseau was probably planning to publish them, a supposition supported by his re-writing of the copies of the *Lettres* in the Bibliothèque publique et universitaire de Neuchâtel:[96] 'JJ avait sûrement pensé à les recueillir pour en faire un petit ouvrage.'[97] This assessment must be considered, however, against Rousseau's explanation that he wrote them at his friend's request: 'Votre idée d'amuser un peu la vivacité de votre fille[98] et de l'exercer un peu à l'attention sur des objets agréables et variés comme les plantes me semble excellente, mais

92. Rousseau to Latourrette, 4 July 1770; *CC*, xxxviii.52-53.

93. Rousseau, *Lettres à Malesherbes*, *OC*, i.1136.

94. The *Lettres élémentaires sur la botanique à M. de L****. were first published in the *Collection complète des œuvres de J. J. Rousseau*, Pierre-Alexandre Du Peyrou and Paul Claude Moultou (eds), vol.7, p.531-88, in-4°; vol.14, p.431-518, in-8 ° and in-12 ° (Geneva, [Société typographique de Genève], [1780-]1782).

95. It is unclear how they could be construed as a 'fiktive Korrespondenz'. Alison Martin, 'Frauenzimmerbotanik: unschuldiger Zeitvertreib und Mode?' *Frauen und Gärten um 1800: Weiblichkeit, Natur, Ästhetik*, ed. Christiane Holm and Holger Zaunstöck (Halle, 2009), p.37-47 (37).

96. BPUN Neuchâtel MsR80.

97. R. A. Leigh, 'Notes explicatives', *CC*, xxxviii.256.

98. He means Marguerite-Madeleine, or 'Madelon' Delessert (1767-1839), Mme Delessert's eldest daughter. The herbarium he made for her is now in Montmorency, France. See Appendix 3.

je n'aurais osé vous la proposer de peur de faire le M. Josse; puisqu'elle *vient de vous* je l'approuve de tout mon cœur et j'y concourrai de même.'[99] Another objection to Leigh's theory is that the *Lettres* contain many personal references to Mme Delessert's mother, husband, and children.[100]

These letters originated not only in Rousseau's interest in botany, but also in his connections with the Roguin and the Boy de la Tour families, with whom he had long-standing connections in Lyon, Yverdon and Neuchâtel. Two of Mme Boy de La Tour's daughters, Julie Emélie Willading-Boy de la Tour (1751-1826) and Madeleine-Catherine Delessert, had close relationships with Rousseau, who fostered their interest in botany, writing letters to one and making a superb herbarium ('Zürich') for the other.[101] He praised his 'Cousine', Madeleine-Catherine Delessert, for her 'grand sens et excellent caractère', and he reminisced fondly about his herborisations with 'Tante Julie' in the Croix de Vague.[102] Indeed, Mme Delessert initiated the *Lettres* project, asking Rousseau to send her 'un petit catalogue des plantes les plus connues avec des marques pour les reconnaître'.[103]

The *Lettres* are usually said to have been directed to an exclusively female audience.[104] Yet Rousseau addressed 'vos enfants'; this could be taken to mean *all* of Mme Delessert's children: seven sons and three girls.[105] Two of her sons became avid naturalists, Etienne (1771-1794), born the year letters began, and Jules Paul Benjamin Delessert (1773-

99. My emphasis; Rousseau to Mme Delessert, 22 August 1771; *CC*, xxxviii.251-52.

100. See Jean-Jacques Rousseau, *Lettres inédites de Jean-Jacques Rousseau à Mmes Boy de la Tour et Delessert, comprenant les lettres sur la botanique*, Philippe Godet and Maurice Boy de la Tour (eds), (Paris and Geneva, 1911), p.iv.

101. 'Je dois à Tante Julie un petit échantillon d'herbier que je n'ai pu préparer encore; mais je tâcherai de réparer en partie cette négligence, et quand vous serez en doute des plantes dont j'aurai à vous parler dans la suite, ce petit recueil vous aidera pour celles qui pourront S'y trouver.' Hence, he invited Mme Delessert to use her sister's herbarium. Rousseau to Mme Delessert, 20 October 1771; see also *CC*, xxxviii.281; Rousseau to Julie Anne Marie Boy de la Tour, 16 April 1772; *CC*, xxxxi.34-35. See Appendix 3.

102. Rousseau, *Confessions*; *OC*, i.590; Rousseau to Mme Delessert, 16 May 1772; *CC*, xxxix.49. See also Thierry Hoquet, 'La bibliothèque botanique de Benjamin Delessert', *Bulletin du bibliophile* 1 (2002), p.102-41 (112).

103. Rousseau to Mme Delessert, 22 August 1771; *CC*, xxxviii.252. Botany was popular in the upper classes. See Daniel Mornet, *Les Sciences de la nature en France au XVIIIe siècle* (Paris, 1911), p.181; Williams, *Botanophilia*. The family's Swiss background probably contributed to this interest.

104. See the claim that Rousseau makes botany a 'Lehrgegenstand für Mädchen und junge Frauen [object of instruction for girls and young women]'. Martin, 'Frauenzimmerbotanik', p.38. George asserts that Rousseau 'vehemently' recommends botany to women: George, *Botany, sexuality and women's writing*, p.67.

105. Rousseau to Mme Delessert, 30 August 1773; *CC*, xxxix.190; see also Rousseau to Mme Delessert, 16 July 1772; *CC*, xxxix.82.

1847), born two years later. Inspired by Rousseau, Benjamin Delessert pursued 'la vocation familiale pour la botanique', founding one of the great private botanical museums of the nineteenth century.[106]

iv. Rousseau applies the natural method

> Cum autem sincerus naturae contemplator, ipsam in plantis explorans, teneatur integram earum organisationem scrutari ac evolvere, omnes ideo prosequitur caracteres, et neglecta methodo pauciores memorante, eam tantum inquirit quae nullos omittens et mutuas omnium deprehendens affinitates.[107]

The *Lettres élémentaires sur la botanique* begin with a conventional discussion of the fructification:

> Une plante parfaite est composée de racine, de tige, de branches, de feuilles, de fleurs, et de fruits: car on appelle fruit en botanique, tant dans les herbes que dans les arbres, toute la fabrique de la semence. Vous connaissez tout cela; du moins assez pour entendre le mot; mais il y a une partie principale qui demande un plus grand examen. C'est la *fructification*, c'est à dire la fleur et le fruit. Commençons par la *fleur*, qui vient la première. C'est dans *cette partie* que la nature a renfermé le sommaire de son ouvrage, c'est par elle qu'elle le perpétue, et c'est aussi de toutes les parties du végétal la plus éclatante pour l'ordinaire, et toujours la moins sujette aux variations.[108]

But the *Lettres* do not in fact explicate the sexual system, offering instead a short course in the spirit of the natural method. Four features of Rousseau's approach make this allegiance clear: (1) he commences each analysis with the corolla, using it to place the plant in the categories of Tournefort's system; (2) he employs the 'premier coup d'œil'[109] to ascertain (3) the 'air de famille', facies propria, 'port', or habit; (4) based on habit, he compares six common plant *families* fundamental to the natural method.

First, based on the system of Tournefort, Rousseau examines the

106. Hoquet, 'La bibliothèque botanique', p.114. See George, *Botany, sexuality and women's writing*, p.43. I discuss this museum in Chapter 9, p.340-42.

107. 'The sincere contemplator of nature, who seeks to know her in plants, must [...] examine and disclose their *entire* organization; to that end, he must investigate *all* characters, and [...] must search in such a fashion that omitting nothing he discovers the mutual affinities of all plants.' My emphasis; Jussieu, *Genera plantarum*, p.xxxv; trans. S. Rosa in Stevens, *Development of biological systematics*, p.354.

108. My emphasis; Rousseau to Mme Delessert, 22 August 1771; *CC*, xxxviii.252-53.

109. The 'premier coup d'œil' exercises the eye; learning to see is a principal goal of Emile's education and of the *Lettres*. For example, 'je crois néanmoins qu'il lui sera toujours utile d'apprendre à bien *voir* ce qu'elle [Madelon] *voit*'. My emphasis; Rousseau to Mme Delessert, 22 August 1771 *CC*, xxxviii.255.

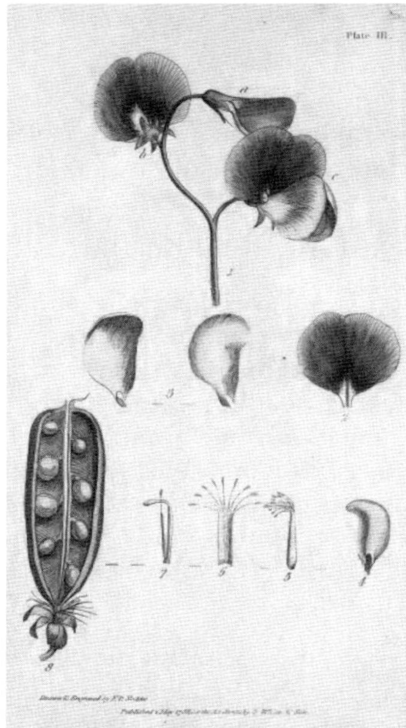

Figure 6.2: Frederick P. Nodder, *Plate III, letter III*, drawing and engraving from Thomas Martyn, *Thirty-eight plates, with explanations intended to illustrate Linnaeus's system of vegetables, and particularly adapted to the letters of the elements of botany* (London, B. & J. White, 1794), courtesy of the Biodiversity Heritage Library (online).

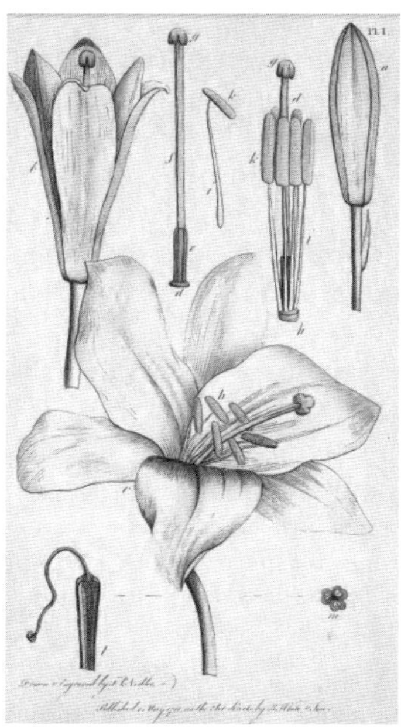

Figure 6.3: Frederick P. Nodder, *Plate I, letter I*, drawing and engraving from Thomas Martyn, *Thirty-eight plates*, courtesy of the Biodiversity Heritage Library (online).

corolla of the pea plant in detail the third letter on the Papilionaceae or Leguminosae (legumes); this exemplifies Rousseau's analysis of the corolla, or petals, of the flower (fig.6.2, p.192):

> Les pois sont à présent en pleine fructification. Saisissons ce moment pour observer leur caractère. Il est un des plus curieux que puisse offrir la botanique. Toutes les fleurs se divisent généralement en régulières et irrégulières. Les premières sont celles dont toutes les parties s'écartent uniformément du centre de la fleur et aboutiraient ainsi par leurs extrémités extérieures à la circonférence d'un cercle. Cette uniformité fait qu'en présentant à l'œil les fleurs de cette espèce, il n'y distingue ni dessus ni dessous, ni droite ni gauche; telles sont les deux familles ci-devant examinées. Mais au premier coup d'œil vous verrez qu'une fleur de pois est irrégulière, qu'on y distingue aisément dans la *corolle* la partie plus longue qui doit être en haut, de la plus courte qui doit être en bas et qu'on connait fort bien en présentant la fleur vis-à-vis de l'œil si on la tient dans sa situation naturelle ou si on la renverse. Ainsi toutes les fois qu'examinant une fleur *irrégulière* on parle du haut et du bas c'est en la plaçant dans sa situation naturelle.[110]

He finds that it is *polypetalous* and irregular, which places it in either class ten or eleven of Tournefort's system, although he does not mention this explicitly.

Second, Rousseau's letters to Mme Delessert stress the importance of the familial habit, which is apprehended by the sense of sight.[111] The habit of a plant[112] – its general aspect – is grasped 'au premier coup d'œil', much as one might intuit a person's gender, occupation or ethnic origin.[113] The habit is a primary quality, while scent, taste and colour are relegated to a secondary role: 'leur couleur, [...] leur goût, leur odeur, leur saveur [...] n'est [...] qu'une étude analogique et *secondaire*, pour éclairer et confirmer celle des *formes*'.[114] Thus, in the second letter on botany Rousseau tells Mme Delessert, '[d]ans ce Calice vous trouverez une corolle composée de quatre pétales dont je laisse à part la couleur, parce qu'elle ne fait pas *caractère*'.[115]

This distinction permeates Rousseau's reflections on painting, music and language, as well as botany, 'belles couleurs bien nuancées plaisent à

110. My emphasis; Rousseau to Mme Delessert, 16 May 1772; *CC*, xxxix.50.

111. Rueff notes that Rousseau accepted Berkeley's focus on size and distance, rejecting colour theory. Rueff, 'Apprendre à voir', p.197. This is consistent with the primary-secondary qualities distinction.

112. See Delaporte, *Le Second règne*, p.140-41, who says Rousseau thinks *flowers* have female faces.

113. For example, compare a Swiss who herds cows or a Corsican who tills soil to a Parisian wig-maker. The examples are mine. Rousseau, *Projet de constitution pour la Corse*, *OC*, iii 905; *Considérations sur le gouvernement de Pologne*, *OC*, iii.960, 966.

114. My emphasis; Rousseau, *Fragments de botanique*, *OC*, iv.1249.

115. My emphasis; Rousseau to Mme Delessert, 18 October 1771; *CC*, xxxviii.277.

la vue, mais ce plaisir est purement de sensation. C'est le *dessin* [...] qui donne à ces couleurs de la vie et de l'âme'. By analogy, melody is to form or 'dessin' what harmony is to colour: 'La mélodie fait précisément dans la musique ce que fait le dessin dans la peinture; c'est elle qui marque les traits et les figures dont les accords et les sons ne sont que les couleurs.'[116]

Rousseau introduces important criteria for recognising plant habit and insists on their significance. Most important, habit can be grasped *without* the fructification, a procedure that clearly distinguishes Rousseau's approach from the sexual system.

With practice, the student comes to identify the familial habit, starting with the lilies (Liliaceae) (fig.6.3, p.192):

> [L]es premiers linéaments des plantes, quoique si légèrement marqués, que votre *œil clairvoyant* sait déjà distinguer *un air de famille* dans les liliacées, et que notre chère petite botaniste s'amuse déjà de corolles et de pétales, je vais vous proposer une autre famille sur laquelle elle pourra derechef exercer son petit savoir.[117]

In the fourth letter he returns to this procedure:

> Je voudrais même que vous tâchassiez d'en déterminer la lignée ou section par la *physionomie*, et que vous vous exerçassiez à juger au simple *coup d'œil* si la fleur en gueule que vous voyez est une labiée ou une personée. La *figure* extérieure de la corolle peut *suffire* pour vous guider dans ce choix, que vous pourrez vérifier ensuite ôtant la corolle et regardant au fond du calice; car si vous avez bien jugé, la fleur que vous aurez nommée labiée vous montrera quatre graines nues, et celle que vous aurez nommée personée vous montrera un péricarpe: le contraire vous prouverait que vous vous êtes trompée, et par un *second examen* de la même plante vous préviendrez une erreur semblable pour une autre fois. Voilà, chère Cousine, de l'occupation pour quelques promenades.[118]

In the sixth, and final, letter on plant families he reiterates the importance of long practice in identifying the habit:

> [J]e crois néanmoins vous en avoir donné une idée suffisante pour pouvoir, après *quelques mois* d'herborisation vous familiariser avec l'idée générale du *port* de chaque famille: en sorte qu'à *l'aspect d'une plante*, vous puissiez conjecturer à peu près si elle appartient à quelqu'une des cinq familles et à laquelle; sauf à vérifier ensuite par l'analyse de la fructification si vous vous êtes trompée ou non dans votre conjecture [...] Or au *simple aspect* de l'ombelle et de la plante qui la porte on doit prendre une idée si nette des ombellifères qu'à la rencontre d'une plante de cette famille on s'y trompera rarement au *premier coup d'œil*.[119]

116. My emphasis; Rousseau, *Essai sur l'origine des langues*, *OC*, v.413.
117. My emphasis; Rousseau to Mme Delessert, 18 October 1771; *CC*, xxxviii.276.
118. My emphasis; Rousseau to Mme Delessert, 19 June 1772; *CC*, xxxix.72.
119. My emphasis; Rousseau to Mme Delessert, 2 May 1773; *CC*, xxxix.151.

Compare a letter to Malesherbes, written during this period:

> [J]e crois qu'il faut commencer à connaître empiriquement un certain nombre d'espèces pour parvenir à déterminer les autres, et je crois que celles que je vous envoie peuvent suffire en les étudiant bien à vous familiariser avec la *famille* et à en distinguer au moins les genres au *premier coup d'œil* par le *faciès* propre à chacun d'eux.[120]

Mme Delessert must distinguish plants without their fructification by collecting branches and leaves, which are necessary to determine genera and species:

> Les arbres et toutes les grandes plantes ne se prennent que par échantillon. Mais il faut que cet échantillon soit si bien choisi, qu'il contienne toutes les parties constitutives du *genre* et de *l'espèce*, afin qu'il puisse suffire pour reconnaître et déterminer la plante qui l'a fourni. Il ne suffit pas que toutes les parties de la fructification y soient sensibles ce qui ne servirait qu'à distinguer le genre, il faut qu'on y voie bien le caractère de la *foliation* et de la *ramification*; c'est-à dire, la naissance et la forme des *feuilles* et des branches, et même autant qu'il se peut quelque portion de la *tige*; car comme vous verrez dans la suite; tout cela sert à distinguer les espèces différentes des mêmes genres qui sont parfaitement semblables par la fleur et le fruit.[121]

Thus, she can distinguish hemlock from plants of a similar appearance – parsley and chervil – 'surtout par le feuillage':

> Quand vous vous serez bien assurée de la ciguë en fleurs, vous vous confirmerez dans votre jugement en froissant légèrement et flairant son feuillage; car son odeur puante et vireuse ne vous la laissera pas confondre avec le persil ni avec le cerfeuil qui tous deux ont des odeurs agréables. Bien sûr enfin de ne pas faire de qui-pro-quo, vous examinerez ensemble et séparément ces trois plantes dans tous leurs états et par toutes leurs parties, surtout par le *feuillage*, qui les accompagne *plus constamment* que la fleur; et par cet examen comparé et répété jusqu'à ce que vous ayez acquis la certitude du coup d'œil vous parviendrez à distinguer et connaitre imperturbablement la ciguë. L'étude vous mène ainsi jusqu'à la porte de la pratique, après quoi celle-ci fait la facilité du savoir.[122]

In a letter written after the initial eight, Rousseau commented on Mme Delessert's specimens: 'Vos plantes sont parfaitement bien conservées [...] mais vous n'avez pas eu soin de dessécher aussi des *feuilles* qui quand le

120. My emphasis; Rousseau to Malesherbes, 19 December 1771; *CC*, xxxviii.301. There is therefore no difference in how Rousseau teaches botany to an erudite male aristocrat and to a female friend. Duris, *Linné et la France*, p.105.

121. My emphasis; Rousseau to Mme Delessert, 11 April 1773; *CC*, xxxix.140.

122. My emphasis; Rousseau to Mme Delessert, 16 July 1772; *CC*, xxxix.88. Cf. Bernhard Kuhn, 'Natural history and the history of the self: geology and autobiography in the works of Goethe and Rousseau', *Colloquium helveticum* 25 (1997), p.41-62 (45). See also Rousseau to Mme Delessert, 11 April 1773; *CC*, xxxix.141.

genre est connu, servent à déterminer l'*espèce*.'[123] He reiterated this point in a later letter: 'Voici, chère Cousine, les noms des plantes que vous m'avez envoyées en dernier lieu. J'ai ajouté un point d'interrogation à ceux dont je suis en doute, parce que vous n'avez pas eu soin d'y mettre des *feuilles* avec la fleur, et que le *feuillage* est souvent nécessaire pour déterminer l'espèce à un aussi mince botaniste que moi.'[124]

Rousseau consulted works on foliage: for example, early in his botanical studies in May 1765, he requested Duhamel du Monceau's treatise on trees[125] and Sauvages's system of classification based on the foliage: 'j'aurai encore besoin du traité des arbres et arbustes de M. du Hamel [du Monceau] chez Guérin et de la Tour, et d'un in-12 intitulé *Methodus Foliorum*, par M. [François Boissier] de Sauvages [de la Croix]'.[126]

However, there can be exceptions to relying on the foliage to identify a species: '*Le Chardon-roland*, au contraire, n'a guère le port d'une ombellifère, et néanmoins c'en est une, puisqu'il en a tous les caractères dans sa fructification.'[127] It is therefore imperative to observe both the fructification and 'port' in order to arrive at the correct conclusion. This process of cross-checking demands concentrated observation and practice – it is not to be learned from a book.

The classificatory basis of the *Lettres* is therefore family resemblance revealed by the habit. In the third letter Rousseau wrote:

> [R]eprenons le fil interrompu de nos *familles* végétales.
> Mon intention est de vous décrire d'abord six de ces *familles* pour vous familiariser avec la structure générale des parties caractéristiques des plantes. Vous en avez déjà deux; reste à quatre qu'il faut encore avoir la patience de suivre. Après quoi laissant pour un tems les autres branches de cette nombreuse lignée et passant à l'examen des parties différentes de la fructification nous ferons en sorte que, sans peut être connaître beaucoup de plantes vous ne serez du moins jamais en terre étrangère parmi les productions du règne végétal.[128]

For Rousseau these were the Liliaceae (lilies), Cruciferae (cabbages), Leguminoseae (beans), Labiateae (mints), Compositae (daisies) and

123. My emphasis; Rousseau to Mme Delessert, 24 May 1773; *CC*, xxxix.162.

124. My emphasis; Rousseau to Mme Delessert, April/March 1774; *CC*, xxxix.234.

125. Henri Louis Duhamel du Monceau, *Traité des arbres et arbustes qui se cultivent en France en pleine terre*, 2 vols (Paris, H. L. Guerin and L. F. Delatour, 1755).

126. My emphasis; Rousseau to Duchesne, 19 May 1765; *CC*, xxv.301-02. The work he requested is Sauvages, *Methodus foliorum*. Sauvages was an important Montpellier botanist. Rousseau sold his annotated copy of this work to Daniel Malthus, and it is part of Thomas Malthus's collection in Jesus College, Cambridge. See Cook, 'Jean-Jacques Rousseau's copy of Albrecht von Haller's *Historia stirpium*'; Henry Cheyron, 'Ray et Sauvages annotés par Jean-Jacques Rousseau', and Appendix 1.

127. Original emphasis; Rousseau to Mme Delessert, 16 July 1772; *CC*, xxxix.86.

128. My emphasis; Rousseau to Mme Delessert, 16 May 1772; *CC*, xxxix.50.

Umbelliferae (carrots). He introduced them to his students in an order that facilitated comparison, starting with the lilies: '[l]e Lis que j'ai choisi [...] est de la saison et aussi à cause de la grosseur de sa fleur et de ses parties qui les rend plus sensibles' (fig.6.2, p.192).[129]

Antoine Laurent de Jussieu likewise considered these same families 'précieuses [...] puisqu'elles sont la base de nos recherches', but he listed them in a different order than Rousseau did and included the grass family,[130] which Rousseau considered 'difficile'.[131] Jussieu considered these seven families both 'très simples et avouées des tous les Botanistes'.[132] They are 'several obviously natural groups, which he variously called families and orders, over whose circumscription and existence there could be no doubt [...]. These taxa [...] were regarded by Jussieu as exemplars of the *continuous* natural order, not as discrete, bounded groups.'[133] The families chosen by Rousseau for the *Lettres* therefore occupied centre stage on the continuous web or map of the plant kingdom as conceived by Jussieu, Linnaeus and others. This conception, in which all gaps would eventually be filled, stood in sharp contrast with the plant kingdom viewed as distinct groups – species, genera, orders, classes, etc. – for example in an artificial system. Furthermore, even the creators of artificial systems ultimately concurred in the continuity of nature, and acknowledged any boundary as itself artificial.[134]

In the *Lettres* Rousseau treated these six principal families before embarking on discussions of fruit trees and herbaria; the last were discussed in what was originally the sixth letter, which he placed at the end of the series 'parce qu'elle interromproit *l'ordre* que je me suis proposé'.[135] In the last letter on plant families (sixth in the final version) Rousseau asserted his allegiance to the natural method, stating that his exposition of plant families *precedes* the introduction of genera and species: 'il ne sera pas question si tôt des genres et des espèces'.[136]

129. Rousseau to Mme Delessert, 22 August 1771; *CC*, xxxviii.254.

130. Jussieu, 'Exposition d'un nouvel ordre des plantes', p.179. Tournefort's seventeen classes include all of these.

131. Rousseau to Malesherbes, 17 April 1772; *CC*, xxxix.39. Rousseau's Neuchâtel herbarium contains many grass specimens; he had Benjamin Stillingfleet (1702–1771), 'Observations on grasses' in *Miscellaneous tracts relating to natural history, husbandry and physick* (London, R. and J. Dodsley, 1762) and Johannes Scheuchzer (1684-1738), *Agrostographia sive graminum, juncorum, cyperorum, cyperoidum, usque affinium historia* (Zürich, Bodmer, 1719). Rousseau to Coindet, 5 July 1767; *CC*, xxxiii.191. Du Peyrou brought him the work in Trye. Du Peyrou to Rousseau, 26 August 1767; *CC*, xxxiv.55.

132. Jussieu, 'Exposition d'un nouvel ordre des plantes', p.178.

133. My emphasis; Stevens, *Development of biological systematics*, p.35-36.

134. Stevens, *Development of biological systematics*, p.256

135. My emphasis; Rousseau to Mme Delessert, 2 May 1773; *CC*, xxxix.151.

136. Rousseau to Mme Delessert, 2 May 1773; *CC*, xxxix.151.

After concluding the six letters on plant families, Rousseau offered Mme Delessert a summary of his exposition:

> Lorsque *vous* me proposâtes [...] de vous donner la connaissance de quelques plantes pour l'amusement de vos enfants, je jugeai qu'on pouvait leur rendre cet amusement utile par une étude un peu *méthodique* qui les accoutumât peu à peu à l'attention, à l'observation et surtout au bon raisonnement. Au lieu qu'une simple *nomenclature* qui ne chargerait que leur mémoire, ne les amuserait pas longtemps, serait bientôt oubliée, et ne leur serait d'aucun profit après cet oubli. Je commençai donc par tâter leur goût et le vôtre par quelques notions générales des parties de la *fructification* où résident les caractères les plus *essentiels* et les plus constants des plantes et par lesquels on vient le mieux à bout de les classer. Je vous offris pour premiers objets cinq ou six *familles* des plus *nombreuses* et des plus *saillantes* du règne végétal, et je tâchai d'accoutumer vos *yeux* à démêler et distinguer leurs parties essentielles en attendant que vous y pussiez reconnaître cet *air de famille* [port] qui les distingue même *sans fructification*, mais qui ne frappe que les yeux suffisamment exercés.[137]

Rousseau's comment seems at first to follow the conventions laid down by Linnaeus for the sexual system, but ends on a very different note, emphasising the habit, the family and how to distinguish plants by their foliage alone. He therefore did not limit himself to the sexual system, but went beyond it to study the habit of six of the most numerous and salient of the hundred or so in the plant kingdom.

He later stated that he selected these families 'au hasard', but nevertheless maintained – with other proponents of the natural method – that the 'véritable étude' of plants entails the study of *all* parts:

> [J]'aurai toujours le même empressement à contribuer à vos amusements et à ceux de la charmante Madelon, mais pour reprendre ce petit travail avec un peu de succès, il faudrait que j'eusse une idée un peu plus précise de vos goûts et de vos progrès, et que je visse de quel point je dois partir pour vous marquer la route que vous devez suivre. Les six familles, dont j'ai essayé de vous décrire la *fructification* pour consulter votre goût et vous familiariser avec quelques *termes*, sont prises pour ainsi dire au hasard et n'ont pas une suite dont on puisse prendre le fil. Cet essai étant fait il en faudrait connaître bien le succès pour commencer au point convenable la véritable étude, qui ne consiste pas seulement dans celle de la fructification mais des plantes dans leur ensemble et dans *toutes leurs parties*.[138]

Rousseau endorsed the families of the natural method not only by employing them as the organising principle for the *Lettres*, but also in the herbarium he made for Madelon Delessert,[139] in which he gave the plants

137. My emphasis; Rousseau to Mme Delessert, 30 August 1773; *CC*, xxxix.190-91.

138. My emphasis; Rousseau to Mme Delessert, 23 August 1774; *CC*, xxxix.265-66. See also Adanson, *Famelles des plantes*, p.clv.

139. See Alexandra Cook and Christopher Kelly, 'Introduction', *The Collected writings of*

their Latin and French names, and situated them in their natural families (for example 'Legumineuses' (Leguminoseae) for *Medicago arabica* L.):

> Cet herbier confirme la position de Rousseau en matière de classification. Ni dans l'herbier, ni dans les lettres, il n'utilise le système [sexuel] de Linné, qui lui aurait fait par exemple placer le Lis de la première lettre dans l'Hexandrie et la Luzerne dans Diadelphie. Il connaît bien sûr ce système et [...] il s'avoue s'être pris pour lui une passion dont il ne s'est pas bien guéri 'même après en avoir senti le vide.[140]

Drouin adds: 'Là encore, derrière la posture simplificatrice, se cache une analyse attentive à la complexité'.[141]

The desideratum of a natural method persists to this day, although the procedures outlined by Rousseau no longer suffice: 'Today, the systematist is expected to observe not only superficial characters, such as the arrangement and shape of leaves, but also such intimate details as the structure and sculpture of the pollen grain, the nodal anatomy, the number of chromosomes and many other data on morphology, anatomy, physiology, cytology, etc.'.[142] Ultimately, '[a]ny system adopted by us will be but one of equally good, or bad, alternatives, the number of these depending on the accidents of evolutionary history in the different groups of organisms, on the number of workers studying them, as well as on the fashion of the day'.[143]

v. The appeal of the non-floral

Rousseau probably started studying cryptogams, such as mosses and lichens, at a relatively early stage; this would have diverted him from the paradigm of flowering plants on which the first twenty-three classes of Linnaeus's sexual system is based. 'Cryptogam' refers to secret marriage, that is to say, unknown means of reproduction, so in cryptogams there

Rousseau (Hanover, NH, 2000), vol.8, p.xxvi. The herbarium for Madelon was acquired by the Musée Jean-Jacques Rousseau, Montmorency, France in 2001 and is briefly described in Appendix 3.

140. Drouin, 'Les herborisations d'un philosophe', in *Rousseau et les sciences*, ed. Bernadette Bensaude-Vincent and Bruno Bernardi (Paris, 2003), p.77-92 (84), quoting Rousseau, *Confessions*, *OC* i.643. 'Sa méthode pratique, raisonnée, logique, compréhensible pour les esprits moins cultivés, contraste avec les procédés nuageux dont un Lamarck n'était point complètement affranchi. Et pourtant ils sont de la même école, qui cherche à *sortir* des cadres *linnéens*.' My emphasis; Godet and Boy de la Tour in Rousseau, *Lettres inédites de Jean-Jacques Rousseau*, p.vii.

141. Drouin, *L'Herbier*, p.78. Rousseau's choice of natural families forced his English translator, Thomas Martyn, to make a clumsy transition to Linnaean classes and orders. See Chapter 9, p.314 for an elaboration.

142. Jacob Lorch, 'The natural system in biology', *Philosophy of science* 28:3 (July 1961), p.282-95 (289).

143. Lorch, 'The natural system', p.294.

are no stamens or pistils to count.[144] Linnaeus relegated them to the residual, twenty-fourth class of the sexual system because of their obscure reproductive organs and functions. This interest in cryptogams therefore displaces the flower from the central place it has been assumed to occupy in Rousseau's study of botany.[145]

As cryptogams are not flowering plants, such as those analysed in the letters to Mme Delessert, what was their attraction for Rousseau? The discovery of plant sexuality gave cryptogam reproduction a mysterious character that attracted many naturalists, besides Rousseau. Rousseau's cryptogam herbarium at the Musée des arts décoratifs (Paris) contains 315 moss specimens and 11 lichen specimens (fig.6.4, p.201), and offers ample evidence of this fascination.[146] His interest in cryptogams took him well beyond the average amateur; it required detailed attention to plants whose means of reproduction were not understood at the time.[147]

Rousseau's cryptogam studies seem to have been stimulated by several factors. First, such plants are accessible in winter when there is little else to study:

> Voici, Monsieur, quelques échantillons de mousses que j'ai rassemblées à la hâte pour vous mettre à portée au moins de distinguer les principaux genres avant que la saison de les observer soit passée. C'est une étude à laquelle j'employai délicieusement l'hiver que j'ai passé à Wootton où je me trouvais environné de montagnes de bois et de rochers tapissés de capillaires et de mousses des plus curieuses. Mais depuis lors j'ai si bien perdu cette famille de vue que ma mémoire éteinte ne me fournit presque plus rien de ce que j'avais acquis en ce genre, et n'ayant point l'ouvrage de Dillenius, guide indispensable dans ces recherches je ne suis parvenu qu'avec beaucoup d'effort et souvent avec doute à déterminer les espèces que je vous envoie.[148]

Second, Rousseau cited health-related reasons for this study; cryptogams grow on rocks well above the ground, so that he did not have to stoop down to study them, as he recounted to the famous Swiss physician, Dr Samuel Tissot: 'Je me promène tous les jours mais doucement, sans me lasser et sans me *baisser*, et ne pouvant plus herboriser par terre vu *mon*

144. Lichens are symbiotic organisms of fungus and algae. In the eighteenth century they were grouped in the genus 'Lichen', following Johann Jakob Dillenius, *Historia muscorum* (Oxford, The Sheldonian theatre, 1741).

145. Cf. David Scott, 'Rousseau and flowers: the poetry of botany', *SVEC* 182 (1979), p.73-86, and Lisa Gasbarrone, 'The book of nature: Rousseau's floral collections and the text', *L'Esprit créateur* 28:1 (1988), p.27-41.

146. Musée des arts décoratifs, Paris: henceforth MdAD. The cryptogam herbarium was probably a gift to Malesherbes, see Appendix 3.

147. Goethe appreciated Rousseau's investigation of mosses. Johann Wolfgang von Goethe, 'Der Verfasser teilt die Geschichte seiner botanischen Studien mit', Dorothea Kuhn (ed.) in *Goethes Werke*, Erich Trunz, *et al.* (eds) (Hamburg, 1955), vol.13, p.159-60.

148. Rousseau to Malesherbes, 19 December 1771; *CC*, xxxviii.300; see also Rousseau to Malesherbes, 11 November 1771; *CC*, xxxviii.289.

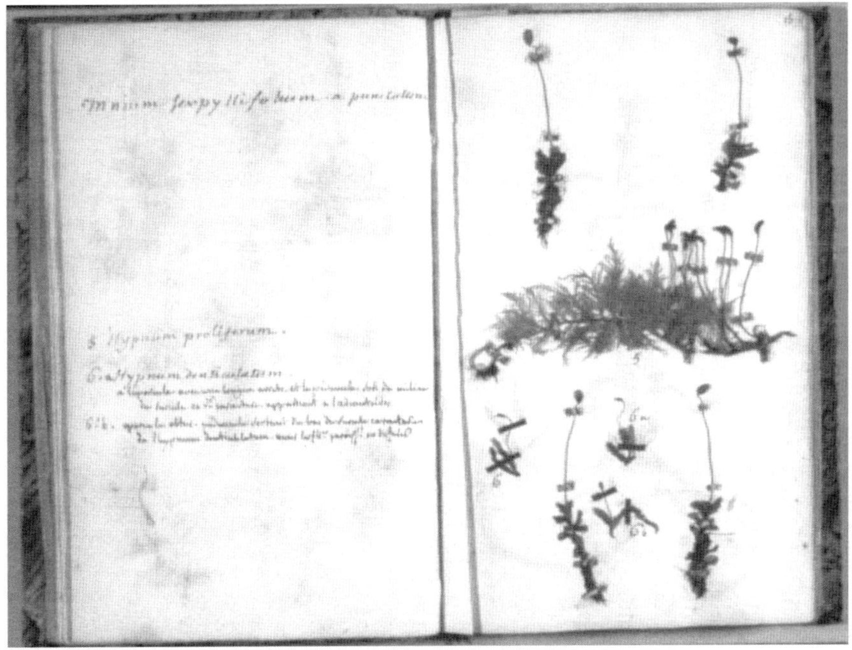

Figure 6.4: Specimens of *Mnium serpyllifolium* L. (= *M. punctatum* Hedw.) *Hypnum proliferum* L. (= *Thuidium tamariscinum* (Hedw.) Schimp.), *Hypnum denticulatum* L. (= *Plagiothecium denticulatom* (Hedw.) Schimp.), from Herbarium of Rousseau, feuille 15, by permission of the Musée des arts décoratifs, Paris.

état et la saison, je m'amuse à observer et déterminer les mousses et les lichens sur les troncs des arbres, et je soupçonne déjà que cette étude pourrait être encore perfectionnée, même après Michelius[149] et Dillenius'.[150]

In his letter to Tissot Rousseau refers to two authoritative works: Micheli's *Nova plantarum genera* (1729), considered the first treatise in mycology,[151] and Dillenius's *Historia muscorum* (1741), a treatise on mosses and other cryptogams. Rousseau's references to their works demonstrate his knowledge of the leading literature in these fields.[152]

Pier-Antonio Micheli (1679-1737) was an eminent authority in the new field of mycology. Inspired by Tournefort's artificial classification system, Micheli divided mushrooms into 900 genera – some new – and studied their structure and reproduction. He discovered spores and contributed to the study of mosses and lichens.[153]

In the letter to Malesherbes, cited above, Rousseau calls Dillenius the 'guide indispensable dans ces recherches'. Dillenius's work formed the basis of everything done in this field for decades to come. Rousseau appears to have owned a copy of *Historia muscorum* in 1768,[154] but later borrowed it from Malesherbes, and in a letter to Liotard he stated that he no longer had his own copy.[155] According to one tradition, Rousseau gave his copy to the Montpellier botanist, Antoine Gouan. In that case, this gift was probably in exchange for plant specimens and a copy of Gouan's treatise on fish.[156] Gouan's copy[157] contains two annotations in

149. My emphasis; Rousseau is referring to Pier-Antonio Micheli, *Nova plantarum genera juxta Tournefortii methodum disposita* (Florence, Bernardo Paperini, 1729).

150. Rousseau to Dr Samuel Auguste André David Tissot, 1 February 1769; *CC*, xxxvii.38.

151. See Rousseau to Guy, 20 January 1766, 8 January 1767, and 20 January 1767.

152. Duchess of Portland, 12 September 1767, *CC*, xxxiv.97; Rousseau to Latourrette, 17 December 1769; *CC*, xxxvii.187.

153. Magnin-Gonze, *Histoire de la botanique*, p.123.

154. Rousseau requested this work from Guy on 25 November 1767, but objected to the price. He obtained a copy in 1768, possibly from Daniel Malthus's cousin, Jane Dalton: 'Je n'ai pas pu la trouver, mais Mademoiselle Dalton qui l'a [...] veut positivement que vous l'acceptiez'. Daniel Malthus to Rousseau, 24 January 1768, *CC*, xxxv.53. His annotations of Regnault in 1777-1778 contain one reference to Dillenius. BdAN [Munch MS], art. 253

155. Rousseau to Liotard, 7 November 1768; *CC*, xxxvi.171; Rousseau to Malesherbes, 17 April 1772; *CC*, xxxix.36-41.

156. Rousseau to Gouan, 26 December 1769; *CC*, xxxvii.194. The work appeared in both French and Latin, but Rousseau simply refers to Gouan's 'Icthyologie': *Historia piscium: sistens ipsorum Anatomen externam, internam, atque Genera in Classes et Ordines redacta: accedunt Vocabularium locupletissimum, Indices latini ac gallici, Experimenta circa Motum natatorium et muscularem, Respirationis mechanismum, Auditus et Generationis organa or Histoire des poissons, contenant la description anatomique de leurs parties externes & internes, et le caractère des divers genres rangés par classes et par ordres. Avec un vocabulaire complet, des tables raisonnées en latin et en françois* (Strasbourg, A. König, 1770).

157. Dillenius, *Historia muscorum*, p.iv. The annotated copy is found in the collections of the

English; both marginalia to Dillenius's 'Praefatio'.[158] The tone is critical: 'For want of using proper microscopes, Dillenius has given very imperfect descriptions of the smaller tribes, particularly of the Confervae.'

Finally, the sheer challenge of the subject matter might have been one of its attractions. Mosses and lichens, with their small size and encrypted reproduction, are, like grasses, difficult to study: 'les mousses et les graminées [sont] deux familles difficiles et nombreuses. Il faudrait qu'il tâchât de les prendre dans l'état de floraison où leurs caractères déterminants sont les plus marqués'.[159]

These difficulties notwithstanding, Rousseau contemplated writing a treatise on mosses in response to a proposal from the famous publisher, Charles-Joseph Panckoucke (1736-1798). In his reply to Panckoucke, Rousseau apologised for his diminished faculties: 'D'abord je suis bien éloigné de me sentir capable de la besogne qu'il vous a pris la fantaisie de me proposer, et quand je serais en état de m'en tirer je n'aurais jamais le courage d'entreprendre dans l'anéantissement où je suis une tâche aussi laborieuse que celle qu'on me propose d'embrasser'.[160] Yet he accepted the proposal, limiting its scope to mosses, while excluding lichens, mushrooms and exotic plants:

> Puisque vous avez songé aux Mousses, les mousses soit: mais n'y ajoutons rien de plus, ni byssus, ni conferva, ni champignons, ni plantes exotiques. A peine ai-je la moindre idée de tout cela, et il est trop tard et j'ai de trop mauvais yeux pour prendre toute cette étude ab ovo. Mon dessein n'est même pas d'embrasser les Lichens; mais je n'aurai déjà que trop de mes mousses pour le peu de forces qui me restent, et je suis bien éloigné d'espérer rien faire qui soit digne d'entrer dans l'ouvrage auquel vous voulez me donner la présomption d'avoir quelque part. Cependant comme la matière est de mon goût, je m'en occuperai sans répugnance comme sans diligence, et quand mes barbouillages ne seraient bon à la fin qu'à jeter au feu, j'aurai toujours eu le plaisir de botaniser en les faisant, de faire auprès de Mr de Buffon acte de zèle et de bonne volonté, et d'obéissance auprès de vous.[161]

Faculty of Medicine, Montpellier (call number 210491; old call number DC 128 4°). I discuss Rousseau's relationship with Gouan in Chapter 8.

158. According to Grasset, it is 'non contestée' that the annotations are Rousseau's. J.-J. Grasset, 'Jean-Jacques Rousseau à Montpellier', *Académie des sciences et lettres de Montpellier: mémoires de la section des lettres* (1847-1854), p.553-85 (573, n.2). Yet Rousseau claims: '[M]on ignorance dans la langue anglaise qui m'empêche d'en entendre la poésie ne me laisse pas partager le plaisir que l'on prend à la lire'. Rousseau to the Duchess of Portland, 19 May 1772; *CC*, xxxix.57. The annotations could be from Jane Dalton, but I have yet to sight her handwriting for verification.

159. Rousseau to Malesherbes, 19 December 1771; *CC*, xxxviii.301. Rousseau also studied grasses, as noted earlier.

160. Rousseau to Panckoucke, 16 October1771; *CC*, xxxviii.274.

161. Rousseau to Panckoucke, 16 October1771; *CC*, xxxviii.274-75.

Rousseau overcame his reluctance with an enthusiasm for the subject, 'la matiére est de mon goût'. He suggested that Buffon and Philibert Guéneau de Montbeillard (1720-1785) were preparing the 'premier volume du régne végétal' to provide 'des régles pour la forme de mon travail'.[162] The fate of this work is not known.

vi. The disappearance of Rousseau's *méthode*

Eighteenth- and nineteenth-century commentators knew that Rousseau was 'au courant du débat sur les classifications qui agite alors les naturalistes, sans être connu du public'.[163] For example, Mme Roland (1754-1793) wrote: 'Quelle différence de tous les précepteurs,[164] au simple et sublime Jean-Jacques! Sans adopter *nul système*, il fait suivre la nature et met à la portée de *choisir* ensuite avec profit celui qui plaira davantage. Les idées qu'il donne sont distinctes, agréables; il fait aimer la science; c'est déjà l'avoir enseignée à moitié.'[165] Similarly, Goethe praised Rousseau's adoption of the natural families:[166] 'Seine methode das Pflanzenreich ins Engere zu bringen, neigt sich [...] offenbar zur Einteilung nach *Familien*;[167] und da ich in jener Zeit auch schon zu Betrachtungen dieser Art hingeleitet war, so machte sein Vortrag auf mich einen desto größeren Eindruck.'[168] Augustin-Pyramus de Candolle, the father of modern plant taxonomy, told his students that the *Lettres* were '[i]nspirées par [...] le génie de Jussieu'.[169] Because they did not

162. Rousseau to Panckoucke, 16 October1771; *CC*, xxxviii.275.
163. Drouin, 'Les herborisations d'un philosophe', p.80.
164. These 'précepteurs' were Lamarck and François Joseph Lestiboudois (d.1815), author of *Botanographie belgique, ou Méthode pour connoître facilement toutes les plantes qui croissent naturellement: ou que l'on cultive communément dans les provinces septentrionales de la France* (Lille, Imprimerie de J. B. Henry, 1781).
165. My emphasis; Jeanne-Manon Roland de la Platière to Jean-Marie Roland de la Platière, 16 January 1782; Jeanne-Manon Roland de la Platière, *Lettres de Madame Roland*, ed. Claude Perroud, vol.1 (Paris, 1900), p.148. Mme Roland also praised Latourrette and Rozier's *Démonstrations élémentaires*.
166. Polianski argues that Goethe's was a 'selektive Wahrnehmungsweise der Rousseauschen Botanik [selective apprehension of Rousseau's botany]'. Indeed, '[b]ezeichnend ist, daß Goethe, dessen metaphysische Botanik im klaren Gegensatz zu Rousseau steht, im Nachhinein diese Haltung Rousseaus vôllig ausblendet [significantly, Goethe, whose metaphysical botany clearly contradicted that of Rousseau, later completely ignores Rousseau's position]'. Igor Polianski, *Die Kunst, die Natur vorzustellen: die Ästhetisierung der Pflanzenkunde um 1800* (Cologne, 2004), p.186.
167. See Bernhard Kuhn, '"A chain of marvels": botany and autobiography in Rousseau', *European Romantic review* 17:1 (January 2006), p.1-20 (7).
168. 'His method of narrowing down the plant world lends itself to the classification of plants according to families [...] and since I too had at that time been led to observations of this kind, his presentation made an even greater impression upon me.' Goethe, 'Der Verfasser teilt die Geschichte seiner botanischen Studien mit', p.159.
169. Candolle, *Histoire de la botanique genevoise*, p.20.

follow the sexual system, the *Lettres* retained their appeal: 'M. Benjamin Delessert avait puisé le goût qui le portait vers l'étude des plantes [...] dans les lettres de Jean-Jacques Rousseau sur la botanique [...] ou [...] l'auteur a su se renfermer dans les choses les plus fondamentales, écrites d'hier, quoique 70 années au moins nous séparent de l'époque où elles ont été rédigées'.[170]

Rousseau's reliance on the natural families in the *Lettres* was deemed revolutionary by the German microbiologist, Ferdinand Cohn in his 1886 review of Albert Jansen's *Jean-Jacques Rousseau als Botaniker*:[171]

> Wir müssen Rousseau's botanische Unterrichtsmethode umso höher stellen, als zu jener Zeit die natürlichen Pflanzenfamilien noch nicht gegründet waren. Zwar hatte, wie schon bemerkt Bernard de Jussieu bereits 1758 [sic] die Klein-Trianon angebauten Gewächse nach natürlichen Familien geordnet, aber er hat nicht gewagt damit vor der Öffentlichkeit zu treten: erst 1789 wurde die natürliche Methode durch seinen Neffen Antoine Laurent in die Wissenschaft eingeführt. Rousseau war eben auch in der Botanik, wie in der Politik, der *Vorläufer einer Revolution*, welche erst ein Jahrhundert nach seinem Tode in der Welt Bahn brechen sollte.[172]

Cohn lamented that botanical teaching in Germany followed the sexual system, while ignoring Rousseau's *Lettres*, which were available in German, French, English and several other languages. Although Cohn probably did not know Thomas Martyn's 1785 English translation of the *Lettres*, we can readily date the Linnaean slant given to Rousseau's botanical thought to this edition, which was translated into French, Dutch and Portuguese (discussed in Chapter 9, p.396). Cohn argued correctly that Rousseau's botanical teaching had been misunderstood:[173]

> Rousseaus botanische Briefe sind eine pädagogische That, deren vollen Bedeutung noch heut[e] nicht genügend anerkannt ist. In der Zeit wo jene Briefe geschrieben wurden, war das Linné'sche System zur unbedingten Herrschaft gelangt: kein wissenschaftliches, kein populäres Buch erschien, in dem die Pflanzen nicht nach Linné geordnet waren. Wir wissen, welch begeisterter Verehrer Linné's gerade Rousseau war; die logische Consequenz

170. Antoine Lasègue, *Musée botanique de Benjamin Delessert* (Paris, 1845), p.43-44.

171. Ferdinand Cohn, 'Jean-Jacques Rousseau als Botaniker', *Deutsche Rundschau* 47 (April-May-June 1886), p.364-85 (382-83). Cohn (1828-1898), a botanist, and founder of microbiology, mentored Robert Koch (1843-1910), who won the Nobel Prize for Medicine in 1905 for his research on anthrax and tuberculosis.

172. 'We must rate Rousseau's botanical instruction method that much higher since at that time the natural plant families were not yet established. Indeed, as already mentioned, Bernard de Jussieu had arranged the plants in the Petit Trianon according to natural families, but he had not dared to bring them before the public: only in 1789 was the natural method introduced into science by his nephew, Antoine Laurent. Rousseau was in botany, as in politics, the *forerunner of a revolution* that would only establish itself a century after his death'. My emphasis.

173. Cohn, 'Jean-Jacques Rousseau als Botaniker', p.382.

des Linné'schen Pflanzenystems, die Genauigkeit seiner Artmerkmale, selbst die energische, poetisch angehauchte Sprache des schwedischen Reformators erregt seine vollste Bewunderung. Aber für den botanischen Jugendunterricht verwerthet Rousseau nicht das künstliche System Linné's, sondern die natürliche Methode der Pflanzenfamilien.[174]

Furthermore, despite Goethe's praise for Rousseau's application of the natural method, this method had been ignored, and botanical teaching in Germany consequently retained a Linnaean cast:

Noch heut[e] haben unsere Schulmänner noch kein richtiges Verständnis fur Rousseau's pädagogische Bedeutung. Noch immer wird auf unseren Schulen beim botanischen Unterricht das Linné'sche System reglementsmäßig zu Grunde gelegt, und dadurch der Jugend die Meinung beigebracht, als bestünde die liebenswürdigste der Naturwissenschaften im Zählen von Staubfäden und im gedächtnismäßigen Erlernen von Classen und Ordnungen.[175]

Recent scholarship presents a strange contrast to these assessments,[176] missing the revolutionary character of Rousseau's taxonomic choice in the *Lettres*,[177] and even denying it outright. Instead of acknowledging Rousseau's debt to the natural method,[178] scholars now commonly

174. 'Rousseau's botanical letters are a pedagogical fact, whose significance even today is not sufficiently recognised. At the time when these letters were written, the Linnaean system had achieved dominance: no scientific, no popular book appeared, in which plants were not arranged according to [the system of] Linnaeus. We know what an enthusiastic admirer of Linnaeus Rousseau was; the logic of the Linnaean plant system, the exactitude of his species characteristics, even the energetic, poetically inclined language of the Swedish reformer aroused [Rousseau's] utter amazement. But for botanical teaching of the young Rousseau did not use the artificial system of Linnaeus, but the natural method of plant families.'

175. 'Even today our school teachers do not have a correct understanding of Rousseau's pedagogical significance. In our schools the Linnaean system is still used as the obligatory basis, and youth thereby given the idea that the most appealing of the natural sciences consists in counting filaments and memorising classes and orders', Cohn, 'Jean-Jacques Rousseau als Botaniker', p.383.

176. Exceptions are Cook and Kelly, 'Introduction', *The Collected writings of Rousseau*, vol.8, p.xxvi and Drouin, 'Les herborisations d'un philosophe', p.84.

177. 'Ces huit lettres sont consacrées à l'étude de six familles de plantes que Rousseau a choisies parmi les plus *courantes*.' My emphasis; Ducourthial, *La Botanique*, p.194; see also p.210.

178. Based on the 1789 French translation of Martyn's Linnaean edition of Rousseau's botany, and on the erroneous belief that Rousseau was acquainted with Linnaeus's pupil, Daniel Solander (1773-1782), Polianski concludes: 'Rousseau, der mit dem Linné-Schüler Daniel Solander befreundet war und die Hoffnung auf ein *Natürliches System* der Pflanzen demütig *aufgab*, gehörte zu den schwärmerischen Verehrer Linnés [Rousseau, who was acquainted with Linnaeus's student, Daniel Solander, and gave up hope for a natural system of plants, belonged to the most enthusiastic fans of Linnaeus]'. My emphasis; Polianski, *Die Kunst*, p.185.

characterise the *Lettres* as 'Linnaean', although what this might mean is usually not explained. For example: 'Fontenelle strove to make astronomers of the fair sex while Rousseau attempted to instruct them in *Linnaean* botany',[179] and 'Rousseau was driven to study plants systematically and on a *Linnaean basis* in spite of his hostility to academic science'.[180] Rousseau has been characterised as 'l'artisan en France de la popularisation des idées *linnéennes*'.[181] 'As Rousseau was a Linnaean, and was published before Jussieu brought out his natural classification in 1789, his popularity helped to ensure that popular botanophilia would remain hitched to the *Linnaean sexual system* into the 19th century and increasingly remote from scientific botany.'[182] In short, recent accounts deny that the natural method played any role in Rousseau's botanical work.

Contrary to this received view, the *Lettres élémentaires* do not refer directly to the sexual system; nor does their classificatory basis rest on that system. They are hybrid in approach, taking their inspiration as much from Jussieu and Adanson, as they do from Linnaeus. This hybridisation has occasioned puzzlement on the part of scholars who propagate the image of Rousseau as Linnaean sexualist: 'It is also curious that [...] Rousseau, Europe's greatest advocate of sexual complementarity, did not indulge in heavy-handed sexual imagery [...] His letters on botany [...] are remarkable for their *neglect* of sexual reproduction in plants.'[183]

vii. Conclusion

Rousseau's taxonomic choices were more contingent and his loyalties less fixed than they have usually been taken to be; I argue that he was, in effect, a taxonomic pluralist who accepted the legitimacy of more than one way of carving up the plant kingdom. However, his botanical work has all too frequently been construed as demonstrating a loyalty to Linnaeus and the sexual system that is not sustained by a careful examination of his famous botanical course; at the same time, his turn toward the natural method has been marginalised or even forgotten. We

179. My emphasis; George, *Botany, sexuality and women's writing*, p.51. See also Williams, *Botanophilia in eighteenth-century France*, p.105, and Schiebinger, *Nature's body*, p.12, who suggests that by praising binomial nomenclature, Rousseau followed the sexual system.

180. My emphasis; George, *Botany, sexuality and women's writing*, p.53. This conclusion is based on a few quotations from the letters that seem 'flirtatious', 'intimate' or sexually initiatory, together with Rousseau's celebration of Linnaeus's binomial names in the so-called 'Introduction' to the *Fragmens pour un dictionnaire*. Yet none of these texts constitute a 'Linnaean basis' for the eight letters.

181. My emphasis; Duris, *Linné et la France*, p.105.

182. My emphasis; Williams, *Botanophilia in eighteenth-century France*, p.95.

183. My emphasis; Schiebinger, *Nature's body*, p.219-20, n.66. *Emile* and *Julie* address sexual complementarity.

should be struck not by Rousseau's devotion to one (sexual) system, but rather, by his flexibility in matters taxonomic as well as political and moral: different approaches suit different circumstances.[184] The natural method suits the exposition of the fundamentals of botany, while the sexual system is useful for organising large numbers of specimens. It was, I argue, not for his sexual system alone that Rousseau prized Linnaeus, but rather for the totality of his insights: the *Philosophia botanica* gave him a 'profit plus réel [...] que tous les livres de morale';[185] 'il était réservé à l'illustre Linnaeus' to make of botany 'une science philosophique'.[186]

It is therefore time to extract the *Lettres* from the framework of the sexual system in which their various editors and commentators have placed them, and to restore them to their true role as a lucid expression of the natural method. If Rousseau had wished to give an explication of the sexual system in the *Lettres*, he would have done so; it is most telling that he did not. Turning the conventional wisdom on its head, I argue that far from pandering to a presumed audience of feminine and juvenile amateurs, the *Lettres* represent Rousseau's most advanced thinking on botany, which took its inspiration not only from Linnaeus, but also from Antoine Laurent de Jussieu and Michel Adanson.[187] Following the Linnaean teaching on habit in *Philosophia botanica* section 168, Rousseau made the 'air de famille' or 'port' the centrepiece of his lessons to Mme Delessert and her children. Following Adanson, he stressed the relevance of all the plant's parts. My conclusion only appears unlikely if we assume that Rousseau did not read his Linnaeus very carefully; yet we have every reason to believe that he read his Linnaeus very carefully indeed.

Once we see Rousseau's taxonomic choices clearly, we can understand the myriad ways in which his study of botany relates to his philosophy as a whole.

First: throughout his works, Rousseau exalted nature, which cannot lie, as a source of truth, as the common mother of humanity and as a refuge from the social passions. In this refuge he could meditate on the beings around him, achieve mastery over his negative feelings and harness his energies to an edifying study that brought him closer to the Author of all

184. See for example Rousseau, *Préface à Narcisse*, *OC*, ii, 971-72; *Du contrat social*, *OC*, iii.414-19.

185. Rousseau to Linnaeus, 21 September 1771; *CC*, xxxviii.267. See also Jean-Marc Drouin, 'Rousseau lecteur de Linné', p.39-57.

186. '[J]e ne conviendrais pas tout à fait [...] que Tournefort soit le plus grand Botaniste du siècle; il a la gloire d'avoir fait le premier de la Botanique une étude vraiment méthodique'. Rousseau to Dutens, 5 February 1767, *CC*, xxxii.99.

187. Duris, *Linné et la France*, p.105. Rousseau's critical comments about Adanson's dismissal of Linnaeus and his nomenclature should not obscure Rousseau's debt to this non-conformist botanist. See Rousseau, 'Introduction', *OC*, iv.1208.

things through the apprehension of final ends and the continuous chain of being.

Second: Rousseau understood that in order to achieve these results, his meditations required a focus and a way to order the material he was studying: to regard the plant realm dumbly as mere spectacle without any idea of system was insufficient. He began studying botany by an artificial system that he ultimately found wanting, and like most thoughtful students, he advanced to more advanced theories, of which the natural method was one. Indeed, it might be argued that for a philosopher who exalted nature as a guide, Rousseau's disillusionment with the artificial sexual system was over-determined. His progression to the natural method implied a conscious decision to adopt a way of viewing the plant kingdom that leading contemporary botanists considered most 'faithful to nature' because it was based on the concept of continuity.[188] Given Rousseau's philosophical principles, nothing short of such a natural method would do.

188. Jussieu, *Genera plantarum*, p.xxxiii, translated in Stevens, *Development of biological systematics*, p.353.

7. Modes of mediation: botanical books and binomial names

> [L]e botaniste ne souffre point d'intermédiaire entre la nature et lui.[1]

i. Botanical mediations

Language is the first social institution, and thus books, like names, are mediations, conventions or 'coinages'.[2] Rousseau held that the name should not be substituted for the thing itself.[3] Yet, these objections notwithstanding, Rousseau believed that names and books are indispensable to botanists. He championed an orderly nomenclatural regime in botany – one and only one set of rigid designators – thereby acknowledging the utility, if not the legitimacy, of this mediation.[4] He likewise recognised that to reject nomenclature would entail renouncing books, as the former is impossible without the latter. This he was also unwilling to do, for botanical texts report the observations of others that readers may confirm or reject:

> Il y a longtemps que j'éprouve les difficultés de la *nomenclature*, et j'ai souvent été tenté d'abandonner tout à fait cette partie; mais il faudrait en même temps renoncer aux *livres* et à profiter des observations d'autrui, et il me semble qu'un des plus grands charmes de la botanique est après celui de voir par soi-même, celui de *vérifier* ce qu'ont vu les autres; donner sur le témoignage de mes propres yeux mon assentiment aux observations fines et justes d'un auteur, me paraît une véritable jouissance: au lieu que quand je ne trouve pas ce qu'il dit, je suis toujours en inquiétude si ce n'est point moi qui vois mal. D'ailleurs ne pouvant voir par *moi-même* que si peu de chose, il faut bien sur le reste en ces choses me fier à ce que d'autres ont vu, et leurs différentes nomenclatures me forcent pour cela de percer de mon mieux le *chaos* de la synonymie.[5]

For example, at Montmorency he discovered that Vaillant had reported findings in *Botanicon parisiense* that could not be verified (fig.7.1, p.213).[6]

1. Rousseau, *OC*, iv.1250.
2. Rousseau, *Essai sur l'origine des langues*, *OC*, v.375; *Introduction*, *OC*, iv.1209.
3. Rousseau, *Emile*, *OC*, iv.434.
4. Today, the object so designated is to be the same in all worlds. Bird and Tobin, 'Natural kinds'.
5. My emphasis; Rousseau to Malesherbes, 17 April 1772; *CC*, xxxix.37.
6. See for example Rousseau to Latourrette, 25 January 1772; *CC*, xxxix.22.

This peer review expresses the critical spirit of the republic of science in which results had to be demonstrated before one's peers. These mediations enabled a global community of botanists to constitute itself.

ii. The problem with books

> L'abus des livres tue la science.[7]
>
> Toujours des livres! Quelle manie. Parce que l'Europe est pleine des livres, les Européens les regardent comme indispensables, sans songer que sur les trois quarts de la terre on n'en a jamais vu.[8]

Rousseau acquired a considerable botanical library and was conversant with the important works of antiquity, the Renaissance, and the early-modern period.[9] It is telling that when considering the sale of his botanical library, he was concerned that it not be dispersed to preserve its integrity for a future owner: 'Ces sortes de collections, quoique petites, perdent beaucoup de leur prix étant dépareillées, vu la peine qu'il en a coûté pour les rassembler'.[10]

Yet the problematic character of books pervades Rousseau's botanical reflections as it does his other writings; he was a critic of books in general and of botanical books in particular. For Rousseau the book was a necessary evil, an artefact of the socialisation of humanity and humanity's tragic imprisonment in relations of mediation. Studying botany constitutes an effort to supersede mediation: 'le botaniste ne souffre point d'intermédiaire entre la nature et lui'.[11]

Rousseau's antipathy toward books seems paradoxical for the author of so many works.[12] In the first *Discours* he condemns the invention of the printing press, citing 'les désordres affreux que l'imprimerie a déjà causés en Europe'; he praises the Caliph Omar for destroying the library at Alexandria on the grounds that books should be burned if they contained anything contrary to the Koran, but if they contained *nothing* contrary to it, they should still be burned.[13]

Rousseau's criticisms of writers and literary activities are legion. He deplored his own entry into the literary life and claimed it commenced his lifelong miseries and sufferings at the hands of a cabal of malicious

7. Rousseau, *Emile*, *OC*, iv.826.
8. Rousseau, *Emile*, *OC*, iv.620.
9. Concerning Rousseau's botanical library, see Appendix 1.
10. Rousseau to Gouan, 26 December 1769; *CC*, xxxvii.196.
11. Rousseau, *Fragments de botanique*, *OC*, iv.1250.
12. For a discussion of Rousseau and authorship, see Christopher Kelly, *Rousseau as author: consecrating one's life to the truth* (Chicago, IL, 2003).
13. Rousseau, *Discours sur les sciences et les arts*, *OC* iii.28.

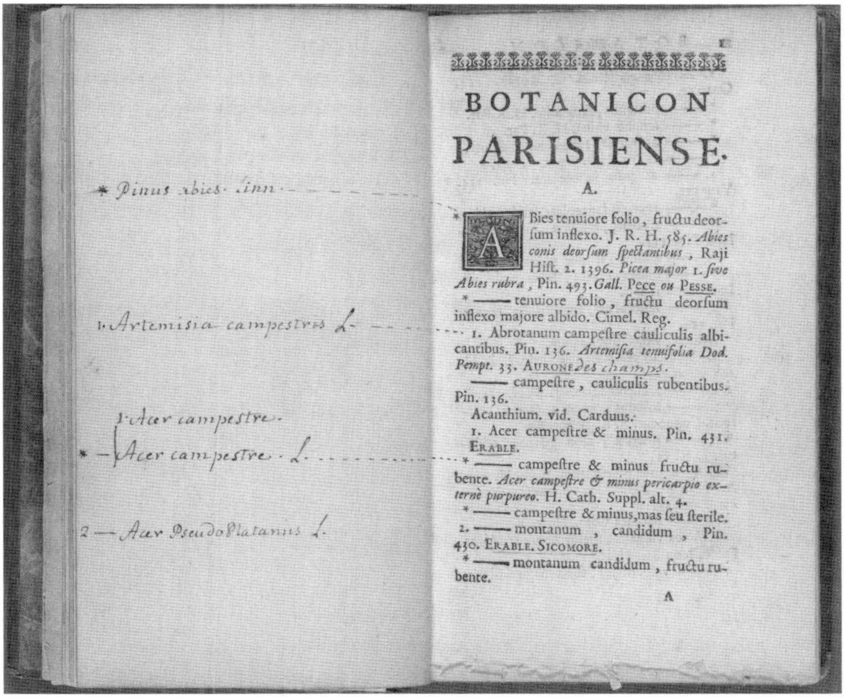

Figure 7.1: Rousseau's annotated copy of Sebastien Vaillant, *Botanicon parisiense:* (Leiden and Paris, Briasson,1743), p.1, Call number 007197381, Houghton Library, Harvard University.

philosophes.[14] He criticised those who study plants merely to parade their knowledge 'pour faire des livres [...] pour devenir auteur ou professeur'.[15]

In botanical books people had traditionally sought pharmaceutical information: 'on ne cherchait pas des plantes mais des simples'.[16] By the seventeenth century another imperative had emerged: the need for ways to order an ever-growing mass of botanical information, the product of 'voyages immenses, coûteux, pénibles, et périlleux'.[17] Rousseau accepted neither of these imperatives, however, since he rejected both pharmaceutical and exotic botany.[18]

Even when studying nature without the aid of books, Rousseau often recommended the one book that is always good to read: the one written not by men, but by nature. This credo derives from his rejection of 'l'autorité des hommes qui sont menteurs', in favour of 'nature[,] qui ne ment jamais'.[19] Rousseau praised the Duchess of Portland and Pierre Clappier for their ability to read this book[20] and he stressed to Mme Delessert that she must first study the book of nature, before using the books of men:

> [J]e vous préviens que si vous voulez prendre des livres et suivre la nomenclature ordinaire, avec beaucoup des noms vous aurez peu d'idées, celles que vous aurez brouilleront, vous ne suivrez bien ni ma marche ni celle des autres, et n'aurez tout au plus qu'une connaissance des mots. Chère Cousine, je suis jaloux d'être votre seul guide dans cette partie. Quand il en sera temps je vous indiquerai les livres que vous pouvez consulter. En attendant ayez la patience de ne lire que dans *celui de la nature*.[21]

Rousseau's method can be learned without books because 'il ne sera pas question si tôt des genres et des espèces'.[22] Only later may she 'prendre les livres' to learn two hundred binomial names of 'Linnaeus, quoique latins'.[23]

14. Rousseau, *Rêveries*, *OC*, i.1077.
15. Rousseau, *Rêveries*, *OC*, i.1069.
16. Rousseau, 'Introduction', *OC*, iv.1201.
17. 'Introduction', *OC*, iv.1209; see also Linnaeus, *Disquisitio*, p.4; *Sexes of plants*, p.3.
18. He used the findings of both.
19. Rousseau, *Fragments de botanique*, *OC*, iv.1250; see also Rousseau, *Discours sur l'origine de l'inégalité*, *OC*, iii.133, Rousseau, *Dialogues*, *OC*, i.833. This is nature perceived and mediated by the senses.
20. Rousseau to the Duchess of Portland, 20 October 1766; *CC*, xxxi.40; Rousseau to Clappier, 17 April 1769; *CC*, xxxvii.84.
21. My emphasis; Rousseau to Mme Delessert, 16 May 1772; *CC*, xxxix.50. Arguably, Rousseau was writing a botanical book, but he declined to act the part of M. Josse.
22. Rousseau to Mme Delessert, 2 May 1773; *CC*, xxxix.151.
23. Rousseau to Mme Delessert, 30 August 1773; *CC*, xxxix.192. See his recommendation to Panckoucke that he learn 'familièrement et empiriquement environ une centaine de plantes'. Rousseau to Panckoucke, 16 October 1771; *CC*, xxxviii.275. Hence, Mme Delessert is to learn *twice* as many plants and their names as a man of learning.

In *Emile* Rousseau directs his pedagogical polemics against teaching children from books. The child does not have to be a Bacon, Descartes or Newton, but he should learn about the phenomena of the natural world so that he can lead a self-sufficient life; by abandoning books and rote learning, the child should learn from the things themselves (mediated by the teacher's construction of circumstances): 'il prend ses leçons de la nature et non pas des hommes'.[24] For the child there is no science of words; there is no science of books. Encountering phenomena is the only lesson; nature is the only teacher. Rousseau allowed Emile only one book: *Robinson Crusoe*. By the time he was writing the *Rêveries*, Rousseau had reduced his own reading to Plutarch's *Vies*, the book he first read with his father as a child in Geneva, and that accompanied him throughout his life.[25]

iii. Rousseau's turn toward illustrated botanical books

Rousseau's general complaint against botany books is that they are written not for novices, but for 'maîtres'.[26]

> Tous les livres du monde ne valent pas un bon guide et n'y sauraient suppléer, parce qu'ils sont tous écrits pour des disciples déjà instruits par des cours sous des professeurs ou démonstrateurs [...] j'ai pris des livres qui supposant des connaissances que je n'avais pas ne m'ont pu mener à l'inconnu que par l'inconnu ce qui n'est pas le moyen d'apprendre.[27]

A truly elementary work of botany was lacking: 'Les livres des botanistes modernes n'instruisent que les botanistes; ils sont inutiles aux ignorants. Il nous manque un livre vraiment élémentaire, avec lequel un homme qui n'aurait jamais vu de plantes pu[t] parvenir à les étudier seul.'[28] The problems posed by contemporary botanical texts can be reduced to three: (1) the difficulties of systems and methods (discussed previously), (2) lack of illustrations, and (3) the difficulties of synonymy (discussed below, p.244-47).

For Rousseau one deficiency of eighteenth-century botanical texts stood out beyond all others: the absence of illustrations. This was perhaps a more serious problem than the advanced systems and methods conveyed in these texts. Important botanical works to which Rousseau

24. Rousseau, *Emile*; *CC*, iv.361.
25. Rousseau, *Rêveries*; *OC*, i.1024. Rousseau read Plutarch, *Les Vies des hommes illustres Grecs et Romains comparée l'une avec l'autre par Plutarque de Chaeronée*, trans. Simon Goulart, *et al.* (Lyon, Paul Frellon, 1611), cited in *OC*, i.1238, n.2.
26. Rousseau to the Duchess of Portland, 12 February 1767; *CC*, xxxii.134.
27. Rousseau to Pierre Clappier, 23 December 1768; *CC* xxxvi.215.
28. Rousseau to the Duchess of Portland, 12 February 1767; *CC*, xxxii.134. Not even the textbook by Rozier and Latourrette was sufficiently elementary. Rousseau to Clappier, 23 December 1768; *CC*, xxxvi.214-16.

often referred, for example Linnaeus's *Species plantarum*, lacked illustrations: 'Quand j'ai vu dans mon Linnaeus la Classe et l'ordre d'une plante qui m'est inconnue, je voudrais me *figurer* cette plante, savoir si elle est grande ou petite, si la fleur est bleue ou rouge, me représenter son *port*? Rien. Je lis une description caractéristique, d'après laquelle je ne puis rien me représenter. Cela n'est-il pas désolant?'[29]

Rousseau's objection to the absence of illustrations from botanical texts is connected with his application of the distinction between primary and secondary qualities to the plant; *form* – its habit or 'port' – is its primary quality, in contrast with secondary qualities such as odour or colour that are merely apprehended in the mind of the observer. This form can be conveyed by illustrations, such as engravings and woodcuts.

The absence of illustrations from his works was no oversight on Linnaeus's part, however; Linnaeus explicitly justifies the exclusion of illustrations in 'Ratio operis', the preface to *Genera plantarum*: 'Icones pro determinandis generibus non commendo, sed absolute rejici'.[30] Pictures may help boys, but the mature brain does not need them. He supports his position with three arguments: (1) an argument cannot be deduced from a picture, (2) it is not always to possible to engrave or depict a plant and (3) there are too many variations in number and shape of characters to depict all of them in pictures, whereas the 26-letter alphabet provides a ready way to do so without any pictures whatsoever. In a plant its primary qualities could be reliably apprehended: for example form, shape, number and position of the parts of the fructification. These were considered objective, while the secondary ones of colour, taste and smell were considered too variable to provide a constant point of reference.

Linnaeus reiterated his rejection of illustrations in his description of botanical education in *Philosophia botanica*: 'Sic plantae omnes uno anno, primo intuitu, absque praeceptore, sine inconibus aut descriptionibus, constanti memoria addiscuntur.'[31] Daston and Galison argue that in the truth-to-nature tradition represented by Linnaeus, the object for the scientist was to 'learn to see the typical', not the details of every particular thing; this was 'the achievement of a lifetime'.[32] According to Linnaeus,

29. My emphasis; Rousseau to the Duchess of Portland, 12 February 1767; *CC*, xxxii.134.

30. Carolus Linnaeus, *Genera plantarum*, 2nd edn (Leiden, C. Wishoff, 1742), p.viii, section 13. 'I do not recommend drawings for determining genera, but absolutely reject them.' Staffan Müller-Wille and Karen Reeds, 'A translation of Carolus Linnaeus's introduction to *Genera plantarum*', p.568.

31. Linnaeus, *Philosophia botanica*, p.97, section 151. '[A]ll plants become known in a single year, *at first sight*, with no instructor and *without pictures* or description by means of stable recollection.' My emphasis; Linnaeus, *Philosophia botanica* (2003), p.111, section 151.

32. Daston and Galison, *Objectivity*, p.82.

picturing individual plants or species was a hopeless endeavour; given the variety of nature, plant descriptions should be exclusively verbal.[33] With his stipulation against pictures, Linnaeus anticipated twentieth-century logical positivism, which held pictures to lack objectivity: 'these ascetics [...] aspired to a higher, purer form of knowing entirely free of pictures, intuitions or indeed any aspect of the senses'.[34]

Linnaeus was not alone in holding this view; his position cohered with the general trend in eighteenth-century taxonomic works. The 'disappearance of images from botanical books in the seventeenth century' is a well-documented fact; illustrations were replaced by 'a precise technical language that, to the uninitiated, conveyed more than woodcuts or engravings could'.[35]

Despite his admiration for the Swedish botanist, Rousseau could not swallow this argument; very early in his botanical studies he become enamoured of illustrated works, especially Renaissance herbals, which he sought at great expense. He also looked for contemporary illustrated works such as florilegia – works intended more for the book connoisseur than for the botanist.[36] There were also many works to help health professionals such as apothecaries and physicians identify plants by sight, rather than from written descriptions and synonymies; physicians and apothecaries needed to be able to identify medicinal plants with certainty in order to avoid harming patients.[37] Rousseau consulted such works for their illustrations despite his criticism of pharmaceutical botany:

> Il me semble que tous les livres qu'on écrit sur la botanique ne sont bons que pour ceux qui la savent déjà. J'ai acquis votre Stillingfleet et je n'en suis pas plus avancé. J'ai pris le parti de renoncer à toute *lecture* et de vendre mes *livres* et mes estampes, pour acheter des *plantes gravées*; sans avoir le plaisir d'apprendre, j'aurai celui d'étudier et pour mon objet cela revient à peu près au même.[38]

33. Linnaeus, 'Ratio operis', *Genera plantarum*, section 18α. Linnaeus's antipathy to illustrations may have been influenced by his poor draughtsmanship. Personal communication, Karen Reeds.
34. Daston and Galison, *Objectivity*, p.259-60. Daston and Galison do not discuss the manifesto of 'Ratio operis.' They are concerned with the Ur-typic character of the uncoloured illustrations of leaves and other plant parts in *Hortus cliffortianus* and *Philosophia botanica*.
35. Brian W. Ogilvie, 'Image and text in natural history, 1500-1700', in *The Power of images in early modern science*, Wolfgang Lefèvre, *et al.* (Basel, 2003), p.141-66 (141, 142).
36. Brian W. Ogilvie, 'The many books of nature: Renaissance naturalists and information overload', *Journal of the history of ideas* 64:1 (January 2003), p.29-40 (39-40).
37. Such a work was Elizabeth Blackwell, *A Curious herbal containing five hundred cuts of the most useful plants which are now used in the practice of physic: engraved on folio copper plates, after drawings taken from the life*, 2 vols (London, J. Nourse, 1739-1751). See Malthus to Rousseau, 24 January 1768, *CC*, xxxv.53-56.
38. My emphasis; Benjamin Stillingfleet, the original bluestocking, wrote a treatise on grasses,

Remarkably, Rousseau was interested in pictorial works of botany as early as December 1764 when he wrote to Nicolas-Bonaventure Duchesne, 'Je ne manque pas de livres de botanique; ce dont j'ai besoin, c'est des figures de *plantes* bien dessinées et bien *gravées*, et je me tourmente pour en avoir sans pouvoir en obtenir aucune d'aucun côté: cela me désole'.[39] Three years later he was still searching for illustrated works: 'S'il paraît quelque chose de nouveau sur la botanique, surtout des *plantes gravées*, je vous prie de m'en donner avis'.[40]

The illustrated works on plants that Rousseau collected, in which he expressed an interest, or that were proposed to him, were many and diverse in authorship, purpose and date. Taken together, these works contained thousands of illustrations and ranged widely over time and space, starting with the principal works of the botanical Renaissance in sixteenth-century Flanders, Germany and England. Illustrated works in-folio such as florilegia and herbals were expensive; while it is not always possible to determine whether he purchased these works, was given them or borrowed them, it appears that Rousseau purchased some of them. This helps explain his complaints about the cost of books: 'Une des choses qui me dégoûteraient de la botanique serait le *prix énorme* de la plupart des livres qui en traitent, et la nécessité toutefois d'avoir tous ces livres ou la plupart, surtout lorsque n'ayant point suivi de cours ni étudié sous aucun maître on est réduit à étudier seul'. Consequently, he had to pass up the *Hortus cliffortianus* and *Amoenitates academicae* of Linnaeus.[41]

Although Rousseau is not known for taking an interest in Renaissance botanical authors, he acquired a number of important illustrated works from this era during which the herbal flourished.[42] Owing to its prosperity and central role in the northern Renaissance, botany and horticulture thrived in the courts and universities of sixteenth-century Flanders. Premier among the Flemish botanical authors was Carolus Clusius (Charles de l'Ecluse) (1526-1609), one of the founders of scientific as well as exotic botany in the sixteenth century.[43] Originally from Arras, Clusius studied with the famous Montpellier physician and botanist, Guillaume Rondelet (1507-1566). As overseer of the imperial medical garden of Vienna (1573-1576), Clusius was able to travel and collect

referred to in Chapter 6 (p.197n), 'Observations on grasses' in *Miscellaneous tracts*. Rousseau to Malthus, 2 January 1767; *CC*, xxxii.2.

39. My emphasis; Rousseau to Duchesne, 16 December 1764; *CC*, xxii.242.

40. My emphasis; Rousseau to Guy, 20 January 1767; *CC*, xxxii.55.

41. My emphasis; Rousseau to Gouan, 6 October 1769; *CC*, xxxvii.152.

42. Arber, *Herbals: their origin and evolution*. I am not aware of any scholarly work on Rousseau's interest in Renaissance botany.

43. See Florike Egmond, *The World of Carolus Clusius: natural history in the making 1550-1610* (Brookfield, VT, 2010).

widely, becoming an expert on eastern European flora. In 1593 Clusius accepted the chair of botany at Leiden. He is known for his exact descriptions, and for having introduced a number of economically important exotic plants to Europe such as the tulip and the potato. The 1109 woodcuts in his *Rariorum plantarum historia* (Antwerp, 1601), included 233 from the Spanish flora and 356 from the Austro-Hungarian flora. Rembert Dodoens and Matthias de Lobel used his illustrations.

Rousseau was keen to acquire Clusius' works, but had difficulty obtaining them. In a letter of 28 December 1767 Rousseau requested Clusius's richly illustrated collected works, *Rariorum plantarum historia* (Antwerp, 1601), in one folio volume,[44] although a year later he still did not have them, complaining that 'la France est si barbare encore en botanique, qu'on n'y trouve presque aucun Livre de cette science; et j'ai été obligé de faire venir à grands frais de Hollande et d'Angleterre le peu que j'en ai, encore ai-je cherché partout ceux de *Clusius* sans pouvoir les trouver'.[45] Yet he must have consulted another work of Clusius at some point, for he wrote in his annotations of the Regnaults' work that the tamarisk 'devient [souvent] beaucoup plus grand, et Clusius assure en avoir vu en Espagne dont un homme avait peine à embrasser le tronc' (fig.7.2, p.220).[46]

Another important Flemish botanist was Matthias de Lobel (de L'Obel or Lobelius) (1538-1616), who came from Lille and, like Clusius, studied in Montpellier with Rondelet; Lobel eventually made his way to England, where he became botanist to King James I. Lobel's *Plantarum seu stirpium icones* (1581)[47] was famous for its 1441 woodcuts, which 'eurent une large répercussion, étant de fait l'ouvrage botanique soi-disant lobélien le plus souvent cité'.[48] Rousseau requested it from Rey under the title, 'Lobelii Plantarum *icones*'.[49]

A letter of January 1770 shows that Rousseau owned several important Renaissance botanical works:

> Je suis ici dans un grand embarras pour le transport de mon bagage consistant en grande partie dans un attirail de Botanique [...] Si même la

44. Rousseau to Rey, 28 December 1767; *CC*, xxxiv.268.
45. My emphasis; Rousseau to Du Peyrou, 19 December 1768; *CC*, xxxvi.211.
46. BdAN [Munch MS]. '115 – Le Tamaris, ou Tamarisc'. This comment suggests that he had seen Carolus Clusius, *Rariorum aliquot stirpium per Hispanias observatarum historia* (Antwerp, Christoph Plantin, 1576), p.104. Clusius called the tamarisk Myrica sylvestris (= *Tamarix gallica* L.). See Linnaeus, *Species plantarum*, vol.1, p.270-71.
47. Matthias de Lobel, *Plantarum seu stirpium icones*, 2 vols in-4° obl. (Antwerp, Christoph Plantin, 1581).
48. Armand Louis, *Mathieu de L'Obel (1538-1616): épisode de l'histoire de la botanique* (Ghent, Louvain, 1980), p.34. For Rousseau's knowledge of Lobel, see BdAN [Munch MS], no. 230. *L'Herbe aux cuillers*. This reference indicates that Rousseau had access to Lobel's work.
49. My emphasis; Rousseau to Rey, 28 December 1767; *CC*, xxxiv.268.

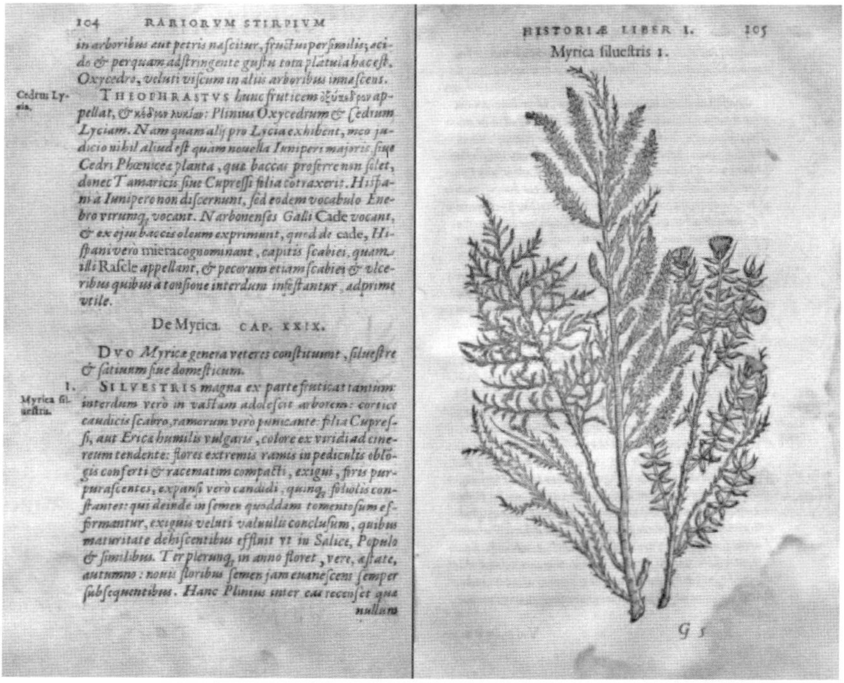

104 RARIORVM STIRPIVM

in arboribus aut petris nascitur, fructus persimilis, acido & perquam adstringente gustu tota plantula hac est. Oxycedro, veluti viscum in aliis arboribus innascens.

Cedrus Lycia. THEOPHRASTVS *hunc fruticem ὀξύκεδρον appellat, & κέδρον λυκίαν: Plinius Oxycedrum & Cedrum Lyciam. Nam quam alij pro Lycia exhibent, meo judicio nihil aliud est quam nouella Iuniperi majoris siue Cedri Phœniceæ planta, quæ baccas proferre non solet, donec Tamaricis siue Cupressi folia cotraxerit. Hispani a Iunipero non discernunt, sed eodem vocabulo Enebro vtrunque vocant. Narbonenses Galli* Cade *vocant, & ex ejus baccis oleum exprimunt, quod de* cade, *Hispani vero* mieracognominant, *capitis scabiei, quam illi* Rascle *appellant, & pecorum etiam scabiei & vlceribus quibus a tonsione interdum infestantur, adprime vtile.*

De Myrica. CAP. XXIX.

DVO *Myricæ genera veteres constituunt, siluestre & satiuum siue domesticum.*

1\. Myrica siluestris. SILVESTRIS *magna ex parte fruticat tantum: interdum vero in vastam adolescit arborem: cortice caudicis scabro, ramorum vero punicante: folia Cupressi, aut Ericæ humilis vulgaris, colore ex viridi ad cinereum tendente: flores extremis ramis in pediculis oblongis conferti & racematim compacti, exigui, foris purpurascentes, expansi vero candidi, quinque foliolis constantes: qui deinde in semen quoddam tomentosum efformantur, exiguis veluti valuulis conclusum, quibus maturitate dehiscentibus effluit vt in Salice, Populo & similibus. Ter plerunque in anno floret, vere, æstate, autumno: nouis floribus semen jam euanescens semper subsequentibus. Hanc Plinius inter eas recenset quæ nullum*

HISTORIÆ LIBER I. 105

Myrica siluestris 1.

G 5

Figure 7.2: Myrica silvestris (= *Tamarix gallica* L.), woodcut from Carolus Clusius, *Rariorum aliquot stirpium per Hispanias observatarum historia* (Antwerp, Christoph Plantin, 1576), p.105, courtesy of the Biodiversity Heritage Library (online).

revue de mon herbier et de mes livres de botanique pouvait vous amuser quelques moments, le tout pourrait être déposé chez vous et vous le visiteriez à votre aise. Je ne doute pas que vous n'ayez la plupart de mes livres. Il peut cependant s'en trouver d'Anglais comme Parkinson et le Gerard émaculé[50] que peut être n'avez-vous pas. Le Valerius Cordus est assez rare; j'avais aussi Tragus, mais je l'ai donné à M. Clappier.[51]

The presence of works by the English botanists, John Parkinson (1567-1650) and John Gerard (1545-1612), in Rousseau's library reflects the importance of English botany and his friendship with the Duchess of Portland, who introduced him to Gerard and John Ray.[52] Rousseau's search for English botanical texts is attested by Daniel Malthus,[53] who recommended major contemporary works to him, including Sir John Hill, *British herbal* (1756), Elizabeth Blackwell, *A Curious herbal* (1739-1751)[54] and Philip Miller, presumably his very popular *Gardener's dictionary*.

Parkinson, apothecary to James I and first botanist to Charles I of England, kept a botanical garden in what is now Covent Garden, London. He authored two works; the first was *Paradisi in sole paradisus terrestris* (London, 1629), an important horticultural treatise on English flower gardens, which described nearly a thousand plants with 109 magnificent, full-page woodcuts specially cut in England, illustrating 780 plants, mostly after Clusius and Lobel.[55] Parkinson's *Theatrum botanicum* (London, 1640) was an ambitious attempt to describe 3800 plants, largely from a medical perspective; it contained unpublished work left by Lobel, who died in England. The *Theatrum botanicum* contains twenty-eight new species never previously mentioned. Rousseau does not specify which work(s) of Parkinson he had, but it is possible he had both.

Rousseau referred several times to *The Herball* of John Gerard, a translation of Rembert Dodoens's *Pemptades sex* (1583). Thomas Johnson's revision of Gerard's *Herball* contained 2821 woodcuts, and this may have

50. The term 'émaculé' may refer to an error-free edition that Daniel Malthus could not locate: 'Je ne sais pas si vous voulez dire par *Gerard emaculatus*, le titre du livre, car vous l'avez souligné comme les autres: nos libraires ne connaissent pas ce titre, et la seule bonne édition que nous avons du Gerard, est Johnson sur *Gerard*.' Malthus to Rousseau, 24 January 1768; *CC*, xxxv.53.

51. Rousseau to Latourrette, 26 January 1770; *CC*, xxxvii.213.

52. Rousseau to the Duchess of Portland, 20 October 1766; *CC*, xxxi.41.

53. Malthus to Rousseau, 24 January 1768; *CC*, xxxv.53-56.

54. Blackwell, *A Curious herbal*.

55. John Parkinson, *Paradisi in sole paradisus terrestris: or A garden of all sorts of pleasant flowers which our English ayre will permitt to be noursed up: with a kitchen garden of all manner of Herbes, Rootes, and fruites, for meate or sause used with us, and an orchard of all sorte of fruitbearing trees and shrubbes fit for our land: together with the right orderinge, planting and preserving of them and their uses and vertues collected by John Parkinson apothecary of London* (London, Humfrey Lownes and Robert Young, 1629).

been the edition Rousseau knew.[56] In April 1767 Rousseau returned a copy of this work to Bernard Granville,[57] but he seems to have had another copy after his return to France, given the statement above in his letter of January 1770 to Latourrette.[58] In his notes to the Regnaults' *La Botanique mise à la portée de tout le monde* he refers to Gerard: 'On a confondu ici deux plantes toutes différentes, quoiqu'elles aient toutes deux les feuilles maculées, ce qui peut être a été cause du quiproquo. La racine et toute la figure (a) appartiennent à l'*Orchis mascula* de Linna'us, ainsi que les synonymes cités de Dodonée[,] de Matthiole[59] et de Gerard' (fig.7.3, p.223).[60]

Another author mentioned in this letter is Valerius Cordus (1515-1544), son of Euricius (Eberwein) Cordus (1486-1535), a botanist and medical professor at Ferrare and Marburg. The young Cordus was a brilliant botanist and physician, who died of malaria at twenty-nine. In addition to his pharmacopeia, *Dispensatorium* (1540), and an herbal in five volumes, *Historia Plantarum* (1544), Cordus synthesised ether by combining ethyl alcohol with sulfuric acid. In his *Annotationes in Pedacii Dioscoridis Anazarbei de medica material libros v* (published posthumously in 1549), Cordus gave 'his own systematic observations of many of the same plants described by Pedanius Dioscorides in the first century CE'.[61] He also described about 500 new species with precision, presenting plant characteristics in a regular pattern, including the features of stem and leaves, the flower and time of flowering, the fruit and seeds, the number of loculi in the seed, the root, whether annual or perennial, taste and smell and habitat. Cordus thus established the basis of scientific plant description.[62] His work was important not only because it described plants new to Europe, but also because it made fine distinctions among them, giving

56. John Gerard, *The Herball, or general historie of plantes*, fol. (London, John Norton, 1597) contained c.2146 woodcuts. Rousseau may have had Johnson's edition, *The Herball or general historie of plantes*, ed. Thomas Johnson (London, Adam Filip, Joice Norton and Richard Whitakers, 1636) which contained 2821 woodcuts! He appears to have borrowed this work from Richard Granville and returned it upon his departure from England. See Rousseau to Granville, 29 April 1767; *CC*, xxxiii.36.

57. Rousseau to Granville, 29 April 1767; *CC*, xxxiii.36.

58. Malthus looked for this work on Rousseau's behalf. Malthus to Rousseau, 24 January 1768; *CC*, xxxv.53.

59. Rousseau is supposed to have known a seventeenth-century edition, Matthioli, *Opera quae extant Omnia*. Matthioli's commentary on Dioscorides was first published in Venice in 1554.

60. Original emphasis; BdAN [Munch MS], art. 14 – 'Le satyrion'. *Orchis maculata* L. has been subdivided into *O. maculata* (L.) L., *O. maculata* Alsch., and *O. maculata* Crantz.

61. http://www.summagallicana.it/lessico/c/Cordus%20Valerius.htm.

62. Valerius Cordus, *Annotationes in Pedacii Dioscoridis Anazarbei de medica materia libros V*, ed. Conrad Gessner (Strasbourg, Josias Rihel, 1561).

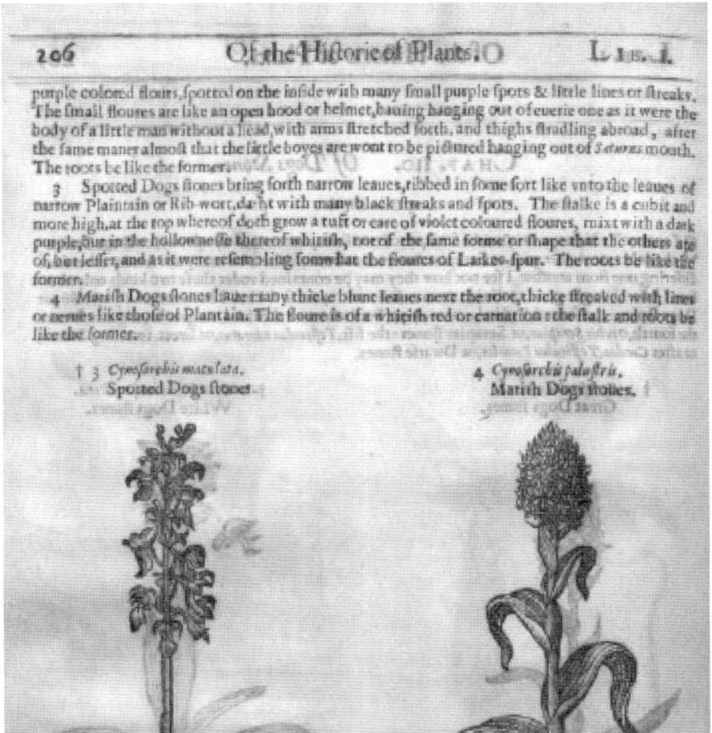

206 Of the Hiſtorie of Plants. LIB. I.

purple colored floures, ſpotted on the inſide with many ſmall purple ſpots & little lines or ſtreaks. The ſmall floures are like an open hood or helmet, hauing hanging out of euerie one as it were the body of a little man without a head, with arms ſtretched forth, and thighs ſtradling abroad, after the ſame maner almoſt that the little boyes are wont to be pictured hanging out of *Saturns* mouth. The roots be like the former.

3 Spotted Dogs ſtones bring forth narrow leaues, ribbed in ſome ſort like vnto the leaues of narrow Plaintain or Rib-wort, daſht with many black ſtreaks and ſpots. The ſtalke is a cubit and more high, at the top whereof doth grow a tuft or eare of violet coloured floures, mixt with a dark purple, but in the hollowneſſe thereof whitiſh, not of the ſame forme or ſhape that the others are of, but leſſer, and as it were reſembling ſomwhat the floures of Larkes-ſpur. The roots be like the former.

4 Mariſh Dogs ſtones haue many thicke blunt leaues next the root, thicke ſtreaked with lines or nerues like thoſe of Plantain. The floure is of a whitiſh red or carnation: the ſtalk and roots be like the former.

† 3 *Cynoſorchis maculata.*
Spotted Dogs ſtones.

4 *Cynoſorchis paluſtris.*
Mariſh Dogs ſtones.

‡ 5 This hath fiue or ſix little leaues: the ſtalk is ſome handfull or better in height, ſet about with ſomewhat leſſe leaues: the tuft of floures at the top of the ſtalke are of a purple colour, ſmall, with a white lip diuided into foure partitions hanging downe, which alſo is lightly ſpotted with purple: it hath a little ſpur hanging down on the hinder part of each floure. The ſeed is ſmal, and contained in ſuch twined heads as in other plants of this kinde. The roots are like the former, but much leſſe. ‡

¶ *The Place.*

Theſe kinds of Dogs ſtones do grow in moiſt and fertil medows. The mariſh Dogs ſtones grow for the moſt part in moiſt and wateriſh woods, and alſo in mariſh grounds. ‡ The fift groweth in many hilly places of Auſtria, Germany, and England. ‡

¶ *The Time.*

They floure from the beginning of May to the end of Auguſt.

¶ The

Figure 7.3: *3. Cynosorchis maculata* (*Spotted Dogs Stones*); *4. Cynosorchis palustris* (*Marish Dogs Stones*) (= *Orchis mascula* (L.) L.), engravings from John Gerard, *The Herball or general historie of plantes*, ed. Thomas Johnson (London, Adam Filip, Joice Norton & Richard Whitakers, 1636), p.206, courtesy of the Biodiversity Heritage Library (online).

up the pretense of his predecessors to match them to plants described by the ancients.[63]

Upon his arrival in London in January 1766, Rousseau had gone to considerable trouble to obtain a work by Cordus[64] in addition to other sixteenth- and seventeenth-century botanical works:

> J'avais eu l'intention de prendre à Strasbourg chez M. Koenig quatre livres de botanique, savoir: *Mathiolus*,[65] *Michelius, Valerius Cordus* et le *Pinax* de *Gaspard Bauhin*; n'ayant pu mettre ces livres dans ma malle, je les laissai. Je vous prie, quand vous écrirez à M. Koenig, de le saluer de ma part, et de le prier d'envoyer ces quatre livres encaissés par la première occasion à Amsterdam, à M. Rey, à qui j'écrirai pour lui marquer la voie par laquelle il doit me les faire parvenir.[66]

Rousseau's purchase of works by Cordus, Matthioli, Micheli and Bauhin attests not only to his serious approach to botany, but also his early preference for pre-Linnaean works, and desire to broaden his botanical horizons beyond the works of the Swedish systematist, which he had already acquired.

The last author whom Rousseau mentioned in his January 1770 letter was Hieronymus Tragus (Jerome Bock), (1498-1554). Tragus, another notable Renaissance botanical author, published his *New Kreütter Buch* (Strasbourg, 1539) without illustrations; however, his *Kreüter Buch* (Strasbourg, 1546 and subsequent editions) was illustrated with up to 537 pictures (fig. 7.4, p.225).[67] Rousseau confirmed that he had once owned this work, but had given it away to the Montpellier physician, Pierre Clappier: 'j'avais aussi Tragus, mais je l'ai donné à M. Clappier'.[68]

Rousseau also looked for illustrations published by Jakob Dietrich (Theodorus) von Bergzabern (c.1520-1590), under his Latin name, Tabernaemontanus, in a companion volume to his *Neuw Kreuterbuch* (Frankfurt, 1588, 1591), the *Eicones plantarum seu stirpium* (1590). The *Eicones* contained 3000 plant illustrations drawn from Leonhard Fuchs (1501-1566), Dodoens, Bock, Matthioli, Lobel and Clusius. Rousseau requested these illustrations from Rey in a letter of 28 December 1767.[69] However, it is not clear if Rousseau ever owned or consulted this work.

63. Ogilvie, 'The many books of nature', p.32.
64. Rousseau does not specify the work in question.
65. The edition Rousseau sought was Matthioli, *Opera quae extant omnia* (1674), referred to above.
66. Original emphasis; Rousseau to Guy, 20 January 1766; *CC*, xxviii.209–10; see also letters of 8 and 20 January 1767; *CC*, xxxii.58-61 and *CC*, xxxii.55-58, respectively, regarding the settlement of the account for these books.
67. Hieronymous Tragus [Jerome Bock], *New Kreütter Buch* (Strasbourg, Josias Rihel, 1546).
68. Rousseau to Latourrette, 26 January 1770; *CC*, xxxvii.213.
69. Rousseau to Rey, 28 December 1767; *CC*, xxxiv.268.

Figure 7.4: Engravings from Hieronymus Tragus [Jerome Bock], *Kreüter Buch* (Strasbourg, Josias Rihel, 1560), Frontispiece and p.21, courtesy of the Biodiversity Heritage Library (online).

Rousseau likewise collected eighteenth-century illustrated works. These included regional florae such as *Traité historique des plantes qui croissent dans la Lorraine* by Pierre Joseph Buc'hoz (1731-1802), a lawyer-turned-physician who produced a total of 300 volumes of botanical texts, mainly pictographic. The folio and quarto editions of his *Traité* comprised ten and twenty volumes respectively.[70] Du Peyrou sent Rousseau the publication notice, which reads:

> Cet ouvrage, dont deux volumes ont déjà paru, en aura 20 in 8°, avec des *Planches gravées* qui se distribuent séparément pour pouvoir être reliées à part. L'ouvrage ne se distribue que par souscription; elle est de 60 livres de France jusqu'au premier janvier prochain, et 72 pour les années postérieures, payable en 4 termes, en recevant le premier, le cinquième, dixième et quinzième volumes.[71]

Rousseau also consulted Pierre-Joseph Garidel (1658-1737), *Histoire des plantes qui naissent aux environs d'Aix* (Aix, 1715), containing 100 plates.[72] He borrowed it from the Marquis de Mirabeau, the famous physiocrat and 'ami des hommes', whose letter advocating the use of horses over oxen was published in the second French translation (1764) of Hirzel's *Le Socrate rustique*.[73] In this, as in other cases, Rousseau used friends' libraries to supplement his own book collection.

A work with a broader geographical reach was that of the French Dominican, Jacques Barrelier (1606-1673), *Plantae per Galliam, Hispaniam et Italiam observatae* (Paris, 1714), edited and published by Antoine de Jussieu of the Jussieu botanical dynasty (discussed in Chapter 5, p.157); this work contains 1327 engravings of the plants of France, Spain and Italy.[74]

On Du Peyrou's recommendation, Rousseau purchased François-Alexandre-Pierre de Garsault, *Les Figures des plantes et animaux d'usage en médecine*, 6 vols (1764-1765) which contains 729 illustrations of plants and animals.[75] An author and illustrator, Garsault (1693-1778) had been captain of the King's stud farm, a post with an obvious zoological dimension. Rousseau ordered Garsault from Duchesne shortly after

70. Pierre-Joseph Buc'hoz, *Traité historique des plantes qui croissent dans la Lorraine et les trois Evêchés: contenant leur description, leur figure, leur noms, l'endroit où elles croissent, leur culture, leur analyse et leurs propriétés, tant pour la médicine, que pour les arts et métiers*, 10 vols (Paris, Fetil, 1770). Buc'hoz served as physician to Stanislas I (1577-1766), the Polish king in exile in France.

71. Du Peyrou to Rousseau, 10 October 1764; *CC*, xxi.237-38.

72. Pierre-Joseph Garidel, *Histoire des plantes qui naissent aux environs d'Aix* (Aix, J. David, 1715).

73. Rousseau had borrowed Garidel from the Marquis de Mirabeau: 'le Garidel qui appartient à M. le Marquis de Mirabeau à qui je vous prie de le faire restituer'. Rousseau to Du Peyrou, 7 June 1770; *CC*, xxxviii.33.

74. Jacques Barrelier and Antoine de Jussieu, *Plantae per Galliam, Hispaniam et Italiam observatae: iconibus aeneis exhibitae* (Paris, Stéphane Ganeau, 1714). Guy to Rousseau, 29 August 1765; *CC*, xxvi.269.

75. Garsault, *Les Figures de Plantes et animaux.*

receiving Du Peyrou's recommendation;[76] and reported using it while a guest of the Marquis de Mirabeau in 1768.[77] This suggests he may have had it shipped back from England.

The illustrations in two important pre-Linnaean works held particular value for Rousseau throughout his botanical studies, and he refers to them frequently. He had

> [U]n extrême besoin de deux livres de botanique, l'un de Tournefort, en trois volumes in-4°, intitulé *Institutiones rei herboriae*; ou [...] le livre du même auteur intitulé *Elemens de botanique*; l'autre livre est de M. Vaillant, intitulé *Botanicon Parisiense* in-folio;[78] ces deux livres doivent être chers à cause des figures: cependant ne laissez pas, je vous en prie, de me les envoyer le plus tôt que vous pourrez en me marquant le prix, parce que, s'il est trop cher pour moi, je trouverai ici à m'en défaire.[79]

As we have seen, Tournefort was one of the leading pre-Linnaean systematists, while Vaillant was an important early sexualist. Rousseau had recourse to both of these works when searching for *Plantago monanthos* in Montmorency with the 'caterve du jardin' in 1771.[80]

It might be objected that, following Linnaeus, a herbarium is better than any picture: 'Herbarium praestat omni Icone, necessarium omni Botanico'.[81] In his commentary on the herbarium given to him by Joseph Dombey, Rousseau seems to concur: it was to be his 'unique bibliothèque'.[82] However, only one day after making this comment, he admitted that Dombey's herbarium made a library necessary: 'on m'a fait présent il y a quelque temps d'un très bel herbier qui vaut mieux pour moi qu'une *bibliothèque*, mais qui me *la* rendrait quelquefois nécessaire'.[83] Hence, despite the beauty and value of Dombey's herbarium (see p.279-81), it could not stand alone as a botanical reference work, and would have to be supplemented by books.

To Rousseau the road to botany through books seemed crooked and perilous; it is easy to lose one's way; un-illustrated works leave the

76. Rousseau to Duchesne, 19 May 1765; *CC*, xxv.301–02.
77. Rousseau to Du Peyrou, 3 March 1768; *CC*, xxxv.156–58.
78. Rousseau's personal copy, containing the bookplate of Jane Dalton (Daniel Malthus's botanist cousin), is at Harvard University. See Hawkins, 'Manuscripts of Jean-Jacques Rousseau at Harvard University', p.215-17.
79. My emphasis; Rousseau to Duchesne, 28 April 1765; *CC*, xxv.197; see also Rousseau to Duchesne, 16 December 1764; *CC*, xxii.273; Rousseau had considerable knowledge of the works of Tournefort and Vaillant.
80. Rousseau to Latourrette, 25 January 1772; *CC*, xxxix.22.
81. Linnaeus, *Philosophia botanica* (1751), p.7, section 11; 'A HERBARIUM is better than any picture, and necessary for every botanist'; Carolus Linnaeus, *Philosophia botanica* (2003), p.18, section 11.
82. Rousseau to Du Peyrou, 10 June 1768; *CC*, xxxv.307.
83. My emphasis; Rousseau to Guy, 11 June 1768; *CC*, xxxv.309-10.

uninitiated without a visual guide – the novice must feel his or her way like a blind person. Yet, books, however dubious, were there to stay.[84]

iv. Uses of the illustrated botanical text

It seems that the luxuriantly illustrated botanical works that Rousseau acquired may have been intended for something more than study or passive visual enjoyment. At an early stage in his botanical study, Rousseau sent the following request to François Coindet of Geneva: 'Je voudrais trouver un recueil de plantes gravées et bien ressemblantes, quand même il faudrait y mettre un certain prix. Ne pourriez-vous point m'aider dans cette recherche? Cela me procurerait encore le plaisir de m'occuper l'hiver à les *enluminer*.'[85] Du Peyrou had told Rousseau the previous autumn that he was giving him 'de l'occupation cet hiver, si je reçois mes planches enluminées. Car je pense que vous vous serez pourvu de couleurs et de pinceaux à moins que vous n'aimiez mieux me laisser ce Soin'.[86] Six months later Du Peyrou recommended Garsault's work to aid Rousseau's artistic endeavours: 'Le plus grand plaisir que vous pouviez me faire était de trouver de votre goût l'Etui avec les couleurs. Pour en faire usage, il ne faut pas être au fait de toutes les plantes. Mais il vous faut un Garsault.'[87]

Rousseau's correspondence in the years 1764-1767 contains numerous references to his procuring paints and painters' supplies (for example paint brushes) to colour engravings ('estampes') of plants, flowers, and landscapes for an *optique* that he wished to acquire:[88]

> L'usage que j'en veux faire est d'enluminer des *plantes* et *fleurs* dans leurs *couleurs naturelles* et d'autres estampes et paysages pour une *optique*. Je connais ces boîtes de Nuremberg; les *couleurs* en sont mauvaises et falsifiées; je voudrais avoir à part une petite provision des couleurs les plus nécessaires comme Carmin, outremer, gomme gutte, vert de Vessie, vert de gris, encre de la Chine etc. et je voudrais surtout que ces couleurs en petite quantité fussent bien choisies et aussi pures qu'il est possible. Joignez-y aussi, je vous prie un peu de gomme arabique; car on ne trouve ici rien du tout.[89]

84. This position contradicts Rousseau's advice in *Emile*, *OC*, iv.361.
85. My emphasis; Rousseau to Coindet, 27 April 1765; *CC*, xxv.187.
86. Du Peyrou to Rousseau, 20 October 1764; *CC*, xxi.281.
87. Du Peyrou notes that this work 'n'est pas d'un grand prix, les 5 volumes ne coûtant que deux Louis'. Du Peyrou to Rousseau, 30 April 1765; *CC*, xxv.211.
88. D'Ivernois to Rousseau, 14 December 1764; *CC*, xxii.230-31; see also d'Ivernois to Rousseau, 4 January 1765; *CC*, xxiii.20-22. Rousseau eventually abandoned this plan. See Rousseau to François Henri d'Ivernois; 20 July 1765; *CC*, xxvi.105-07. For an earlier reference to Rousseau's knowledge of this device, see *Julie*, *OC*, ii.77: 'Ajoutez à tout cela les illusions de l'optique'.
89. My emphasis; Rousseau to Deluc, 20 December 1764; *CC*, xxii.256. See also Rueff, 'Apprendre à voir'.

In response to Rousseau, Deluc describes the optique, a popular home entertainment device of the period:

> On appelle communément aujourd'hui *une optique*, cette machine dans laquelle un miroir et un grand verre lenticulaire redressent des estampes couchées horizontalement et en détachent les objets. S'il s'agit de cette machine, voulez-vous la boëte, ses verres et un assortiment d'estampes; celles-ci varient beaucoup de prix suivant leur degré [de] perfection. Vous en avez peut-être et la boëte vous suffirait. En un mot, Monsieur, indiquez-moi ce que vous souhaitez et je ferai tout ce qui me sera possible.[90]

Rousseau's reply shows a nuanced appreciation of the different types of optiques, and he explains his need for this type of diversion:

> C'est effectivement une *optique* telle que vous le décrivez dont nous avions parlé. Je n'aime point celles qui restant toutes ouvertes laissent de toutes parts entrer la lumière et présentent avec l'image les objets environnants. Vous me parlâtes d'une manière d'enclore tellement l'image dans la boëte par une espèce de Quadre noir qu'on ne vit absolument que l'estampe. Voilà, Monsieur, ce que je désirerais, et si vous trouviez à Genève de bons verres, que vous connussiez quelque bon ouvrier que vous voulussiez bien diriger pour faire la boîte, je vous serais obligé de vouloir bien y donner quelques soins. Ces sortes de machines sont fort plates à moins qu'elles ne fassent tout à fait *illusion*, mais quand elles la font elles sont très *amusantes*, et je sens que dans mon état, enfermé plus de six mois tous les ans j'ai très grand besoin d'amusements qui fassent diversion aux excursions de ma tête et l'empêchent de me consumer dans ma prison.
> Cherchant la plus grande *illusion*, il s'ensuit qu'il faut aussi les meilleures estampes: Celles d'Angleterre me paraissent les plus propres à la chose; il est vrai qu'elles sont extrêmement chères; si vous en pouviez rassembler une douzaine ou quinzaine de bien bonnes, cela me suffirait en attendant que j'en fisse chercher à Paris. Vous savez qu'il faut des lointains des *perspectives*, des allées, avenues, *Galeries*, marines, en un mot tout ce qui chasse et prolonge l'espace.
> *L'architecture* avec des cours et avant-cours, colonnades etc. fait très bien aussi.[91]

Like the television or DVD of today the *optique* provided an alternative source of entertainment to books. The *optique* could display images with various degrees of reduction or enlargement and lighting. 'The heyday of optical boxes [...] spanned from the second half of the eighteenth century to its close [...] Their sides were often papered with engravings [...]

90. My emphasis; Deluc to Rousseau, 15 December 1764; *CC*, xxii.240-41. This was the same Jean-André Deluc (1727-1817) who was a pioneer in the fledgling science of geology, discussed in Chapter 3, p.64-65, 68-70.

91. My emphasis; Rousseau to Jean André Deluc, 20 December 1764; *CC*, xxii.256-57; see also Rousseau, *Rêveries*, *OC*, i.1070; Juan Calatrava, 'Rousseau et l'architecture', *Annales de la société Jean-Jacques Rousseau: Jean-Jacques Rousseau et les arts visuels*, Actes du colloque de Neuchâtel (20-22 septembre 2001) 45 (2003), p.81-111.

Instructions for their assembly could also be found in the optics section of every major dictionary and encyclopedia.'[92] There was a 'pan-European marketing of scientific baubles and luminous effects', including peep boxes, perspective games, visual knick-knacks, slides, mobile plaques, projection mania.[93] These provided adult education, scientific popularisation and above all the distraction needed to perk up a weary or strained mind: in short a leisure occupation in the Aristotelian sense of engaged - not couch potato - leisure.[94] Prints lit from behind 'domesticated credulity [...] They accustomed the viewer to visual replacements.'[95]

Thus, in 1764 the illusion provided by the optique was already linked in Rousseau's mind to botany, and the satisfaction provided by the optique was not just a passing fancy; it later provided a metaphor for expressing the pleasures provided by his herbarium: 'Cet herbier est pour moi un journal d'herborisations qui me les fait recommencer avec un nouveau charme et produit l'effet d'une *optique* qui les *peindrait* derechef à mes yeux.'[96]

Six months later, Rousseau abandoned the optique for botany, but his interest in colouring plant engravings persisted.[97] Du Peyrou, Coindet, Jean André Deluc and Francois-Henri d'Ivernois helped Rousseau obtain painting supplies, by no means a straightforward matter. Du Peyrou had provided Rousseau with a case ('étui') containing 'couleurs', while Coindet, Deluc and d'Ivernois searched for additions to this collection.[98] In late 1764 Deluc was assisting d'Ivernois in purchasing supplies (presumably in Geneva), but confessed he did not know what use Rousseau wanted to make of these 'couleurs', and requested further specification:

> Voici, Mon cher Monsieur, une boëte qui contient à peu près toutes les *couleurs* qu'on emploie ordinairement dans la *miniature* et six bons pinceaux; il y manque de l'outremer, mais il n'est essentiel que pour les portraits. Ne sachant pas *précisément quel usage* vous voulez faire des couleurs que vous

92. Barbara Maria Stafford, *Body criticism: imaging the unseen in Enlightenment art and medicine* (Cambridge, MA, 1991), p.371.
93. Stafford, *Body criticism*, p.366.
94. Stafford, *Body criticism*, p.368. For Aristotle leisure is not simply an entitlement of a certain class of people, it is a *privilege*, like citizenship itself, and should be used to the noblest ends. See Aristotle, *Politics*, 1338a10-25.
95. Stafford, *Body criticism*, p.377.
96. Rousseau, *Rêveries*, *OC*, i.1073.
97. Rousseau to François Henri d'Ivernois, 29 July 1765; *CC*, xxvi.105-107. As an enthusiastic student of botany at the time, Du Peyrou was presumably the other person referred to in this passage as having 'le même goût que moi'.
98. François Henri d'Ivernois to Rousseau, 15 December 1764; *CC*, xxii.239; 9 January 1765; *CC*, xxiii.66-68; 15 January 1765; *CC*, xxiii.106-107; Coindet to Rousseau, 11 October 1767; *CC*, xxxiv.141-42.

souhaitez, j'ai préféré cette boëte qui vous servira d'échantillon. Quand vous l'aurez examinée vous pourrez m'indiquer ce qui vous manque ou celles des couleurs dont vous croirez de faire un plus grand usage. Le nom de la couleur ou un morceau de la coquille suffira, on l'achètera alors séparément chez un droguiste.[99]

Coindet wrote to Rousseau, 'Je m'occuperai avec grand plaisir à rechercher quelques recueils de plantes tels que vous les desirez [...] Quand vous voudrez *enluminer* je vous enverrai si vous le voulez les *utensiles* et les *couleurs* nécessaires pour cela'.[100] A letter from Coindet explained in some detail how to prepare the 'couleurs':

Je vous envoyai [...] des pinceaux et les *couleurs* que vous vouliez avoir en plus grande quantité. La gomme gutte, et le vert de vessie se délayent tout Simplement Sans aucune préparation.[101] Il faut bien broyer le bleu de Prusse et le vert distillé, et le délayer peu à peu avec de l'eau dans laquelle vous aurez fait fondre un peu de gomme arabique, il faut moins de gomme dans le vert que dans le bleu. Je n'ay pas encore pu vous faire faire la petite machine pour broyer les *couleurs* et je vous ay envoyé la mienne en attendant. Je vous enverrai par la première occasion de la gomme arabique, de l'encre de la Chine, et de la flanelle d'Angleterre.[102]

The precision of Coindet's instructions, the range of colours and supplies involved, Rousseau's own comments on these points, all show a clear understanding of which colours artists considered 'les plus nécessaires'.[103] The duration of these discussions over three years likewise gives us some idea of the effort, cost and interest Rousseau must have had in these artistic activities. For instance, an accounting sent by d'Ivernois shows that by early January 1765 he had spent a total of 17.6 French pounds (livres) on brushes and paints for Rousseau.[104]

These findings may seem all the more surprising given Rousseau's reputation as a philosopher who took little or no interest in the pictorial arts.[105] This impression is reinforced by his having left no treatise on art or aesthetic judgement (see Kant's account of the sublime and aesthetic

99. My emphasis; Jean-André Deluc to Rousseau, 15 December 1764; *CC*, xxii.240.
100. My emphasis; Coindet to Rousseau, 15 May 1765; *CC*, xxv.281.
101. 'Vert de vessie: attestée dès 1640: voir Littré, *vert*, 12°: 'nom donné dans le commerce à une sorte de couleur verte qu'il reçoit ordinairement enfermée dans des vessies, et qui est extrait des baies mûres du nerprun des teinturiers (rhamnées)'. Rousseau to d'Ivernois, 7 January 1765; *CC*, xxiii.45, note (c).
102. My emphasis; Coindet to Rousseau, 4 November 1767; *CC*, xxxiv.165.
103. Rousseau to Jean-André Deluc, 20 December 1764; *CC*, xxii.256.
104. Francois Henri d'Ivernois to Rousseau, 4 January 1765; *CC*, xxiii.20-22.
105. For this reason, Rousseau's positions on artistic representation in engravings, painting, and sculpture have to be reconstructed out of many sources. See *Annales de la société Jean-Jacques Rousseau* 45 (Geneva, 2003), *passim*.

judgement in *Kritik der Urteilskraft* [*Critique of judgement*]), and by his outright hostility toward the arts in the first *Discours*.

Yet Rousseau used painting as a key analogy for his theories in music: 'La musique [...] *peint* tout, même les objets qui ne sont que visibles [...] elle semble mettre l'œil dans l'oreille'.[106] In the *Essai sur l'origine des langues* and the *Dictionnaire de musique* he analogises form ('dessin') to melody and colour to harmony, relying, as we have seen, on the philosophical distinction between primary and secondary qualities. Secondary qualities such as colour are therefore lifeless without the primary ones of number and shape.

Rousseau therefore preferred the uncoloured engraving to a painting: 'Si je n'aime pas beaucoup les tableaux, j'aime extrêmement les estampes – elles laissent quelque chose à faire à mon imagination. Elle les colore d'abord à sa fantaisie, et il me semble que je vois alors les objets tout comme ils sont dans la Nature'.[107] The engraving already contained the essence of what the viewer needed: 'L'intérêt et le sentiment ne tiennent point aux couleurs; les traits d'un tableau touchant nous touchent encore dans une *estampe*'.[108] Hence, paintings held virtually no interest for him, and there was only one, the 'Déluge' of Nicolas Poussin (1594-1665), that he is supposed to have admired: 'je n'ai aucun goût pour ces choses-là. Je n'aime que les Etres vivants [...] Il n'y a qu'un tableau qui m'ait frappé en ma vie – c'est celui de Poussin qui représente le Déluge. Je le regardais une heure entière et je ne pus le quitter quoiqu'il remplit toute mon âme de la plus vive amertume.'[109]

Similarly, the form, habit or 'port' of plants is their primary quality, while colour 'n'est [...] qu'une étude analogique et secondaire, pour éclairer et confirmer celle des *formes*'.[110] This relegation of colour to a secondary quality makes Rousseau's aesthetic concerns in botany – expressed in his desire for illustrated works, and for paints to 'enluminer des plantes et fleurs dans leurs couleurs naturelles'[111] – all the more striking.[112] In fact, this desire for coloured representations of flowers and landscapes was not new. In the *Confessions* Rousseau professed a love of drawing that included the use of colour:

> Le lavis des mappes [...] m'avait aussi rendu le goût du dessin. J'achetai des couleurs et je me mis à faire des fleurs et des paysages. C'est dommage que je me suis trouvé peu de talent pour cet art; l'inclination y était tout entière. Au milieu de mes crayons et mes pinceaux j'aurais passé des mois entiers sans

106. My emphasis; Rousseau, 'Imitation', *Dictionnaire de musique*, *OC*, v.860-61.
107. Jakob Heinrich Meister to Johann Heinrich Meister, 1 June 1764; *CC*, xx.127.
108. My emphasis; Rousseau, *Essai sur l'origine des langues*; *OC*, v.413.
109. Jakob Heinrich Meister to Johann Heinrich Meister, 1 June 1764; *CC*, xx.127.
110. My emphasis; Rousseau, *Fragments de botanique*, *OC*, iv.1249.
111. Rousseau to Deluc, 20 December 1764; *CC*, xxii.256.

> sortir. Cette occupation devenant pour moi trop attachante, on était obligé de m'en arracher. [...] je ne vois plus rien au monde que l'amusement dont je suis occupé. L'âge ne me pas guéri de ce défaut [...] et maintenant que j'écris ceci, me voilà comme un vieux radoteur, engoué d'une autre étude inutile où je n'entends rien.[113]

In herbaria he appreciated specimens in which the colour was well preserved, an art that Mme Delessert had cultivated: 'Vos plantes sont parfaitement bien conservées, les *couleurs* surtout, ce qui était difficile pour la fleur bleue'.[114]

At the same time that Rousseau was pursuing botany, investigating the *optique* and colouring botanical plates, he enjoyed warm friendships with artists such as Claude-Henri Watelet (1718-1786) and Barthélémy-Augustin Blondel d'Azaincourt (1719-1794). Blondel d'Azaincourt was an engraver who followed François Boucher and was inspired by the Flemish art collected by his father, and the rustic scenes it depicted.[115] Watelet was a prominent artist and patron of the arts, whose wealth derived from tax farming. Coindet visited Watelet and his mistress, the pastellist Marguerite Le Comte, at Watelet's experimental farm, where many physiocratic innovations were on display. Watelet and Madame Le Comte promised to convey some of their artistic work to Rousseau: 'Je fus Dimanche au Moulin Joly, Mr Watelet avait reçu votre lettre, il vous répondra incessement. Madame Le Comte et lui doivent me remettre un rouleau de leurs ouvrages que je vous enverrai Soigneusement'.[116] Watelet proclaimed himself a disciple of the gardening precepts of Rousseau's Elysium in his *Essai sur les jardins* (1774), and Du Peyrou proposed Watelet to supervise engravings for the posthumous *Collection complète*.

Much research remains to be done on the linkages among Rousseau's theory of representation, his foray into painting and his botanical interests. Based on my findings, it seems he believed the student of nature not only needed to see the plant's primary quality – for example its port, habitus, or form – but also to experience its secondary qualities such as colour, which should be represented, not only in the herbarium, but also in books for the uninitiated.

112. Linnaeus excluded colour from generic characters. *Philosophia botanica* (2003), p.130, section 167; Latourrette and Rozier, *Démonstrations élémentaires*, p.35.

113. This 'autre étude inutile' is, of course, botany. Rousseau, *Confessions*, *OC*, i.180.

114. My emphasis; Rousseau to Mme Delessert, 24 May 1773, *CC*, xxxix.162; in the *Rêveries* he speaks of 'vives couleurs', *OC*, i.1063.

115. Madeleine Pinault Sørenson, 'Rousseau et l'art du paysage', *Annales de la société Jean-Jacques Rousseau: Jean-Jacques Rousseau et les arts visuels*, Actes du colloque de Neuchâtel (20-22 septembre 2001) 45 (2003), p.271-300 (278).

116. Coindet to Rousseau, 15 May 1765; *CC*, xxv.281.

v. What's in a name? Mediation and the 'Babel' of botanical names

> [L]a nomenclature [de Linnaeus] paraît tenir tellement à son système, qu'on ne s'avise guères de l'en séparer.[117]

According to Genesis, Adam named the plants and animals at God's command; Adam was therefore the father of mediation, which had visited so many ills upon humanity.[118] Linnaeus considered naming like Adam in the Garden of Eden an essential part of the botanist's job. In fact, Linnaeus saw himself as a second Adam[119] for whom the practice of botany was synonymous with *naming* plants: 'Botanici (veri) ex fundemento genuino Botanicam intelligunt, et vegetabilia omnia nomine intelligibili nominare sciant'.[120] He was responding to two major problems confronting eighteenth-century botany, namely, information overload and unwieldy phrases to designate plants.[121] Nomenclature had become a very messy affair; long, descriptive phrases doubled as plant names, as in a case cited by Rousseau, a grass species named and described by the seventeenth-century English botanist, Leonard Plukenet (1642-1706): '*Gramen* myloicophorum caroliniarum seu gramen altissimum, panicula maxima speciosa, è spicis marjoribus compressiusculis utrinque pinnatis blattam molendariam quodam modo referentibus, composite, foliis convolutus mucronatis pungentibus'.[122] As a result, 'la nomenclature se chargeait'.[123]

From the sixteenth century onwards the immense influx of new plants into Europe as the result of exploration, exacerbated the naming problem, leading to an 'information explosion' in natural history.[124] Linnaeus

117. Rousseau, 'Introduction', *OC*, iv.1207.
118. *The Bible*, Genesis, 2:19-20.
119. Müller-Wille, 'Collection and collation', p.545.
120. Linnaeus, *Philosophia botanica* (1751), p.4, section 7, 'The [true] Botanists should have a real basic understanding of botany, and should know how to *name* all vegetables with intelligible names'. Linnaeus, *Philosophia botanica* (2003), p.15, section 7.
121. For a discussion of how Renaissance botanical nomenclature differs from Linnaean nomenclature, see Selosse, 'The underlying pattern of the renaissance botanical genre pinax'.
122. 'A cockroach-bearing Carolinian *grass*, or a very tall grass with a large showy panicle composed of rather large, rather compressed spikes, pinnate on both sides in a manner recalling the grinding cockroach, with convolute, mucronate, pungent leaves rolled up in one another, terminating in a hard, sharp point', original emphasis, Leonard Plukenet, *Almagestum botanicum* (London, the Author, 1696), p.173, quoted in Rousseau, 'Introduction', *OC*, iv.1205.
123. Rousseau, 'Introduction', *OC*, iv.1205.
124. Mary M. Slaughter, *Universal languages and scientific taxonomy in the seventeenth century* (Cambridge, 1982), p.10, quoted in Cook, 'Rousseau and the languages of music and botany', p.78. The number of known plant species leapt from approximately 1000 in the

developed an ingenious solution to this information overload: a so-called *nomen triviale* ('trivial name') that he noted in the margin next to the generic name and polynomial species designation; the combination of the generic and trivial names is the binomial name.[125] Linnaeus laid down his rules for generic and specific Latin names in *Critica botanica* of 1737, and reiterated them in *Philosophia botanica*, 1751: 'Nomina generica, quae ex graeca vel Latina lingua radicem non habent, rejicienda sunt'.[126] He also specified the sources from which Latin names had to be derived, excluding 'primitive' or 'barbarous' names, as well as any names of uncertain derivation: 'Nomina generica primitiva nemo sanus introducit. *Barbara* vocabula omnia sunt nobis uti primitiva, quum lingua eorum ab eruditis non intelligitur. *Dubia* vocabula plantarum, de quibus vix constet e qua lingua orta sunt'.[127] He rejected names imposed by anyone other than 'Botanicis genuinis [genuine botanists]': 'Idiotae imposuere nomina absurda'.[128] The best names conveyed an essential feature of the genus, but as a last resort, generic names could memorialise botanical heroes: 'Nomina generica, ad Botanici optime meriti memoriam conservandam constructa, sancte servanda sunt'.[129]

This solution, which garnered Rousseau's admiration, has proven remarkably robust.[130] In its current form, the official plant name consists of three parts: generic name, species name (formerly trivial name) and the authority (that is, the conventional abbreviation of the name of the botanist to whom the name is attributed, for example 'L.' for Linnaeus); so the white lily is called *Lilium candidum* L. The *Species plantarum* of 1753, which systemically used binomial names, was enshrined in 1905 by the

sixteenth century to 18,000 in John Ray's *Methodus plantarum nova* (London, Henry Faitborne & John Kersey, 1682). A. C. Crombie, *Styles in scientific thinking* (London, 1994), vol.2, p.1263.

125. Although the classes and orders of the sexual system had double names, for example Monandria Monogynia (Class I); these names should on no account be confused with binomial names.

126. Linnaeus, *Philosophia botanica* (1751), p.163, section 229. 'Generic names, which do not have Greek or Latin root, are rejected.' Translation mine.

127. Original emphasis; Linnaeus, *Philosophia botanica*, p.160, section 220. 'No sane person introduces primitive generic names. All barbarous names are regarded by us as primitive, since they are from languages not understood by the learned. [So are] doubtful appellations of plants, when it is hard to decide what language they are derived from', Linnaeus, *Philosophia botanica* (2003), p.172, section 220.

128. Linnaeus, *Philosophia botanica*, p.158, section 211. 'Private individuals have applied absurd names', Linnaeus, *Philosophia botanica* (2003), p.169, section 211.

129. Linnaeus, *Philosophia botanica*, p.171, section 238. 'Generic names that have been formed to perpetuate the memory of a botanist who has done excellent service should be religiously preserved', Linnaeus, *Philosophia botanica* (2003), p.185, section 238.

130. Rousseau, 'Introduction', *OC*, v.1205-208.

international community of botanists as the starting-point of modern botanical nomenclature.[131]

Yet Linnaeus's own bold pronouncements regarding the botanist's prerogative to name plants belied the incremental way in which the binomial naming system grew out of his work on the flora of Öland and Gotland (Sweden), which he visited in 1741. In *Öländska och Gothländska Resa* (1745),[132] binomial nomenclature 'began as an indexer's paper-saving device' that Linnaeus developed as he worked; consequently, it did not arise from a taxonomic master plan.[133] In his *Philosophia botanica*, for example, Linnaeus's application of 'the technique [of binomial nomenclature] was scattershot and at the mercy of hurry and chance'.[134]

Indeed, Linnaean binomial nomenclature neither directly depended upon, nor arose from, the sexual system. According to Linnaeus's own vivid metaphor, his generic names were simply 'coinages', that is to say, a means of exchange in the botanical marketplace: 'The generic name has the same value on the market of botany, as the coin has in the commonwealth, which is accepted at a certain price – without necessitating a metallurgic examination – and is received by others on a daily basis, as long as it has become known in the commonwealth'.[135] The Linnaean binomials could be used in conjunction with any system or method, as was demonstrated by Antoine-Laurent de Jussieu in the Jardin du roi in 1773-1774.[136] Taxonomy is an ongoing project in which species can, of course, be shifted to other genera.

Thus, the Linnaean binomial is a mere *convention*, and nothing more. Unlike the Renaissance botanical name, or Linnaeus's polynomial species names, the binomial – consisting of genus and trivial name – does not convey anything *essential* about the species being named. Generic names can be entirely arbitrary and derive from a variety of sources, which Linnaeus enumerated in *Critica botanica* and *Philosophia botanica*. These include names bestowed by the ancients, or names of famous botanists; the best generic names, he conceded, in *Philosophia botanica* section 240, would

131. Linnaeus retained traditional polynomials as the complete names of species, consisting of a description together with the generic name. For examples, see Linnaeus, *Species plantarum*, *passim*.

132. Linnaeus, *Öländska och Gothländska Resa*. Given that this work was written in Swedish and printed in Fraktur, it would have been difficult for Rousseau to use it, but he nevertheless requested it, which suggests he was aware of its signficance; there is no letter confirming receipt. Rousseau to Rey, 28 December 1767, *CC*, xxxiv.268.

133. Åsberg and Stearn, 'Linnaeus's Öland and Gotland Journey 1741', p.6.

134. Koerner, *Linnaeus*, p.53.

135. Linnaeus, *Critica botanica*, quoted in Staffan Müller-Wille, 'Nature as marketplace: the political economy of Linnaean botany', *History of political economy* 35:1 (2003), p.154-72 (158).

136. Jussieu, 'Sixième notice historique sur le Muséum', p.11-12; Jussieu, 'Exposition', p.178, 193. See also Muller-Wille, 'Collection and collation', p.541-42.

show character or habit.[137] The trivial name, on the other hand, is subject to no laws: 'legibus etiamnum caret'.[138]

Rousseau was a thorough-going nominalist: 'Il est vrai que les noms sont arbitraires', and they might not even be necessary: 'un homme intelligent pourrait être un *excellent Botaniste*, quoiqu'il ne connût pas une seule plante par son *nom*'.[139] However, for Rousseau as for Linnaeus, the name, once assigned, served as a rigid designator: the plant so designated was always and everywhere supposed to be the same one.

vi. Rousseau's defence of Linnaean nomenclature

Linnaeus's reform of botanical nomenclature encountered opposition from the outset, and continues to be criticised to this day for its alleged 'linguistic imperialism'.[140] Linnaeus's own contemporaries were often sharply critical of his 'reform': the French botanist, Michel Adanson, the Swiss botanist, Johann Amman of St Petersburg and the creole priest, José Antonio Alzate (1737-1799), among others, argued for retaining indigenous names against the adoption of Linnaean binomial nomenclature.[141] While the Latin botanical names enshrined in the International Code of Botanical Nomenclature tend to conceal indigenous names for genera, Linnaeus *never* prescribed the elimination of indigenous names or indigenous knowledge about plants: 'In his Latin works [Linnaeus] listed vernacular plant names. He made his students do the same, arguing that local names reflect unknown uses or properties of plants'.[142] Linnaeus insisted that only botanists needed Latin names established according to his strict principles; everyone else was free to use vernacular names.[143] In fact, many of Linnaeus's generic names incorporate indigenous names and/or references to traditional medicinal, artisanal or alimentary uses.[144]

137. Linnaeus, *Critica botanica* (1938), p.79, section 240; Linnaeus, *Philosophia botanica* (2003), p.191.

138. Linnaeus, *Philosophia botanica*, p.202, section 257; 'is still free from any laws', Linnaeus, *Philosophia botanica* (2003), p.219, section 257; see also Linnaeus, *Critica botanica* (Leiden, Conrad Wishoff, 1737), p.204, section 284; see also Müller-Wille, 'Collection and collation', p.560.

139. My emphasis; 'Introduction', *OC* iv.1209.

140. Schiebinger, *Nature's body*, p.209, 267, n.29; *Plants and empire*, p.194-225; Joseph Needham, *et al.*, *Science and civilisation in China*, vol.6 *Biology and biological technology*, part 1: *Botany* (Cambridge, 1986), p.19, 168.

141. Adanson, *Familles des plantes*, p.clxxii-iv; Jean-Marc Drouin, 'Linné et la dénomination des vivants: portrait du naturaliste en législateur', in *Fondements de la botanique: Linné et la classification des plantes*, ed. Thierry Hoquet (Paris, 2005), p.54

142. Koerner, *Linnaeus: nature and nation*, p.100.

143. Linnaeus, *The 'Critica botanica' of Linnaeus*, trans. Hort (1938), p.37, 38, section 229.

144. A. Cook, 'Linnaeus and Chinese plants: a test of the linguistic imperialism thesis', *Notes and records of the Royal Society* 64 (June 2010), p.121-38.

Rousseau ignored the 'violence symbolique' of Linnaean binomial nomenclature, treating it as an unproblematic means of communication.[145] He advocated Linnaean binomial nomenclature, even when he did not use the sexual system, for example in his advice to Mme Delessert to learn 200 Latin names by heart.[146] Various reasons have been advanced for Rousseau's advocacy of Linnaean binomial nomenclature. Drouin, for example, has stressed the social consequences of this nomenclature for botany:[147] 'Le *Philosophia botanica* est bien en effet un livre de morale: elle donne la primauté à l'étude désintéressée de la nature sur la recherche de l'utilité, elle nous suggère de la dépouiller de ses beautés d'emprunt et enfin elle nous montre la constitution d'un *collectif* qui est à la connaissance ce que le corps politique est à la volonté.'[148]

While Drouin's suggestion is well-founded, there were other reasons for Rousseau to advocate Linnaean names, deriving from his desire for authenticity and his critique of mediation in virtually every domain of human life.[149] Just as he sought unmediated political relations in *Du contrat social*,[150] Rousseau sought unmediated relations with nature: 'le botaniste ne souffre point d'intermédiaire entre la nature et lui'.[151] This desire for the unmediated moment vis à vis the Other would seem to preclude the intervention of nomenclature or classification. Yet even though Rousseau held that *any* mediation between humanity and nature is undesirable, he acknowledged the necessity of mediation by books and names for doing botany.

Rousseau's principal consideration of Linnaean nomenclature is an essay on the history of botany published posthumously – and possibly

145. Drouin, 'Linnaeus et la dénomination des vivants', p.53.

146. Rousseau to Mme Delessert, 30 August 1773; *CC*, xxxix.192.

147. Linnaeus appointed himself Legislator (in the ancient sense of Lawgiver, e.g. Solon or Lycurgus) of names in this 'commonwealth'. Linnaeus, *Critica botanica*, p.xix.

148. My emphasis; Drouin, 'Rousseau lecteur de Linné', p.57.

149. Cook, 'Rousseau and languages of music and botany', p.75; Starobinski, *Transparence*, *passim*. Roche writes, 'botany [...] authorised a relationship with nature and with the world that was richer and more sensitive, an unmediated knowledge whose fervour encompassed the wisdom of the greatest, such as Linnaeus'. Daniel Roche, 'Natural history in the academies', in N. Jardine, *et al.*, *Cultures of natural history* (Cambridge, 1996), p.127-144 (143).

150. Intermediaries such as political factions impede the expression of the general will. Rousseau, *Du contrat social*, II.3; *OC*, iii.372.

151. Rousseau, *Fragments de botanique*, *OC*, iv.1250; Derrida derides this notion: 'Que la botanique soit le supplément de la société, c'est là plus qu'une catastrophe. C'est la catastrophe de la catastrophe. Car dans la nature, la plante est ce qu'il y a de plus *naturel*. C'est la vie naturelle. Le minéral se distingue du végétal en ce qu'il est nature morte et utile, servile de l'industrie de l'homme.' Jacques Derrida, *De la grammatologie* (Paris, 1967), p.212.

erroneously – as an introduction to his so-called *Fragmens pour un dictionnaire des termes d'usage en botanique*.[152] The 'Introduction' probably was, as Cheyron has argued, just one of the many fragments on botany found among Rousseau's manuscripts.[153] This document could have been finalised no earlier than Antoine-Laurent de Jussieu's introduction of Linnaean names into the Ecole de botanique of the Jardin du roi in 1773-1774,[154] to which Rousseau refers in his text.

Rousseau offered a series of justifications for Linnaean nomenclature, but his reasons were purely practical: assigning rigid designators to plants. Like classification, this is an information management technique. For Rousseau the sole utility of binomial names is to serve as rigid designators; the utility of plant names has no elective affinity with other considerations of utility such as commercial or other material interests, including those of the body, that detract from the true task of botany: 'Le premier Malheur de la Botanique est d'avoir été regardée dès sa naissance, comme une partie de la Médecine [...] on négligea la connaissance des plantes mêmes [...] On ne cherchait des plantes que pour trouver des remèdes, on ne cherchait pas des plantes mais des simples'.[155]

Binomial nomenclature is just as *useful* to natural philosophers as it is to those who exploit plants for profit.

In botany, as in music, the language of mediation should be serviceable and 'utile'; that is to say, simple, accessible and precise.[156] The Spartan terseness, economy and parsimony of Linnaean nomenclature was an advantage; the names contain 'pas un mot qui ne soit nécessaire'.[157] This nomenclature is 'aussi commode et nécessaire aux Botanistes que celle de l'Algèbre aux Géomètres'.[158] It also pleases the ear: 'Ces mots cependant sont tous grecs ou latins, expressifs, courts, sonores, et

152. Contrary to much scholarship on this text, it is an essay on nomenclature, not classification. This was acknowledged by Deville, who placed it after the other botanical texts, and entitled it, 'Réflexions sur la nomenclature botanique', p.293ff. See George, *Botany, sexuality and women's writing*, p.69.

153. Cheyron, 'Ray et Sauvages annotés par Jean-Jacques Rousseau', p.86.

154. Rousseau, 'Introduction', *OC*, iv.1207. Antoine Laurent de Jussieu's conversion of the Jardin du roi from Tournefort's system used 'les familles et les divisions primitives de Trianon', i.e. those set in place by Bernard de Jussieu in the Jardin du Trianon (Versailles) in 1759 (discussed in Chapter 6, p.178). Jussieu also used Linnaean species, genera and binomial nomenclature. Jussieu, 'Sixième notice historique sur le Muséum', p.11-12; Jussieu, 'Exposition', p.193.

155. Rousseau, 'Introduction', *OC*, iv.1201; see also *Rêveries*, *OC*, i.1065-66.

156. Rousseau, 'Projet concernant de nouveaux signes pour la musique', *OC* v.130;Cook, 'Rousseau and the languages of music and botany', p.80.

157. Björnståhl to Carl Christoffer Gjörwell, 1 September 1770; *CC* xxxviii.94.

158. Rousseau, 'Introduction', *OC*, iv.1206.

forment même des constructions élégantes par leur extrême précision'.[159]

Like Antoine-Laurent Jussieu, Rousseau used the natural method, while advocating Linnaean names. This separation of the names from the system with which they are often associated may seem to contradict Rousseau's statement that Linnaeus's 'nomenclature *paraît* tenir tellement à son systême, qu'on ne s'avise guères de l'en séparer'.[160] Yet Rousseau's choice of the verb 'paraître' suggests that Linnaeus's nomenclature only *seemed* tied to his sexual system, but in fact was not.

Rousseau justified binomial names on the basis of their (1) universality, (2) tidiness, and (3) accessibility.

First, Linnaeus's binomial Latin names were on the verge of becoming universal; they were accepted throughout practically all of Europe:[161] 'La grande commodité de cette nouvelle nomenclature et son utilité que l'usage a fait connaître, l'on fait adopter Presque universellement dans toute l'Europe [...] et même à Paris. M. [Antoine-Laurent] de Jussieu vient de l'établir au Jardin du Roi'.[162] Of this adoption, Jussieu wrote: 'A la nomenclature de Tournefort, seule admise jusqu'alors dans le jardin, il [Jussieu] substitua celle de Linnaeus, plus abrégée, plus commode, et généralement adoptée dans toute l'Europe: ce qui mit en ce point le jardin de Paris en harmonie avec les autres jardins de botanique étrangers'.[163] In Rousseau's eyes, Jussieu had achieved a real triumph, considering chauvinistic French attitudes toward a 'système étranger':[164] 'Je sais avec quel mépris on affecte en France de traiter ce grand Naturaliste [Linnaeus], mais le reste de l'Europe l'en dédommage, et la postérité l'en vengera'.[165]

Mastery of the rigid designators facilitates *communication*, as Rousseau

159. Rousseau, 'Introduction', *OC* iv.1206. But '[n]ous n'avons aucune idée d'une langue sonore et harmonieuse qui parle autant par les sons que par les voix', Rousseau, *Essai sur l'origine des langues*, *OC*, v.390.

160. My emphasis; Rousseau, 'Introduction', *OC*, iv.1207; see Jussieu, 'Exposition d'un nouvel ordre des plantes', p.193; see Drouin *L'Herbier*, p. 74. 'La nécessité de distinguer les problèmes théoriques de la nomenclature et ceux de la classification [...] est souligné aujourd'hui par les ethnobotanistes.'

161. It has been claimed that given the many rival systems at the time, it was impossible to predict, as Rousseau did, the eventual supremacy of Linnaean binomials. Schiebinger, *Plants and empire*, p.223-24.

162. Rousseau, 'Introduction', *OC*, iv.1207.

163. Jussieu, 'Sixième notice', p.11-12. Contrary to Jussieu's positive assessment of Linnaeus's contributions to botany, it has been asserted that the reorganisation of the Jardin du roi resulted from '[a]nti-Linnaean feeling', and that Linnaean names were *not* introduced into the Jardin. Schiebinger, *Plants and empire*, p.223-24. See also Jussieu, 'Exposition', p.178, 193-94.

164. Rousseau, *Introduction*, *OC*, iv.1207.

165. Rousseau to Dutens, 5 February 1767; *CC*, xxxiv.99.

repeatedly counselled Mme Delessert: 'je vous conseille de tenir une note fidèle des noms latins et de tacher même de les retenir; car c'est le seul moyen de *s'entendre* avec les botanistes';[166] 'Ce n'est pas en vain que je vous donne ceux de Linnaeus, quoique *latins*. Ce sont les seuls admis dans toute l'Europe et par lesquels on est sûr de *s'entendre* avec les botanistes de toutes les nations'.[167] But nomenclature plays a limited role in botany: 'Ce n'est pas que j'aie changé d'avis sur la nomenclature; je ne la crois pas plus utile que quand je vous en ai parlé à la botanique qu'on veut étudier de soi-même; mais, encore une fois, pour *s'entendre* avec quelqu'un qui est absent, il faut bien *convenir* des noms qu'on donne aux objets dont on parle'.[168]

Second, Linnaeus had brought order to the chaos of botanical classification and nomenclature

> Enfin M. Linnaeus [...] forma le projet d'une refonte générale dont tout le monde sentait le besoin, mais dont nul n'osait tenter l'entreprise. Il fit plus, il l'exécuta, et après avoir préparé dans son *Critica botanica* les règles sur lesquelles ce travail devait être conduit, il détermina dans *Genera plantarum* ces genres des plantes, ensuite les espèces dans son *Species*; de sorte que gardant tous les anciens noms qui pouvaient s'accorder avec ces nouvelles règles et refondant tous les autres, il établit enfin une *nomenclature* éclairée, fondée sur les vrais principes de l'art qu'il avait lui-même exposés.[169]

Species polynomials were retained, but the trivial name could stand for the entire phrase: '[i]l fallut pour cela créer [...] à la Botanique une nouvelle langue qui épargnait ce long circuit de paroles qu'on voit dans les anciennes descriptions'.[170]

Finally, Linnaeus's rules for naming make the study of botany easier and more accessible: names were to be short and clear, easy to pronounce and derived from a discrete set of sources. These features were very important to Rousseau, aligning closely with his own efforts to increase the accessibility of music and botany through compiling dictionaries and creating symbolic keys and systems of notation.[171]

vii. Le 'disciple' critique le 'maître'

Rousseau agreed with Linnaeus that plant names needed to be subjected to an orderly nomenclatural regime, and he eulogised binomial nomenclature and its originator. Yet he ultimately rejected Linnaeus's view

166. My emphasis; Rousseau to Mme Delessert, 24 May 1773; *CC*, xxxix.163.
167. My emphasis; Rousseau to Mme Delessert, 30 August 1773; *CC*, xxxix.192.
168. My emphasis; Rousseau to Mme Delessert, 30 August 1773; *CC*, xxxix.192.
169. My emphasis; Rousseau, 'Introduction', *OC*, iv.1205-06.
170. Rousseau, 'Introduction', *OC*, iv.1206.
171. Cook, 'Rousseau and the languages of music and botany', p.80.

of the botanist as a new Adam, the master of nomenclature. Adam, after all, was simply the first apothecary, that is, he used plants for their properties, without regarding them for themselves.[172]

Rousseau stated on several occasions that mastery of names does not constitute the study of botany: 'Il ne s'agit donc pas encore de la nomenclature qui n'est qu'un savoir d'*herboriste.* J'ai toujours cru qu'on pouvait être un *très grand* botaniste sans connaître une seule plante par son *nom*'.[173] In the so-called 'Introduction' to the *Fragmens pour un dictionnaire,* Rousseau reiterated this position: 'un homme intelligent pourrait être un *excellent Botaniste,* quoiqu'il ne connût pas une seule plante par son *nom*'.[174] The study of plant structure should take precedence over names:

> On prétend que la botanique n'est qu'une Science de mots qui n'exerce que la mémoire et n'apprend qu'à nommer des plantes. Pour moi, je ne connais point d'étude raisonnable qui ne soit qu'une science de mots; et auxquel des deux, je vous prie, accorderai-je le nom de botaniste, de celui qui sait cracher un nom ou une phrase à l'aspect d'une plante, sans rien connaitre à sa structure, ou de celui qui connaissant très bien cette structure ignore néanmoins le *nom très arbitraire* qu'on donne à cette plante en tel ou en tel pays? Si nous ne donnons à vos enfants qu'une occupation amusante, nous manquons la meilleure moitié de notre but qui est, en les amusant, d'exercer leur intelligence et de les accoutumer à l'attention. Avant de leur apprendre à nommer ce qu'ils voient, commençons par leur apprendre à le voir.[175]

In the next letter he returns to this theme in a more polemical vein: 'ce n'est pas une *nomenclature des perroquets* qu'il s'agit d'acquérir, mais une *science réelle,* et l'une des sciences les plus aimables qu'il soit possible de cultiver'.[176]

This standpoint agrees with that articulated by Antoine-Laurent de Jussieu in a contemporary work: 'La nomenclature ne doit pas être négligé, mais la recherche des *caractères* est une partie plus important de la Botanique'.[177] In a later work, Jussieu expanded on this point: 'Botanicum diù dilusit errans aut vaga nimis defintio; scientiae quae dum plantis tantummodo designandis ac nominandis incibuit[...] apud vulgum visa fuit scientiae merae nomenclationis, non menti sed memoriae exercendae praeprimis idonea'.[178]

172. Rousseau, *Rêveries, OC,* i.1064.

173. My emphasis; Rousseau to Mme Delessert, 22 August 1771; *OC,* iv.1152.

174. My emphasis; 'Introduction', *OC,* iv.1209.

175. My emphasis; Rousseau to Mme Delessert, 16 July 1772; *CC,* xxxix.82.

176. My emphasis; Rousseau to Mme Delessert, 2 May 1773; *CC,* xxxix.151.

177. My emphasis; Jussieu, 'Examen de la famille des renoncules', p.218.

178. 'For a long time the definition of botanical science has been incorrect, deceptive, or at the very least, vague; it has concentrated on designating and naming plants [...] being seen by laymen as a science of pure nomenclature, more especially suitable to the exercise of

Yet even while conceding the pointlessness of a botany fit for parrots, Rousseau did not accept the attack on nomenclature advanced by Louis Jean Marie Daubenton, who condemned naming as a complete waste of time; the object should be the study of *properties*.

La nomenclature des plantes n'est donc pas nécessaire pour la découverte de leurs propriétés; cela est si vrai qu'il serait ridicule de l'avoir mis en question, s'il n'était prouvé par l'état présent de la *Botanique* et par l'expérience du passé, que l'on s'est appliqué à la nomenclature par préférence aux autres parties de cette science. On fait plus d'observations et on tente plus de combinaisons pour parvenir à réduire la nomenclature des plantes en système, qu'il ne faudrait peut-être faire d'expériences et acquérir de faits pour découvrir quantité de nouvelles propriétés utiles dans ces mêmes plantes. Ce défaut de conduite dans l'étude de la *Botanique*, est un obstacle à l'avancement de cette science, parce qu'il nous éloigne de son principal objet. Il est même à craindre que si on continuait à marcher dans cette fausse route, on ne vînt à le perdre de vue. Pour s'en convaincre il faut examiner quelle est l'utilité que l'on a retirée de la nomenclature des plantes, poussée au point de perfection que les *Botanistes* se sont efforcés de lui donner; à quoi cette nomenclature peut servir dans la *Botanique*; et à quoi elle peut nuire, en supposant que cette connaissance soit réduite en système constant et même infaillible.[179]

Rousseau mounted two objections against Daubenton: (1) against the view that the identification of properties, virtues or 'simples' is the proper object of botanical study, he maintained the importance of studying the plants themselves; and (2) against the rejection of nomenclature he deplored the waste of a vast store of information. Rousseau replied to Daubenton's position, if not to Daubenton himself:[180]

Je demande à tout lecteur sensé comment il est possible de s'attacher à l'étude des plantes en rejetant celle de la nomenclature [...] Il s'agit de savoir si trois cents ans d'études et d'observations doivent être perdus pour la botanique, si trois cent volumes de figures et de descriptions doivent être jetés au feu, si les connaissances acquises par tous les savants [...] doivent être inutiles à leurs successeurs, et si chacun partant toujours de zéro pour son premier point, pourra parvenir lui-même aux mêmes connaissances qu'une longue suite de recherches et d'études a répandues dans la masse du genre humain.[181]

Hence, Rousseau depicted nomenclature as a necessary, but not sufficient, condition for doing botany. For Mme Delessert this stipulation

---

the memory than to that of the mind', Jussieu, 'Introductio', *Genera plantarum*, p.xxxiv-xxxv; trans. Susan Rosa in Stevens, *Development of biological systematics*, p.354.

179. Original emphasis; Louis Jean Marie Daubenton, 'Botanique', para.5 in Diderot and D'Alembert, *Encyclopédie*.

180. The two men were acquainted, as attested by a letter of Rousseau to Marc-Antoine-Louis Claret de Latourrette of 4 July 1770; *CC*, xxxviii.53.

181. Rousseau, 'Introduction', *OC*, iv.1209.

had special significance; by mastering Latin plant names she could enter the 'commonwealth' of botany.[182] She was therefore not a merely passive female student under the tutelage of a male preceptor,[183] but rather, 'seule de [son] sexe avec Madame la Duchesse de Portland au très petit nombre des vrais botanistes'.[184] Mme Delessert and the Duchess of Portland thus belonged to an elite band of true savants.[185]

viii. Latin and synonymy

Linnaeus held that Latin, under threat from vernacular tongues, should remain the learned language of Europe. He himself knew only Latin and Swedish; he had a high regard for Latin and a strong grounding in it. He took '[c]lassicism or pseudo-classicism' for granted 'as far as generic names were concerned'.[186] While his preference for the classical languages had its roots in his own provincial Swedish background, Linnaeus was also defending Latin as a universal language against linguistic fragmentation.[187] Stearn argues that while Latin still is '[t]he property of no one nation or linguistic group, Latin has, in consequence of its neutrality, become world-wide'.[188] At the same time, however, use of Latin not only in published works, but also in private correspondence (for example between Albrecht von Haller and Johannes Gessner, both German Swiss) excluded literate women, who were not, as a rule, taught Latin.[189] Some scholars therefore question whether Latin could have served as a universal language even in the eighteenth century, given that women and less well-educated men usually could not read it.[190]

Rousseau knew Latin names could be difficult, for he wrote to Mme Delessert:

> On pourrait faire des mots plus francisés: mais il me parait avantageux de vous tenir toujours le plus près qu'il se pourra des termes admis dans la

182. Linnaeus referred to the 'commonwealth' of botany in *Critica botanica*, p.xix.
183. See Gasbarrone, 'The book of nature', p.34-35.
184. Rousseau to Mme Delessert, 2 May 1773; *CC*, xxxix.158.
185. Like true philosophers, the number of 'vrais botanistes', is small; many botanists teach simply to make a name for themselves: 'quoique ce fut maintenant un usage assez général de commencer à enseigner les autres avant de s'instruire soi-même, il [Rousseau] ne voulait point le suivre'. Björnståhl to Gjörwell, 1 September 1770; *CC*, xxxviii.94; see also Rousseau, *Rêveries*, *OC*, i.1069. On Rousseau's elitism in science, see *Discours*, *OC*, iii.29.
186. S. F. Blake, 'Byways of nomenclature', *American journal of botany* 36:1 (January 1949), p.8-9 (8).
187. Koerner, *Linnaeus*, p.47.
188. William T. Stearn, 'The background of Linnaeus's contributions to the nomenclature and methods of systematic biology', *Systematic zoology* 8 (1959), p.4-21 (9).
189. William T. Stearn, *Botanical Latin: history, syntax, terminology and vocabulary*, 4th edn (Newton Abbot, 1992), p.7.
190. Schiebinger, *Plants and empire*, p.200.

> botanique, afin que sans avoir besoin d'apprendre ni latin ni grec, vous puissiez néanmoins entendre passablement le vocabulaire de cette science, pédantesquement tiré de ces deux langues, comme si pour connaître les plantes il faillait commencer par être un savant grammairien.[191]

In contrast with Rousseau, Michel Adanson rejected Linnaeus's imposition of Latin and Greek names on plants: 'On parle le langaje de la Botanike en François, en Anglois, en Allemand, en Italien, etc. comme en Grec et en Latin [..] quele régle de latinité impose à tous les noms une terminaison en *a*, en *ia*, en *um* ou en *us* exclusivement?'[192] Latin and Greek names from antiquity would only be acceptable if validated by scientific authority: 'Nous pensons comme M. Ludwig (*Instit.* ¶213) que les noms Grecs ou Latins, reçus en Médecine et en Botanike ne doivent pas être chanés pour leur substituer un autre synonyme Grec ou Latin'.[193] Since important authors such as Haller and Adanson did not accept Linnaean nomenclature, it was difficult to relate their works to a consistent set of names. Rousseau preferred Haller to Adanson because the former, in contrast to the latter, cited 'toujours les genres et quelquefois les phrases des espèces de Linnaeus'.[194]

Even though Rousseau was a self-taught autodidact who only briefly attended school, learning and using Latin seems to have posed no problem to him either practically or theoretically. Furthermore, he seems to have been unconcerned by the mediation of the human-nature relationship through Latin. The threshold problem from his point of view was mediation itself; as a linguistic mediation, Latin is no worse than any other language, and perhaps it is even better.

Effective communication in botany requires that the plant name function as a rigid designator, that is: that it always and invariably refers to the same plant. Rousseau understood that it was indispensable to link competing names with rigid designators through synonymy: 'La Synonymie est la partie [...] de la Botanique [...] sans laquelle [...] il est impossible à l'observateur de profiter des observations des *autres*; chose pourtant très nécessaire à qui ne peut tout voir de ses propres yeux'.[195] So, for example, he wrote to the Duchess of Portland: 'J'ai hasardé de donner des noms du *Species* de Linnaeus à celles [plantes] qui n'en avoient point'.[196]

In the few extant botany books owned by Rousseau we find extensive

191. Rousseau to Mme Delessert, 16 July 1772; *CC*, xxxix.84.
192. Adanson, *Familles des plantes*, p.clxxiii-clxxiv. Adanson's idiosyncratic orthography has been retained.
193. Adanson, *Familles des plantes*, 1966, p.clxxii-iii. Adanson referred to *Institutiones historico physicae* by Christian Gottlieb Ludwig (1709-1773).
194. Rousseau, 'Introduction', *OC* iv.1208.
195. My emphasis; Rousseau to Clappier, 26 May 1769; *CC*, xxxvii.93.
196. Rousseau to the Duchess of Portland, 29 April 1767; *CC*, xxxii.35.

annotations, usually cross-references or synonymies to the works and names of Linnaeus. This constitutes an important testimonial of Rousseau's conviction of the utility of Linnaean names: 'Il a fallu pour ne pas m'y perdre tout rapporter à une nomenclature particulière; et j'ai choisi celle de Linnaeus [...] parce que ses noms, composés seulement de deux mots me délivrent des longues phrases des autres.'[197] In a letter to Du Peyrou he explains the value of this exercise:

> [J]'ai fait une assez considérable collection de livres de Botanique [...] j'ai fait sur la plupart de ces livres un grand travail par rapport à la Synonymie, en ajoutant à la plupart des descriptions et des figures le nom de Linnaeus. Il faut s'être essayé sur ces sortes de concordances pour comprendre la peine qu'elles coûtent, et combien celle que j'ai prise peut en éviter à ceux à qui passeront ces mêmes Livres s'ils en veulent faire usage.[198]

He annotated several important works in this fashion, including Sébastien Vaillant, *Botanicon parisiense* (Leiden, 1743),[199] Dominique Chabrey, *Omnium stirpium sciagraphia et icones* (Geneva, 1678),[200] Albrecht von Haller, *Historia stirpium indigenarum Helvetiae inchoata* (Berne, 1768),[201] John Ray, *Synopsis methodica stirpium britannicarum*, edited by J. J. Dillenius (London, 1724) and François Boissier de Sauvages de la Croix, *Methodus foliorum* (The Hague, 1751); the last contains nearly 2000 annotations.[202] Sometime during 1777-1778 Rousseau annotated a copy of Nicolas and Geneviève de Nangis Regnault, *La Botanique mise à la portée de tout le monde* (Paris, 1774). He also composed a table of the 349 plants in *La Botanique*, following the sexual system.[203]

Rousseau attested to how time-consuming it was to add synonymies to books by pre-Linnaean authors and authors who did not use Linnaean names.[204] Yet synonymy, while difficult, is necessary in order to unite the

197. Rousseau to Malesherbes, 17 April 1772; *CC*, xxxix.37. Note that Rousseau stated his preference for Linnaeus's 'système', presumably the sexual system, in the past tense.
198. Rousseau to Du Peyrou, 15 November 1769; *CC*, xxxvii.175; see also Rousseau to Du Peyrou, 3 March 1768; *CC*, xxxv.156-58, par.4.
199. Hawkins, 'Manuscripts of Jean-Jacques Rousseau at Harvard University'. Rousseau expressed an 'extrême besoin' for Vaillant. Rousseau to Duchesne, 28 April 1765; *CC*, xxv.197; for his use of Vaillant, see for example Rousseau to Gouan, 26 December 1769; *CC*, xxxvii.196.
200. This book is in the Lloyd Library, Cincinnati, Ohio. See Takuya Kobayashi, 'Seven botanical books annotated by Rousseau', *Lloydiana* 11 (2007), p.10-11.
201. Cook, 'Jean-Jacques Rousseau's copy of Albrecht von Haller's *Historia stirpium*'.
202. Cheyron, 'Ray et Sauvages'.
203. Rousseau to the abbé de Pramont, 13 April 1778, *CC*, xl.209-12; Cheyron, 'L'amour de la botanique: les annotations de Jean-Jacques Rousseau sur la Botanique de Regnault', *Littératures 4* (1981), p.53-95. An imperfect transcription of these annotations was published by Musset-Pathay as 'Notes inédites de J. J. Rousseau sur la botanique'. Leigh published an extract from them in *CC*, xl.A666. See BdAN [Munch MS].
204. Rousseau to the Duchess of Portland, 12 February 1767; *CC*, xxxii.134.

descriptions of plants provided by authors employing different systems of nomenclature. The work of synonymy could also help would-be gardeners such as Du Peyrou:

> J'ai aussi réduit [...] les phrases de Sauvages à la *nomenclature triviale* de Linnaeus qui est très *commode*. Si le plaisir d'avoir un Jardin vous rend un peu de goût pour la botanique, je pourrai vous épargner beaucoup de travail pour la *Synonymie* en vous envoyant pour vos exemplaires ce que j'ai noté dans les miens, et il est absolument nécessaire de débrouiller cette *partie critique* de la botanique pour reconnaître la même plante, à qui souvent chaque Auteur donne un nom différent.[205]

Even though Rousseau's source of rigid designators was Linnaeus, he often returned to the seventeenth-century Swiss author, Caspar [Gaspard] Bauhin, to settle questions of synonymy:

> Il a fallu pour ne pas m'y perdre tout rapporter à une nomenclature particulière, et j'ai choisi celle de Linnaeus[206] [...] Pour y rapporter sans peine celles de Tournefort il ne faut très souvent que recourir à l'Auteur commun que tous deux citent assez constamment savoir Gaspard Bauhin. C'est dans son Pinax[207] que je cherche leur concordance. Car Linnaeus me parait faire une chose convenable et juste, quand Tournefort n'a fait que prendre la phrase de Bauhin, de citer l'auteur original et non pas celui[208] qui l'a transcrit, comme on fait très injustement en France.[209]

ix. Beyond names – Rousseau's botanical pasigraphy

Pierre Prévost of Geneva, who visited Rousseau in Paris near the end of his life, wrote, 'Comme il avait imaginé précédemment une méthode nouvelle pour noter la musique, il s'occupait alors à inventer une *écriture abrégée* pour la botanique. J'ai vu écrite de sa main, avec ces nouveaux *caractères*, une partie des genres et espèces de Linnaeus, qu'il rassemblait dans un fort petit volume, pour pouvoir les *porter* plus aisément avec lui dans ses promenades solitaires'.[210]

205. My emphasis; Rousseau to Du Peyrou, 3 March 1768; xxxv.156-58. By 'nomenclature triviale' Rousseau means the nomen triviale, or specific epithet, rather than the generic name coupled with the polynomial, which constituted the species name for Linnaeus.
206. I.e. the Linnaean binomial nomenclature.
207. 'Pinax' refers to a 'tablet; a list or register inscribed on a tablet; a catalogue or index.' *OED*, 'pinax'.
208. That is, Bauhin.
209. Rousseau to Malesherbes, 17 April 1772; *CC*, xxxix.37. In other words, the French have treated the Swiss Bauhin, unjustly. Similarly, 'les phrases de Tournefort étant presque toutes tirées de Gaspard Bauhin, tout l'honneur qu'en cela les Français veulent faire à leur compatriote remonte à *un Suisse* en toute équité'. My emphasis; BdAN [Munch MS], Article 190.
210. My emphasis; Prévost (1751-1839) was tutor to Mme Delessert's children in Paris. Pierre Prévost, Appendice 659, *CC*, xl.267.

At the BPU Neuchâtel these 'caractères' survive in the same cahier of manuscripts that holds the eight letters on botany to Mme Delessert (fig. 7.5, p.249). They constitute what is perhaps one of Rousseau's greatest, but most neglected, innovations: a botanical symbol system or pasigraphy.[211] This pasigraphy was to supply a new, symbolic language for describing and classifying plants; it would have superseded Linnaean binomials and polynomials in succinctness and brevity and would have avoided the problems associated with the use of Latin. It would also have been portable, like Rousseau's small herbaria (discussed in the next chapter).

Rousseau's pasigraphy continued a tradition initiated by Linnaeus's hijacking of alchemical symbols as abbreviations to which he accorded novel meanings.[212] This pasigraphy applied the principles of accessibility, simplicity and utility that Rousseau articulated in his system of musical notation, introduced to the Académie royale des sciences in 1742. This reform was supposed to make music 'plus simple', 'plus précise' and 'plus aisé à apprendre' by reducing – 'et c'est ici le point le plus important [...] le nombre des signes et leur combinaisons'. Hence, Rousseau wrote, '[j]e fonde l'utilité de mon système sur son extrême facilité'.[213] Like his notation system for music, Rousseau's pasigraphy exemplifies the kind of universal scientific language extolled by seventeenth-century savants such as Leibniz: 'an ideographic writing system similar to Chinese characters [...] Egyptian hieroglyphics, algebraic symbols [...] and the like'.[214]

For Leibniz, Chinese might be the ideal universal language, while Rousseau relegated ideographic languages to a lower standing than alphabetic languages in his hierarchy of languages.[215] Yet he nevertheless experimented with an ideographic scientific language for botany that promised to be more fit for purpose than alphabetic languages.[216] Joseph Needham, the great sinologist, concluded two hundred years later that in botany Chinese would be the ideal master code to replace Latin. Needham reached his conclusion in ignorance of Rousseau's experimentation with botanical ideographs in the eighteenth century; the only

211. This system has recently been published alphabetically in Ducourthial, *La Botanique*, p.435-532.

212. William T. Stearn, 'Signs used by Linnaeus', in Carolus Linnaeus, *Species plantarum*, facsimile edn with an introduction by William T. Stearn (London, 1953), vol.1, p.162-63.

213. Rousseau, *Projet concernant de nouveaux signes pour la musique*, *OC*, v.130.

214. Slaughter, *Universal languages and scientific taxonomy in the seventeenth century*, p.85.

215. Rousseau, *Essai sur l'origine des langues*, *OC*, v.384-85.

216. Cook, 'Languages of music and botany', p.82-84. Ironically, Rousseau's system of musical notation was exported to China via Japan and is used in schools in China today. Personal communication, Yin Shoufu, 2009.

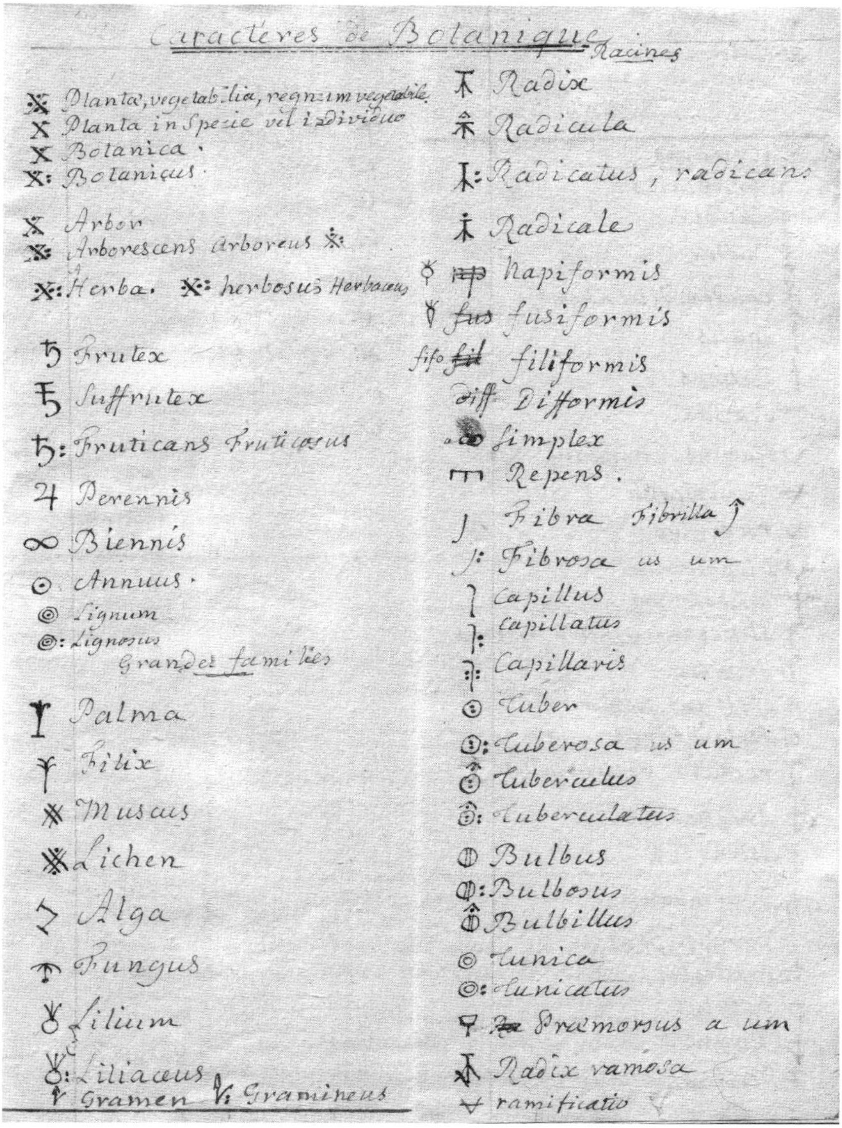

Figure 7.5: Jean-Jacques Rousseau, *Caractères de botanique*, by permission of the Bibliothèque publique et universitaire de Neuchâtel, MsR21, f.49.

predecessors of which Needham was aware were the alchemical symbols that Linnaeus borrowed from chemistry.[217]

Bernardin de Saint Pierre (1737-1814) is perhaps the sole source for how Rousseau regarded his botanical symbol system. Bernardin, best known for his pastoral novel, *Paul et Virginie*, cultivated a friendship with Rousseau in Paris in the 1770s based on their mutual interest in botany; he later served briefly as intendant of the Jardin des plantes (1792-1793). Bernardin writes of Rousseau's pasigraphy:[218]

> Il avait imaginé quelque chose de très ingénieux: c'étaient des formules botaniques pour exprimer le caractère des plantes: par exemple la feuille par un signe: si en cœur, un cœur. Les pétales par un autre; les anthères par un autre; de sorte qu'au moyen de huit ou dix caractères il exprimait la description d'une plante que plusieurs lignes n'auraient pas décrit.[219]

According to Bernardin, Rousseau was disillusioned by these efforts: 'quelque ingénieuse et expéditive que fut cette méthode, il me dit qu'il y avait renoncé, parce qu'elle ne lui présentait que des squelettes'.[220] An advocate for this invention did not appear until Needham made his plea two hundred years later.

x. Conclusion: Rousseau's verdict on Linnaean nomenclature

Linnaeus's binomial names revolutionised botany, and were widely adopted within twenty years after their publication in the *Species*; their utility forms the cornerstone of Rousseau's brief for preferring Linnaeus to all other nomenclatural authorities. Yet his advocacy of Linnaean binomials did not mean that Rousseau felt an exclusive commitment to Linnaeus's other major invention, the sexual system. Rousseau was, like Linnaeus himself, a 'promiscuous realist', who believed that there is more than one way to carve up nature.[221]

Like other promiscuous realists of his day, Rousseau was indebted to Linnaeus for his binomial names, whatever their shortcomings. Even the

217. Needham, *et al.*, *Science and civilisation in China*, vol.6, part 1, p.178-82.

218. Jacques Henri Bernardin de St Pierre, *La Vie et les ouvrages de Jean-Jacques Rousseau*, ed. Maurice Souriau (Paris, 1907), p.163.

219. 'Les uns [...] représentaient les formes des fleurs; d'autres, celles des feuilles; d'autres, celles, des fruits [...] Il n'employait que neuf ou dix de ces signes, pour former l'expression d'une plante. Il y en avait de placés les uns au-dessus des autres, avec des chiffres qui exprimaient les genres et les espèces de la plante, en sorte que vous les eussiez pris pour les termes d'une formule algébrique.' Jacques Henri Bernardin de St Pierre, *Etudes de la nature*, 11 (no publication information), quoted in Roger de Vilmorin, 'Introduction: lettres sur la botanique, fragments pour un dictionnaire de botanique', *OC*, iv.cci.

220. 'Notes et variantes, *OC*, iv.1830.

221. Bird and Tobin, 'Natural kinds'.

most committed proponents of the natural method such as Antoine-Laurent Jussieu used Linnaean binomials. Despite various deficiencies of the Linnaean nomenclatural reform, therefore, Rousseau preferred it to any other; the alternative was that botanists 'retombe[nt] dans les phrases de Tournefort et Gaspard Bauhin'.[222] Rousseau could find no nomenclature superior to that of Linnaeus in his time. Hence, binomial names did for botany, what Rousseau's numerical system of musical notation did for music, reducing the 'nombre des signes', and simplifying them; this nomenclatural reform made botany, already becoming an accessible and popular science, more manageable, and easier to learn. It was for a later age to ponder whether to give up Linnaean binomials in favour of an identification system for organisms better aligned with what we now know about nature.

222. Rousseau, 'Introduction', *OC*, iv.1207.

8. The herbarium as boundary object

> Science [...] seule distingue le vrai botaniste de l'herboriste ou nomenclateur. Il s'agit donc ici d'apprendre à préparer, dessécher et conserver des plantes ou échantillons des plantes de manière à les rendre faciles à reconnaitre et à déterminer. C'est en un mot un *herbier* que je vous propose de commencer.[1]

i. Evocations

In his *Dictionnaire de musique*, Rousseau described the Swiss cowherds' song, the 'ranz des vaches',[2] as 'cet Air si chéri des Suisses qu'il fut défendu sous peine de mort de le jouer dans leurs Troupes parce qu'il faisait fondre en larmes, déserter ou mourir ceux qui l'entendaient tant il excitait en eux l'ardent désir de revoir leur pays'. This peculiar song, devoid of 'accents énergiques', nevertheless recalls to the homesick Swiss soldiers 'leurs anciens plaisirs, leur jeunesse, et toutes leurs façons de vivre, excitent en eux une douleur amère d'avoir perdu cela'. The 'ranz des vaches' 'n'agit point précisément comme *Musique*, mais comme signe mémoratif', evoking feelings of extreme homesickness, a 'douleur 'amère'.[3] According to Rousseau, the homesickness evoked by the 'ranz des vaches' was so overwhelming that its performance was punishable by death in the Swiss mercenary regiments serving eighteenth-century European states in the thousands.[4] In Rousseau's semiotics, therefore, this melancholy song is less music than a sign standing for the homeland.

The herbarium could take on a similar evocative power. Rousseau described his herbarium in the septième promenade of the *Rêveries* as a

1. My emphasis; Rousseau to Mme Delessert, 11 April 1773; *CC*, xxxix.139.
2. This song has many versions in French and German; its German name is 'Kuhreihen'.
3. Original emphasis; Rousseau, 'Musique', *Dictionnaire de musique*, *OC*, v.924. The ranz des vaches was a 'simple air sans paroles' that originated in the German part of Switzerland, from which it spread to the Suisse romande. Fifty versions of these improvisational songs were reported in 1813. Jost, *Jean-Jacques Rousseau Suisse*, vol.1, p.400.
4. Original emphasis; The German word for homesickness is 'Heimweh', which Rousseau writes as 'Hemé' or 'Hemve'. Rousseau to the maréchal de Luxembourg, 20 January 1763; *CC*, xv.53. In 1710 the Basel physician, Theodor Zwinger III (1658-1724), connected the illness of nostalgia (proposed by the Swiss physician Johann Hofer in 1688) to the ranz des vaches. Zwinger also seems to have been the first to speak of the interdiction of this melody among Swiss troops in foreign service. Leigh, *CC*, xv.67, note (g).

'journal d'herborisations' that, like the 'ranz des vaches', brought back a flood of nostalgic memories:

> Toutes mes courses de botanique, les diverses impressions du local des objets qui m'ont frappé, les idées qu'il m'a fait naître, les incidents qui s'y ont mêlés, tout cela m'a laissé des impressions qui se renouvellent par l'aspect des plantes herborisées dans ces mêmes lieux. Je ne reverrai plus ces beaux paysages, ces forêts, ces lacs, ces bosquets, ces rochers, ces montagnes dont l'aspect a toujours touché mon cœur: mais maintenant que je peux plus courir ces heureuses contrées je n'ai qu'à ouvrir mon herbier et bientôt il m'y transporte Les fragments des plantes que j'y ai cueillies suffisent pour me rappeler tout ce magnifique spectacle. Cet herbier est pour moi un journal d'herborisations qui me les fait recommencer avec un nouveau charme et produit l'effet d'une optique[5] qui les peindrait derechef à mes yeux.[6]

Without referring directly to Rousseau's characterisation of the 'ranz des vaches' as a 'signe mémoratif', Jean Starobinski writes that 'la plante desséchée devient le *signe mémoratif*', that is to say that the specimen is no longer just a dried plant any more than the 'ranz des vaches' is merely music; the plant's role has been transposed, shifted into another domain – that of deep feelings.[7]

A close reading of two works that Rousseau wrote almost contemporaneously, the *Considérations sur le gouvernement de Pologne* (1770-1771) and the *Lettres sur la botanique* (1771-1774), shows that the herbarium could take on an emotive power like that of the ranz des vaches – patriotism. What better way to attach the citizens to the soil of the homeland than by having them learn the plants that grow upon it, including those most useful to the life of man? The ancient Lawgivers 'cherchèrent les liens qui attachassent les Citoyens à la patrie', establishing 'les institutions douces qui les attachaient [...] tous à leur *sol*'.[8] Rousseau tells the Poles: '[j]e veux qu'en apprenant à lire il lise des *choses* de son pays, qu'à dix ans il en connaisse toutes les *productions*'.[9] The ten-year old is to acquire an inventory of the productions of his country: its plants, animals, minerals, foods, etc.[10] A herbarium could therefore educate the future citizen in the productions of his country: 'un herbier [...] S'augmentera sans cesse avec vos connaissances, et contiendra enfin l'histoire de toute la végétation du pays'.[11]

5. I discuss Rousseau's interest (c.1764) in the optique, an eighteenth-century stereoscope, in Chapter 7, p.228-30.
6. Rousseau, *Rêveries*, *OC*, i.1073.
7. Original emphasis; see Starobinski, *Jean-Jacques Rousseau: la transparence et l'obstacle*, p.281.
8. My emphasis; Rousseau, *Considérations sur le gouvernement de Pologne*, *OC*, iii.957. Gourevitch translates 'sol' as 'fatherland', not soil.
9. My emphasis; Rousseau, *Considérations sur le gouvernement de Pologne*, *OC*, iii.966.
10. Poland had no significant industry at this date.
11. Rousseau to Mme Delessert, 11 April 1773; *CC*, xxxix.142.

ii. 'C'est en un mot un herbier que je vous propose de commencer'

While the evocative quality of the herbarium for Rousseau is well-known, little attention has been paid to the long and complex process – the intense and protracted expenditure of expertise, effort and energy that was needed to make herbaria of the calibre of Rousseau's. What, in other words, did he have to know, understand and be able to do in order to make herbaria? Did he have other reasons to make a herbarium besides evoking pleasant memories or patriotic sentiments?

I place Rousseau's herbaria in a new light by showing how he (1) adopted and (2) perfected a *pratique* advocated and taught by the major botanical authorities of the time, (3) how he *modified* this pratique, and (4) used it to advertise botany and the study of nature to a broad audience, thereby rendering it the 'boundary object' par excellence. A 'boundary object' is defined as

> [A]n analytic concept of those scientific objects which both inhabit several intersecting social worlds [..] and satisfy the informational requirements of each of them. Boundary objects are objects which are both plastic enough to adapt to local needs and the constraints of the several parties employing them, yet robust enough to maintain a common identity across sites [...] They have different meanings in different social worlds but their structure is common enough to more than one world to make them recognizable, a means of translation.[12]

The hortus siccus, or herbarium of dried plants had been developed as a research and teaching tool in Renaissance Italy, and was a well-established part of botany when Rousseau took up plant studies.[13] Rousseau faithfully followed Linnaeus's dictum that a 'Herbarium praestat omni Icone, necessarium omni Botanico'.[14] Linnaeus required that every beginner must *himself* collect, dry and mount specimens;[15] hence the

12. Susan Leigh Star and James Richard Griesemer, 'Institutional ecology, "translations" and boundary objects: amateurs and professionals in Berkeley's Museum of Vertebrate Zoology, 1907-1939', *Social studies of science* 19 (1989), p.387-420 (393).
13. 'A herbarium is a collection of plant specimens that usually have been dried and pressed, carefully mounted on sheets of quality paper, identified and labelled with important information [...] and stored and arranged on shelves in cabinets in the sequence of an accepted classification. These specimens [...] serve as a permanent reference to the diversity of plants [...] They also serve as a definitive reference for the identification and correct naming of newly collected plants [...] analogous to a library of carefully preserved plants where the specimens themselves and the labels [...] provide invaluable information.' (http://www.afcd.gov.hk/english/conservation/con_flo/con_flo_hkh/con_flo_hkh.html, accessed 18 May 2012.)
14. Linnaeus, *Philosophia botanica* (1751), p.7, section 11; 'A HERBARIUM is better than any picture, and necessary for every botanist', Linnaeus, *Philosophia botanica* (2003), p.18, section 11.
15. Linnaeus, *Philosophia botanica* (2003), p.329.

botanist must become a kind of artisan, and making a herbarium an integral part of the science of botany.

This important aspect of botanical practice has generally been underappreciated, as has been artisanship in the sciences generally. Scholars have recently begun to investigate seriously the relationship between artisanship and science in the Renaissance and early-modern periods. Noted artisans such as Dürer and da Vinci contributed to the sciences:[16] this was hardly a rote, mechanical process. Rather, 'science and art are both image-making and knowledge-producing activities'.[17] While Linnaeus argued that the herbarium was 'better than any picture', there can be no doubt (as I discuss below), that preparing the herbarium was also very much an image-producing activity, and remains so today.

Collecting plants and making herbaria might appear to have been a mere pastime for Rousseau because it had become dangerous for him to philosophise publicly after the arrest warrants issued against him in 1762. Moreover, Rousseau's own depiction of his botanical practice makes it sometimes seem practically frivolous.[18] I argue, however, that Rousseau's herbarium project was truly original, and that it exerted a lasting influence on the popularisation and dissemination of botany beyond the networks of savants, physicians and chemists in which he moved.[19]

We have seen that in his chemical studies Rousseau showed a keen awareness of the importance of apparatus, tools, and instruments; as a botanist he showed the same awareness. Whereas other botanists had sought principally to collect and preserve specimens simply for their own satisfaction or further study, Rousseau saw the herbarium could be more useful if made available in different sizes and systems of classification. Specimens could also be presented in various ways.

Rousseau also foresaw the utility of the herbarium not only for the savant, but also its attraction for the *amateur*, whose interest in botany could be aroused by seeing an aesthetically appealing rendering of naturalia.[20] These *amateurs* could include women and children, who were not practitioners in natural philosophy, chemistry or medicine.[21]

16. See for example Bacon, *The New organon*, p.66-67.
17. Pamela H. Smith, 'Art, science and visual culture in early modern Europe', *Isis* 97 (2006), p.96, 99; see also Pamela H. Smith, *The Body of the artisan* (Chicago, IL, 2004).
18. My emphasis; Rousseau, *Rêveries*, *OC*, i.1069; see also A. Cook, 'The septième promenade of the *Rêveries*'.
19. A. Cook, 'Jean-Jacques Rousseau et les réseaux d'échange botanique'.
20. 'Sans avoir l'ambition des herbiers classiques, l'herbier de l'amateur peut aussi ne servir qu'à la reconnaissance des végétaux, en devenant une sorte de fichier illustré', Marc Pignal, 'Techniques particulières', *Orchidées: culture et protection* 23 (1995), p.43-47 (46). 'Amateur' in this context means one who loves something.
21. Dubourg evinced a similar desire to place botany 'à la portée de tout le monde, sans

Rousseau's herbaria expressed his desire to propagate botany as a salutary science by advertising 'le plus riche et les plus varié des trois' realms of nature.[22]

In promoting herbaria, the author of *Emile* likewise promoted artisanship; every man should be able to live by the work of his hands: 'de toutes les occupations qui peuvent fournir la subsistence à l'homme, celle qui le rapproche le plus de l'état de Nature est le travail des mains'.[23] He himself chose to live by this creed after his 'reform'.[24] But for Rousseau botanical artisanship aims beyond the virtues of hard work and self-sufficiency to advertise botany as an *edifying*, and therefore salutary, science.

For Rousseau the goal was not only to make herbaria that would attract people to nature study, but also to teach others how to make them: the student selects and prepares the dried plant samples and sends them to the teacher, who provides the identifications. The student should be able to 'Classer, ranger, et nommer [...] celles que vous verrez pour la première fois'.[25] In fact, while important, the herbarium's use for beginners is only secondary: 'Les herbiers servent de mémoratifs pour celles qu'on a déjà connues; mais ils font mal connaître celles qu'on n'a pas vues auparavant'.[26] The virtue of herbaria lies therefore in reinforcing information that one has already acquired.

Rousseau's herbaria displayed not only a clear understanding of the botanical enterprise, but also bridged different worlds – the worlds of the philosopher, the botanist, the collector, the craftsman and the nature lover.[27] Presented individually on the page, bordered in red, sometimes with notes, each specimen dramatically testifies to the philosopher's love of nature and his desire to share this love of nature with others (fig.8.1, p.258).[28]

---

exception des Herboristes, ni des Gens de la campagne, des Femmes, ni des Enfans'. Dubourg, *Botaniste*, p.x. Dubourg does not invoke the larger psychological and theological context in which Rousseau places botanical study.

22. Rousseau, *Fragments de botanique*, *OC*, iv.1249; see also Rousseau, *Rousseau juge de Jean-Jacques: dialogues*, *OC*, i.832.

23. Rousseau, *Emile, ou De l'éducation*, *OC*, iv.470. Rousseau's artisanal background as an apprentice engraver may have given his botanical pursuits a strong orientation toward the herbarium. He also picked up other artisanal skills: music copying, lace-making, and book-binding.

24. Rousseau, *Confessions*, *OC*, i.363.

25. Rousseau to Madame Delessert, 11 April 1773; *CC*, xxxix.139.

26. Rousseau to Mme Delessert, 11 April 1773; *CC*, xxxix.139.

27. 'It is normally the case that objects of scientific inquiry inhabit multiple social worlds, since all science requires intersectional work', Star and Griesemer, 'Institutional Ecology', p.392.

28. However, as several of these herbaria are lost, any generalisation is potentially misleading.

Figure 8.1: Jean-Jacques Rousseau, specimen of *Viola odorata* L., Herbarium for Julie Emélie Willading, née Boy de la Tour, by permission of the Zentralbibliothek Zürich.

iii. Herbaria: a brief history

Since the Renaissance, herbaria have provided a practical and convenient 'way for botanists to escape the tyranny of the seasons',[29] hence the alternative name for herbaria, 'hortos hyemales'[30] or 'winter garden'. By the eighteenth century herbarium fabrication was a well-established, indeed integral, part of botany. Rousseau thus drew on a tradition dating from the Renaissance, although the origins of the modern herbarium remain obscure. Over the course of time the term 'herbarium' has variously referred to herbals, *historia plantarum*, *materia medica* or collections of dried medicinal plants.[31]

The modern herbarium of dried plants, carefully preserved and annotated for future use, emerged in sixteenth-century Italy, the homeland of early-modern botany. 'The literature on the history of herbaria is full of debates about the inventor of the technique. A Silesian physician, Thomas von Sarepta (1297-1378), is supposed to have kept a collection of dried plants and prepared a herbal from them in the late fourteenth century'.[32] Most modern scholars[33] award priority to Luca Ghini (c.1490-1556), professor of botany at Bologna and later director of the botanical garden of Pisa (1544-1546) for two reasons: he sent dried plants to fellow botanist, Pietr Andrea Matthioli, and his students made herbaria.

Yet while Ghini, the widely accepted originator of the modern herbarium, is reputed to have preserved his plant collections for later consultation, none of his herbaria survive. Ghini inspired others to make herbaria, even if he himself never made one;[34] he sent dried specimens to Mattioli to illustrate the latter's commentary on Dioscorides.[35] As Reeds notes, '[t]here is no doubt [...] that Luca Ghini [...] succeeded in making the technique popular among university students and teachers beginning in the 1530s and [15]40s. His students taught their own students and friends and so the idea quickly spread throughout Europe'.[36] Findlen adds, 'Ghini and his students were among

29. Karen M. Reeds *Botany in medieval and Renaissance universities* (New York and London, 1991), p.35.
30. Adriaan van den Spiegel, *Isagoges in rem herbariam libri duo* (Padua, Paolo Meietti, 1606), p.79.
31. Jean-Baptiste Saint-Lager, *Histoire des herbiers* (Paris, 1885), p.4, 6. Herbaria may have existed in antiquity as collections of medicinal plants used by pharmacists to prevent confusions between medicinal and poisonous plants. Saint-Lager, *Histoire des herbiers*, p.2.
32. Reeds, *Botany in medieval and Renaissance universities*, p.35.
33. Ernst H. F. Meyer, *Geschichte der Botanik* (Königsberg, 1857), vol. 4, p.272; Arber, *Herbals*, p.139.
34. Meyer acknowledges that '[v]on einem Herbarium Ghini's wissen wir zwar nichts Bestimmtes, allein von Pflanzen, die er an Mattioli geschickt, hörten wir bereits [of Ghini's herbarium we know indeed nothing certain, we only know something about plants that he sent to Mattioli]'. Meyer, *Geschichte der Botanik*, p.271.
35. Saint-Lager, *Histoire des herbiers*, p.16-17.
36. Reeds, *Botany in medieval and Renaissance universities*, p.35.

the first to collect specimens. Rather than leaving nature where they found it [...] they brought home samples. The growth of herbaria at the hands of scholars such as Aldrovandi and Cesalpino gives testimony to the facility with which these techniques spread'.[37] Ghini allegedly passed his technique on to his student, Ulisse Aldrovandi (1522-1605), who 'seems to have been the first to aim at a collection of dried plants [5000 specimens], which should include those of the whole world'.[38] However, neither Aldrovandi nor Andrea Cesalpino (another Ghini student) left any account of their herbaria or how they learned to make them.

In his *Histoire des herbiers* J. B. Saint-Lager argues, on the other hand, that a sixteenth-century Englishman, John Falconer, was the first to create a herbarium;[39] Falconer is supposed to have shown this work to other naturalists in Ferrara, Italy in 1545. Saint-Lager[40] relies for this information on contemporary testimony from the Jewish-Portuguese physician, João Rodrigues de Castelo Branco (1511-1568), also known as Amatus Lusitanus.[41] Lusitanus resided in Italy, including Ferrara, for a number of years, and described Falconer's herbarium, presumably from direct knowledge.[42]

Ernst Meyer argued against awarding Falconer priority for creating the first modern herbarium, contending that it would have been impossible for a sixteenth-century Englishman to have had a sufficiently advanced scientific understanding or training to originate the herbarium; because of the allegedly 'traurigem Zustande' of medicine and the natural sciences in sixteenth-century England![43]

37. Paula Findlen, *Possessing nature: museums, collecting, and scientific culture in early modern Italy* (Berkeley, CA, 1994), p.166.
38. Arber, *Herbals*, p.140.
39. Saint-Lager, *Histoire des herbiers*, p.17. Even Meyer acknowledges that Falconer's was the first known herbarium: 'sein Buch [...] ist nicht allein Eins der ersten, sondern geradezu das erste, nicht allein in England, sondern in der ganzen Welt das erste der Art, wovon ich bestimmte Nachricht finden konnte [his book [...] is not only one of the first, but actually the first, not only in England, but also the first of its kind in the entire world of which I could find a definite report]'. Meyer, *Geschichte der Botanik*, p.271.
40. Saint-Lager, *Histoire des herbiers*, p.18.
41. Lusitanus, a Christianised Jew of Spanish extraction, disagreed not only with Mattioli's interpretation of Dioscorides' plant identifications, but also with identifications offered by Otto Brunfels, Leonhard Fuchs and Jean Ruel. Matthioli attacked Lusitanus in *Apologia adversus Amathum Lusitanum* (1558). Pope Paul VI's persecution of Christianised Jews drove Lusitanus to the Jewish enclave of Salonica, where he died of plague contracted from one of his patients.
42. Amatus Lusitanus, *In Dioscorides Anazarbei de medica materia libros quinque ennarationes*, (Venice [Gualterum Scotum], 1553), p.322; 'il avait formé une collection varié d'un grand nombre d'échantillons admirablement préparés et collés sur des feuilles de papier réunies en volume' (Saint-Lager, *Histoire des herbiers*, p.18); see also Otto A. J. Penzig, *Contribuzioni alla storia della botanica* (Milan, 1905), p.2.
43. Meyer, *Geschichte der Botanik*, p.271. Saint-Lager notes that had this been the case, Shakespeare could not have written plays. Saint-Lager, *Histoire des herbiers*, p.19.

Saint-Lager suggested the inventor of the herbarium was not one person, but many; pharmacists, physicians and others probably dried plants between the pages of books.[44] The twin developments of book-binding and paper-making in the Middle Ages were crucial to the presentation of the herbarium in book form. Also very important was the reduction in the price of paper in the wake of the invention of printing in the fifteenth century. Irrespective of its precise origin, however, the herbarium was an established, and central, part of botanical study by the 1760s, when Rousseau took it up.

The impact of herbaria on botany cannot be overestimated. Herbaria not only allowed botanical work to be continued during the winter months, but also enabled botanists to preserve specimens for later consultation, comparison, and determination of species.

> A *hortus siccus* had a great many advantages over a real garden [...] It was compact, it required no gardener, it travelled well, it could easily be enlarged (although the practice of mounting plants in bound volumes rather than on loose sheets made it hard to rearrange them). It enabled men to collect plants on long trips and made it easy to exchange botanical knowledge. Dried plants could be sent with fewer mishaps than living ones; distant friends could readily see whether or not they were talking about the same plant.[45]

In the late eighteenth century, Lamarck justified the importance of herbaria to botany: they provide more plants than one can see in years of voyaging and visiting gardens and they permit comparisons among specimens collected in different seasons and in different climates. In the herbarium one can view more plants at once than one can see in the best-stocked garden; moreover, the loss of plants under cultivation reduces the utility of botanical gardens. Finally, despite being dried, the fructification can be made visible by means of rehydration.[46] Other contemporary sources likewise stressed the large information storage capacity of herbaria.[47]

Herbaria have recently taken on a new significance as a taxonomic tool to address a range of problems. Once thought to be 'an out-dated, minor discipline of little importance in the modern world', taxonomy has shown its value 'in the present struggle to reclaim deserts and degraded landscapes, breed pest resistance in our crops, and find new

44. Saint-Lager, *Histoire des herbiers*, p. 20.

45. Original emphasis; Reeds, *Botany in medieval and Renaissance universities*, p.35.

46. Jean Baptiste Pierre Antoine de Monet de Lamarck, 'Herbier (Herbarium)', in *Encyclopédie méthodique: botanique* (Paris, Panckoucke, 1789), vol.3, p.111-16 (112).

47. Jean Baptiste François Bulliard [Pierre], *Dictionnaire élémentaire de la botanique, ou, exposition par ordre alphabétique des préceptes de la botanique, et de tous les termes, tant françois que latins, consacrés à l'étude de cette science* (Paris, chez l'auteur and Didot, Barrois and Belin, 1783), p.99.

sources of energy, food, medicines and useful materials'.[48] Herbaria 'constitute the primary database for the world's plant diversity', and supply the only examples of plants that have become extremely rare or even extinct. 'Worldwide there are 2639 herbaria, containing more than 272,800,000 specimens.'[49] The national herbarium of France alone contains over 8 million specimens.[50] It has thus been said that the herbarium is an 'instrument d'étude indispensable au botaniste'.[51]

iv. Early-modern sources on herbaria

Making herbaria is a craft; it is therefore principally learned by doing, rather than from books. There were, however, significant accounts of the practice in print by Rousseau's lifetime. The earliest was *Isagoges in rem herbariam* by Adriaan van den Spiegel (1578-1635),[52] which describes the so-called 'winter garden' or 'hortos hyemales' for the winter months, and explains how to make one. 'Spieghel realises the great importance of herbaria, and observes that the labour expended in making one is deserving of high praise.'[53] However, there is no evidence that Rousseau knew this work.

An important source on herbaria was Tournefort's *Elemens de botanique* (1694), probably in the Latin version, *Institutiones rei herbariae*,[54] which Rousseau owned.[55] Tournefort (1656-1708) used the term herbarium for the first time: 'C'est proprement un amas de plantes sèches que l'on conserve dans des boites ou dans des livres; afin de les pouvoir examiner avec soin dans toutes les saisons de l'année: on l'appelle en Latin *Herbarium*, ou *Hortus siccus*.'[56] Tournefort's brief instructions are unremarkable, with the exception of the mercuric chloride insecticide he concocted to protect his collections. This insecticide was used well

48. Diane Bridson and Leonard Forman, *The Herbarium handbook*, revised edn (Kew, 1992), p.1.
49. Brian M. Boom, 'Societal and scientific information needs from plant collections', in *Sampling the green world: innovative concepts of collection, preservation, and storage of plant diversity*, ed. T. F. Stuessy and S. H. Sohmer (New York, 1996), p.19.
50. For comparison: Tournefort's herbarium contains 4000 specimens, Lamarck's 19,000, and that of the average nineteenth-century amateur, 8-10,000.
51. Abderrazak Marouf and Jean Vallade, *Dictionnaire de botanique: les phanérogames* (Paris, 2000), 'Herbier'.
52. Van den Spiegel, *Isagoges in rem herbariam libri duo*, p.79.
53. Arber, *Herbals*, p.142. The orthography of this name varies. Rousseau never refers to Van den Spiegel, but Linnaeus mentions him in the dissertation, *Instructio musei*, defended in 1753. Carolus von Linnaeus, *Amœnitates academicae*, vol.3, p.458.
54. Tournefort, *Institutiones rei herbariae.*
55. Rousseau owned this text at his death, as attested by the marquis de Girardin, and by a list sighted by the author at the Musée Jacquemart-André, Chaalis, France, 25 August 2004; see *CC*, xlii.4-22, and Richebourg, *Les Lectures de Rousseau*, p.184.
56. Original emphasis; Joseph Pitton de Tournefort, *Elemens de botanique*, vol.1 (Paris, Imprimerie royale, 1694), p.547; see also Arber, *Herbals*, p.142.

into the twentieth century and has created serious toxicity problems, with 'the potential for irreversible nervous system damage'.[57]

Rousseau was undoubtedly aware of chemical treatments, which seventeenth- and eighteenth-century sources discuss, but we do not known how Rousseau regarded these treatments.[58] A text purportedly by Rousseau that was published posthumously first by the French mycologist, Jean Baptiste François (usually known as Pierre) Bulliard (1752-1793) and subsequently by others, prescribes an anti-insect treatment with alum: 'on aura attention, pour les garantir des insectes, d'arroser un peu les tiges de chaque sujet, et le papier blanc, qui les renfermera, avec la solution de sublime [corrosif dissous avec du sel ammoniac cru dans de l'eau], citée plus haut [p.155-56] pour les graines'.[59] This is a safer method than that advocated by Tournefort, although this was probably not known at the time.[60] Rousseau may have used a chemical treatment, because (1) he used glue, which attracts insects, yet (2) his herbaria are well-preserved; generalisations are impossible, however, given that (3) some of his herbaria are not extant or have yet to be found.[61]

Other authorities on herbaria may have also been known to Rousseau, for example from his time in Switzerland. The Genevan apothecary, Jean-Baptiste Tollot, whom we have already encountered decrying the state of Swiss botany, provided a set of instructions on how to make a herbarium in 1741. He seems to follow Tournefort closely:

> [C]omment il faut s'y prendre pour en faire un Herbier; c'est- à-dire une petite Bibliothèque portative,[62] dont un Botaniste ne saurait se passer; par

57. Linda Rader and Celia Ison, 'Legacy of mercuric chloride', in *Managing the modern herbarium: an inter-disciplinary approach*, ed. D. A. Metsger and S. C. Byers (Washington, DC, 1999), p.354.
58. For example Tournefort, *Elemens de botanique*, p.547.
59. Bulliard, *Dictionnaire élémentaire*, p.100; Lebreton, *Manuel de botanique, à l'usage des amateurs et des voyageurs: contenant des principes de botanique, l'explication du système de Linné, un catalogue des différens végétaux étrangers, les moyens de transporter les arbres et les semences; la manière de former un herbier, etc* (Paris, Prault, 1787), p.162; Louis Claude Marie Richard, *Dictionnaire élémentaire de botanique de Bulliard, revu et presqu'entièrement refondu par Louis Claude Richard*, 2nd edn (Paris, 1802), p.75-79. The authenticity of this text in all its versions is doubtful.
60. Alum is '[a] whitish transparent mineral salt, crystallising in octahedrons, very astringent, used in dyeing, tanning skins and medicine, also for sizing paper and making materials fire-proof; chemically a double sulphate of aluminium and potassium' (*OED*, 'Alum'). Alum is used to remove phosphates from lake water; while its toxicity is uncertain, the precipitate it forms with phosphates is filtered out before human consumption.
61. Alum was apparently widely used; an entry for a 'pound of fine rock alum' is found on the list of supplies for the Spanish botanical expedition in which Joseph Dombey, participated. Hipólito Ruiz, 'Travels of Ruiz, Pavón and Dombey in Peru and Chile (1777-1788)', translated by B. E. Dahlgren, official documents and epilogue A. J. Barreiro, *Botanical series: Field museum of natural history* 21:467 (28 March 1940), p.5-327 (297).
62. Note Tollot's reference to the herbarium as portable library, an idea that Rousseau took up.

ce moyen, il peut faire son Cours chaque Année et sans sortir de son Cabinet. Pour faire cet Herbier avec succès, il faut cueillir les Plantes dans leur vigueur, et les placer sans confusion, en écrivant à côté le nom de chacune: Dans l'arrangement il faut suivre autant qu'il est possible, l'ordre établi par Mr. de *Tournefort*; il faut prendre garde que les Plantes ne prennent pas un mauvais pli; on doit conserver autant qu'on le peut, leur port, leur figure naturelle, et surtout celle de leurs Fleurs. Il faut ensuite conserver cet Herbier dans un lieu qui ne soit ni trop sec ni trop humide; trop sec les Plantes se blanchissent et se réduisent peu à peu en poussière; trop humide, elles risqueraient de perdre leur couleur et de se pourrir.

Tollot adds that one must revisit the herbarium frequently to retain plant names, since names are arbitrary, and some plants have more than one; this must be a repeated study in the manner of Erasmus's study of grammar.[63]

Rousseau also undoubtedly knew Linnaeus's definition of the herbarium in his *Philosophia botanica*: 'Plantae methodo data collectae in Herbarium vivum [...] disponuntur secundum Systema, ut prompte evolvantur'.[64] Linnaeus issued sixteen succinct instructions for preparing herbaria, which have changed little since 1751:[65] (1) plants should be dry, (2) no parts should be removed, (3) the parts should be displayed spread out (4) but not bent, (5) the fruit-body should be present, (6) the specimens should be dried between dry papers (7) as soon as possible (8) in a press, (9) glued down with fish glue, (10) kept on folio sheets, (11) placed only *one* per page (12) on unbound sheets (not books, see Tournefort), (13) the genus written above, (14) species and 'Historica [explanation]' on the back, (15) all plants of the same genus should be held within a band of linden bark.[66]

Linnaeus's sixteenth and last stipulation is to arrange plants *methodically*; the arrangement may either be by his own system or that of someone else.[67] Tournefort also used a system, but does not mention methodical arrangement, perhaps because the proliferation of systems occurred later. Linnaeus shows how the herbarium cupboard should be laid out: large enough to accommodate 6000 dried plants ordered

63. My emphasis; [Jean-Baptiste Tollot], 'Réflexions sur l'étude des plantes', p.1154-55. On the epistemic connections between taxonomy and grammar in the eighteenth century, see Foucault, *Les Mots*, p.71.
64. Linnaeus, *Philosophia botanica* (1751), p.291. 'Plants that have been collected by a given method[...] into a herbarium made from living plants, which[...] are arranged according to a system, so that they may be promptly produced.' Linnaeus, *Philosophia botanica* (2003), p.329.
65. Metsger and Byers, 'Introduction', in *Managing the modern herbarium*, p.1-18 (8).
66. Linnaeus, *Philosophia botanica* (2003), p.18 section 11.
67. Linnaeus, *Philosophia botanica* (2003), p.329.

according to the twenty-four classes of his artificial system of sexual classification.[68]

Two aspects of Linnaeus's view of herbaria specimens deserve closer examination: (1) his preference for the herbarium over pictures and (2) his isolation of the specimen, one per page.

Linnaeus maintained that the herbarium is *better* than any picture,[69] but given his view of pictures, this assertion is difficult to interpret. Recall that in *Genera plantarum* and *Philosophia botanica* he argues against using pictures in learning and doing botany.[70] Herbaria seem to have been exempt from the objections that Linnaeus made against pictures, which were seconded by Alphonse de Candolle in the nineteenth century: 'No illustration, including a photograph, could in his mind, compete with the authenticity of a herbarium specimen, however flat and faded'.[71] Yet the question remains: are herbaria implicated in the controversies surrounding objectivity of images, the validity of the senses, 'picturability' and representation in the sciences? Are herbaria true to nature, as Linnaeus would have it? Are they objective, as de Candolle would have it? For his part, Rousseau did not pose such questions; the value of both the herbarium and pictures was self-evident.

Linnaeus's logical positivist objection to depictions of plants notwithstanding, he was dedicated to promoting herbaria in which, inevitably, the specimens, once part of the natural world, *represented* nature just as pictures do; they *represent* nature because the botanist selects, prepares and arranges the specimen, making a series of complicated and critical *choices* about which specimen(s) to use, in what state, and in what way. Eighteenth-century practitioners made these decisions in various ways, with some taking great care to present the specimen as intact as possible with the parts of the fructification, as well as roots; such a practitioner was Rousseau (see fig.8.1, p.258).[72] Others, such as Haller, seem to have been more concerned to amass a large number of specimens, taking less care over their presentation or their condition. Yet the herbarium

68. Linnaeus, *Philosophia botanica* (2003), p.329-30, tab.11. One of the original cupboards is in the Linnean Society, Burlington House, Piccadilly, London.

69. Linnaeus, *Philosophia botanica*, reprint edn (1751), p.7, section 11; 'A HERBARIUM is better than any picture, and necessary for every botanist', Carolus Linnaeus, *Philosophia botanica* (2003), p.18, section 11.

70. Müller-Wille and Reeds, 'A translation of Carolus Linnaeus's introduction to *Genera Plantarum*', p.568. Linnaeus, *Philosophia botanica* (2003), p.111, section 151.

71. Daston and Galison, *Objectivity*, p.109, quoting Alphonse de Candolle, *La Phytographie* (Paris, 1880).

72. Art historian Hans Belting examines the 'Bild [picture]' not as a work of art, but as it is used by actual historical subjects, for example the icon's function as an object of veneration. Belting claims, however, that scientific illustrations can and should have no aesthetic implications, a view with which my discussion implicitly takes issue. Klaus Sachs-Hombach, *Wege zur Bildwissenschaft: interviews* (Cologne, 2004), p.124-45.

constitutes a presentation even if the botanist abstains from the decorations favoured by eighteenth-century practitioners: borders, urns and other flourishes.

This is supported by my second point: Linnaeus's stipulation that only one specimen should be attached to each page. This stipulation has important aesthetic as well as scientific significance; for example, the specimen can be centred and arranged on the page as a *composition*, thereby focusing the viewer's attention on one specimen for a detailed analysis and separating it from anything else, such as other specimens of the same species or genus, or elements of the environment in which it was found.

The process of preparing and presenting the herbarium specimen is therefore neither random nor neutral; rather, it can be a carefully orchestrated *presentation* and the result may still be authentic, but it can no longer be said to be mere nature. Semiotically, the herbarium specimen has a two-fold function: it operates as both signified and signifier at the same time.

Followers of Linnaeus such as Rousseau and James Edward Smith adhered to this principle, but this was hardly a universal practice. Like Linnaeus, Rousseau usually presented his specimens one to a page, following known principles of composition, such as centring, and placing heavier elements on the bottom and lighter ones on top. Where a specimen was too long to be presented vertically, he bent it to fit inside the frame, creating a very pleasing result (fig.8.2, p.267). This attention to the composition of the herbarium page is consistent with Rousseau's stated aim of advertising botany (which is not a stated aim of Linnaeus). Yet Rousseau's comments on arranging the plant on the page prioritise the concerns of the botanist and teacher over those of the artist; indeed, he hardly refers to the aesthetics of presentation.[73] For example, he specified that the specimen should lie in its 'situation naturelle', in order to 'conserver des plantes ou échantillons des plantes de manière à les rendre faciles à reconnaître et à déterminer'.[74] For teaching purposes he preferred 'un rameau desséché et *non collé*, qui eût feuilles et fleurs'.[75]

Rousseau usually adhered to Linnaeus's prescription to attach only one specimen per page, but he departed from the master's rules by gluing down small bands ('bandelettes') to attach the specimen to the

73. A rare example is Rousseau's comment to Mme Delessert: 'Vos plantes sont parfaitement bien conservées, les couleurs surtout, ce qui était difficile pour la fleur bleue.' 24 May 1773; *CC*, xxxix.162.

74. Rousseau to Madame Delessert, 11 April 1773; *CC*, xxxix.141, 139.

75. My emphasis; Rousseau to Madame Delessert, 30 August 1773; *CC*, xxxix.191. Lamarck preferred the specimen not to be glued down.

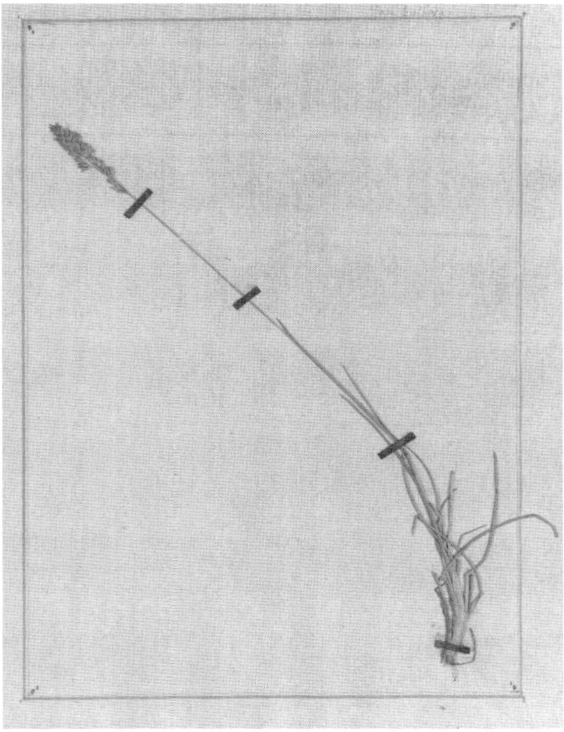

Figure 8.2: Jean-Jacques Rousseau, specimen of *Poa bulbosa* L., Herbarium for Julie Emélie Willading, née Boy de la Tour, by permission of the Zentralbibliothek Zürich.

Figure 8.3: Herbarium of Julie Emélie Willading, née Boy de la Tour, by permission of the Zentralbibliothek Zürich.

page, not the plant itself.[76] This method leaves the plant intact and detachable for inspection from all angles. Rousseau used both the older and more modern methods of storing specimens; while his small, portable herbarium (MdAD) is bound in a book, he kept other herbaria unbound in custom-made containers (fig.8.3, p.267). Storage in books has since been abandoned as such collections have to be unbound in order to rearrange the specimens or to examine them closely:[77] 'Ce n'est pas le lieu d'un savoir figé mais celui de la science en train de se faire.'[78]

Linnaeus's requirement that only one specimen should be placed on the herbarium sheet contrasts dramatically with the herbaria of notable contemporaries such as Haller, who crammed their herbaria with specimens, rather than isolating one specimen on each page; while perhaps useful for comparison, the visual effect of Haller's herbarium is one of excess. In this respect his herbarium resembles eighteenth-century rococo shell collections in which quantity and variety of specimens played a paramount role (see fig.3.4, p.85)[79]

Rousseau criticises the motives, if not the aesthetic preferences, of collectors more concerned to fill their cabinets than to learn something about nature. In the *Rêveries* he writes: 'Il est aisé, je l'avoue d'aller ramasser du sable et des pierres, d'en remplir ses poches et son cabinet et de se donner avec cela les *airs* d'un naturaliste: mais ceux qui s'attachent à ces sortes de collections sont pour l'ordinaire de *riches ignorants* qui ne cherchent à cela que le *plaisir* de l'étalage.'[80]

Hence, Rousseau sided with scientific systematisers such as Linnaeus, Buffon and Daubenton against the 'window-dressing' assembled by collectors such as Antoine-Joseph Dezailler d'Argenville (1680-1765),[81] for whom a principal criterion of a collection's worth was the number of rare, precious, and beautiful specimens it contained.[82]

76. The MdAD herbarium has multiple specimens per page.
77. Rousseau's major herbaria at Zürich, Neuchâtel and Montmorency are *unbound* in the modern fashion. The MdAD and Carnavalet herbaria (the latter's attribution to Rousseau may be spurious) are bound; the herbarium at Chaalis (also possibly spurious) is unbound, but too fragmentary to assess. See Appendix 3.
78. Thierry Hoquet, 'La bibliothèque botanique de Benjamin Delessert', p.104.
79. See Emma Spary, 'Scientific symmetries'; Bettina Dietz, 'Mobile objects: the space of shells in eighteenth-century France', *British journal for the history of science* 39:142 (September 2006), part 3, p.363-82. Haller's herbarium is, however, much more chaotic than the collections of a Dezailler d'Argenville, discussed below.
80. My emphasis; Rousseau, *Rêveries*, *OC*, i.1067.
81. Dezailler possessed important scientific credentials as a 'free associate' of the Montpellier Société royale des sciences, who 'had done primary work on the classification of mollusks'. Livesey, 'London by the light of Montpellier', p.94.
82. Dietz, 'Mobile objects', p.371, 382. Yet even in these collections, 'systematic claims combined with an aesthetic approach to produce a characteristic constellation reflected in the ordering of collections'. Dietz, 'Mobile objects', p.374. Thus, '[a] categorical

The utility of the herbarium specimen to teaching and research is enhanced by information added by the collector. In the sixteenth-century Gherardo Cibo (1512-1600) already 'scrupulously followed advice about the need to localize descriptions. Each entry recorded 'the location, the day and the hour in which it was gathered, as well as the names of the persons who had accompanied him'.[83] Yet, in *Philosophia botanica* Linnaeus does not spell out what information the label ('Historica') should contain, and his own practice in this regard was erratic, ranging from careful inscriptions on the backs of specimens to none at all.[84]

The Linnaean dissertation, *Instructio musei*, defended in 1753,[85] is somewhat more specific, stating: 'In chartae aversa parte annotentur, quae ad historiam plantae pertinent, ut differentia, locus proprietas, usus etc'.[86] Rousseau probably never read this passage, since he was 'forcé [...] de renoncer [...] aux amoenitates academicae'.[87] However, he prepared careful documentation of his specimens, following such as models as Latourrette (discussed below, p.284-89). In a modern herbarium the label includes the complete name (genus, species) with its author (e.g. *Rosa canina* L.), the location, as well as the date, of collection, and any other information regarding the environment in which the specimen was found.

Another important contemporary textbook that taught how to make a herbarium was Rozier and Latourrette's[88] *Démonstrations élémentaires de botanique pour l'usage de l'école royale vétérinaire* (Lyon, 1766).[89] Of all the texts considered so far, this provided the most detailed discussion,

distinction between "scientific" collections [...] and purely "amateur" collections [...] misrepresents the permeable character of the field of natural history well into the 1780s'. Dietz, 'Mobile objects', p.375.

83. Findlen, *Possessing nature*, p.169.

84. Compare specimens 26 and 27 from Linnaeus's herbarium (Linnean Society, London), *Veronica incana*, with the lichens, for which there are very few inscriptions: '26.Veronica 7 incana 5 (//Veronica spicata languinosa et incana, fl. Caeruleo Amm. Gerb. Tanais. 323. Habitat in desertis tanacaesibus ubiq: p[er] Ucraniam'. Spencer Savage, *A Catalogue of the Linnaean herbarium* (London, 1945), p. 2; see p.197. The symbol (denotes the western edge of Asia.

85. Linnaeus wrote the dissertation, which the student defended.

86. 'On the reverse side of the page is to be noted, what pertains to the history of the plant, in what way it is differentiated, place [,] property, use, etc.' Linnaeus, 'Instructio Musei', *Amoenitates academicae*, p.460.

87. Rousseau to Gouan, 6 October 1769; *CC*, xxxvii.152.

88. They botanised with Rousseau in the summer of 1768, and on other occasions as well.

89. Haller attributed sole authorship of this work to Rozier; yet Latourrette determined the 'forme' of the work and wrote its Introduction, 'véritable chef d'œuvre de concision et de clarté'. From the second edition onwards, Latourrette should be considered its 'principal auteur'. Antoine Magnin, *Claret de la Tourrette*, p.11.

running to over ten pages of quarto text.[90] This work, published in revised editions in 1773, 1787 and 1796, came highly recommended: 'l'illustre Bernard de Jussieu et son neveu [Antoine-Laurent de Jussieu] en ont constamment recommandé la lecture à ceux qui fréquentaient le jardin des plantes de Paris'.[91] It was cited as a fundamental reference work in the instructions issued by the King of Spain to the botanical expedition of Hipólito Ruiz López (1754-1816), José Antonio Pavón Jiménez (1754-1840) and Dombey that departed for South America in 1777.[92] However, this text presents a strange lacuna when compared with the texts on herbaria by Linnaeus, as it contains no references to labels or other descriptive information. As for methodical arrangement, the choice is left up to the practitioner.

The authors focused on conservation: gathering, drying, preventing spoilage and presenting well-preserved specimens. They accordingly opposed gluing down the specimens because glue attracts insects; they thereby reduce, if not eliminate, the need for insecticide. Furthermore, 'L'herbier doit être tenu dans lieu sec, renfermé, garanti de l'air extérieur; on doit le visiter de tems en tems, pour détruire les *larves* d'insectes qui s'y introduisent'.[93] The authors also stressed colour retention, which Linnaeus, Lamarck and other authors ignored; they offered the unusual suggestion of treating the flowers with essences so they might retain their odour![94]

Despite his friendship with Latourrette and Rozier, Rousseau rarely referred to the *Démonstrations élémentaires de botanique*, and it is not known if he owned a copy. To Latourrette he noted that 'les applaudissements avec lesquels vos premiers essais ont été reçus du public vous sont garants de ceux avec lesquels il accueillerait un travail plus considérable'.[95] On the other hand, Rousseau asserted the work was written for the learned, and was not elementary enough for beginners.[96] This was the same complaint he made against all botanical works, theirs not excepted.

v. The botanical field trip

Herbaria require the collection and identification of specimens; this is the first step in constituting a herbarium. Collection in turn entails trips to the field: herborisations. Botanists must examine plants not just in gardens, but most especially, uncultivated in the wild; for Rousseau, as we

90. Latourrette and Rozier, *Démonstrations élémentaires*, p.234-44.
91. Magnin, *Claret de la Tourrette*, p.11.
92. Ruiz, 'Travels of Ruiz, Pavón and Dombey in Peru and Chile (1777-1788)', p.284.
93. My emphasis; Latourrette and Rozier, *Démonstrations élémentaires*, p.235.
94. Latourrette and Rozier, *Démonstrations élémentaires*, p.243.
95. Rousseau to Latourrette, 26 January 1770; *CC*, xxxvii.212.

have seen, the uncultivated species display the beauty and variety, as well as utility, bestowed on them by the Author of nature. For this reason, he criticised Linnaeus as a 'grand observateur [...] qui ait vu jusqu'ici la botanique en naturaliste et en philosophe; mais il l'a trop étudiée dans des herbiers et dans des jardins et pas assez dans la nature elle-même'.[97] Rousseau was not denigrating herbaria; rather, he was pointing out that the botanist must start from the living plant 'sur pied', not from a dried specimen which may be missing parts, may have faded, or deteriorated: 'on herborise inutilement dans un herbier [...] si l'on n'a commencé par herboriser sur la terre. Ces sortes de recueils doivent servir seulement de mémoratifs mais non pas d'instruction première'.[98] Rousseau understood therefore that botany is no armchair hobby, and that every botanist must travel.

The earliest herbarium-makers in sixteenth-century Italy – Ghini, Aldrovandi, Cesalpino and Cibo – made many collecting field trips; Ghini in particular seems to have exploited the pedagogical possibilities of field trips early on, establishing 'the field trip as a standard part of student training [...] Botanizing further strengthened the ties between mentors and disciples, as an essential rite of inclusion'.[99]

Rousseau's comments about Linnaeus, cited above, are somewhat unfair because Linnaeus was in no doubt as to the importance of botanical field trips.[100] In his *Philosophia botanica* he specified procedures and details of the botanical field trip, down to the clothing that should be worn: 'levissimus, laxissimus, Botanicis propius'![101] He stipulates the tools to be used, provides for a division of labour, the length of the journey ('Milliaria ad summum duo cum dimidio'[102]), what is to be collected, recorded and taught. The professor should give 'DEMONSTRATIO singula a Professore fiat intra horae'.[103] He promoted scientific travel by his so-called 'Apostles' to far-flung places

96. Rousseau to Clappier, 23 December 1768; *CC*, xxxvi.214-16.

97. Rousseau, *Confessions*, *OC*, i.643. This criticism targeted Linnaeus's advocacy of scientific 'gardens of paradise' that supposedly provide the conditions for plants to flourish according to their natures. See Staffan Müller-Wille, *Botanik und weltweiter Handel*, p.289-90.

98. Rousseau to Malesherbes, 19 December 1771; *CC*, xxxviii.301.

99. Findlen, *Possessing nature*, p.166.

100. See Hanna Hodacs, 'Linnaeans outdoors: the transformative role of studying nature "on the move" and outside', *British journal for the history of science* 44 (June 2011), p.183-209.

101. 'Very light and very loose CLOTHING proper to botanists'. Linnaeus, *Philosophia botanica* (2003), p.330.

102. 2.5 miles at most.

103. Linnaeus, *Philosophia botanica* (1751), p.293; 'single DEMONSTRATION [...] lasting not longer than half an hour'. Linnaeus, *Philosophia botanica* (2003), p.331.

worldwide,[104] and he led many botanical trips closer to home, cataloguing the flora of the homeland, especially for its useful properties.[105]

We have seen that Rousseau's field trips took many forms – from solitary walks, to group expeditions with or without a guide; some lasted one day, others several days. On all these expeditions, organisation mattered just as much as it did to Linnaeus, who decreed: 'LEGES Serovientis, Discendentis, Absentis. Divisionis, Pransionis hora 2, Quietis in h. 4, Secretarii'.[106] In fact, Rousseau captained the 'collège de botanique' that explored the Val de Travers: 'Rousseau, comme le plus âgé, était le capitaine de la petite troupe, chargé de la discipline du corps, et d'y maintenir l'ordre et la subordination'.[107] In France Rousseau continued to take the organisational initiative; in July 1768 he was ready to 'aller herboriser à la grande Chartreuse avec belle et bonne compagnie botaniste que j'ai trouvée et recrutée en ce pays.'[108]

These expeditions put Rousseau in a good mood, which he experienced again when opening his herbarium: he recalled 'l'aisance et la gaité des voyages pédestres', and reminded Du Peyrou, 'nous n'étions point maussades à Brot [Val de Travers] ni les uns ni les autres'.[109] Rousseau employed terms of endearment for his companions on these expeditions, for example 'Caravanne' and 'Caterve'. To the abbé Baurin, he wrote of their upcoming expedition to Mont Pilat (discussed below, p.273), 'nous nous aiderons tous les uns les autres, et nous reviendrons de petits Linnaeus [...] Je vous salue, Monsieur, et embrasse cordialement; puisque nous allons être camarades de voyage[,] permettez que j'en prenne d'avance la familiarité.'[110]

The object of the expedition was to collect and preliminarily prepare specimens; this entailed making records, a point to which we shall return. Rousseau's care to record important information about specimens right away in the field was thrown into relief when he feared a herbarium he had made had been lost: 'vous trouverez que je n'ai pas laissé d'y prendre

104. Mariette Mantelkow and Kenneth Nyberg, 'Linnaeus's apostles and the development of the *Species plantarum*', *Symb. bot. Upsal.* 33 (2005), p.73-80.

105. These were recorded in his travel reports on Lapland, Öland and Gotland. See for example his *Öländska och Gothländska Resa.*

106. Linnaeus, *Philosophia botanica* (1751), p.293; 'There are RULES for those who come late, depart early or are absent. And for the division [of labour], lunch at 2, rest at 4, and for a secretary'. Linnaeus, *Philosophia botanica* (2003), p.331.

107. D'Escherny, *Mélanges de littérature*, p.66.

108. This 'compagnie' included Claret de Latourrette, the leader, abbé de Grange-Blanche and abbé Rozier. Rousseau to Du Peyrou, 6 July 1768, *CC*, xxxvi.8; see also Rousseau to Luc Antoine Donin de Champagneux, 12 August 1769, CC, xxxvii.126-27.

109. Rousseau to Du Peyrou, 16 September 1769; *CC*, xxxvii.142. Similarly, 'mes voyages pédestres ayant été jusqu'ici tous très gais, faits avec des camarades d'aussi bonne humeur que moi'. Rousseau to the comte de Laurencin, 10 October 1769; *CC*, xxxvii.158.

110. Rousseau to the abbé Baurin, 8 August 1769, *CC*, xxxvii.119.

quelque soin. C'est une perte qui quoique petite ne me serait pas facile à réparer promptement, surtout à cause d'un catalogue accompagné de divers petits éclaircissements écrits *sur le champ*, et dont je n'ai gardé aucun double'.[111]

A good guide made all the difference between a productive or an unsatisfactory expedition. In Switzerland Rousseau and his friends enrolled Abraham Gagnebin as their guide to the flora of the Jura, while in August 1769 Rousseau's expedition to Mont Pilat with a Dr Meynier,[112] the abbé Baurin and Luc Antoine Donin de Champagneux lacked a guide, and found little to collect:[113]

> Un voyage de botanique [...] que j'ai fait au mont Pilat [...] a été désastreux, toujours de la pluie; j'ai trouvé peu de plantes, et j'ai perdu mon chien blessé par un autre, et fugitif; je le croyais mort dans les bois de sa blessure, quand à mon retour je l'ai trouvé ici bien portant, sans que je puisse imaginer comment il a pu faire douze lieues et repasser le Rhône dans l'état où il était.[114]

The botanical garden and the field trip were closely linked. Demonstrators from the Chelsea Physick Garden and the Jardin du roi routinely took students to the field. These field trips started very early, shortly after dawn and lasted most of the day, depending on the place visited. In Paris, Bernard and Antoine Laurent de Jussieu led popular herborisations so crowded that Rousseau found little value in attending![115]

Rousseau was privileged instead to join select groups from the Jardin du roi guided by Bernard and Antoine Laurent de Jussieu, as discussed earlier (p.188). In May 1771, for example, Antoine Laurent de Jussieu and André Thouin joined Rousseau in Montmorency to locate *Plantago monanthos* (= *P. uniflora* L., *Littorella lacustris* E. H. L. Krause), that had been sighted by various botanists in the vicinity.[116] Letters from

111. My emphasis; Rousseau to Mme Delessert, 16 July 1772; *CC*, xxxix.81. Rousseau refers to the herbarium he made for Julie Emélie Willading née Boy de la Tour. See Chapter 4, fig.8.1, 8.2 and 8.3.

112. Meynier 'faisaient semblant d'aimer la Botanique, et qui désirant de me cajoler, je ne sais pourquoi, S'imaginèrent qu'il n'y avait rien de mieux pour cela que de me faire bien des façons', Rousseau to Du Peyrou, 16 September 1769; *CC*, xxxvii.142.

113. 'On a beaucoup raisonné sur [...] la composition du petit groupe qui accompagna JJ dans la fameuse expédition au mont Pilat. Or, JJ déclare lui-même qu'il ne fut accompagné que de "trois messieurs". Il y eut donc, outre Baurin et Meynier un troisième, peut-être Donin de Champagneux (voir le nº 6599).' Rousseau to abbé Baurin, 8 August 1769, *CC*, xxxvii.119, editor's note.

114. Rousseau to Laliaud, 27 August, 1769, *CC*, xxxvii.129-30; see also Rousseau to Du Peyrou, 16 September 1769; *CC*, xxxvii.141-44.

115. Rousseau's presence may have made these excursions 'tumultueuse[s]'. Rousseau to Malesherbes, 17 April 1772; *CC*, xxxix.38.

116. For example Bernard de Jussieu and Tournefort; Rousseau to Latourrette, 25 January 1772; *CC*, xxxix.22; Rousseau to Malesherbes, 17 April 1772; *CC*, xxxix.36.

Rousseau to Thouin[117] and a document by Antoine-Laurent de Jussieu confirm Rousseau's frequent participation in these expeditions: '[p]endant cinq des dernières années de sa vie, il [Rousseau] assista *régulièrement* aux herborisations particulières que M. de Jussieu le neveu faisait toutes les semaines, dans l'été, avec M. Thouin et un petit nombre d'amis ou élèves choisis'.[118]

Fieldwork can be not only trying, but also dangerous, as Rousseau discovered during on Mount Pilat in 1769:

> Nous avons eu mauvais tems presque durant toute la route [...] nous avons trouvé un très mauvais gîte. Pour lit du foin ressuant et tout mouillé, hors un seul matelas rembourré de puces [...] Le quatrième [article], des accidents de toute espèce; un de nos Messieurs a été mordu d'un chien sur la montagne. Sultan a été demi-massacré d'un autre chien, il a disparu, je l'ai cru mort de ses blessures ou mangé du Loup[119] [...] Le cinquième article et le pire est que nous n'avons presque rien trouvé, étant allés trop tard pour les fleurs, trop tôt pour les graines, et n'ayant eu nul guide pour trouver les bons endroits.[120]

vi. How to make a herbarium

The object of botanical expedition is to gather specimens for a herbarium. Making a herbarium is as much an artisanal as a scientific undertaking; it is one for which Rousseau, with his craft background, seems to have had a particular talent. The work requires time, concentration, attention to detail, knowledge of preservation techniques, and skill. As Rousseau relates in *Rousseau juge de Jean-Jacques: dialogues*:

> Dans de grandes et fréquentes herborisations il a fait une immense collection de plantes; il les a desséchées avec des soins infinis; il les a collées avec une grande propreté sur des papiers qu'il ornait de cadres rouges. Il s'est appliqué à conserver la figure et la couleur des fleurs et des feuilles, au point de faire de ces herbiers ainsi préparés des recueils de miniatures. Il en a donné, envoyé à diverses personnes, et ce qui lui reste suffirait pour persuader à ceux qui savent combien ce travail exige de tems et de patience qu'il en fait son unique occupation.[121]

For Rousseau, as for Linnaeus, botany was synonymous with the collection of specimens that can later be examined to reinforce the memory, both of the plants themselves and of the circumstances of

117. Rousseau to Thouin, 16 July 1771; *CC*, xxxviii.239.

118. My emphasis; Jussieu, 'Sixième notice', p.14, note (1).

119. Sultan was Rousseau's dog. Not only did Sultan survive, he reappeared at his master's home 'tranquille et parfaitement guéri', after crossing the Rhône. Rousseau to Du Peyrou, 16 September 1769; *CC*, xxxvii.142.

120. Rousseau to Du Peyrou, 16 September 1769; *CC*, xxxvii.142.

121. Rousseau, *Dialogues*, *OC*, i.832. Rousseau speaks of himself in the third person.

collection. In the eighth and last letter on botany to Madeleine-Catherine Delessert of 30 August 1773, Rousseau gave explicit instructions for making herbaria. He prefaced these instructions with the comment, '[j]e crains que nous ne l'ayons [la botanique] traitée jusqu'ici d'une manière trop abstraite; en n'appliquant nos idées sur des objets déterminés'.[122] He thereby acknowledged the concrete, material character of botanical work: 'Science [...] seule distingue le vraie botaniste de l'herboriste ou nomenclateur. Il s'agit donc ici d'apprendre à préparer, dessécher et conserver des plantes ou échantillons des plantes de manière à les rendre faciles à reconnaître et à déterminer. C'est en un mot un herbier que je vous propose de commencer.'[123]

Herbaria are formal scientific collections, and Rousseau's were no exception, their evocative role and aesthetic character notwithstanding. Rousseau's stipulations were for the most part standard practice, following the *Philosophia botanica*:

1. Pick a dry day and collect plants at the driest time of day;
2. Select plants 'en pleine fleur';
3. Include roots 'qu'autant qu'elles ont singularités remarquables';
4. Include all the 'parties constitutives du genre et de l'espèce';
5. Flatten and dry the plants in a press; change the paper (not necessarily every day, the modern method);
6. Use white paper for pressing with gray sheets in between 'assez fort et bien collé';
7. Store the specimens in the driest room in the house, not on the ground floor;
8. Keep the herbarium 'bien serré et un peu en presse';[124]
9. The student does not make the identifying labels; that is the teacher's task.

In a letter of 25 January 1772 Rousseau refined his instructions to include gluing the plants down as they dried.[125]

In order to be used, the plants collected must be organised in some way. 'J'y ai vu seulement des cartons remplis des rameaux de plantes dont je viens de vous parler, et des graines distribuées dans de petites boëtes classées, comme les plantes qui les fournissent, selon le système [sexuel] de Linnaeus'.[126] This statement may refer to the 11-volume quarto herbarium he made for himself (the so-called 'Berlin' or 'kleines' her-

122. Rousseau to Mme Delessert, 11 April 1773; *CC*, xxxix.139.
123. Rousseau to Mme Delessert, 11 April 1773; *CC*, xxxix.139.
124. Rousseau to Mme Delessert, 11 April 1773; *CC*, xxxix.142.
125. Rousseau to Latourrette, 25 January 1772; *CC*, xxxix.20.
126. Rousseau, *Dialogues*, *OC*, i.833.

barium destroyed during WWII). Dombey's herbarium, in the traditional folio size, would also have been arranged following the sexual system, in all likelihood.[127] In the herbarium for Madelon Delessert, Rousseau used an alphabetical ordering of genera with family names of the natural method (Montmorency); he may have also employed a chronological system in other herbaria, such as the herbier portatif (MdAD). This does not exclude the possibility that he used other systems or methods in herbaria that are no longer extant.

vii. Rousseau's herbaria and the republic of science

> L'herbier n'est pas le lieu d'une collection qu'on admire. C'est un lieu de travail et de *sociabilité*, où l'on *s'oblige mutuellement*, les dons de spécimens rétribuant les identifications des plantes. C'est enfin un lieu où les rivalités se cristallisent, et où les *paternités* sont difficiles à reconnaître.[128]

Rousseau worked hard to learn botany, and claimed to have done so entirely on his own: 'Toujours *seul* et *sans autre maître* que la nature[,] j'ai mis des efforts incroyables à de très faibles progrès'.[129] This statement is belied, however, by Rousseau's contacts with many leading botanists, who contributed specimens and/or know-how to his herbaria.[130]

Rousseau's correspondence and other writings indicate who his botanical contacts were, but give few indications about how or from whom he learned to make herbaria. This may seem odd, given how important a role herbaria played in his botanical studies. From the standpoint of modern science, we would expect Rousseau to acknowledge his teachers and his sources. Yet, viewed from the perspective of an early-modern craft, Rousseau's silence about his masters does not seem at all odd. Craftsmen usually passed on their trade (and their trade secrets) by example and word of mouth, rather than through textual media:

> Skills are not always learned and disseminated in ways that can be described by algorithmic models, such as the mere reading of the methods section of a scientific article; rather they are acquired through direct contact and personal observation of experimental techniques. Because skills are clearly associated with manual dexterity, they had been ignored by earlier groups of historians of science more interested in the 'high theory' of science. In

127. See Appendix 3.

128. My emphasis; Hoquet, 'La Bibliothèque botanique de Benjamin Delessert', p.104.

129. My emphasis; Rousseau to Malesherbes, 19 December 1771; *CC*, xxxviii.301; see also Rousseau to Clappier, 23 December 1768; *CC*, xxxvi.214-16.

130. Alexandra Cook, 'Jean-Jacques Rousseau et les réseaux d'échange botanique', p.93-114; Alexandra Cook, 'Botanical exchanges: Jean-Jacques Rousseau and the Duchess of Portland', *History of European ideas* 33 (2007), p.142-56.

addition [...] artisanal knowledge was often seen as antithetical to scientific knowledge.[131]

The first person to introduce Rousseau to herbaria may have been Abraham Gagnebin.[132] Herbaria formed part of the apparatus of the field trips in the Swiss Jura and Creux du Van guided by Gagnebin, for 'M. Du Peyrou avait soin des herbiers'.[133]As I note in my earlier discussion of Gagnebin, the remounting of his herbarium specimens makes it difficult to observe his presentation of specimens, apart from his often voluminous notes about the features and properties of plants.

Rousseau had the beginnings of a herbarium by October 1765 when he departed Switzerland; he left his books and engravings in the care of Du Peyrou,[134] to whom he wrote in January 1766: 'Je vous recommande mes *plantes sèches*. Ce recueil fait en *Suisse* me sera bien *précieux* en Angleterre où j'espère m'en occuper'.[135] Thus, he was already acquainted with dried plant specimens, but it is not known if he had prepared them himself or how he may have arranged them.

For answers to where, how and from whom Rousseau acquired the skills to make herbaria, we must turn to the correspondence and exchange networks of the Enlightenment republic of science in which gifts of books, specimens, and even entire herbaria circulated reciprocally among its citizens functioning as currency.[136] In 1766 Rousseau began a botanical exchange with the Duchess of Portland, who gave him English botany books.[137] In return, Rousseau offered her and other correspondents herbaria that he made himself in a range of sizes for a variety of recipients; the principle of reciprocity at work in the republic of science may have motivated him to undertake this project.[138]

131. Myles Jackson, 'Review article: labor, skills and practices in the scientific enterprise: recent works in the cultural history of science', *Journal of modern history* 71:4 (1999), p.902-13 (902).

132. Frédéric Eigeldinger concurs in this inference; see Rousseau, *Deux lettres*, p.96.

133. D'Escherny, *Mélanges de littérature*, p.66.

134. Rousseau insisted to Du Peyrou that his botany books and herbaria be kept separate from his other books and papers. Rousseau to Du Peyrou, 17 November 1765; *CC*, xxvii.274-76.

135. My emphasis; Rousseau to Du Peyrou, 1 January 1766; *CC*, xxviii.146, Du Peyrou made the arrangements for shipping these precious items to Rousseau in England: 'J'ai accompagné cet envoi d'un Catalogue spécifique du contenu de chaque Caisse [...] Il y a 7 Caisses tant grandes que petites numérotées d'un à sept, emballées et marquées J. J. R. Le N° 1 renferme tous les livres de Botanique, Estampes et *Portefeuilles* [herbiers]'. My emphasis; Du Peyrou to Rousseau, 1 June 1766; *CC*, xxix.242-43.

136. Cook, 'Rousseau's copy of Albrecht von Haller's *Historia stirpium*', p.149-56; Cook, 'Botanical exchanges', p.142-56.

137. For example James Petiver and John Ray. Rousseau to the Duchess of Portland, 12 February 1787; *CC*, xxxii.134.

138. The Duchess gave Rousseau four works, one of which, the *Herbarium amboinense* of Rumphius, he returned; he gave her two portable herbaria. See Cook, 'Botanical exchanges', *passim*.

Upon his return to France from England in 1767, Rousseau became involved with botanical exchange networks centred on Montpellier and Lyon. These networks distinguished themselves from Paris, with its 'star network' centred on the Jardin du roi and its intendant, Georges Louis Leclerc, comte de Buffon,[139] as 'grands fiefs linnéens français'.[140] Montpellier was the more important of these two 'fiefs', while Lyon was 'le second bastion du linnéisme en France après Montpellier'.[141]

Between 1760 and 1790 the Montpellier botanical network was dominated by two important botanists: François Boissier de Sauvages de la Croix[142] and Antoine Gouan, both distinguished professors in the famous medical faculty of that city. Sauvages, author of a system of classification on the basis of leaves, corresponded with Linnaeus. Many members of this botanical network were graduates of the medical faculty of Montpellier as well as members of the Société royale des sciences of Montpellier.[143] In both their theoretical commitments (for example Linnaean systematics) and organisational ethos, these botanists followed foreign models.[144] Originally an academy under royal patronage like that of Paris, the Montpellier Société royale des sciences was reconstituted in the 1740s on the scientific society model, in which 'self-organising groups of scholars and scientists [...] recruited external members[,] who in turn acknowledged and internalised their scholarly values'.[145]

The Montpellier botanists played an important role in the republic of science; they were in close contact with royal societies in England (London) and Sweden (for example Uppsala). Linnaeus was elected an 'associé étranger' of the Montpellier Société royale des sciences in 1743. Gouan, together with four other Montpellier botanists, was a founding member of the Linnean Society in London in 1788.

Gouan was a distinguished botanist to whom Rousseau turned with questions, and whom Rousseau credited with reawakening his interest in botany.[146] Gouan and Rousseau exchanged gifts of books and specimens, in accordance with the norms of the Enlightenment republic of science.[147] Gouan sent Rousseau an advance copy of his *Historia piscium*

139. James Livesey, 'Botany and provincial Enlightenment in Montpellier: Antoine Banal père et fils 1750-1800', *History of science* 43 (2005) p.57-76 (71).
140. Duris, *Linné et la France*, p.39.
141. Duris, *Linné et la France*, p.44.
142. Sauvages was elected to the Swedish Royal Society in 1740. He authored *Methodus foliorum*.
143. Cook, 'Jean-Jacques Rousseau et les réseaux d'échange botanique', p.93-114.
144. Livesey, 'Botany and provincial Enlightenment in Montpellier', p.71.
145. Livesey, 'London by the light of Montpellier', p.93.
146. Rousseau to Gouan, 6 October and 26 December 1769, *CC*, xxxvii.152-55, 194-97.
147. For example, several specimens in Rousseau's quarto herbarium ('Berlin') were identified by, and may have come from, Gouan. Albert Jansen, *Jean-Jacques Rousseau als Botaniker*, (Berlin, 1885), p.283, 287, 289, 292. See Appendix 3.

(*'Icthyologie'*),[148] while Rousseau reciprocated Gouan's gifts with plant specimens, as the norms of the republic of science dictated:

> J'ai reçu, Monsieur, avec bien de la reconnaissance votre Ichthyologie,[149] qui m'a été envoyée par M. de la Tourrette, et les plantes, que vous avez eu la bonté de me faire passer par un négociant. Que puis-je, dans ma misère, vous *rendre* pour ces *cadeaux*, si ce n'est des actions de grâces? Je vous les fais de tout mon cœur, et pour ne pas *renvoyer* votre cahier tout-à-fait *vide*, j'y ai mis le peu de *carex* que j'ai recueillis en ce pays et qui consistent en huit ou dix espèces. Il est vrai que j'en aurais pu ramasser davantage, mais dans mes courses nonchalantes, j'aime à observer sur la terre et suis trop paresseux pour recueillir.[150]

In addition to sending specimens, Rousseau may have reciprocated with Dillenius's *Historia muscorum*, that is, the copy of this work in Gouan's collection at the Montpellier medical faculty.[151]

The network around Gouan likewise facilitated Rousseau's botanical pursuits. It included, in order of relative importance: (1) Joseph Dombey (1742-1794), a student of Gouan and explorer of South American flora (1778-1784), (2) Marc-Antoine-Louis Claret de Latourrette (1729-1793) counsellor of the Cour des monnaies of Lyon (1749-1769), botanical author and collector who was a prime mover in establishing the botanical garden of the Ecole vétérinaire of Lyon, (3) abbé Jean Baptiste François Rozier (1734-1793), who collaborated with Latourrette and published an important journal, and (4) Pierre Clappier (1740-1818), a Montpellier physician who, like Dombey, had studied under Gouan.

Joseph Dombey, one of Gouan's prize students, botanised with his professor in the Pyrenees circa 1767. He served Rousseau as a botanical guide in the Lyon region, circa 1768-1769:

> C'est en herborisant ainsi qu'il [Dombey] avait rencontré à Bourgoin, en Dauphiné, Jean-Jacques Rousseau qui commençait à s'occuper de

148. Rousseau to Gouan, 26 December 1769; *CC*, xxxvii.194-97.

149. Rousseau means Antoine Gouan, *Historia piscium*. A French edition was published the same year: *Histoire des poissons* (Strasbourg, A. König, 1770). Rousseau exchanged this book along with ginseng root, for coffee that Bernardin de St Pierre had brought back from the Ile de Bourbon. Bernardin de St Pierre, 'Essai sur J.-J. Rousseau', *Œuvres complètes*, p.28. This incident is discussed in Chapter 4.

150. My emphasis; Rousseau to Gouan, 26 December 1769; *CC*, xxxvii.194-95. Note that Rousseau received Gouan's *Historia piscium* via Latourrette, whom I discuss below.

151. J.-J. Grasset, 'Jean-Jacques Rousseau à Montpellier'. The copy of Dillenius in Montpellier is that of Oxford, not one of the London editions (1763 or 1768) that seem to be discussed by Daniel Malthus and Rousseau in early 1768. Gouan's book is annotated in English, although this is not an argument against Rousseau's having owned it, since Malthus's letter suggests that his cousin, Jane Dalton, may have given her copy to Rousseau. Malthus to Rousseau, 24 January 1768, *CC*, xxxv.53-56. See the discussion of Dillenius in Chapter 6 p.202-203.

botanique.[152] Il lui servit plusieurs fois de guide, lui fit présent d'un herbier et Jean-Jacques lui adressant ses remerciements pour ce don précieux lui dit [...] *que ce serait là qu'il étudierait ses leçons* [...] Rousseau reconnaissant recommandait peu après à Buffon le jeune naturaliste dont il avait apprécié à la fois le savoir et le caractère.[153]

Dombey's gift to Rousseau of 1500 rare and exotic dried plants in 1768 undoubtedly conveyed something about making herbaria, as Rousseau declared this collection would be his only library:

> Je suis occupé maintenant à mettre en ordre un très bel herbier, dont un jeune homme est venu ici me faire présent, et qui contient un très grand nombre de plantes étrangères et rares parfaitement belles et bien conservées. Je travaille à y fondre mon petit herbier que vous avez vu et dont la misère fait mieux sortir la magnificence de l'autre. Le tout forme dix grands Cartons ou volumes in folio qui contiennent environ quinze cent plantes, prés de deux mille en comptant les variétés. J'y ai fait faire une belle caisse pour pouvoir l'emporter partout commodément avec moi. Ce sera désormais mon unique bibliothèque et pourvu qu'on ne m'en ôte pas la jouissance je défie les hommes de me rendre malheureux désormais.[154]

Dombey's herbarium did indeed serve Rousseau as an important reference work. Rousseau planned to augment it with plants he had collected, and he used it as the basis for collections he was making for the Duchess of Portland. However, it was not a self-sufficient library: 'on m'a fait présent il y a quelque temps d'un très bel herbier qui vaut mieux pour moi qu'une bibliothèque, mais qui me la rendrait quelquefois nécessaire'.[155]

Dombey's gift to Rousseau evolved into his so-called 'grand herbier'; this designation could have referred both to its folio dimensions (the largest paper size) and/or the number of species and specimens that it contained. A portion of these may in fact constitute the Rousseau herbarium at the BPUN of Neuchâtel.[156] This herbarium was therefore an important model for Rousseau's work, even though he seems to have returned it to Dombey circa 1770.[157]

We know of Dombey's accomplishments from a number of sources. By his early thirties, he was sufficiently renowned that the French Crown recommended him to accompany an important Spanish botanical ex-

152. In fact, Rousseau had pursued botany since 1764.

153. Original emphasis; Ernest Théodore Hamy, *Joseph Dombey: médecin, naturaliste, archéologue, explorateur du Pérou, du Chili et du Brésil 1778-1785: sa vie, son œuvre, sa correspondance* (Paris, 1905), p.xv-xvi.

154. Rousseau to Du Peyrou, 10 June 1768; *CC*, xxxv.307.

155. Rousseau to Guy, 11 June 1768, *CC*, xxxv.309.

156. See Rousseau to the Duchess of Portland, 2 July 1768; *CC*, xxxvi.3; Rousseau to Du Peyrou, 10 June 1768; *CC*, xxxix.308, and Appendix 3.

157. Rousseau to Du Peyrou, 7 June 1770; *CC*, xxxviii.32-33; See Appendix 3.

pedition to Peru and Chile, one of whose goals was to collect detailed information on the species of *Cinchona* native to the region.[158] The Spanish Crown's instructions to the botanists explicitly refer to Dombey's knowledge and skills in herbarium preparation: 'For the formation of herbaria they shall consult their teachers before leaving, inspect those that these men have made for their use, and ask instructions from Doctor Dombey, who is said to be particularly experienced in this matter'.[159] Another document relating to the organisation of the expedition states: 'This Frenchman [...] because of his knowledge and botanical experience [...] deserves to be chief of the expedition',[160] but political considerations precluded a leader of foreign origin. In the end, Dombey's considerable scientific contributions were omitted from the Spanish publication of the expedition's findings, and the specimens were confiscated from him at Cadiz in 1785.[161]

Dombey, for his part, esteemed Rousseau highly; unaware of Rousseau's death in July 1778, Dombey wrote to Thouin in December 1778 stating that he intended to send Rousseau a plant bearing Rousseau's name, and he later praised Marie-Antoinette's visit to Rousseau's tomb at Ermenonville.[162]

The Linnaean network in Lyon also included Latourrette and Rozier, collaborators in the teaching of botany at the Ecole vétérinaire of Lyon and co-authors of the *Démonstrations élémentaires*, an important textbook referred to in earlier chapters. They botanised with Rousseau at the Grande Chartreuse in July 1768; prior to this visit, Rousseau wrote to Latourrette: 'J'ai le coeur plein de vos bontés, de celles de M. l'Abbé Rozier [...] A Jeudi donc, Monsieur, j'attends avec impatience ce jour qui commencera pour moi une époque aussi douce qu'utile, si elle m'attire de votre part autant de bienveillance que d'instruction' (fig.8.4, p.282).[163]

158. Trees and shrubs of the genus *Cinchona*, native to the Andes, are natural sources of quinine, which is used to treat malaria; this drug was a prerequisite to the tropical colonial enterprise, and all the colonial powers were therefore interested in it. The genus takes its name from Francisca Henríquez de Ribera (1576-1639), Countess of Cinchòn, who was supposedly cured of malaria by it.

159. Ruiz, 'Travels of Ruiz, Pavón and Dombey', p.283-84.

160. Ruiz, 'Travels of Ruiz, Pavón and Dombey', p.292.

161. Dombey was forbidden by the Spanish Crown from publishing his work, and Buffon forced him to surrender his herbarium to L'Héritier de Brutelle, who shipped it clandestinely to London to keep it out of the hands of the Spanish. Lasègue, *Musée botanique*, p.244ff. The French part of his collection, numbering c.1500 specimens, including 60 new genera, was deposited in the Museum national d'histoire naturelle, Paris. Dombey died in a Corsair prison at Montserrat in 1794.

162. Rousseau to Du Peyrou, 10 June 1768; *CC*, xxxv.308, n.(d). Dombey, a partisan of the French Revolution, was no monarchist.

163. Rousseau to Latourrette, 5 July 1768; *CC*, xxxvi.7; see also Rousseau to the Duchess of Portland, 2 July 1768; *CC*, xxxvi.4.

Figure 8.4: Abbé Jean-Baptiste François Rozier, specimen of *Arnica* 'chartreuse!', courtesy of the Fonds anciens, Bibliothèque municipale de Lyon.

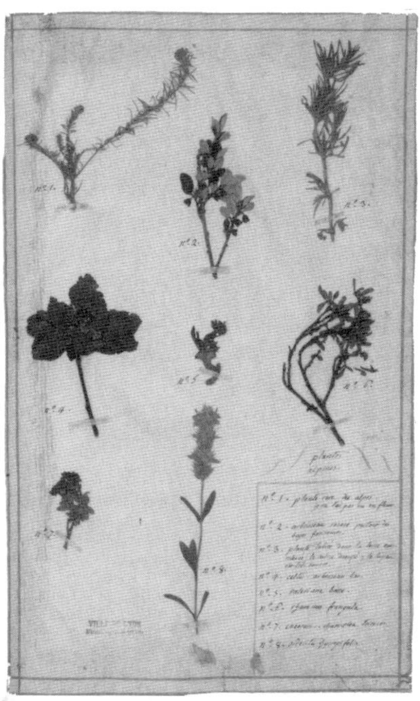

Figure 8.5: Abbé Jean-Baptiste François Rozier, specimens of 'Plantes Alpines', courtesy of the Fonds anciens, Bibliothèque muncipale de Lyon.

Despite his ecclesiastical office, Rozier devoted a large portion of his time to natural science. He briefly held the post of director of the Ecole vétérinaire of Lyon (Alfort) 1765-1766,[164] and was well-known for his journal, *Observations sur la physique, sur l'histoire naturelle et sur les arts*,[165] also called *Journal de physique*. This was a path-breaking publication with a varied audience and contributors, a wide scope of content and rapid turnaround times.[166] Rozier 'deliberately set out to create an alternative to publication in learned society proceedings';[167] he 'addressed his Journal directly to the most serious and advanced scientific researchers of his day, and he rejected completely any ties to the popular or the dilettantish'.[168]

When he came to Lyon in 1768, Rousseau sought Rozier's acquaintance and visited him frequently, consulting his herbarium and taking botany lessons from him.[169] Rozier confirms this in a letter he wrote to the marquis de Girardin upon hearing of Rousseau's death; he had Rousseau's 'portrait en medaillon, dessiné et modelé pendant qu'il *feuilletait mon herbier* lors du séjour qu'il fit à Lyon'.[170]

This herbarium exemplifies the kind of models that were available to Rousseau. In its final form, Rozier's herbarium is truly magnificent, displaying a 'grande qualité esthétique'.[171] Rozier unabashedly combined science with aesthetics, as did other collectors of the time. Some pages are decorated with borders, the sky and the sun. The herbarium comprises ten volumes; each page contains between one and fifteen specimens, attached with bands of gummed paper and surrounded by red borders in a manner similar to that employed by Rousseau (fig.8.5, p.282).

Rozier's herbarium followed Linnaean norms for naming and ordering his specimens. Most are labelled with Linnaean binomial names (generic name and species epithet or trivial name), and sometimes also with the order and class of the plant in the Linnaean sexual system. The herbarium is organised alphabetically, according to the first letter of the generic name. When Rousseau consulted it, Rozier's herbarium was

164. A. J. Dugour, 'Notice sur la vie et les écrits de l'abbé Rozier', in *Cours complet d'agriculture: théorique, pratique, économique, et de médecine rurale et vétérinaire ou dictionnaire universel d'agriculture* (Paris, 1800), vol.10, p.i-xvi (iv).

165. The journal was published in Paris until 1823.

166. James E. McClellan III, 'The scientific press in transition: Rozier's journal and the scientific societies in the 1770s', *Annals of science* 36 (1979), p.425-49.

167. McClellan, 'The scientific press in transition', p.433.

168. McClellan, 'The scientific press in transition', p.438.

169. Dugour, 'Notice sur la vie et les écrits de l'abbé Rozier'.

170. My emphasis; Rozier to the marquis de Girardin, 17 July 1778; *CC*, xli.33.

171. Frédéric Danet, 'Rozier Jean-Baptiste François', in *Herbiers de la région Rhône-Alpes 2e partie: catalogue des herbiers*, ed Andrine Faure (Lyon, 2006), p.300-301. The herbarium is conserved at the Bibliothèque municipale of Lyon.

probably still a work in progress, although the absence of collection dates and names of collectors makes it difficult to date the specimens or the herbarium itself.

While there are no extant letters between them, Rousseau's correspondence with Latourrette attests that Rozier sent Rousseau specimens, and perhaps some of these found their way into herbaria that are no longer extant.[172] According to Rozier's obituary, he and Rousseau were close during Rousseau's time in the Lyon region:

> Ces deux hommes dont les goûts simples les rapprochaient de la nature, concurrent l'un pour l'autre, en dissertant sur les productions du règne végétal, une estime mutuelle, une amitié qu'aucun événement n'a troublé. Leur liaison devint si étroite qu'ils faisaient presque *tous les jours* des excursions pour herboriser dans les riches campagnes du Lyonnais, qu'ils les poussèrent même jusques dans les montagnes du Dauphiné, ou la grande Chartreuse les accueillit pendant quelques jours.[173]

Neither Rozier nor Rousseau seem to have actively pursued their friendship later on,[174] and in the quatrième promenade Rousseau painted his former friend in a less than favourable light.[175] However, Rozier, who was undoubtedly ignorant of this last testimonial, wrote to the marquis de Girardin shortly after Rousseau's death attesting not only to his high regard for his friend, but also to how they had botanised together:

> Vous avez eu la triste consolation de fermer les yeux de l'homme unique, du vrai philosophe dont je regrette vivement la perte. Autrefois son compagnon *sur les Alpes* dans ses *excursions botaniques*, j'ai appris à le connaître, à juger l'homme lorsque la masse du public ne le jugeait que par ses ouvrages ou par les diatribes imprimées contre lui. Ma vénération pour cet homme qui méritait d'être mieux connu a toujours été extrême et sa mort ajoute à sa vivacité.[176]

Rozier's collaborator, Latourrette, played a particularly important role in inspiring Rousseau's interest in herbaria. As Rousseau writes,

172. Rousseau to Latourrette, 25 January 1772; *CC*, xxxix.20.

173. My emphasis; Dugour, 'Notice sur la vie et les écrits de l'abbé Rozier'. It seems odd that if their relationship had been so close they did not correspond after their face-to-face contact ended.

174. Rozier, who published his journal in Paris, may have visited Rousseau bearing plant specimens from Latourrette: 'J'avais chargé l'abbé Rozier de vous en parler quelquefois, mais livré a beaucoup d'occupations et logé fort loin de vous, il ne peut surement pas vous voir autant que je le désirerais pour lui et pour moi. Il a dû vous remettre de ma part, dans le courant de l'été dernier, un petit paquet de nos plantes sèches', Latourrette to Rousseau, 14 January 1772; *CC*, xxxix.8.

175. Rousseau, *Rêveries*, *OC*, i.1024-25.

176. My emphasis; Rozier to the marquis de Girardin, 17 July 1778; *CC*, xli.33. Rozier wrote the marquis a mere thirteen days after Rousseau's death, a rapid response by eighteenth-century standards.

'vous, Monsieur, dont la riche collection rassemble en petit presque toutes les productions de la nature, et prouve par son bel assortiment combien M. l'abbé Rozier a eu raison de dire qu'elle est l'ouvrage de choix et non du hasard'.[177] Latourrette cultivated plants on his estate (called La Tourrette); his herbarium contained 7000 plant specimens from France, Europe and the New World, including Canada and the Caribbean. His herbarium included specimens sent by botanists such as Dombey, Carlo Allioni (1728-1804) and Philippe Isidore Picot de Gouan (1744-1785). La Peyrouse (fig.8.6, p.287). Latourrette was a lichen specialist, author of *Chloris lugdunensis* (1785), a treatise on cryptogams and constituted an important collection of 1500 lichen specimens; the latter collection still exists as a distinct part of the herbarium of the Jardin botanique of Lyon, while his phanerogams have been incorporated into the herbarium's general collection. My research in the general herbarium of the Jardin botanique of Lyon has turned up a specimen of *Gentiana acaulis* L. that Rousseau apparently sent to Latourrette in 1777 (fig.8.6, p.287).[178] This gift would have occurred during the last revival of Rousseau's interest in botany, after he had sold his botany books (presumably to Daniel Malthus; see Appendix 1) and returned his herbarium to its creator (presumably Dombey).

Rousseau obviously esteemed Latourrette both as a naturalist and as a person: 'S'il m'était donné de me choisir une vie égale et douce, Je voudrais tous les jours de la mienne passer la matinée au travail soit à ma copie soit sur mon herbier; Diner avec vous [...] puis me promener tête à tête avec vous le reste de la journée en herborisant et philosophant selon notre fantaisie.'[179] In another letter Rousseau tells Latourrette, 'Les détails de vos herborisations et de vos découvertes m'ont fait battre le cœur d'aise. Il me semblait que j'étais à votre suite, et que je partageais vos plaisirs; ces plaisirs si purs si doux que si d'hommes savent goûter, et dont parmi ce peu-là moins encore sont dignes.'[180]

Magnin's study of Latourrette provides a clue to his direct influence on Rousseau's herbaria

> Les notes et observations de l'herbier méritent d'autant plus d'être signalées, qu'à cette époque, la plupart des naturalistes attachaient peu d'importance à ces constatations minutieuses de localités, de stations, etc.; c'est pourquoi l'on voit tant de collections, si riches à d'autres point de vue, n'être d'aucune utilité pour les recherches ultérieures. Il n'en est pas ainsi de l'herbier de La Tourrette, dont tous les échantillons sont accompagnés des notes explicites

177. Rousseau to Latourrette, 25 January 1772; *CC*, xxxix.21.
178. I found this specimen in the General Herbarium of the Jardin botanique de Lyon circa 4 April 2008.
179. Rousseau to Latourrette, 4 July 1770; *CC*, xxxviii.53.
180. Rousseau to Latourrette, 25 January 1772; *CC*, xxxix.21.

sur les localités, les stations, la date de la récolte, en un mot, de tous les renseignements qui peuvent servir à l'histoire de la plante.[181]

Latourrette and Rousseau engaged in a lively botanical correspondence[182] in which two important commonalities emerge: (1) both made notes 'minutieuses de localités, [et] de stations' in their herbaria. This important form of scientific documentation is exemplified by Rousseau's notes in the Zürich herbarium;[183] in the 'catalogue' of this herbarium he specifies the locations where specimens were collected,[184] how rare they were (for example *Drosera rotundifolia* L. (fig.8.7, p.287), and *Erica tetralix* L.),[185] and the environments in which they grew.[186] For example, in his note on *Anthriscus caucalis* M. Bieb. (burr chervil) (fig.8.8, p.288) he observes that it grows in a wide range of habitats, from sandy riverbanks to mountain tops: this plant 'aime le sable et les rivages; je l'ai cependant trouvée aussi tout au haut du Mont Pilat'.[187] In the miniature moss and lichen herbarium (MdAD) his description of 'n' notes essential points concerning the plant's environment, season, and habit: 'Trouvée à Monquin[188] au printemps en abondance couchée dans le limon d'une eau courante selon la direction du fil de l'eau. Feuillage assez âpre et rigide au toucher, sans aucun vestige de fructification' (fig.8.9, p.288).[189] He therefore did not merely collect for the sake of collecting, or to fill empty slots in his herbarium.

Both men collected lichens; lichens are fungi parasitic upon algals, whose form is modified by the influence of the parasite; these plants appealed particularly to Rousseau, who probably became interested in them during 1764-1765 when he could have observed them in profusion

181. Magnin, *Claret de la Tourrette*, p.49. Magnin's observations about the documentation habits of eighteenth-century collectors are curious in light of the practice of Renaissance naturalists such as Cibo, who kept scrupulous notes about locale, environment and circumstances of collection (see above).

182. Rousseau, 'Botanical writings', p.214-27.

183. This is the herbarium Rousseau made for Mme Delessert's sister, Julie Emélie Willading-Boy de la Tour (1751-1826). Rousseau to Mme Delessert, 20 October 1771; *CC*, xxxviii.281.

184. See Théophile Dufour, 'Petit herbier pour Mademoiselle Julie Boy de la Tour', *Annales de la société Jean-Jacques Rousseau* 2 (1906), p.260-68.

185. Appendix 3; Dufour, 'Petit herbier', p.263.

186. Dufour, 'Petit herbier'; Schneebeli-Graf, *Das Zürcher Herbar von Jean-Jacques Rousseau*. 'Rare' should be understood occurring infrequently.

187. Mont Pilat (1432 m.) lies between the Loire and Rhone valleys southwest of Vienne in the department of the Loire.

188. Rousseau lived at the Ferme de Maubec from January 1769 to April 1770, located in Monquin, above Bourgoin-Jallieu (formerly Bourgoin) in the département d'Isère, in the Rhône-Alpes region of France. It is still standing.

189. MdAD, f.23A-24.

Figure 8.6: Specimens of *Gentiana acaulis* L. given to Latourrette by Philippe-Isidore Picot de Lapeyrouse (upper left), by Rousseau (upper right) and by Dombey (lower left); from the General Herbarium, Jardin Botanique de Lyon, by permission of Jardin botanique de Lyon.

Figure 8.7 Jean-Jacques Rousseau, specimen of *Drosera rotundifolia* L., Herbarium of Julie Emélie Willading, née Boy de la Tour, by permission of the Zentralbibliothek Zürich.

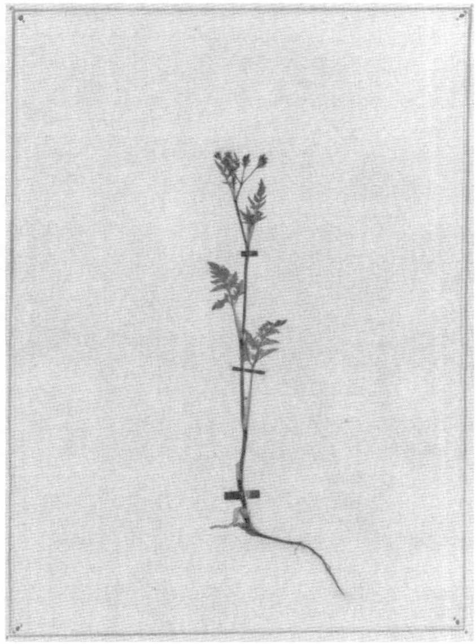

Figure 8.8: Jean-Jacques Rousseau, specimen of *Anthriscus caucalis* M. Bieb., Herbarium of Julie Emélie Willading, née Boy de la Tour, by permission of the Zentralbibliothek Zürich.

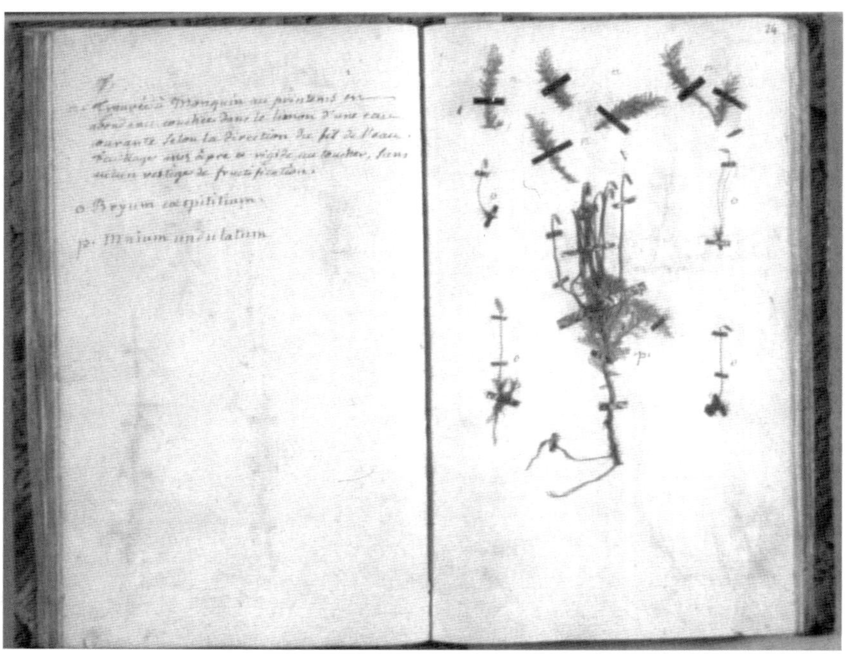

Figure 8.9: Jean-Jacques Rousseau, specimens of 'n': unidentified, 'o': *Bryum caespiticium* Hedw. and 'p': *Mnium undulatum* L. (= *Atrichum undulatum* (Hedw.) P. Beauv., feuille 23A-24, Herbarium of Jean-Jacques Rousseau, by permission of the Musée des arts décoratifs, Paris.

on rocks in the Jura and Creux du Van.[190] He studied cryptogams in winter when other plants were not available.[191]

Latourrette's extensive knowledge and collections in this field may have given Rousseau inspiration and encouragement to study a group of plants that is difficult to identify and collect successfully. For example, on 16 September 1769 Rousseau told Du Peyrou that he had acquired a specimen of *Lichen islandicus* or 'Iceland moss' (*Cetraria islandica* L.) during his trip to Mont Pilat the previous August.[192] Rousseau's miniature herbier portatif (MdAD), contains eleven lichen specimens, of which three are *Cetraria islandica* L. He was also interested in mosses, represented by over 300 specimens in the herbier portatif (MdAD).

The physician Pierre Clappier belonged to the Montpellier botanical network, and seems to have educated Rousseau about herbaria. In a letter of 23 December 1768 Rousseau praised Clappier's skills as a botanical teacher,[193] while in another letter of 26 May 1769 to Clappier, Rousseau praised his correspondent's specimen preservation skills: 'Vous joignez au savoir de les bien connoître l'art particulier de les bien conserver.'[194]

In addition to these provincial contacts, Rousseau received specimens from Jean-Baptiste Christophe Fusée Aublet (1720-1778),[195] the first European to describe the flora of French Guyana; he and Rousseau probably were acquainted, but to what extent is unknown.[196] There were several Fusée Aublet specimens in Rousseau's small ('kleines') quarto herbarium that was destroyed in Berlin during WWII.[197] In May 1778 Rousseau apparently acquired one of Fusée Aublet's herbaria, in which

190. Rousseau to Tissot, 1 February 1769; *CC*, xxxvii.37-40.

191. Rousseau to Malesherbes, 11 November 1771; *CC*, xxxviii.289; Rousseau to Malesherbes, 19 December 1771; *CC*, xxxviii.301.

192. Rousseau to Luc Antoine Donin de Champagneux, 12 August 1769; *CC*, xxxvii.126-27; to Du Peyrou, 16 September 1769; *CC*, xxxvii.143; to Laliaud, 27 August, 1769, *CC*, xxxvii.128-30; to Gouan 26 December 1769, *CC*, xxxvii.195.

193. Rousseau to Clappier, 23 December 1768; *CC*, xxxvi.215.

194. Rousseau to Clappier, 26 May 1769; *CC*, xxxvii.92.

195. Jean-Baptiste Christophe Fusée Aublet, *Histoire des plantes de la Guiane françoise*, 4 vols (Paris, Pierre-François Didot jeune, 1775); see also Mark J. Plotkin, Brian M. Boom and Maloyre Allison, *The Ethnobotany of Aublet's Histoire des plantes de la Guiane françoise (1775)* (St Louis, MO, 1991).

196. While the relationship between Rousseau and Fusée Aublet is unclear, Fusée Aublet moved in the same 'group of enlightened magistrates, courtiers and authors which included Malesherbes, Rousseau [...] and several others'. Stafleu, *Linnaeus and the Linnaeans*, p.282.

197. This herbarium was apparently destroyed during the bombardment of Berlin in WWII. The only record of its contents was made by Jansen, who mentions several Fusée Aublet specimens in *Jean-Jacques Rousseau als Botaniker*, p.278 (*V. teucrium*), p.280-82 (*Psyllium ciliaris*, *P. indicum* and *P. aegyptium*, *Cornus racemosa*) and possibly *B. rubra* 'de Cayenne', p.285.

he made notations.[198] Since Rousseau died on 4 July 1778, he probably received Fusée Aublet's herbarium too late for it to have served as a model for his own herbaria; however, its presence among his collections may attest to his friendship with this important botanist, and that botanist's belief that Rousseau would appreciate such a legacy. Rousseau was interested in Caribbean flora, as evidenced by his notes on the Campeche tree and other Caribbean trees taken from Jacquin's *Selectarum stirpium americanarum historia*.[199] Guyanese specimens also figure among Rousseau's herbarium fragments preserved at the Musée Jacquemart-André, Chaalis, France and in his seed collection, which came into the hands of Louis-Philippe ('Philippe-Egalité'), duc d'Orléans (1747-1794).[200]

viii. A botanical tool: the portable herbarium

Sometime in the early 1770s Rousseau conceived the novel project of creating small and even miniature herbaria, or *herbiers portatifs*, to give or sell to friends and others interested in learning botany.[201] His experience travelling with specimens may have inspired these efforts, and he may have been aware of other botanists who used *herbiers portatifs*.[202]

Rousseau described this project to the Duchess of Portland on 17 April 1772:

> Pour faire une diversion de mon goût à mes occupations, je me suis proposé de faire des herbiers pour les naturalistes et amateurs qui voudront en acquérir [...] J'ai pensé que de petits herbiers bien choisis et faits avec soin pourraient favoriser le goût de la botanique, et je vais travailler cet été à des collections que je mettrai j'espère en état d'être distribuées un an d'ici.[203]

The Duchess could have been expected to appreciate this way of making botany accessible and attractive: 'Le régne végétal, le plus riant des trois

198. The herbarium is in the Laboratoire de phanérogamie of the Muséum national d'histoire naturelle, Paris.

199. Neuchâtel BPUN MsR81; Rousseau used the same vernacular names as Fusée Aublet for Guyanese plants when other names were available. Cook, 'Jean-Jacques Rousseau and exotic botany', p.188.

200. These are recorded in the list of seeds allegedly given to the Duke by Rousseau; the list was drawn up in 1781 by the Duke's physician, Jean Etienne Guettard (1715-1786). It includes sugar apple (279), Campeche wood (396) and Guianese rosewood (678). Muséum national d'histoire naturelle (Paris), Ms 324. How Rousseau's seeds became part of the Duke's collections is unknown, but Rousseau knew Guettard's works. See for example Rousseau to Malesherbes 17 April 1772; *CC*, xxx.38. Rousseau and Guettard are supposed to have been acquainted. Charles Gilbert Romme to M. Dubruel, 15 November 1774; *CC*, xxxix.274.

201. These would be quarto, octavo or smaller, like the MdAD herbarium. Herbaria are usually folio. See Appendix 3

202. See J.-B. Tollot, who proposed 'une petite Bibliothèque portative'. 'Réflexions sur l'étude des plantes et sur leur usage', p.1154.

203. Rousseau to the Duchess of Portland, 17 April 1772; *CC*, xxxix.42.

et peut-être le plus riche, est très négligé et presque oublié dans les cabinets d'histoire naturelle, où il devait briller par préférence'.[204]

Rousseau had already begun a similar project for his friend, the jurist and royal censor, Chrétien-Guillaume Lamoignon de Malesherbes, as related in a letter written around April 1772:

> [J]e trouverais convenable de me faire durant cet été de bonnes fournitures, de les préparer[,] coller et ranger durant l'hiver, et de ne livrer que dans un an d'ici ce que j'aurais alors rassemblé, après quoi continuant de même, d'année en année les personnes pour lesquelles j'aurais commencé des herbiers [...] jusqu'à ce que j'eusse tout ce que je pourrais fournir.[205]

While eighteenth-century herbaria were usually folio-sized, Rousseau found a smaller size to be more convenient and convenience – or user-friendliness – trumped tradition. He first broached this issue when preserving some of Malesherbes' plant collections: 'L'expérience m'ayant appris que les grands herbiers sont plus incommodes qu'utiles, j'ai réduit cet échantillon à une forme plus portative'.[206] He raised the question of paper size with Malesherbes again in a subsequent letter: 'Mon idée serait de faire le fond de chaque herbier sur du papier à lettre tel que celui-ci, c'est ainsi que j'en ai commencé un pour mon usage,[207] et je sens chaque jour mieux que la commodité de ce format compense amplement l'avantage qu'ont de plus les grands herbiers'.[208]

The question of paper size is not banal, and Rousseau's innovative use of different paper sizes stands at the core of his project. In a letter of 22 October 1773 he offered the Duchess of Portland 'deux échantillons d'herbiers portatifs qui me paraissent plus commodes et presque aussi utiles que les grands'.[209] However, herbaria could be even smaller than letter size (quarto), that is to say, octavo or smaller:

> J'ai le projet aussi d'une forme de petits herbiers à mettre dans la poche qui ne sont pas les moins curieuses, et je n'y ferais entrer néanmoins que des plantes qui pourraient y tenir entières, racine et tout; entre autres, la plupart des mousses, les Glaux, Peplis, Montia, Sagina, Perchepiert, etc. Il me semble que ces herbiers mignons pourraient devenir charmants et précieux en même temps.[210]

204. Rousseau to the Duchess of Portland, 17 April 1772; *CC*, xxxix.42.
205. Rousseau to Malesherbes, 17 April 1772; *CC*, xxxix.36.
206. Rousseau to Malesherbes, 21 October 1771; *CC*, xxxviii.282; Malesherbes to Rousseau, 28 October 1771; *CC*, xxxviii.283.
207. This would have been the Berlin herbarium destroyed in World War II; see Appendix 3.
208. Rousseau to Malesherbes, 17 April 1772; *CC*, xxxix.38.
209. Rousseau to the Duchess of Portland, 22 October 1773; *CC*, xxxix.204. In this letter he laments the decline in his interest in botany, and expresses the hope that the Duchess's pleasure in this gift might revive it.
210. Rousseau to Malesherbes, 17 April 1772; *CC*, xxxix.37.

Rousseau thus proposed two herbarium sizes: letter size (quarto) and pocket-sized, or portable ('portatif'). Whenever possible, he preferred the paper dimensions to correspond to the plant size.[211] He likewise collected larger specimens on folio-sized paper because 'il y a des plantes d'une certaine grandeur qui ne peuvent conserver leur port dans un petit espace, et des échantillons si parfaits que ce serait dommage de les mutiler'.[212]

Rousseau's experiments with herbarium format were not only original, but also significant for the dissemination and popularisation of botany.[213] The small herbarium facilitates botanical study, while portable herbaria provide plant samples ready to hand for comparison in the field and may thus relieve the collector of having to carry large botany texts for reference or to check plant identities after the collecting expedition. Smaller herbaria are also easier to study, and to store. Convenience had inspired the herbarium at its inception in the sixteenth century; Rousseau took this a step further.

By 18 April 1773 Rousseau feared, however, that he could not produce what he promised:

> J'ai fait des préparatifs admirables, j'ai ramassé une grande quantité de plantes. On m'en a donné plusieurs. J'ai acheté des boites, du papier, du Carton, des portefeuilles; j'ai fait des cadres. Ce n'est que quand j'ai voulu venir enfin à l'emploi de tout cela que j'ai senti mon insuffisance. J'aurais certainement mieux réussi si j'avais moins entrepris, mais aussi je n'y aurais pas mis le même zèle. La plupart de mes préparatifs sont perdus parce qu'ils ont été malentendus [...] ce que j'en ai conservé se réduit presque à rien.

Yet he did not give up the project:

> Cependant pour donner encore quelque objet à mes courses champêtres je ne renonce pas tout à fait aux herbiers [...] J'ai commencé trois ou quatre petits herbiers en miniature auxquels je me bornerai; ce que j'y ferai chaque année ne sera ni fatigant ni coûteux et entretiendra toujours ce reste de goût que je conserve pour l'herborisation.[214]

Indeed, it is likely that the miniature herbarium of mosses, algae, and lichens at the Musée des arts décoratifs ('MdAD'), Paris, was one of the herbiers portatifs that Rousseau offered to Malesherbes in letters of 21 October 1771[215] and 17 April 1772.[216] Rousseau probably never

211. Rousseau to Malesherbes, 19 January 1772; *CC*, xxxix.17.

212. Rousseau to Malesherbes, 17 April 1772; *CC*, xxxix.37.

213. Rousseau to Malesherbes, 18 April 1773; *CC*, xxxix.147. On Rousseau's role in popularising botany, see Cook, 'Propagating botany: the case of Jean-Jacques Rousseau' in the *Transmission of culture in Western Europe, 1750-1850: papers celebrating the bi-centenary of the Bibliothèque britanique*, ed. D. Bickerton and J. Proud (Berne, 1999), p.69-94.

214. Rousseau to Malesherbes, 18 April 1773; *CC*, xxxix.147.

215. Rousseau to Malesherbes, 21 October 1771; *CC*, xxxviii.282-83.

216. Rousseau to Malesherbes, 17 April 1772; *CC*, xxxix.36-37; see also 'Notices

completed this ambitious project, but the remaining examples of his herbaria attest to his dedication and care in preparing them.

ix. Lamarck's anti-aesthetic stance

Jean Baptiste Pierre Antoine, chevalier de Lamarck (1744-1829), has come down to us as author of the discredited theory of inheritance of acquired characteristics, and is depicted pejoratively in modern biology textbooks as misguided.[217] He was, however, a prodigious botanist; his botanical masterpiece, the *Flore françoise* (1789), is consulted by botanists to this day.[218] In mid-career Lamarck was assigned to invertebrate zoology, which he pursued with similar skill. His herbarium is far less well-known than his other works, but nonetheless constitutes an important reference work, comparable to the herbarium of Antoine-Laurent de Jussieu, and superior to that of Michel Adanson.[219]

Lamarck is reputed to have botanised with Rousseau circa 1776 in Paris and its environs.[220] It has been argued that Rousseau influenced him in a number of ways: in his alienation from his fellow savants, his idiosyncratic views on chemistry and Newton and his assumption of the guise of the 'promeneur solitaire'. [221]

The eight volumes on botany prepared by Lamarck and his collaborator, Jean Louis Marie Poiret (1755-1834), for Panckouke's *Encyclopédie methodique* (1783-1808) are magisterial. In 'Herbier (Herbarium)' Lamarck provided a comprehensive discussion of the topic; like Rousseau, he followed Linnaeus's dictate regarding the fundamental role of the herbarium as 'une collection que doit se former un Botaniste, ainsi que toute personne que veut se livrer entièrement à l'étude des végétaux'.[222] Lamarck considered assembling a herbarium fundamental to botanical training: 'c'est en recueillant, préparant et desséchant soi-même les

bibliographiques: XII. Herbiers', *OC*, iv.1890-91. An analysis of this herbarium will be published in the *Œuvres complètes de Jean-Jacques Rousseau* (Paris, forthcoming).

217. But see Pietro Corsi, *Lamarck: genèse et enjeux du transformisme: 1770-1830*, trans. D. Menard (Paris, 2001).

218. Personal communication, Professor Peter Stevens, Harvard University Herbaria and Missouri Botanical Garden.

219. Gérard G. Aymonin, 'L'herbier de Lamarck', *Revue d'histoire des sciences* 34:1 (1981), p.25-58 (48).

220. Aymonin, 'L'herbier de Lamarck', p.32.

221. See Jacques Roger, 'Lamarck et Jean-Jacques Rousseau', *Gesnerus* 42 (1985), p.369-81 (372-73). Roger gives conflicting information about Lamarck's view of Rousseau: 'Jusqu'à présent nous n'avons pu faire état d'aucune référence explicite de Lamarck à of Rousseau' (p.374), but '[d]ans le *Système analytique*, Lamarck fait explicitement un grand éloge de Jean-Jacques Rousseau, "le plus profond de nos moralistes"'. Roger, 'Lamarck et Jean-Jacques Rousseau', p.377.

222. Lamarck, 'Herbier', p.111.

plantes, qu'on se forme un Herbier en bon état, et qu'on devient vraiment Botaniste; parce que [...] ce moyen forme insensiblement le coup d'œil de celui qui le met en usage, et le rend capable de reconnoître au premier aspect les plantes qui s'offrent a sa vue'.[223] As mentioned earlier in this chapter, Lamarck justified herbaria on grounds that Rousseau could have endorsed.

Like Rousseau, Lamarck paid particular attention to the presentation of specimens, stressing the need to arrange the flowers so that all their parts are visible and to prevent damage by other plant parts.[224] Lamarck did not refer to Rousseau's herbaria directly, but implicitly questioned Rousseau's prescriptions and practices. He stated that dry collection is not necessary; plants may need to be collected despite bad weather, but proper drying techniques can compensate for this. Lamarck also opposed gluing or attaching plants in any way, even in the careful way Rousseau employed, as it is best to be able to pick them up and examine them from all angles; furthermore, glue attracts insects.[225]

Lamarck concluded that the herbarium should not be an 'objet de luxe [...] en formant des cadres, des guirlandes, etc. sur les papiers, autour des plantes qui y sont posées'. Furthermore, the 'unique utilité' of the herbarium 'est un moyen d'instruction pour le Possesseur qui fait le consulter et y travailler sans cesse'.[226] It is not known if Lamarck intended these comments as a criticism of Rousseau's herbaria, but they can be taken as Lamarck's warning against aestheticising nature in the herbarium.

x. Conclusion

Rousseau's herbaria served a variety of purposes that lay outside the immediate scope of the science of botany, strictly understood. Yet even though he presented his specimens as compositions and his herbaria are decorated, albeit with restraint, Rousseau made scientific notations about the specimens, and used them to teach and popularise botany. He was not primarily concerned with an aesthetic presentation as were, for example, the proprietors of opulent collections of rarities; indeed, he was sharply critical of their pretensions to science, siding in his *Rêveries*[227] not with the shell collector, Dezailler d'Argenville, but rather, with the rigorist, Buffon.[228] Rousseau's herbaria are therefore a far cry from

223. Lamarck, 'Herbier', p.112-13. Rousseau likewise stresses the importance of educating the sight. Rousseau to Mme Delessert, 30 August 1773; *CC*, xxxix.190-91.

224. Lamarck, 'Herbier', p.114.

225. Lamarck, 'Herbier', p.116.

226. Lamarck, 'Herbier', p.116.

227. Rousseau, *Rêveries*, *OC*, i.1067.

228. Dietz, 'Mobile objects', p.382.

eighteenth-century rococo *naturalia* collections in which symmetry and variety took precedence.[229] They likewise bear little resemblance to Haller's chaotic, if vivid, collections.

Jean-Jacques thus gave a new direction to the use of the herbarium; a dried plant collection could serve not only to house plants for collectors and to provide material for scientific colleagues to study, but also to advertise botany to a wider, less specialised audience. He did this work himself, following the Linnaean dictate that every botanist should make herbaria, as well as his own dictate that every man should be able to work with his own hands.

Rousseau was, I argue, the first to recognise the potential of the herbarium to further goals outside the immediate purview of natural history, botany and medicine; this entailed transforming the herbarium into a portable *tool*, sized and adapted to the needs of the collector or student, complete with Latin and vulgar names, clear presentation of specimens and notes on locale and environment.[230] The herbarium could be as much a tool for the botanist as were alembics, beakers and distillation apparatus for the chemist.

The herbarium also played a personal role for Rousseau as 'un journal d'herborisations qui [...] produit l'effet d'une optique qui les peindrait derechef à mes yeux'.[231] Rousseau's herbaria therefore bring together the personal and the practical, the aesthetic and the scientific; they serve multiple purposes, and are therefore boundary objects par excellence.

Together with providing science with a teaching and research tool of inestimable value in documenting biodiversity, the herbarium continues to exercise aesthetic appeal today. This appeal has been powerfully captured in a collection of exquisite photographs from the herbarium of the Natural History Museum, London, entitled simply, *Flora*. The volume's editor writes: 'The specimens we have chosen to reproduce here have been chosen not for their scientific value, but for their beauty and aesthetic appeal [...] we have tried to lead the reader down one of the many avenues that emanate from a simple pressed flower.'[232] This is the spirit in which Rousseau made his herbaria.

229. Dietz, 'Mobile objects', p.371; Spary, 'Scientific symmetries', p.30.

230. Tollot described the herbarium as a 'bibliothèque portative', foreshadowing Rousseau's idea. Tollot, 'Réflexions sur l'étude des plantes'.

231. Rousseau, *Rêveries*, *OC*, i.1073.

232. Sandra Knapp, 'Introduction', in Nick Knight, *Flora* (New York, 2001), unpaginated. At this writing, Dr Knapp is Botanical Secretary of the Linnean Society, London.

9. A forged legacy: the strange fates of Rousseau's botanical works

i. Introduction

Rousseau's literary executors, Paul Moultou and Pierre Alexandre Du Peyrou, were keen to advertise the botanical aspect of Rousseau's work; they published the majority of Rousseau's botanical writings posthumously in the *Collection complète des œuvres de J.-J. Rousseau*, vol.14 (Geneva, [1780-]1782), comprising the eight *Lettres élémentaires sur la botanique*, the so-called *Fragmens pour un dictionnaire des termes d'usage en botanique*, and letters that Rousseau wrote to his more important botanical correspondents such as the Duchess of Portland, Malesherbes and Du Peyrou.[1]

As Raymond Birn has persuasively argued, Rousseau's executors made a number of editorial decisions that were intended to promote their preferred image of Rousseau. In this chapter I show that these decisions affected not only the major works, but also the botanical opus.[2] I argue, moreover, that Rousseau's botanical legacy was forged not only by Du Peyrou and Moultou, but also most especially by Rousseau's English translator, the Reverend Thomas Martyn, who gave the late eighteenth century a Linnaean Rousseau; translations of Martyn's version into French, Dutch and Portuguese consolidated Rousseau's Linnaean image.

In preceding chapters, I have argued that Rousseau did not base his botanical pedagogy on Linnaeus's sexual system, but rather on the natural method expounded by Adanson, Jussieu and others. In this chapter I track the posthumous transformation of Rousseau into the Linnaean author *par excellence*: 'l'artisan en France de la popularisation des idées linnéennes'.[3]

1. *Fragmens pour un Dictionnaire des termes d'usage en botanique*, *Lettres élémentaires sur la botanique à M. de L***.*, *Lettres à M. de M***. sur la formation des herbiers'*, *Collection complète des œuvres de J. J. Rousseau*, vol.7, p.477-598, in-4°; vol.14, p.351-534, in-8 ° and in-12 °; 'Lettre à M. D[u Peyrou]'. *Collection complète*, vol.24, p.492-93, in-8 ° and in-12 °; 'Quinze lettres relatives à la botanique, adressées à Madame la Duchesse de Portland', 'Neuf lettres relatives à la botanique, adressées à M. de la Tourette', in *Supplément à la collection complète*, vol.13, p.310-94, in-4°; vol.26, p.506-64, in-8° and in-12 °, Pierre-Alexandre Du Peyrou, ed., (Geneva, [Société typographique de Genève], 1782).
2. The definitive work on the posthumous 'forging' of Rousseau's image makes only slight reference to the botanical writings. See Raymond Birn, *Forging Rousseau: print, commerce and cultural manipulation in the late Enlightenment* (Oxford, 2001), p.88, 129, 161-62.
3. Duris, *Linné et la France*, p.105.

ii. A 'forged' text? *Fragmens pour un dictionnaire des termes d'usage en botanique*

The first forging of Rousseau's botanical legacy occurred with the posthumous publication of the *Fragmens pour un dictionnaire des termes d'usage en botanique* by Du Peyrou and Moultou in the *Collection complète*. These *Fragmens* are regarded by scholars as Rousseau's own work; to my knowledge, their authenticity has never been challenged. No less a scholar than R. A. Leigh confidently considered the *Fragmens* Rousseau's answer to the deficiencies of contemporary botanical works.[4]

In fact, compelling evidence points to this text or texts having been cobbled together – 'forged' – after Rousseau's death.[5] My argument for this assertion relies on the following basis: (1) the absence of a *manuscript* of the *Fragmens*; (2) the absence of the *Fragmens* from both the 'Traité' (publication contract) and the 'Prospectus' for the *Collection complète*; (3) the participation of an un-named third party in preparing the materials for publication; and (4) Rousseau's own protests that he was not working on such a text.

First, there is no manuscript of the *Fragmens*, and there is no evidence that one ever existed. Conversely, if the manuscript ever existed, it disappeared without a trace. This would have been odd, given the care with which other manuscripts in the care of Du Peyrou were preserved and eventually housed in the archives in the BPU of Neuchâtel. The disposition of Rousseau's manuscripts was stipulated between 'les amis de Rousseau' – René Louis, marquis de Girardin, Paul Claude Moultou and Pierre Alexandre Du Peyrou – and the Société typographique of Geneva in a 'Traité' dated 18 September 1779: 'Et l'Edition parachevée, les *Manuscrits originaux*, ainsi que l'exemplaire imprimé, corrigé et apostillé par l'Auteur lui-même, seront déposés par les Parties contractantes dans une *Bibliothèque publique*, pour constater *l'authenticité et la fidélité* de l'Edition'.[6]

Charly Guyot believed he had found evidence of the existence of the manuscript of the *Fragmens*. He asserted that the manuscript was inventoried in Du Peyrou's study at the time of the latter's death in 1794.[7] Yet a close examination of the inventory of Du Peyrou's study in a the Archives de l'Etat of Neuchâtel shows that the text in question was *Dictionnaire élémentaire de botanique*, most likely the well-known work of that name by Pierre Bulliard (discussed below).[8]

4. Leigh in Rousseau to Latourrette, 26 January 1770; *CC*, xxxvii.214, note (c).
5. I rely here on the double sense of the term 'to forge': (1) the blacksmith's work at his forge, and (2) the legalistic sense of the term: to falsify a document or signature.
6. My emphasis; 'Traité entre les Amis de Rousseau et la Société typographique de Genève by René Louis de Girardin, marquis de Girardin, Paul Claude Moultou, Pierre Alexandre Du Peyrou and Société typographique de Genève', 18 September 1779; *CC*, xliv.25-26.
7. Guyot, *Un Ami et défenseur de Rousseau*, p.156.
8. The inventory of Du Peyrou's home drawn up by judicial officers of the principality of

Second, neither the 'Prospectus' of the *Collection complète* nor the publication contract, the 'Traité', refer explicitly to the *Fragmens*. The botanical writings referred to in the Prospectus are '*Lettres sur la Botanique*, Mss, adressées à une Dame de ses amies, dans le but de lui rendre plus agréable l'étude de cette partie de l'histoire naturelle',[9] while the Traité lists 'Lettres et Mémoires sur divers sujets, en particulier *Lettres sur la botanique*, adressées à Madame de Lessert'.[10] At best it could be argued that the phrase 'Mémoires sur divers sujets' leaves open the possibility of the inclusion of the work later designated 'Fragmens'.

Third, Du Peyrou's correspondence with Moultou the following year indicates he possessed botanical material from Rousseau that was far from ready for publication: 'Il me reste encore [...] quelques *fragments* destinés à former une *espèce* de Dictionnaire, mais ils sont très *informes*, très *incomplets* et demandent une *révision* et un *choix sévère* des *morceaux* dont on pourra faire usage'.[11] Du Peyrou stated the 'fragmens' were 'destinés' to become a kind of dictionary, but does not say by whom they were destined for this purpose. In a subsequent letter to Moultou he described the planned preparation of the material by Moultou's friend, Jean-Laurent Garcin (1733-1781), like his father, Laurent Garcin of Neuchâtel (discussed in Chapter 4), a keen botanist (p.111-119):

> Vous trouverez, Monsieur soit dans le texte, soit au titre des notes, des *observations* que j'ay cru nécessaires pour *faciliter* le travail de *Mr Garcin*; veuillez, Monsieur, tout revoir avant que de lui envoyer ces originaux et la copie; joindre *vos observations aux miennes*, et lui recommander le tout, pour que ces *fragments* ne s'égarent pas. Après son travail, si vous le croyez nécessaire [...]. Renvoyez-moi le tout, et je ferai tirer une nouvelle copie dégagée de toutes les *notes* qui *surchargent* celle que je vous ai envoyée. Faites encore un avertissement pour être mis à la tête de ces fragments.[12]

This passage makes it obvious that Du Peyrou had added considerably to the material left by Rousseau, and that Moultou was to add even more, perhaps; all of this occurred *before* Garcin even took up the project.

Neuchâtel, states that a 'Dictionnaire élémentaire de botanique avec 40 cahiers de plantes figurées au naturel' was found in Du Peyrou's 'cabinet d'affaires', possibly Pierre Bulliard's *Dictionnaire élémentaire de botanique*. See 'Inventaire du mobilier de l'Hôtel Du Peyrou', *Archives de l'état, plumitif civil de la cour de justice de Neuchâtel*, November 1792-August 1798, cahier 58, p.154, n.207. See also R. A. Leigh, 'Les manuscrits disparus de J.-J. Rousseau: quelques observations et quelques fragments retrouvés', *Annales de la société Jean-Jacques Rousseau* 34 (1956-1958), p.31-81 (46).

9. Prospectus [de l'édition de Genève: version annotée par Du Peyrou] Société typographique de Genève and Pierre Alexandre Du Peyrou, 8 February 1779; *CC*, xliii.99.

10. 'Traité entre les Amis de Rousseau et la Société typographique de Genève by René Louis de Girardin, marquis de Girardin, Paul Claude Moultou, Pierre Alexandre Du Peyrou and Société typographique de Genève', 18 September 1779; *CC*, xliv.24.

11. My emphasis; Du Peyrou to Moultou, 22 April 1780; *CC*, xliv.212.

12. My emphasis; Du Peyrou to Moultou, 10 May 1780; *CC*, xliv.231.

Jean-Laurent Garcin of Neuchâtel was the only child of Laurent Garcin, who introduced Linnaean botany to Neuchâtel in the 1740s, and botanised with d'Ivernois and Gagnebin. Garcin fils was certainly known to some of Rousseau's Neuchâtel friends, but was absent from the region during Rousseau's stay, and did not return until well after Rousseau's departure, so the two men never met. He studied theology in Geneva, was ordained in the Reformed church in 1757, and served the parish of Fleurier in the Val de Travers, where his preaching was esteemed; however, he soon left Fleurier, and by 1759 had found employment in Amsterdam as a private tutor. In Holland he preached sermons in Amsterdam and The Hague; at the same time he moved in literary circles, becoming enamoured of the liberally educated poet and novelist, Belle van Zuylen, who accepted him as a literary confidant, but nothing more.[13] Garcin published poetry as well as a verse translation of the 150 psalms. He accompanied his students to Leiden and later to Paris, where he was received by Mme Necker (née Suzanne Curchod, 1737-1794; wife of the financier, Jacques Necker) in her famous salon.[14]

Upon his advantageous marriage in 1771 to a Mlle Sturler of Bern, Garcin retired to a Swiss country house in the canton of Fribourg, styling himself 'Seigneur de Cottens', and followed his father in cultivating botany. He worked on a never-completed *Flore helvétique* to supplement Haller's *Historia stirpium*, which he considered inadequate.[15] His 'magnifique herbier de plantes indigènes', together with his manuscripts, was snatched up by Lord Bute (see p.311) in 1781 after Garcin's early demise at the age of 48.[16] Some of these specimens, having passed through the hands of several owners, are now in the British Museum of Natural History, where they are distributed in the general herbarium, and are therefore virtually impossible to locate.[17]

13. Belle van Zuylen (1740-1805) was born Isabella Agneta Elisabeth van Tuyll van Serooskerken in the castle of Zuylen, near Utrecht. After a scandalous love affair with a married Swiss officer, Baron Constant d'Hermenches, she married Charles-Emmanuel de Charrière de Penthaz of Colombier, Neuchâtel and was known henceforth as Isabelle de Charrière. Her friends included Du Peyrou, who also had roots in aristocratic Dutch society. She wrote several epistolary novels broaching controversial topics and was a confidante of d'Hermenches's nephew, Benjamin Constant.
14. Bridel, *Le Conservateur suisse*, vol.13, p.108-15.
15. Bridel, *Le Conservateur suisse*, p.120-27.
16. Bridel, *Le Conservateur suisse*, p.118. The catalogue of the sale of Bute's library does not specify, however, which herbarium was Garcin's, perhaps number 1254, 'A French Hortus Siccus, contained in many bundles and volumes'. George Leigh and John Sotheby, *A Catalogue of the botanical and natural history part of the library of the late John, Earl of Bute* (London, Leigh and Sotheby, 1794), p.64.
17. Concerning Garcin's important herbarium of Swiss plants, see Hortense S. Miller, 'The Herbarium of Aylmer Bourke Lambert: notes on its acquisition, dispersal, and present whereabouts', *Taxon* 19:4 (August 1970), p.489-553 (524).

The foregoing suggests that the *Fragmens* may have had up to three other editors and/or co-authors in addition to Rousseau himself. Yet in their 'Avis' to the published version, the editors gave no credit to Garcin, and reveal virtually nothing of the process that produced these so-called 'Fragmens':

> Il *paraît* par ces Fragmens, que le projet de M. Rousseau était de faciliter l'intelligence des termes usités chez les Botanistes: il est fâcheux qu'il n'ait laissé sur ce sujet intéressant que des *brouillons*, peut-être aussi *incomplets* par les articles qu'il a ébauchés, que par ceux qu'il n'a point traités. Mais nous avons pensé que, malgré leur *imperfection*, ces fragments méritent de voir le jour, et, quelque *défectueux* qu'ils puissent être, nous n'avons voulu essayer, ni de *suppléer* aux articles qui manquent, ni de *corriger* ou *finir* ceux qui sont faits: tout au plus avons-nous osé nous permettre de *faire disparaître* quelques obscurités, ou quelques *défauts de style* qui avaient échappé à la première composition.[18]

Another indication of how the editors manipulated the *Fragmens* into existence is found in a note they added to the entry 'Plantes': 'Cet article ne paraît pas achevé non plus que beaucoup d'autres, quoiqu'on ait rassemblé, dans les trois paragraphes ci-dessus qui composent celui-ci, trois morceaux de l'Auteur sur autant de chiffons'.[19] These comments by the editors, taken together, suggest that in the absence of the manuscript of the *Fragmens*, it is impossible to ascertain what Rousseau wrote, and what others wrote, deleted or modified.

A dubious origin for the *Fragmens pour un dictionnaire* was intimated by the 'doyen' of Swiss letters, Philippe-Sirice Bridel (1757-1845), who referred to the work as 'un chaos de notes incohérentes du Timon genevois qui furent publiées sous le nom pompeux et trompeur de Dictionnaire de botanique'.[20] Bridel could speak with some authority since he was a friend and protégé of Garcin, the actual editor of the *Fragmens*. Bridel related a further intriguing indication as to the condition of the text prior to publication, quoting Garcin as having said that the *Fragmens* were '*Rudis et indigesta moles* qui ne m'a valu ni gloire, ni plaisir'.[21] As it has so far proven impossible to locate any of Garcin's manuscripts, attempts to follow up on this comment have been fruitless.[22]

18. My emphasis; Rousseau, *Collection complète*, vol.14, p.326 in 8° and in-12°. This editorial note was not reproduced in later editions such as the *OC* of the Bibliothèque de la Pléiade.
19. Rousseau, *Collection complète*, vol.14, p.412, n.(c); see also p.381, n.(a) and 383, n.(b), in-8° and in-12° concerning the entry 'Corolle' that had either disappeared or that Rousseau never wrote in the first place.
20. Bridel, *Le Conservateur suisse*, p.118.
21. Bridel, *Le Conservateur suisse*, p.118.
22. Jean-Laurent Garcin's papers and herbaria were sold to Lord Bute in 1781; however, my inquiries with the Bute family archive (Mount Stuart, Isle of Bute), the British Library, the British Museum, the Natural History Museum, London and the National Museum of Wales have turned up no trace of Garcin's papers.

This brings me to my fourth contention: lack of authorial intention. Rousseau repeatedly distanced himself from the preparation of a botanical 'livre', 'dictionnaire' or traité'. By 1769 it was well-known that Rousseau was a student of botany, and rumour had it that he was preparing a botanical treatise, perhaps on the model of his already-published *Dictionnaire de musique*. Rousseau denied that he was preparing a dictionary or other botanical work: 'C'est pour s'amuser à m'impatienter qu'on fait courir le bruit que je travaille à un dictionnaire de botanique, à un opéra, à que sais-je quoi. J'ai pris le parti de les laisser dire sans me fâcher et sans même prendre la peine de les démentir'.[23] Third parties such as the Uppsala professor, Jacob Jonas Björnståhl (1731-1779), reported similar comments by Rousseau:

> On m'avait dit que Rousseau avait le dessein de publier un livre sur la botanique, et je lui demandai si cela était vrai. Il me répondit que non, ajoutant qu'il y avait une grande différence à être maître ou à être disciple, et que, quoique ce fut maintenant un usage assez général de commencer à enseigner les autres avant de s'instruire soi-même, il ne voulait point le suivre.[24]

Charles-Gilbert Romme[25], recounted that the botanist, Jean Etienne Guettard,[26] queried Rousseau in the garden of the Palais Royal in 1774 about whether he was writing a botanical dictionary: 'M. Guettard, le rencontrant un jour au Jardin du Palais-Royal lui demanda si le public jouerait bientôt du dictionnaire de Botanique auquel il travaillait. J. J. lui répondit qu'on devait s'instruire avant d'écrire sur une partie et qu'il ignorait absolument la Botanique'.[27] In a similar vein, Bernardin de St Pierre reported: 'Un jour, une très aimable dame, chez lui, avec son frère, lui dit: 'Vous vous occupez de botanique: apparemment vous nous en donnerez un traité'; 'On croit, dit-il, qu'on ne s'applique aux choses que pour en donner des leçons. Je cultive la botanique pour la botanique même'.[28] Hence, Rousseau maintained a consistent position on this question: one should not compose a treatise before one is well-versed

23. Rousseau to Rey, 2 September 1770; *CC*, xxxviii.97.
24. Björnståhl to Gjörwell, 1 September 1770; *CC*, xxxviii.94.
25. Romme (1750-1795), a mathematician, created the republican calendar; he was a committed Jacobin who chose to commit suicide rather than face execution as a montagnard.
26. Etienne Guettard (1715-1786), physician to Louis-Philippe, duc d'Orléans, authored *Observations sur les plantes*, 2 vols (Paris, Durand, 1747). Rousseau's acquaintance with Guettard may explain the Duke's acquisition of Rousseau's collection of 739 seeds for his cabinet d'histoire naturelle, the circumstances of which remain unclarified. The seed list, possibly drawn up by Guettard, is Ms 324, Muséum national d'histoire naturelle (Paris). Rousseau mentions this collection several times in his correspondence. Rousseau to the Duchess of Portland, 17 April 1772; *CC*, xxxix.41-43; Rousseau to Malesherbes, 17 April 1772, *CC*, xxxix.36-41; Rousseau to Latourrette, 7 January 1773; *CC*, xxxix.128-29.
27. Romme to Dubruel, 15 November 1774; *CC*, xxxix.274.
28. Bernardin de St Pierre, *La Vie et les ouvrages de Jean-Jacques Rousseau*, p.166.

in the subject, and one should not study a subject only to teach it (and he implied, to indulge one's vanity). One should not use botany to enhance one's reputation by giving lectures or writing books.[29]

So what might have been the origin of the 'fragments' that Du Peyrou sent to Moultou? One possibility is that these were Rousseau's reading notes, since he states in the septième promenade of the *Rêveries* that he took notes from books because he was not able to replace those he had sold: 'Hors d'état de racheter des livres de botanique je me suis en devoir de transcrire ceux qu'on m'a prêtés [...]'.[30] Pierre Prévost (1751-1839), a fellow Genevan and tutor of Mme Delessert's children, confirmed this:

> Il se procurait divers livres de botanique, surtout d'anciens auteurs, tels que Ray, Bauhin, etc. dont il fait des extraits écrits et rangés avec un soin et un ordre recherchés. Ce travail à la fin de sa vie, prit la place des courses de botanique, auxquelles il disait avoir renoncé par lassitude et par ennui, parce que les environs de Paris ne lui offraient plus rien de piquant.[31]

Perhaps some of this painstaking effort contributed to the work that Du Peyrou and Moultou later dubbed *Fragmens pour un dictionnaire*. This hypothesis is supported by recent research that has revealed that many of the definitions in the *Fragmens* closely follow a few, well-known botanical texts such as Tournefort's *Elemens de botanique* (1694), Linnaeus's *Philosophia botanica* (1751), and Adanson's *Familles des plantes* (1763-1764).[32]

While these works were all available by the mid-1760s, a later work may have also influenced Rousseau: the two-volume, folio work, *La Botanique mise à la portée de tout le monde* (1774),[33] written by Nicolas François Regnault (1746-1810) and illustrated with magnificent engravings by his wife, Geneviève de Nangis Regnault (1746-1802).[34] However, some

29. Rousseau, *Rêveries*, *OC*, i.1069-70.
30. Rousseau, *Rêveries*, *OC*, i.1061.
31. Pierre Prévost, 'Souvenirs de Pierre Prévost', A659, *CC*, cl.267-68. Prévost, who knew Rousseau c.1776-1778, accompanied the Delessert family to Paris as their children's tutor. Prévost's son recounts that his father last saw Rousseau as the latter was departing for Ermenonville. Jean-Louis Prévost, 'Rousseau et Pierre Prévost', A660, *CC*, cl.274. See also Elizabeth A. Foster, 'Le dernier séjour de Jean-Jacques Rousseau à Paris, 1770-1778', *Smith College studies in history* 6:1-2 (October 1920-January 1921), p.36-37, 86.
32. Takuya Kobayashi, 'Fragments pour un dictionnaire', p.4.
33. Kobayashi, 'Fragments pour un dictionnaire', p.11-17. The full title is *La botanique mise à la portée de tout le monde ou collection des plantes d'usage dans la médecine, dans les alimens et dans les arts: avec des notices instructives puisées dans les auteurs les plus célèbres, contenant la description, le climat, la culture, les propriétés et les vertus propres à chaque plante, précédé d'une introduction à la botanique, ou dictionnaire abrégé des principaux termes emploiés dans cette science*, 2 vols (Paris, chez l'auteur, 1774).
34. While it was not unheard of for women in the early-modern period to work in scientific illustration (for example Maria Sibylle Merian (1647-1717) and Madeleine Basseporte (1701-1780) of the Jardin du roi), the express acknowledgement of their participation was fairly rare. In this case, Mme Regnault's contribution is expressly acknowledged in the title of the work: 'Executé et publié par les Sr et D[am]e Regnault'.

entries in the *Fragmens* that *appear* to have been drawn from Regnault may actually derive from Linnaeus, a principal authority, as we have seen, for Rousseau's botanical studies.[35]

The copy of the Regnaults' work to which Rousseau had access as late as April 1778 belonged to the abbé de Pramont, an obscure figure about whom little is known; he had occasionally brought music to Rousseau for copying and, presumably in 1777 or 1778, asked Rousseau to arrange the book's loose plant engravings according to the Linnaean artificial system of sexual classification.[36] This Rousseau did, adding marginalia to the text as well.[37] In addition, Rousseau supplied a list of the plants in the Regnaults' work in Linnaean order, 'Table des Plantes gravées dans cet ouvrage, rangées Suivant le Système de Linnaeus'. On 13 April 1778 Rousseau asked the abbé to collect his rearranged book, and Pramont apparently did so, because the book was sold with the abbé's library in October 1786 after his death.[38]

So one very real possibility is that the 'fragments' of which Du Peyrou speaks in his letter of 22 April and 10 May 1780 to Moultou were Rousseau's reading notes, as described by Pierre Prévost, and perhaps nothing more. It may then have been left to Garcin to construct them into a publishable form; this could explain the view on the work expressed by Bridel. Such a possibility has to be considered, given Du Peyrou's letters to Moultou regarding the editorship of Garcin, who in turn produced a text whose manuscript either never existed or mysteriously disappeared.

Given these uncertainties, I believe it is unwarranted to assume that Rousseau intended the texts gathered together posthumously as the

35. The entry 'Pistil', which appears to follow Regnault, likewise closely follows Linnaeus, as Kobayashi concedes. See Takuya Kobayashi, 'Seven botanical books annotated by Roussseau', p.11. Given his laudatory comments to Linnaeus about the *Philosophia botanica*, it is possible that Rousseau used Linnaeus, not Regnault.

36. This is inferred from Rousseau's reply; Rousseau to the abbé de Pramont, 13 April 1778; *CC*, xl.209. It was common practice for publishers to leave engravings unbound so the owner or reader could arrange them as he or she wished.

37. Rousseau notes to the abbé that he has only 'rarement' given in to the temptation to annotate the text. Rousseau to the abbé de Pramont, 13 April 1778; *CC*, xl.209. These marginalia were published by Musset-Pathay in 1825 in *œuvres inédites*. This version contains many inaccuracies; see BdAN [Munch MS]. See also Henry Cheyron, 'L'amour de la botanique: les annotations de Jean-Jacques Rousseau sur la Botanique de Regnault', *Littératures* 4 (Autumn 1981), p. 55-95.

38. Rousseau to the abbé de Pramont, 13 April 1778 *CC*, xl.209; see *CC* xl.210 for the sale catalogue description of the copy annotated by Rousseau. Jansen, *Rousseau als Botaniker*, p.250, believed that Pramont never collected his book from Rousseau: 'Trotz Rousseau's Bitte, wurde das Werk nicht abgeholt, und so nahm er es bis zum 20. Mai noch wiederholt unter seine Feder [Despite Rousseau's request, the work was not collected, and he therefore retained it until 20 May]'. It is now found in the library of Assemblée Nationale, Paris; Rousseau's notes are found in BdAN [Munch MS].

Fragmens as a formal presentation of botanical terminology; the body of texts that were published as the *Fragmens* may simply have been definitions from sources that Rousseau recorded in his last botanical phase, as attested in the septième promenade. Notes from botanical authors and lists of botanical terms are found among Rousseau's unpublished manuscripts in the BPUN archives.[39] That many of the *Dictionnaire*'s entries appear to be based on a limited range of authors – Tournefort, Adanson and the Regnaults – is consistent with this theory. At best, therefore, we may take the definitions in the *Fragmens* – for example 'Fleur' – as preparatory work, an indication of the sources Rousseau was reading, and the topics in which he was interested. More than that, I argue, should not be imputed to the text(s) as published.

To my knowledge no scholar has broached, much less addressed, the obvious questions about the provenance of the *Fragmens* raised by the correspondence between Moultou and Du Peyrou; it is particularly striking that Jean-Laurent Garcin's role in assembling and editing myriad 'chiffons' and 'brouillons' into the *Fragmens* has not surfaced before now in the annals of Rousseau scholarship.[40] Indeed, scholarly opinion seems comfortable with viewing the *Fragmens* not only as a legitimate, if unfinished, text, but also as a text *intended* by its author to see the light of day.

Scholars have based their confidence in Rousseau's authorship of this text principally on two grounds: (1) the simple fact of its publication in the *Collection complète* edited by Du Peyrou and Moultou,[41] and (2) Rousseau's comments to Latourrette in a letter of 26 January 1770: 'C'est à vous Monsieur qu'il faut renvoyer toutes les exhortations que vous me faites sur l'entreprise d'un *Dictionnaire de botanique* dont il est étonnant que ceux qui cultivent cette science sentent si peux la nécessité.' Later in the same letter he returned to the problems he encountered in studying botany and what could be done about them:

> J'avoue pourtant que les difficultés que j'ai trouvées dans l'étude des plantes m'ont donné quelques idées sur les moyens de la faciliter et de la rendre utile aux autres, en suivant le fil du Système végétal par une méthode plus graduelle et moins abstraite que celles de Tournefort et de tous ses

39. Ducourthial, *La Botanique selon Jean-Jacques Rousseau*, p.427-29, Annexe 13; MsR80, 'Extraits de livres botanique'.

40. Even Birn, *Forging Rousseau*, does not take up the mystery surrounding this part of Rousseau's botanical corpus.

41. See for example Louis Claude Marie Richard (ed.), *Dictionnaire élémentaire de botanique de Bulliard de Bulliard, revu et presqu'entièrement refondu par Louis Claude Richard*, (Paris, A. J. Dugour & Durand, 1798), who is cited (no page provided) by Roger de Vilmorin as taking the *Fragmens pour un dictionnaire* to be Rousseau's work. *OC*, iv.ccxviii. I have not been able to locate this reference. Richard's *Dictionnaire* was a revision of that of Pierre Bulliard, discussed below. In preparing the botanical writings for the *Collected writings of Rousseau*, vol.8, I likewise accepted the authenticity of the *Fragmens*.

> successeurs sans en excepter Linnaeus lui-même [...] Nous en causerons si vous voulez quand j'aurai l'honneur de vous voir. Si vous la trouviez digne d'être adoptée et qu'elle vous tentât d'entreprendre sur ce plan des institutions botaniques je croirais avoir beaucoup plus fait en vous excitant à ce travail que si je l'avais entrepris moi-même.[42]

Commenting on this passage, Leigh leapt to an unjustified conclusion: 'En dépit des raisons alléguées ici pour se dérober à cette tâche, Jean-Jacques devait s'y atteler un jour: voir ses *Fragmens pour un dictionnaire des termes d'usage en botanique*, qui parurent pour la première fois dans l'édition de Genève, et dont M. Roger de Vilmorin assigne la rédaction aux années 1771-1773 (*Pléiade* 4, p.cci)'.[43] Interpreting Rousseau's statements to Latourrette in this way took liberties with the clear meaning of his words. Rousseau stated explicitly that he was *not* volunteering to prepare a dictionary of botany, and instead pressed Latourrette, an accomplished botanist, to undertake this project.

On the other hand, little, if any, attention has been paid to Rousseau's statement in 1766 to Du Peyrou that: 'Quoiqu'il en soit, nous avons [...] assez de goûts communs pour les cultiver ensemble avec agrément, et je ne doute pas qu'un jour ou l'autre l'entreprise du *Dictionnaire de botanique* ne se réveille et ne nous fournissent pour plusieurs années les plus agréables occupations'.[44] This, to my knowledge, is Rousseau's only clear statement of authorial intention with respect to the text(s) that became the *Fragmens*, but it occurs at an early date, and is later contradicted; it therefore cannot form the basis for any firm claim for the authenticity of the *Fragmens* as published.

I have argued that the *Fragmens* do not appear to be an authentic text authored by Rousseau himself. However, if we accept the traditional thesis that this text was intended, if not completed, by Rousseau, there still remains the problem of dating it. Most estimates of the work's date(s) of composition rely on the date of May 1774 given by L. J. Courtois in 1923. Courtois bases this dating on one detail: Rousseau's reference to Antoine-Laurent de Jussieu's introduction of Linnaean binomials into the Jardin du roi in the 'Introduction' to the *Fragmens pour un dictionnaire*.[45]

However, caution is warranted here as well; to accept Courtois's reliance on this statement in the 'Introduction', we must accept that the 'Introduction' forms an integral part of the *Fragmens*. Yet, as Henry

42. Rousseau to Latourrette, 26 January 1770; *CC*, xxxvii.212.
43. Leigh in Rousseau to Latourrette, 26 January 1777; *CC*, xxxvii.214, note (c).
44. My emphasis; Rousseau to Du Peyrou, 31 May 1766, *CC*, xxix.236.
45. L. J. Courtois, 'Chronologie critique de la vie et des œuvres de Jean-Jacques Rousseau', *Annales de la société Jean-Jacques Rousseau* 15 (1923), p.226. R. A. Leigh relies on Vilmorin (*OC*, iv.cci) to date the *Fragmens pour un dictionnaire* as having been composed between 1771 and 1773, but these dates do not appear on the page Leigh cites. *CC*, xxxvii.214, n.(c).

Cheyron has persuasively argued, the 'Introduction' may be, like the *Fragmens*, just another fragment – a fragment on the history of botany that Garcin, Moultou and/or Du Peyrou placed at the front of the *Fragmens*: 'Le terme d'Introduction ne me semble pas convenir aux pages qu'on imprime en tête du *Dictionnaire*.'[46] Departing from this tradition, Sénelier offers the years 1763-1765 as the time frame for the composition of the *Fragmens*, but does not justify this dating.[47]

If the rumours about the *Fragmens* being in preparation in the early 1770s were true, it could have been prepared over a number of years; the close resemblance between many of the entries in the *Fragmens* and the Regnaults' work suggests that Rousseau may have still been working on it as late as 1777-1778 when he had access to *La Botanique mise à la portée de tout le monde*. It is of course also possible, if not attested, that Rousseau could have had seen this book on a previous occasion, since it was published in 1774.

As for its subsequent dissemination, the *Fragmens* lagged well behind the letters, which achieved great success not only in French, but also in English; they were translated into Danish, Dutch, German, Portuguese and Russian. The *Fragmens* were not translated by Martyn, and only a partial translation appeared into English in 1979. The *Fragmens* never became as well-known as its sister text, even though it was published in some subsequent editions of Rousseau's works, and in some editions of his botanical writings.[48] The translation in the *Collected writings of Rousseau* is the first complete translation into English.[49]

Nonetheless, scholars continue to take the *Fragmens* seriously, crediting it as an authoritative text; this trend started with the commentary by the *Göttingische Anzeigen von gelehrten Sachen* in 1781 (discussed below, p.321).[50] Pascal Duris, an important scholar of the history of French botany, sees the *Fragmens* as a formal presentation of botany to a learned, male audience, one of two allegedly distinct botanical discourses adopted by Rousseau; the other discourse is supposed to be for amateurs, that of the *Lettres* to Mme Delessert.[51] Similarly, Lisa Gasbarrone argues the *Fragmens*

46. Cheyron, 'Ray et Sauvages annotés par Jean-Jacques Rousseau', p.86; see also Cheyron, 'Fragments pour un dictionnaire des termes d'usage en botanique', in *Dictionnaire de Jean-Jacques Rousseau*, ed. R. Trousson and F. S. Eigeldinger (Paris, 2001), p.351.

47. Jean Sénelier, *Bibliographie générale des œuvres de J.-J. Rousseau* (Paris, 1950), p.170.

48. See for example J.-J. Rousseau, *La Botanique de J.-J. Rousseau: contenant tout ce qu'il a écrit sur cette science; l'exposition de la méthode botanique de M. de Jussieu; la manière de former les herbiers par M. Haüy* (Paris, 1802); Rousseau, *La Botanique de J.-J. Rousseau*, ed. Albéric Deville, p.257ff. Sénelier, *Bibliographie générale*, p.172-73.

49. J.-J. Rousseau, 'Botanical writings', trans. and ed. Cook.

50. 'Genf', *Göttingische Anzeigen von gelehrten Sachen* 2:145 (29 November 1781), p.1162-67 (1164).

51. Duris, *Linné et la France*, p.105. I have discussed the problems with this thesis elsewhere in this work.

are 'more deliberate than inspired'; they constitute 'a closed, self-reflexive system' that exhibits 'a strong authoritarian streak', and makes an 'appeal to the authority of reason over sentiment'.[52]

iii. Rousseau's botany in English: lost in translation?

Rousseau had spent a botanically formative period in England in 1766-1767 botanising with the Duchess of Portland and Daniel Malthus; within twenty years he was well-known in the English-speaking world as a botanical populariser, due largely to the efforts of one man: the Reverend Thomas Martyn (1735-1825), Regius professor of botany at Cambridge (fig.9.1, p.309). In 1785 Martyn published a highly popular translation of the eight letters to Mme Delessert, accompanied by twenty-four letters in the style of Rousseau explicating the Linnaean sexual system. This translation went through eight editions, with the last being published in 1815.

The timing of Martyn's translation was opportune, for the 1780s were a botanically important decade in Great Britain, during which the ideas of Linnaeus were coming into their own. In 1760 Linnaean classification had been introduced to Great Britain by Sir John Hill (1714-1775) in his *Flora britanica*, which re-organised Dillenius's edition of John Ray's *Synopsis methodica stirpium britannicarum* (1690) according to the artificial sexual system.[53] Hill's is usually cited as the first Linnaean work in Great Britain, although another Linnaean work appeared in the same year, James Lee's *An Introduction to botany*, a loose translation of Linnaeus's *Philosophia botanica*.[54]

The Linnaean gospel was propagated in person by Daniel Solander (1733-1782), the Swedish botanist and Linnaean disciple, who accompanied Sir Joseph Banks (long-time president of the Royal Society) on Lt James Cook's first *Endeavour* voyage (1768-1771), a voyage that generated a number of important scientific results.[55] The *Endeavour*

52. Lisa Gasbarrone, 'Blindness or oversight? A closer look at Rousseau's *Dictionnaire de botanique*', *French forum* 11:1 (January 1986), p.5-17 (9,15).

53. Sir John Hill, *Flora britanica, sive, Synopsis methodica stirpium britanicarum sistens arbores et herbas, indigenas et in agricultas in classes et ordines, genera et species redactas secundum systema sexuale, additis nonnullis noviter detectis, cum classium, ordinum et generum characteristicis, specierum descriptionibus, et virium epitome, tabulis aeneis illustrata, post tertiam editionem synopseos rainae, opere Dillenii concinnatam, nunc primum ad celeberrimi Caroli Linnaei methodum disposita* (London, J. Waugh, 1760). Hill was a prolific author, who wrote many works of botany, and gained the patronage of Lord Bute. He was also a gossip columnist and playwright. His knighthood was Swedish, not British.

54. James Lee, *An introduction to botany containing an explanation of the theory of that science, and an interpretation of its technical terms: extracted from the works of Dr Linnaeus [...], with twelve copper plates, and two explanatory tables: to which is added an appendix; containing upwards of two thousand English names of plants* (London, J. and R. Tonson, 1760).

55. These naturalists wished to accompany Cook on his second circumnavigation on the *Resolution*, but the Admiralty refused to grant their requests for substantial alterations to the ship to accommodate their entourage.

Figure 9.1: Coloured engraving by J. Russell (artist) and [?] Vendramini (engraver), *Revd Thomas Martyn, B. D., F. R. S.*, by permission of the Linnean Society of London.

circumnavigated the earth, observed the transit of Venus over Tahiti in 1769, mapped present-day Australia (then known as New Holland) and New Zealand. The voyage's botanical results were especially significant; it brought back 110 genera and 1300 species of plants previously unknown in Europe.[56] The artificial sexual system proved its value in this undertaking, as in so many other ventures in colonial botany, because it allowed for rapid organisation of the collections. Even though Solander and Banks used these efficient Linnaean principles of classification, they failed to publish the planned *Florilegium* of the voyage.

The incursions by Linnaean systematics on British botany were, as Rousseau's correspondence with the Duchess of Portland shows, in fact a fairly slow process; the Duchess, a proficient botanist, was still using pre-Linnaean reference works by John Ray and James Petiver (who annotated Ray's *Synopsis*) when she and Rousseau began corresponding in 1766. Rousseau persuaded her to study Linnaeus, which she had previously not done.[57]

Charles Darwin's grandfather, Dr Erasmus Darwin (1731-1802), served as midwife to the British adoption of Linnaeus. In the 1780s the Botanical Society of Lichfield, of which Darwin was a leading light, brought out English translations of Linnaeus's works; this undertaking attests to the widespread appeal of botany at the time, especially to educated women, who usually did not receive instruction in Latin. The British botanical author, Maria Elisabeth Jacson (1755-1829), described these translations as 'useful'.[58] Darwin himself further popularised Linnaeus's botanical ideas with an enthusiasm and anthropomorphising rakishness that might have even surprised their originator, in 'The loves of the plants', the second part of his *Botanic garden: a poem in two parts* (1791).[59]

Botany was very popular in the highest levels of British society, a trend that had started in the reign of William and Mary, who brought Dutch gardeners and gardening to Great Britain; the Duchess of Portland's father-in-law, Hans Willem Bentinck (1649-1709), first Duke of Portland, came from the Netherlands with William of Orange and supervised the King's gardens until the end of William's reign (1689-1700). Under the Hanoverians, the Dowager Princess of Wales, Augusta (1717-1772),

56. William T. Stearn, 'A Royal Society appointment with Venus in 1769: the voyage of Cook and Banks in the *Endeavour* in 1768-1771 and its botanical results', *Notes and records of the Royal Society of London*, 24:1 (1969), p.64-90.
57. Cook, 'Botanical exchanges', p.147.
58. [Maria Elizabeth Jacson], *Botanical dialogues between Hortensia and her four children* (London, J. Johnson, 1797), p.64.
59. Erasmus Darwin, *Botanic garden: a poem in two parts* (London, J. Johnson, 1791). See Ann B. Shteir, 'Iconographies of Flora: the goddess of flowers in the cultural history of botany', in *Figuring it out: science, gender and visual culture*, ed. Ann B. Shteir and Bernard Lightman (Hanover, NH, 2006), p.3, 19.

continued her husband Frederick's interest in gardens; she lived at Kew and contributed considerable resources to the development of its gardens, employing such botanical luminaries as the Reverend Stephen Hales (author of *Vegetable staticks*, an early investigation of plant physiology) and Sir John Hill; buildings both useful and decorative were provided by the Swedish-British architect, Sir William Chambers (1735-1789), also a progenitor of the Anglo-Chinese style in gardening. John Stuart, third Earl of Bute (1713-1792), a botanical enthusiast and mentor of George III, oversaw the evolution of the gardens.[60]

In 1784, Lord Bute brought out his *Botanical tables*, of which there were only twelve copies; Queen Charlotte (Augusta's daughter-in-law and consort of George III), the Duchess of Portland and Catherine the Great of Russia numbered among those honoured with copies. Foreshadowing the preface that Martyn was to give his edition of Rousseau's letters the following year, Bute wrote that his work was 'composed solely for the Amusement of the Fair Sex'. The nine privately printed volumes prepared for Queen Charlotte were housed in a cabinet painted with flowers.[61] The Queen, together with the Duchess of Portland and Mary Delaney, studied both the science and illustration of plants; they employed the best instructors, such as Francis Bauer (1758-1840), and Georg Dionysius Ehret (1708-1777), engraver of the famous 'Classes S. Literae', illustrating the artificial sexual system in Linnaeus's *Systema naturae* (1735) (see fig.5.7, p.159). Under royal patronage by both the Queen and George III (known as 'farmer George') Kew was developed in the late eighteenth century as a site of serious scientific collecting, cultivation and study, with a growing emphasis on exotic or colonial botany – the acclimatisation of non-European plants, usually for economic or medical purposes intimately connected with the imperial project.

A prime mover in the effort to make Kew 'the great exchange house of the empire' was Sir Joseph Banks.[62] While their long-awaited *Florilegium* of the plants collected on the first *Endeavour* voyage was never published, Banks and Solander wielded enormous influence over the natural history establishment (especially the Royal Society and the British Museum) and its role in the nation-state and the empire. Banks promoted exotic botany, using a large network of collectors or 'agents of empire' to appropriate non-European plant life for European profit, while Solander controlled classifying and naming from his post at the British

60. Lord Bute was rumoured to be Augusta's lover. Ray Desmond, *The History of the royal botanic gardens Kew*, 2nd edn (Kew, 2007), p.28.

61. On the Royal Collection, see online (http://www.royalcollection.org.uk/eGallery/object.asp?category = 279&object = 37017&row = 55&detail = about).

62. Hector C. Cameron, *Sir Joseph Banks K.B., P.R.S: the autocrat of the philosophers* (London, 1952), p.63-64.

Museum.[63] They were instrumental in making intra-tropical plant transfers – for example tea and breadfruit – central to the growth and prosperity of the British Empire.[64]

In 1785, a year after Bute's *Tables* appeared, Martyn brought out his English translation of Rousseau's eight letters and 'introduction' to botany – a translation heavily coloured by the Linnaean sexual system. Martyn, who entered the Anglican priesthood in 1759, succeeded his father, John Martyn, F. R. S. (1699-1768), as Regius Professor of Botany at Cambridge in 1762; he remained in this post until his death in 1825, but stopped actively teaching in the 1790s due to a purported dearth of student interest in the subject. Like his father, John Martyn, Thomas Martyn was a committed disciple of Linnaeus.[65] His father, who had been apprenticed to learn medicine, rather than taking a medical degree, occupied the Chair in Botany at Cambridge as a sinecure from 1733 to 1762; Martyn senior had met Linnaeus during the Swede's brief visit to Great Britain in 1736.[66] Autographed copies of Linnaeus's works graced the family library, and Linnaeus's teachings played an important role in the Martyns' botanical commitments. Shortly after his appointment at Cambridge, Thomas Martyn settled on teaching Linnaeus's sexual system, even though he was agnostic about its longevity.[67]

Martyn, the faithful follower of Linnaeus, painstakingly annotated his personal copy of the *Species plantarum* (1762-1763), interleaving the text with blank pages for his notes. He was careful to record horticultural and other local uses. Next to the entry 'Trifolium' in the *Species plantarum* we

63. David Mackay, 'Agents of empire', in *Visions of empire: voyages, botany, and representations of nature*, ed. Philip Miller and Peter Hans Reill David (Cambridge, 1996), p.47.

64. Lucille Brockway, *Science and colonial expansion: the role of the British botanic gardens* (New York, 1979).

65. Martyn corresponded with well-known English botanists and naturalists such as Richard Pulteney, Erasmus Darwin and Mendes da Costa; he followed in the footsteps of his father, exercising his primary profession alongside his professorial duties, although he ceased teaching in 1796. Oddly, Martyn's *Sketch of a tour through Swisserland* barely acknowledges Rousseau's connection to Switzerland.

66. John Martyn's works include *Tabulae synopticae plantarum officinalium* (London, G. Strahan, W. & J. Innys, C. Rivington, 1726); *Historia plantarum rariorum* (London, Richard Reily, 1728); translations of Virgil's *Georgics* and *Bucolics*, and *Methodus plantarum circa Cantabrigiam nascentium* (London, Richard Reily, 1727). He was involved with the Chelsea Physick Garden, famous for the acclimatisation of exotic plants by its superintendant, Philip Miller (1691-1771), and was a founding member of the first botanical society in Britain, if not the world. S. M. Walters, *The Shaping of Cambridge botany: a short history of whole-plant botany in Cambridge from the time of Ray into the present century* (Cambridge, 1981), p.30.

67. 'It is of no great importance what system we adopt, so that we all agree to talk the same language. That of Linnaeus will probably stand the test of ages, whatever may become of the sexual system.' Thomas Martyn, 'Introduction', in *Letters on the elements of botany addressed to a lady by the celebrated J.-J. Rousseau; translated into English, with notes, and twenty-four additional letters, fully explaining the system of Linnaeus* (London, B. White, 1785), p.15n.

read, for example, what Martyn has gleaned about clover (known as 'shamrock' in Ireland):

> This is a great improvement of clay lands. It is naturally biannual, but cutting makes it continue something longer; as however it soon wears out, it is seldom suffered to continue above two years. The Irish call this Chamrock, and wear this, or the white clover in their hats on St. Patrick's Day. They also put the flowers into their bread.[68]

A Fellow of the Royal and Linnean societies, Martyn authored and edited a number of botanical works, including not only the translation and amplification of Rousseau's *Letters on the Elements of botany*, but also a revised ninth edition of Philip Miller's hugely influential *Gardener's dictionary* (1807),[69] *Plantae cantabrigensis* (1763), and contributions on natural history to the *Gentleman's magazine*.

Martyn chose to use the artificial sexual system, even though it was not the organisational foundation of Rousseau's own text, which, as we have seen, relied on the natural method. The reasons Martyn gives for his choice of method or system (he uses the terms interchangeably) seems strangely under-motivated and unscientific: 'The system I propose to you is not the *French* one by Tournefort, which is very beautiful, and has great merit; but the *Swedish* one by Linnaeus. I prefer this, because it is most complete, and most in fashion.'[70] This comment may appear odd, coming from a dedicated Linnaean, but we have to place it in the context of a work written for females or less-educated males, as indicated by Martyn's comment that Rousseau's letters, 'might be of use to such of my fair countrywomen and unlearned countrymen as wished to amuse themselves with Natural History'.[71] Martyn suggests that 'the unlearned', confronted by 'translations of Linnaeus's works would only reap disgust from a study that is perhaps more capable of affording pleasure than any

68. This comment is found in Martyn's annotated, interleaved copy of Carolus Linnaeus, *Species plantarum*, 2nd edn, 2 (Stockholm, Laurentius Salvius, 1762-1763), p.1082, at the Linnean Society, London.
69. Philip Miller, *The Gardener's and botanist's dictionary*, 9th edn, ed. Thomas Martyn, 4 vols (London, 1807). First published in 1731, the fourth edition of the *Gardener's dictionary* in 3 volumes appeared in 1754 a few months after Linnaeus's *Species plantarum* (1753); Miller only added Linnaean binomial names in his eighth and last edition in 1768. Haller considered it the most important work of its kind. Albrecht von Haller, *Bibliotheca botanica*, 2 vols (Zürich, Orell, Gessner, Fuessli, 1771-1772), vol.2, p.220. See Frans A. Stafleu, 'Review of *The Gardener's dictionary* by Philip Miller', *Taxon* 18:6 (December 1969), p.713-15 (714).
70. My emphasis; Jean-Jacques Rousseau, *Letters on the elements of botany: addressed to a lady by the celebrated J.-J. Rousseau; translated into English, with notes, and twenty-four additional letters, explaining the system of Linnaeus*, trans. and ed. Thomas Martyn, 2nd edn (London, B. White, 1787), p.86.
71. Martyn, *Letters* (1785), p.v.

other'.[72] However, his dedication of the work is most telling: 'To the Ladies of Great Britain: no less eminent for their elegant and useful accomplishments, than admired for the beauty of their persons'. He clearly had a female audience in mind, one that he placated with references to elegance and beauty rather than brains; it is therefore not so strange that he referred to 'fashion' in justifying his choice of Linnaean taxonomy, even though he certainly had other arguments at his disposal. This association of women with fashion and trivial pursuits was deplored by Mary Wollstonecraft in her *Vindication of the rights of women* (1792).

The natural method was first used in a published botanical work by Antoine-Nicolas Duchesne in his *Manuel de botanique* (Paris, 1764), although Antoine-Laurent de Jussieu did not publish his complete version of it until 1789. Yet in 1785 Martyn clearly knew about this system: 'I will now open the whole mystery [...] To do this you must learn a *system*; in which however you are not to expect that all vegetables are arranged in *natural classes*, such as [Rousseau has] hitherto explained to you, but after an *artificial* method'.[73]

Martyn did not scruple to reject the natural method chosen by Rousseau, even though he could have continued the analysis of natural families well beyond the initial six, since they were then thought to number one hundred. He instead adopted the artificial sexual system of classification starting with the ninth letter, the first that he added to the series. He did not seem to have been troubled by the harm this did to the integrity of Rousseau's work; nor, as we shall see, did this shift disturb Martyn's French, Dutch or Portuguese translators, who transmitted his expanded, Linnaean version of Rousseau's letters on botany. Martyn likewise introduced binomial nomenclature into the letters: for example, in the first letter to Mme Delessert he added the Latin binomial for lily ('lis'), *Lilium candidum*, L. in a footnote.[74] Recall that in the *Lettres* Rousseau did not use Latin names, and saved the exercise of learning two hundred of the most important plant names for Mme Delessert's later botanical education.

Martyn framed his choice of botanical classification in cultural or nationalistic terms, telling the reader that Swedish botany had vanquished French botany. This goes against the spirit in which Rousseau pursued botany, for he accused French botanists of prejudice against 'un système étranger',[75] and asserted,'[j]e sais avec quel mépris on affecte en France de traiter ce grand Naturaliste [Linnaeus], mais le reste

72. Martyn, *Letters* (1785), p.vii.

73. My emphasis; Martyn, *Letters* (1787), p.85.

74. Martyn, *Letters* (1785), p.22, n. (a).

75. Rousseau, *Fragmens pour un dictionnaire*, *OC*, iv.1207.

de l'Europe l'en dédommage, et la postérité l'en vengera'.[76] Martyn's position should therefore be clearly distinguished from the intimate connection we have noted between Rousseau's botany and Switzerland: while botany had strong Swiss associations for Rousseau, he did not regard it as 'Swiss botany'; nor did he insist on using a 'Swiss' system like that of Haller. Botany's patriotic significance lies rather, in its ability to provide a way to connect the naturalist with his *patrie*; this can readily be done while using a 'Swedish' system, that is: the Linnaean sexual system, or an indigenous one.

Martyn's Linnaean recasting of Rousseau's botany course was acclaimed by no less a reader than Erasmus Darwin:[77] 'Your continuation of Rousseau's botanical letters has given me much pleasure and improvement: and I dare say will much contribute to propagate the science, which is my principle [sic] design in [...] our translation of Linnaeus.'[78]

Martyn's twenty-four additional letters were adopted for a French edition of the 'Lettres' in 1789, together with F. P. Nodder's illustrations, added to the second edition of 1787. The French editors' commentary on their decision to use Martyn's additions provides an important insight into the value placed both on Rousseau's original work and Martyn's recasting of it:

> Les lettres élémentaires sur la botanique furent particulièrement goûtées en Angleterre. M. Martyn, professeur de l'université de Cambridge, jaloux de rendre hommage à la mémoire de M. Rousseau, a publié vingt-quatre lettres familières sur la méthode de Linné, pour servir de suite aux six [sic] de M. Rousseau. Chaque lettre est accompagnée d'une gravure coloriée pour faciliter l'intelligence des caractères qui servent à designer les classes. Cet ouvrage a été très accueilli, et mérite de l'être. Nous en présentons ici la traduction Françoise, suivie des mêmes gravures, et nous avons lieu d'en espérer les mêmes succès en France. Nous n'avons aucun ouvrage élémentaire où l'on puisse espérer de trouver plus de moyens d'approfondir la méthode linnéenne, qui est aujourd'hui la plus accréditée.[79]

76. Rousseau to Dutens, 5 February 1767; *CC*, xxxiv.99.

77. Letter of Erasmus Darwin to Thomas Martyn, 6 February 1786, Natural History Museum, London, Banksian MS 103, f.573. The letters to John and Thomas Martyn were presented to Joseph Banks by Thomas Martyn in 1817; Banks states that he received them from 'my old and esteemed friend, Professor Martyn[,] March 12 1817. JB'.

78. Darwin refers to the Botanical Society of Lichfield's editions of Linnaeus, *A System of vegetables* (Lichfield, J. Johnson, 1783) and *The Families of plants with their natural characters*, 2 vols (Lichfield, J. Johnson, 1787), 'translated from the last edition [...] of the *Genera plantarum*, and of the *Mantissae plantarum* of the elder Linnaeus; and from the *Supplementum plantarum* of the younger Linnaeus, with all the new families of plants from Thunberg and L'Héritier'.

79. Jean-Jacques Rousseau, *Œuvres complètes, de J. J. Rousseau. Nouvelle édition, classée par ordre de matières, et ornée de quatre-vingt-dix gravures*, Louis Sébastien Mercier, *et al.* (eds), vol.5 (Paris, Poinçot, 1789), p.4-5.

Moreover, Martyn had, in the view of the French editors, acquitted himself well: 'L'auteur de ces lettres a suivi de *très près* la marche de Rousseau, et le traducteur n'a rien négligé pour en rendre le véritable sens.'[80] In 1801 there followed an anonymous Portuguese translation of Martyn's edition (fig. 9.2, p.317) by Leonor de Almeida Portugal de Lorena e Lencastre, the Marquesa da Alorna (1750-1839), a botanical populariser and well-known author in her native country.[81] A Dutch bilingual edition published in 1798 likewise followed Martyn.[82] Like the Marquis de Girardin (1735-1808), Du Peyrou and Moultou before him, Martyn accomplished a 'forging' of Rousseau's botanical legacy that was to have a long-lasting effect.[83]

It is my contention that the Martyn edition and its translations bear much of the responsibility – both directly and indirectly – for creating the image of a Linnaean Rousseau that has had a remarkably resilient life, despite its obvious inaccuracy; the result has been the virtual eclipse of Rousseau's adoption of the natural method in the eight letters to Mme Delessert, and his espousal of its principles to correspondents. Martyn forged the Linnaean Rousseau about whom we still read today.[84]

Yet the fame of the Martyn translation still rested with Rousseau, as Martyn's son-in-law, George Cornelius Gorham, attested:

> It is, perhaps, to be regretted, that the modest author [Martyn] did not make his volume entirely original; for it may be unhesitatingly said that Rousseau's part of the work (however elegant in style) is the *least* valuable portion of the volume. The name of the Swiss, however, generally usurps the place of his infinitely more learned Translator and Continuator; this work being, absurdly enough, usually called '*Rousseau's* Letters on Botany!'[85]

Gorham failed to acknowledge, however, that had Rousseau's name not appeared in conjunction with Martyn's work, the latter probably would

80. My emphasis; Rousseau, *Œuvres complètes*, p.4-5.
81. J.-J. Rousseau, *Cartas sobre os elementos de botanica, com as notas e addiçôes de Thomaz Martyn; (quarta ediçaô em Londres)*, trans. by a Lady of the Court (Lisbon, 1801).
82. Jean-Jacques Rousseau, *Inleyding tot de kruidkunde, een werk geschikt tot gebruik van hen, welke zig in deeze wetenschap willen oeffenen; introduction à la botanique*, trans. J. van Noorden, vol.1 (London, Vernor & Hood; Rotterdam, L. Bennet, 1798).
83. For an elaboration, see Birn, *Forging Rousseau.*
84. Recent examples include Duris, *Linné et la France*, p.105; Williams, *Botanophilia in eighteenth-century France*, p.105 and Fontes da Costa who, following Williams, states, 'Rousseau's *Elementary letters on botany* [...] contributed greatly [...] to the dissemination of the Linnaean system of classification'. Palmira Fontes da Costa, 'Women and the popularization of botany in early nineteenth-century Portugal: the Marquise of Alornás *Botanical recreations*', in *Popularizing science and technology in the European periphery, 1800-2000*, ed. Faidra Papanelopoulou, Agustí Nieto-Galan and Enrique Perdiguero (Farnham, 2009), p.43-63 (54-55).
85. My emphasis; George Cornelius Gorham, *Memoirs of John Martyn, F.R.S. and of Thomas Martyn, B.D., F.R.S., F.L.S. professors of botany in the University of Cambridge* (London, 1830), p.169.

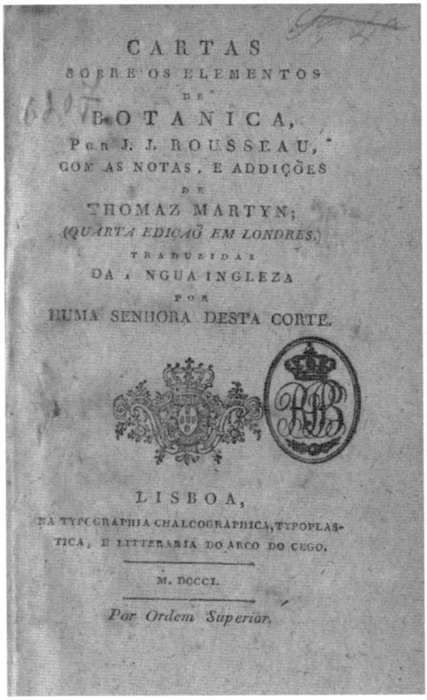
CARTAS
SOBRE OS ELEMENTOS
DE
BOTANICA,
POR J. J. ROUSSEAU,
COM AS NOTAS, E ADDIÇÕES
DE
THOMAZ MARTYN;
(QUARTA EDIÇÃO EM LONDRES.)
TRADUZIDAS
DA LINGUA INGLEZA
POR
HUMA SENHORA DESTA CORTE.

LISBOA,
NA TYPOGRAPHIA CHALCOGRAPHICA, TYPOPLASTICA, E LITTERARIA DO ARCO DO CEGO.
M. DCCCI.
Por Ordem Superior.

Figure 9.2: Title page from Thomas Martyn, *Cartas sobre os elementos de Botanica / por J.-J. Rousseau; com as notas, e addições de Thomaz Martyn (quarta edição em Londres)* (Lisbon, 1801), courtesy of the Bibliotheca Nacional de Portugal.

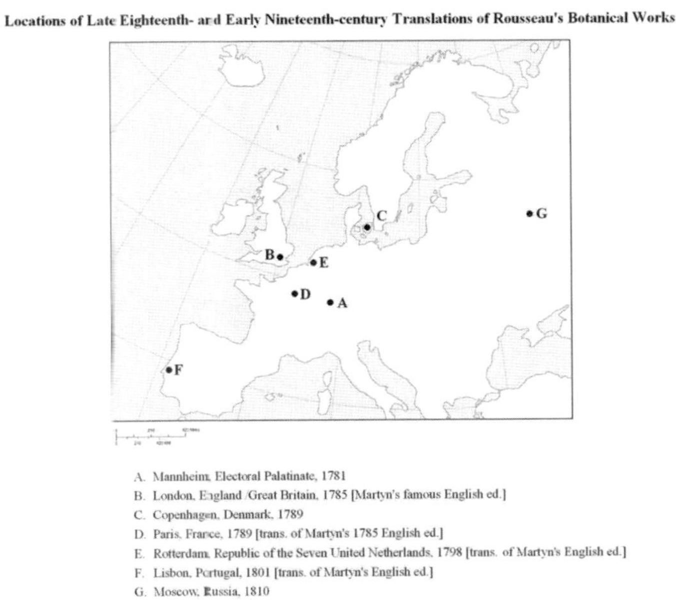

Figure 9.3: Kenneth Yung, Locations of late eighteenth- and early nineteenth-century translations of Rousseau's botanical works, 2011, Hong Kong.

not have enjoyed the success it did. Rousseau's popularity may have likewise eclipsed the role played by authors such as Barbeu Dubourg, who helped shape his view of the botanophile enterprise.

Not all translators of Rousseau's eight letters to Mme Delessert chose to use Martyn's translation and extensive Linnaean additions, however. There are at least two other translations into languages other than English that do not use Martyn's twenty-four additional letters: one into Danish (1789) by Odin Wolff (1760-1830),[86] and the other into Russian (1810) by Vladimir Vasil'evich Izmailow; while the Russian edition contains an explanation of the Linnaean method and the title refers to the 'immortal Linnaeus', it also offers explications of the methods of Tournefort and Antoine Laurent de Jussieu.[87] An anonymous German translation was published in 1781, and therefore pre-dated Martyn: *J. J. Rousseaus Botanik für Frauenzimmer, in Briefen an die Frau von L*** (1781) (fig.9.3, p.317).[88]

iv. Rousseau's English-speaking followers

The effect of the Rousseau/Martyn edition of the letters is difficult to assess, but it certainly seems to have been considerable. Working from Martyn's translation, English women writers such as Priscilla Wakefield (1751-1832)[89] and Maria Jacson (1755-1829) imitated Rousseau's epistolary, and highly accessible, approach to botanical instruction.[90] Wakefield was a particularly successful author, bringing out many works for young

86. J.-J. Rousseau, *Botanik for fruentimmere i breve til Frue de L*** [Botany for women in the letters to Mme L**], trans. Odin Wolff (Copenhagen, S. Poulsen, 1789).

87. J.-J. Rousseau, *Russovy pis'ma o Botanikie, s dopolneniem ego Botanicheskago Slovania, s ob"iaseniem trekh luchshikh metod Turneforta, Linneia, i Zhus'e: I s Botanicheskami chasami, izobriretennymi bezsmertnym Linnem* [Rousseau's letters about botany with an explanation of the methods of Tournefort, Linnaeus, and Jussieu: and with the botanic clock, invented by the immortal Linnaeus], trans. Vladimir Vasil'evich Izmailow (Moscow, 1810).

88. Jean-Jacques Rousseau, *J. J. Rousseaus Botanik für Frauenzimmer, in Briefen an die Frau von L*** (Mannheim, in der schwanischen Buchhandlung, 1781). The anonymous translator claimed to have had the original texts, although this seems unlikely. Rousseau, *J. J. Rousseaus Botanik*, p.1. Sources that discuss this translation are rare, and none I have found shed any light on the translator's identity. See for example Sina Kuchler and Volker Wissemann, '"Botanik für Frauenzimmer": Untersuchungen zur Popularisierung von Botanik im Raum Weimar/Jena um 1800', *Haussknechtia: Mitteilungen der thüringischen botanischen Gesellschaft* 10 (2004), p.299-330.

89. Priscilla Wakefield, *An Introduction to botany, in a series of familiar letters, with illustrative engravings*, 2nd edn (London, E. Newbery, 1798); [Maria Elizabeth Jacson], *Botanical dialogues*, and *Botanical lectures: by a lady: altered from 'Botanical dialogues for the use of schools', and adapted to the use of persons of all ages by the same author* (London, 1804), p.4, 52, 55.

90. Ann Shteir, 'Botanical dialogues: Maria Jacson and women's botanical writing, *Eighteenth-century studies* 23:3 (1990), p.301-17; George, *Botany, sexuality and women's writing*, p.49ff.

people on natural history including *An Introduction to botany, in a series of familiar letters* (1796), which ran to eleven editions and was translated into French. Following Martyn, these authors adopted Linnaeus's artificial sexual system, which formed the backbone of Martyn's additions to Rousseau's original presentation of six families of flowering plants in the mode of Antoine Laurent de Jussieu. Jacson made her allegiance to Rousseau explicit by quoting his fifth letter on botany on the title page of her *Botanical dialogues* (1797).[91] She likewise made her allegiance to Linnaeus explicit throughout the work, calling for the use of Linnaean nomenclature and the learning of classification from the master himself. The contribution of these and other women writers to the 'feminisation' of botany owes an enormous debt to Martyn's Linnaean re-packaging of Rousseau's eight letters.[92]

Martyn's edition likewise gained Rousseau followers in the young United States. Starting in 1788, Benjamin Waterhouse (1754-1846), the first professor of 'physic',[93] delivered a course of lectures on natural history at Harvard University for the first time in the new country's history.[94] Also known as the 'American Jenner', Waterhouse expounded Rousseau's ideas, using Martyn's translation. Waterhouse's promotion of botany and natural history went well beyond the confines of Harvard to the larger Boston community. In 1786 and 1787 he delivered two sets of

91. She quotes Jean-Jacques Rousseau, *Letters on the elements of botany: addressed to a lady by the celebrated J.-J. Rousseau; translated into English, with notes, and twenty-four additional letters, explaining the system of Linnaeus*, trans. and ed. Thomas Martyn 4th edn (London, B. and J. White, 1794), p.49. Jacson was a cousin of Rousseau's friend, Sir Brooke Boothby, whom Rousseau entrusted with a copy of his *Dialogues* in 1776. See also Ann Shteir, *Cultivating women, cultivating science: Flora's daughters and botany in England 1760-1860* (Baltimore, MD, 1996), p.108ff.

92. George, *Botany, sexuality and women's writing* and Shteir, *Cultivating women* both study the phenomenon of botanising women and the 'feminisation of botany' in late eighteenth-century Britain. George emphasises the Linnaean character of this phenomenon, mirrored by the French Linnaean societies, frequented by women. George rightly corrects Shteir's assertion (p.20) that Rousseau took no interest in nomenclature, but George relies on Martyn's version of Rousseau's botanical thought (see Shteir, *Cultivating women*, p.19), leading her to some erroneous interpretations.

93. 'Physic' has many connotations, but the most appropriate one in this context seems to refer to all branches of medicine except surgery. *OED*, 'physic, n[oun].', definition 3(b).

94. Waterhouse, grandnephew of the famous English physician John Fothergill, studied at Edinburgh 1775-76, and obtained the MD degree at Leiden in 1780. He met Benjamin Franklin through Fothergill and in Holland became a lifelong friend of John Adams, whom he attended on his deathbed. He 'left Europe as the best-educated American-born physician until that time'. He became the first professor of medicine at Harvard, and was well-known for his involvement in smallpox vaccination, cooperating closely with Thomas Jefferson. See Philip Cash, *Dr Benjamin Waterhouse: a life in medicine and public service (1754-1846)* (Sagamore Beach, MA, 2006), p.59.

natural history lectures that 'were open to both sexes of the public'.[95] He gave his first course on natural history starting in 1788, introducing Linnaean taxonomy to the University. Waterhouse was sensitive to the potential of the artificial sexual system to offend: 'The botanical phraseology sometimes embarrasses the teacher. We hope however to parry this difficulty, if not entirely surmount it'.[96] Indeed, 'some parents objected to his teaching "youth of both sexes" the Linnaean system of classifying plants by their sex organs. In response to these objections, Waterhouse attempted to "drop the Linnaean metaphor of generation and substituted that of *nutrition*." His actions brought a scolding from Sir Joseph Banks for failing to use the agreed-upon terminology of science'.[97] Waterhouse himself conceded that his attempt to sanitise Linnaean botany was unsuccessful.[98]

While Waterhouse advocated Linnaean names and classification, he took care to state Rousseau's view that botany is more than a mere science of names:

> It is of importance [...] that one universal language should be adopted by botanists; but it is wrong to make that, and classification the primary object. Agreeably to this doctrine is the sentiment of Rousseau [sic], who, in his *Letters on the Elements of botany*, says, 'I have always thought one could be a very great botanist without knowing so much as one plant by name'.[99]

Unfortunately, Waterhouse's work in promoting botanical study at Harvard ended with his dismissal from that institution after a stormy dispute in 1812. His departure had nothing to do with his teaching, which was acclaimed; it had everything to do with his being a Quaker in Congregationalist Massachusetts, and with his vocal expression in the press of political views ('Democratic-Republican' in the parlance of the time) that set him at variance with many influential people.

Rousseau's botanical legacy in the Wakefieldian tradition lived on in the work of the American writer and educator, Almira Hart Lincoln Phelps (1793-1884), another female botanical author who adopted the epistolary mode for transmitting botanical knowledge. In her *Familiar lectures on botany*, first published in 1829, which went through many editions, she proved herself to be the American Priscilla Wakefield.[100]

95. Cash, *Dr Benjamin Waterhouse*, p.87.

96. Benjamin Waterhouse, *The Botanist: being the botanical art of a course of lectures on natural history, delivered in the University of Cambridge [Harvard]; together with a discourse on the principle of vitality* (Boston, MA, 1811), p.24.

97. Original emphasis; Cash, *Dr Benjamin Waterhouse*, p.88; see also p.274.

98. Cash, *Dr Benjamin Waterhouse*, p.274.

99. Waterhouse, *The Botanist*, p.xiii-xiv; see also p.17.

100. See for example Almira Hart Lincoln Phelps, *Familiar lectures on botany*, 18th edn (New York, 1843).

v. Rousseau's botany in and for itself

> In Rousseaus Wercken finden sich ganz allerliebste Briefe über die Botanick, worinn er diese Wissenschafft auf dass fasslichste und zierlichste [Weise] einer Dame vorträgt. Es ist recht ein Muster wie man unterrichten soll und eine Beylage zum Emil.[101]

Rousseau was well-known in the late eighteenth and early nineteenth centuries as a populariser of botany – iconography of the period frequently shows him botanising (fig.9.4, p.322). The image of Rousseau as botanist was particularly popular during the French Revolution, as attested by the proceedings of the 'fête' that celebrated the transfer of his remains from Ermenonville to Paris on 11 October 1794. Botanists were in attendance to honour Rousseau's devotion to their science: 'on en voyait en grand nombre, au milieu desquels on portrait des fleurs, des plantes et des fruits'.[102] His popularisation of botany was well-known, not only through his writings but also, simply, by his practice of it; the great and not-so-great used a sometimes feigned interest in botany to meet the philosopher, and as d'Escherny showed, there was no better way to do so than to herborise with Jean-Jacques.[103]

When Rousseau's botanical works were posthumously published, they received glowing praise well beyond the immediate confines of the Francophone world – France and the Suisse romande; the *Göttingische Anzeigen von gelehrten Sachen*, once the preserve of Rousseau's political adversary, Albrecht von Haller, gushed over the botanical writings:

> Wir waren sehr begierig zu sehen, wie weit sich R. ins Studium der Pflanzen eingelassen hatte [...] Aber wir finden auch hier Tizians goldne Linie, die ihren Meister verrieth, und entsinnen uns nicht, einen angenehmeren und zugleich so scharfsinnigen botanischen Schriftsteller gelesen zu haben.

101. 'In Rousseau's works there are the most delightful letters concerning botany, in which he explains this science to a lady in the most understandable and elegant way. It is truly a model of how to teach and a supplement to Emile.' Goethe to Duke Karl August of Saxe-Weimar, 17 June 1782; Johann Wolfgang von Goethe, *Briefe*, ed. Karl Robert Mandelkow and Bodo Morawe (Hamburg, 1962), vol.1, p.398.

102. 'La fête de Rousseau à Paris: récit', reprinted from the *Moniteur* (15 October 1794); *CC*, xlviii.84. In 1795 Dominique Villars and Pierre Liotard were proposed for a list of pensioners being recognised for their talents and civic contributions, note being made of their botanical connections with Rousseau. R. A. Leigh, 'Remarques', *CC* xlviii.170. See also Vincent Ponçet, *L'Herbier Dominique Villars (1745-1814): témoin de la flore du Dauphiné*, preface by G. G. Aymonin (Grenoble, 1999); Roger L. Williams, *The Letters of Dominique Chaix, botaniste-curé, International archives of the history of ideas* 151 (Dordrecht, 1997).

103. See for example d'Escherny, 'Eloge de J.-J. Rousseau'.

Figure 9.4: Johann Fr. Mayer (artist) and W. (engraver; further information unknown), *J.-J. Rousseau à Ermenonville*, water-coloured engraving, by permission of the Bibliothèque publique et universitaire de Neuchâtel.

Zumal die Definitionen in den Fragmenten des Wörterbuchs sind Muster von Bestimmtheit der Begriffe und Präcision im Ausdruck [...].[104]

Rousseau's legacy to the world of botany – both amateur and professional – spread in many directions: to Germany, where the poet Goethe read him in the original French, to the library of Thomas Jefferson (1743-1826) in Virginia – which contained a copy of Thomas Martyn's 1785 translation – and Harvard University, where Benjamin Waterhouse expounded Rousseau's ideas according to Martyn. Rousseau's renown as a botanist even spread to Russia, where a translation of his botanical writings appeared in 1810; this was remarkable testimony to Rousseau's appeal as a botanist, given that the Russian elite could read his works in the original French, and his other works had yet to be translated into Russian.[105]

Rousseau's botanical activities inspired educational reforms during the French Revolution,[106] and in German education, although German educators did not remain loyal to his natural method.[107] This botanical legacy is now little-known, however, overshadowed by (1) Rousseau's influence in other fields such as political philosophy and literature, (2) undermined by the scholarly neglect of his botany,[108] and (3) marginalised by the presumed march of scientific progress.

vi. Rousseau's botanical legacy in France

Rousseau's name was often on the lips of French botanical authors of the late eighteenth and early nineteenth centuries; vernacular works of botanical popularisation that Rousseau influenced included Pierre Bulliard's *Dictionnaire élémentaire de la botanique* (1783), and its revision

104. 'We were very curious to see how far Rousseau had let himself become absorbed with the study of plants [...] But we also find here Titian's golden line, which betrays its Master, and we cannot recall having read such an agreeable and at the same time such a penetrating botanical author. The more so, as the definitions in the Fragments of the Dictionary are models of conceptual exactitude and precision of expression.' 'Genf', p.1164. The author was not aware of the questionable origins of the *Fragmens*.

105. Rousseau, *Russovy pis'ma o Botanikie*. Botany and Rousseau were popular in early nineteenth-century Russia; pers. comm. Vladimir Dorofeyev, Komarov Botanical Institute, St. Petersburg. On Linnaeus's floral clock, see Linnaeus, *Philosophia botanica* (1751), section 335, p.274-75, and Delaporte, *Le Second règne de la nature*, p.197.

106. Hans-Christian and Elke Harten, *Die Versöhnung mit der Natur* (Reinbek, 1989), p.85ff.

107. Jansen, *Rousseau als Botaniker*, p.264; Cohn, 'Jean-Jacques Rousseau als Botaniker', p.383; Alexandra Cook, 'Propagating botany', p.88-89.

108. 'The works on botany [by Rousseau] were published posthumously and caused – remarkably, in the career of so controversial a writer – absolutely no stir.' Lisa Gasbarrone, 'Blindness or oversight?'. However, Rousseau's eight letters on botany are now being noticed by scholars interested in the popularity of botany among women in the late eighteenth and early nineteenth centuries. Shteir, *Cultivating women*.

by Louis Claude Marie Richard (1798).[109] Dictionaries in natural history were nothing new: for example, Jacques-Christophe Valmont de Bomare's popular *Dictionnaire raisonné universel d'histoire naturelle* appeared in 1764 in five volumes.[110]

Other works that reflect Rousseau's influence are Bulliard's *Histoire des champignons de la France* (1791), and *Manuel de botanique à l'usage des amateurs et des voyageurs* (1787) by François Lebreton. Unlike the works of Linnaeus, all of these were in the vernacular, and aimed at a broad audience, literate in its mother tongue, but not necessarily in Latin.

Another important vehicle of dissemination of Rousseau's botanical writings were the many editions of his *Œuvres complètes* in French. In addition to the French appropriation of Martyn in the 1789 *Œuvres complètes* referred to above, the writings were often reproduced in later editions of Rousseau's works. In addition, special editions gave pride of place to the botanical writings; a case in point was *La Botanique de J.-J. Rousseau* (Paris, 1802), containing an exposition of Antoine Laurent de Jussieu's natural family system – a belated acknowledgement of the importance of this system for Rousseau's work; it also contained a disquisition on preserving plants in herbaria by the mineralogist and botanist, René Just Haüy (1743-1822).[111] The 'Avis de l'éditeur' suggests that the contemporary reputations of both Rousseau and Jussieu in botany were still robust, and that Haüy's lessons on plant preservation suited the spirit of Rousseau's work: 'Des noms pareils [Rousseau, Jussieu, Haüy] justifient notre confiance, et garantissent un accueil favorable'.[112]

In 1805 connoisseurs of special editions of Rousseau's botany were treated to a special edition of the *Lettres* beautifully illustrated by the famous Pierre-Joseph Redouté (1759-1840), an exacting artist and botanist.[113] The illustrator was reluctant to participate in the project, which

109. Bulliard, *Dictionnaire élémentaire de la botanique*; Louis Claude Marie Richard (ed.), *Dictionnaire élémentaire de botanique de Bulliard, revu et presqu'entièrement refondu par Louis Claude Richard* (Paris, A. J. Dugour & Durand, 1798).

110. Jacques Christophe Valmont de Bomare (1731-1807) was a popular natural history teacher in Paris, curator of the natural history cabinet of the Dukes of Bourbon at Chantilly and pharmacist to the king. His botanical work had a strongly horticultural slant. There were several subsequent editions of this work, and each edition comprised different numbers of volumes. It was translated into several languages.

111. The abbé Haüy, a priest and botanist, became interested in mineralogy when he accidentally broke a specimen of calcareous spar. He is known as the father of the geometrical law of crystallisation, elaborated in his *Traité de minéralogie*, 5 vols (Paris, 1801); during the Revolution he was saved from the mob by Geoffroy Saint-Hilaire, who made him professor of mineralogy at the Muséum national d'histoire naturelle, Paris.

112. Rousseau, *La Botanique de J.-J. Rousseau.. par M. Haüy*, p.vi.

113. J.-J. Rousseau, *La Botanique de J.-J. Rousseau ornée de soixante-cinq planches, d'après les peintures de P. J. Redouté* (Paris, 1805, 1821). Redouté started his career at the Jardin du roi in 1782,

was proposed to him by the botanist, Charles l'Héritier de Brutelle (1746-1800); Redouté only agreed to produce the illustrations when he learned that he lived on the very land previously occupied by Rousseau – the former estate of the Marquis de Mirabeau. Like Rousseau, Redouté preferred native plants over exotic and cultivated ones. The 1805 edition was widely known and admired; Goethe used it to help him compose his reflections on Rousseau as botanist, starting in 1824, and completed in 1830-1831.[114]

History did not consign Rousseau's herbaria to the dustbin; they were well known to aficionados. Late eighteenth-century French botanical authors were keen to associate their discussions of herbaria with Rousseau's name, and readily referred to his herbaria. Cases in point are Pierre Bulliard's *Dictionnaire élémentaire de botanique*[115] and François Lebreton's *Manuel de botanique*.[116]

Jean Baptiste François Bulliard (1752-1793), known as Pierre Bulliard, was a leading figure in mycology and developed revolutionary engraving techniques to depict mushrooms with a very high degree of realism and accuracy.[117] His great multi-volume work, *Herbier de la France*, contains his *Dictionnaire élémentaire*, in which he published a purported, possibly spurious, text of Rousseau on herbaria, discussed below. While Bulliard's claim to botanical fame was his skill as an engraver, especially of fungi, the *Dictionnaire* was so popular that it was later edited and re-published in a new version by Louis Claude Marie Richard (1754-1821), a student of Bernard de Jussieu, an important botanical explorer, and professor of botany at the Ecole de médecine (Paris), 1795-1821. He was the grandson of Claude Richard (1705-1789), gardener of the Petit Trianon garden

became a disciple of Linnaeus, and professor of 'iconographie végétale' in 1792. He was painting instructor to Marie Antoinette and the Empress Josephine, and served as a judge of the Salon.

114. Cook, 'Propagating botany' p.83; Elise von Keudell, *Goethe als Benutzer der Weimarer Bibliothek*, reprint edn (Leipzig, [1931] 1982), p.334-35, 343; Goethe, 'Der Verfasser teilt die Geschichte seiner botanischen Studien mit', p.159. These reflections are discussed below, p.334-36.

115. Bulliard, *Dictionnaire élémentaire*, 'Herbier', p.99ff.

116. Lebreton quotes Rousseau's purported discussion of herbarium technique found in Bullliard's *Dictionnaire*. Lebreton proposes that collectors in the 'islands' could easily use Rousseau's method. Lebreton, *Manuel de botanique*, p.163-66.

117. Bulliard was a talented engraver as well as botanical author, ornithologist and medical practitioner; he perfected methods for depicting mushrooms that were unique in their time. He never sought or enjoyed patronage, subsidising production of his ambitious works himself. His work was known to Parisian botanists such as Antoine-Laurent de Jussieu, L'Héritier de Brutelle and L. C. M. Richard. He met a premature, mysterious end, having been possibly shot accidentally while exploring the forest. See E. J. Gilbert, 'Un esprit – une œuvre: Jean Baptiste François Bulliard, dit Pierre (1752-1793)', *Bulletin trimestriel de la société mycologique de France* 68:1 (1952), p.5-131 (9-11).

(Versailles), collaborator of Bernard de Jussieu, and correspondent of Linnaeus. Rousseau was acquainted with Claude Richard, a fact that he relayed with pride in his correspondence.[118] The younger Richard did not claim any direct connection to Rousseau, although Rousseau claimed his uncle as a horticultural contact.[119] The *Dictionnaire* of Bulliard was published in English by the American botanist and educator, Amos Eaton (1776-1842).[120]

Bulliard's *Dictionnaire* adhered to Rousseau's desire to clarify sciences such as music and botany, to make them more accessible and easier to learn.[121] Bulliard appealed both to the utility of botany, and to Rousseau's example as a botanical practitioner who made herbaria; these were somewhat contradictory, given Rousseau's critique of applied botany, especially pharmacology: 'Le premier Malheur de la Botanique est d'avoir été regardée dès sa naissance, comme une partie de la Médecine'.[122]

Bulliard held up botany as a useful science, rather than as an edifying one:

> vous que la fortune aveugle d'avoir comblé ses faveurs, mais dont les trésors immenses étoilent aussitôt dévorés par le luxe et la vanité; ouvrez les yeux, et voyez si les richesses que la Nature vous offre ne sont pas d'un plus grand prix que celles auxquelles vous attachiez d'abord tant d'importance. Bientôt on va vous voir préférer la *vie champêtre* à ces plaisirs bruyants de nos grandes villes; confondus avec des citoyens vos égaux, on vous entendra parler avec mépris de ces passions qui tourmentent les hommes, et les avilissent; l'étude des différentes branches de l'Histoire Naturelle, celle de la *Botanique surtout*, comme la plus *utile*, va fixer votre attention; vous encouragez le *cultivateur*; vous lui apprendrez à raisonner sur son art; peu-à-peu il vous devra de s'être écarté de cette routine aveugle qui oppose depuis si longtemps une funeste barrière aux progrès de *l'Agriculture*; vous protégerez l'orphelin; vous soulagerez le vieillard infirme; vous aiderez l'ouvrier indigent: vous serez béni, adoré; et fussiez-vous même au-dessous de l'aisance, vos jouissances seront plus douces et plus multipliées que, lorsqu'au faîte de la fortune et des

118. Richard's most well-known work is *Démonstrations botaniques, ou Analyse du fruit, considéré en général* (Paris, 1808). For Rousseau's acquaintance with Richard's grandfather, see Rousseau to Latourrette, 4 July 1770; *CC*, xxxviii.53.

119. Rousseau to Latourrette, 4 July 1770; *CC*, xxxviii.53.

120. Louis Claude Marie Richard, *A Botanical dictionary*, trans. and ed. Amos Eaton, *et al.* (New Haven, CT, 1816). This popular work was published in subsequent editions in 1817 and 1819. Eaton, who started his career as a lawyer, studied botany and geology while serving time in jail for forgery. He exerted a considerable influence on botany in the United States, publishing a flora of the northern U.S. He co-founded the Rensselaer Polytechnic Institute in 1824.

121. Cook, 'Languages of music and botany', p.80.

122. Rousseau, 'Introduction', *OC* iv.1201.

rangs, vous viviez dans la mollesse et l'oisiveté, principales causes de tous les désordres.[123]

This view of botany as an aid to agriculture was not foreign to Rousseau, who approved of the utility of agronomy in *Julie*, in his correspondence with Swiss reformers and physiocrats such as Hirzel, Tscharner and Jean-Baptiste-Louis-Théodore Tschudi (1734-1784); he likewise praised Kleinjogg, the Swiss peasant embraced by poets and princes. Even in the *Rêveries*, where he espoused a view of botany as an edifying science, Rousseau lauded Linnaeus for having 'un peu tiré la botanique des écoles de pharmacie pour la rendre à l'histoire naturelle et aux usages économiques'. He likewise referred to the 'règne végétal' as a 'magasin d'aliments donnés par la nature à l'homme et aux animaux.' [124]

Bulliard differed from Rousseau, however, in rejecting the edifying function of botany in favour of its utility: 'Tout le monde s'occupe des moyens d'étendre l'empire de la Botanique; moi, c'est ce dont je m'occupe le moins; je n'envisage cette science que du côté de son *utilité*; mon objet est de mettre sur la voie des découvertes importantes qu'il reste à faire dans cette partie de l'Histoire naturelle'.[125] Elsewhere, he reiterated the utility of botany:

> Nous sommes arrivés enfin à ce terme, où nous devons songer à ne plus regarder la Botanique, que comme un moyen auxiliaire, et comme le fil qui doit nous conduire à un objet plus utile et plus intéressant: cet objet est l'art de nous *approprier* les plantes, et de les faire *servir utilement* à nos différents besoins. Il reste encore à faire dans l'Agriculture, dans la Médecine, dans les Arts, des milliers de découvertes importantes, qui sont en notre pouvoir.[126]

Bulliard's reduction of botany to a useful science and nothing more did not prevent him from seeing the value of Rousseau's method of botanising and his skill in making herbaria. Hence, despite their very different understandings of the ultimate role of botany, Bulliard considered Rousseau's *method* of botanising exemplary:

> Faites, comme dit J. J. Rousseau p.99 [*Dictionnaire* de Bulliard], et comme nous savons qu'il a fait lui-même: lorsque vous trouverez en fleur une plante que vous ne connaîtrez pas, décrivez-la exactement dans les termes de l'art; n'oubliez rien surtout de ses détails caractéristiques; ramassez-en, s'il se peut, deux ou trois échantillons, et conservez-les de votre mieux; lorsque vous en aurez un certain nombre, amusez-vous à les classer suivant une *méthode*

123. My emphasis; Jean Baptiste François Bulliard, *Histoire des champignons de la France, ou Traité élémentaire, renfermant dans un ordre méthodique les descriptions et les figures des champignons qui croissent naturellement en France* (Paris, chez l'auteur, Barrois le jeune, Belin, Croullebois, Bazan, 1791), p.xv-xvi.

124. Rousseau, *Rêveries*, *OC*, i.1064.

125. Bulliard, *Dictionnaire élémentaire*, p.iii.

126. My emphasis; Bulliard, *Dictionnaire élémentaire*, p.155.

quelconque; adressez-vous ensuite à un Botaniste instruit [...] Cette marche, *lente en apparence*, est bien la plus courte et la plus sure de toutes; elle seule vous servira *plus* que les ouvrages les plus savants et les plus *méthodiques*; elle ne manquera pas de vous conduire à de nouvelles découvertes, parce qu'elle vous forcera au besoin d'une attention soutenue et d'une exactitude scrupuleuse; vous vous trouverez bientôt en état de jouir du fruit de vos recherches [...].[127]

Bulliard was also aware of Rousseau's proficiency in herbarium fabrication, to which he devoted a considerable discussion – quoting at length from a text on herbaria purportedly authored by Rousseau (discussed below, p.330-31). Bulliard even reported on the condition of two of Rousseau's herbaria, which he had presumably seen at the home of Chrétien-Guillaume Lamoignon de Malesherbes. For Bulliard, Rousseau's herbaria serve as a prime example of best practice in this exacting work:

On sait que Jean-Jacques Rousseau aimait passionnément la Botanique, et qu'il travaillait même à faire dans cette science quelques réformes avantageuses (1).[128] Il s'est longtemps occupé de l'art de la dessiccation des plantes; il nous a laissé plusieurs herbiers de différents formats. Parmi les livres rares et précieux qui composent la bibliothèque du savant M. de Malesherbes, on trouve deux petits herbiers de Jean-Jacques, faits avec tout le soin et tout l'art possible ; l'un est de *format in*-8°, et ne renferme que des cryptogames; et l'autre de *format in*-4°, est composé de plantes à fleurs distinctes.

M. le Chevalier de Tourmevel[129] ayant appris que j'étais sur le point de faire imprimer cet Ouvrage, a bien voulu concourir de la manière la plus obligeante, à en augmenter l'utilité, en me communiquant un manuscrit du Philosophe Genevois, sur la nécessité d'un herbier, et sur les moyens les plus simples et les plus avantageux en même temps de travailler à s'en faire un.[130]

Another author who claimed a botanical legacy from Rousseau was François Lebreton, a writer on agriculture and botany, who has sunk into obscurity for want of historical records; we do not know his year of birth, while his death year of 1790 is merely conjectural.[131] Of particular

127. My emphasis; Bulliard, *Dictionnaire élémentaire*, p.155.
128. Referring to Rousseau's botanical pasigraphy, Bulliard's footnote reads: '(1) Je ne sais ce qu'est devenu le projet qu'il avait de fixer la nomenclature des plantes; il serait malheureux que cela fût perdu'. Bulliard, *Dictionnaire élémentaire*, p.99.
129. This individual, later termed 'le C. Tourmevel' by Richard in the revised *Dictionnaire élémentaire*, p.76, has proven thus far impossible to identify. During the Revolution, it would have been safer for Richard to use the ambiguous 'C', which could refer to 'citoyen', rather than the discredited aristocratic title, 'chevalier'.
130. Original emphasis; Bulliard, *Dictionnaire élémentaire*, p.99. This manuscript remains unknown at this writing.
131. According to Pierre Zert, librarian of the Académie d'Agriculture de France (successor to the Société royale d'agriculture, Paris), biographical information on Lebreton is not

interest to our discussion is his *Manuel de botanique* (1787).[132] The title page of the work announces that he belonged to the Royal Academy of Sciences of Uppsala (the Swedish city that had been home to Linnaeus) and was a corresponding member of the Société royale d'agriculture (Paris), which vetted the work for publication.[133] He is also described on the title pages of his works as Inspecteur-General des Remises des Capitaineries Royales.

He wrote two works of interest to our discussion: a botany manual, *Manuel de botanique*, which discusses the logistics of acclimatisation, and a treatise on sugar, *Traité sur les propriétés et les effets du sucre*.[134] In his treatise on sugar he presented the most up-to-date information on the cultivation of sugar cane (a key cash crop for the French colonies and those who lived from colonial trade), but deplored the slave labour practices used in producing it, echoing Rousseau's condemnation of slavery in *Du contrat social*.[135] The testimonial appended to the work by the Société royale d'agriculture singled out Lebreton's anti-slavery stand for praise: 'il prouve d'après les recherches de M. Young, que le travail des Nègres est excessivement désavantageux'.[136]

In the *Manuel de botanique* Lebreton provided an explication of herbarium making à la Rousseau, which is nothing more than Rousseau's purported text on herbaria lifted almost whole, and without attribution, from Bulliard's *Dictionnaire*. Lebreton suggested in addition that he received information about making herbaria *directly* from Rousseau: 'L'instruction sur les soins que ce travail exige, nous a paru mériter ici une place: nous l'avons reçue de J.-J. Rousseau, dont le nom seul devient une présomption favorable pour toutes les productions qui ont un objet d'utilité, et *avec qui* nous avons eu plusieurs *conférences* à ce sujet'.[137] When

available due to the loss of records during the Revolution. Personal communication, 20 June 2006.

132. Lebreton, *Manuel de botanique*.

133. Inquiries made to the archivists of these two bodies have yielded no further information about Lebreton. Pers. comm. Pierre Zert, Société d'agriculture, Paris, 20 June 2006.

134. François Lebreton, *Traité sur les propriétés et les effets du sucre* (Paris, Royez, 1789)

135. Lebreton, *Traité sur les propriétés et les effets du sucre*; Rousseau, *Du contrat social*, *OC*, iii.355-58.

136. Lebreton, *Traité sur les propriétés et les effets du sucre*, p.188. 'M. Young' is presumably the English agricultural writer, Arthur Young (1741-1820), who is most well-known for his *Travels in France*, 2 vols (London, W. Richardson, 1792). It is not clear, however, to which of Young's 'recherches' the Academy is referring. The testimonial is dated 8 May 1789; the Tennis Court Oath, marking the beginning of the French Revolution, took place on 20 June 1789.

137. My emphasis. The same text, accompanied by further material at the beginning, was first presented under 'Herbiers' in *Dictionnaire élémentaire de botanique* of Bulliard, revised by L. C. M. Richard. It is quoted in *Œuvres complètes de J. J. Rousseau*, vol. 6 (Paris, 1891), no.4 p.58-60 and possibly in earlier editions of *Dictionnaire élémentaire de botanique* of Bulliard,

and where they are supposed to have met is unknown; there is also no recorded correspondence between them. No contact between Rousseau and Lebreton has been established.[138]

The work also provided also a catalogue of exotic plants, and guidelines for collecting trees and seeds of exotic species in foreign climes and transporting them to France. Lebreton takes travelling botanists as his audience, especially those venturing to, or already resident in, the East or West Indies. Lebreton, it must be noted, was not the first to do this – the requirements of travelling natural history collectors had already been addressed in works by Duhamel du Monceau, Etienne François Turgot (1721-1789), elder brother of the royal minister and by a French translation of a work by an English physician, J. C. Lettsom, *Le Voyageur-naturaliste.*[139] For this audience, methods of collecting, preserving and transporting specimens are critical to the enterprise of exotic botany. Lebreton includes a list of exotic species from both Asia and the Americas that provide edible fruits or other useful products. Their names are recorded in French transliterations of their indigenous names.

As I have argued elsewhere, Rousseau was no friend of the exotic botany promoted by Lebreton and many other botanists of the era.[140] The exotic botany agenda contradicted Rousseau's most deeply held convictions concerning the sanctity of the order of nature, which human beings should not disturb by transplantation; according to the principles outlined in the opening lines of *Emile*, we should eat local, seasonal produce and study the plants of our own locale.

Furthermore, Rousseau's priorities for his herbaria were primarily aesthetic and didactic, while Lebreton's were eminently practical, intended for the travelling collector. Yet Lebreton nonetheless appealed to Rousseau as an authority on herbaria. Rousseau's mastery of technique serves as an example to anyone who wants to preserve plant specimens, irrespective of their motives for doing so.

Following Bulliard, Lebreton reproduced a text on herbaria, allegedly

---

edited by Richard. Bulliard asserted he received the text from a M. le chevalier de Tourmevel, of whom nothing is known. Richard later claimed his source for the text was M. le C. Tourmevel, discussed above (note 129).

138. Lebreton, *Manuel de botanique*, p.xv.

139. John Coakley Lettsom, *Le Voyageur-naturaliste, ou Instructions sur les moyens de ramasser les objets d'histoire naturelle et de les bien conserver* (Amsterdam and Paris, Lacombe, 1775). Yves Laissus, 'Les cabinets d'histoire naturelle', in *Enseignement et diffusion des sciences en France au XVIIIe siecle* ed. R. Taton, (Paris, 1964), p.659-70 (667). Lettsom, (1744-1815), a well-known English physician, authored a treatise on tea in which he criticised its large-scale consumption as potentially injurious to health. Lettsom, *The Natural history of the tea tree.*

140. Cook, 'Jean-Jacques Rousseau and exotic botany'.

from Rousseau,[141] that Bulliard says he received from a 'M. le Chevalier de Tourmevel'.[142] Lebreton does not mention this intermediary, leaving the reader to conclude that Lebreton received it himself from the hand of the philosopher.

> Nous ne parlerons point des avantages qu'il y a de recueillir les plantes sèches, et de former ce qu'on appelle un herbier, personne ne les conteste et ne les ignore; mais ce qui est peut-être moins connu, c'est la meilleure manière de le faire. L'instruction sur les soins que ce travail exige, nous a paru mériter ici une place: nous l'avons reçue de J.-J. Rousseau, dont le *nom seul* devient une présomption favorable pour toutes les productions qui ont un objet d'utilité, et avec qui nous avons eu plusieurs conférences à ce sujet.[143]

The text allegedly from Rousseau is predictable in all respects except one: its advocacy of chemical treatment against insects. 'Pour garantir votre herbier des ravages qu'y feraient les insectes, il faut tremper le papier sur lequel vous voulez fixer vos plantes, dans une sorte de dissolution d'alun.'[144] This sentence renders the authenticity of the text suspect because Rousseau never mentioned such chemical treatments in his writings on herbaria, including the comprehensive letter on herbaria to Mme Delessert. Nor did he mention it in any other attested text either. Yet, Rousseau's herbaria are in excellent condition, which suggests that they could have been treated.[145]

Rousseau's mastery of the art of plant preservation seems to have been well-known among aficionados. Lebreton claimed that Rousseau's were the best-preserved herbaria he had ever seen: 'J'ai vu un de ces herbiers, il y a neuf ou dix ans [c.1777], préparé de [cette] sorte, qui m'a paru être le mieux conservé de tous ce que j'ai vus; je crois que ce même herbier appartient aujourd'hui à M. de Malesherbes'.[146] We are unable to identify this herbarium with certainty, although Bulliard's description suggests it could be the miniature moss and lichen herbarium now at the Musée des arts décoratifs, Paris.[147]

141. Bulliard, *Dictionnaire élémentaire*, p.99. This purported MS from Rousseau on herbaria is unknown. Was the MS in Bulliard's possession when he died in 1793; did his widow pass it on to Richard? Or did Bulliard's publisher acquire it with Bulliard's other papers, as Gilbert speculates? See Gilbert, 'Un esprit – une œuvre', p.104. Bulliard's papers have likewise vanished. Dénis Lamy, Muséum national d'histoire naturelle (Paris), personal communication, 12 April 2006.
142. Bulliard, *Dictionnaire élémentaire*, p.99.
143. My emphasis; Lebreton, *Manuel de botanique*, p.xv.
144. Bulliard, *Dictionnaire élémentaire de la botanique*, p.100; Lebreton, *Manuel de botanique*, p.165.
145. The solution this text recommends is certainly less toxic than the mercuric chloride treatment prescribed by Tournefort, and widely used into the twentieth century.
146. Lebreton, *Manuel de botanique*, p.165-66.
147. See Appendix 3. Bulliard was aware of these herbaria, and praised their quality, but did not characterise their state of preservation: 'Parmi les livres rares et précieux qui

It is strange that Bulliard, Lebreton and Richard, all of whom wrote after the appearance of the posthumous botanical writings of Rousseau in [1780-]1782, professed no awareness of these texts; their praise for Rousseau's contributions to botany relied entirely on his herbaria. This is a significant and neglected moment in Rousseau's botanical legacy, in which the objects and tools of botany – the dried plants composed in a kind of database, the herbarium – occupied centre stage, while the famous written texts remained, exceptionally, at the margins.

In contrast with French authors such as Bulliard, Lebreton and Richard, English writers such as Priscilla Wakefield and Maria Jacson seem to have had little or no awareness of, or interest in, Rousseau's herbaria; on this subject, English readers would probably only have known the material contained in the eighth letter on botany. One exception to this blind spot in the English reception of Rousseau's botany was James Edward Smith, who purchased Linnaeus's collections in 1784, and founded the Linnean Society of London in 1788.[148] In *A Sketch of a tour on the continent*, he recounted visiting Mme Delessert, who showed him Rousseau's 'dried plants neatly pasted on small writing paper, and accompanied with their Linnaean names and other particulars'. 'Botany', Smith wrote, 'seems to have been his [Rousseau's] most favourite amusement in the latter part of life; and his feelings, with respect to this pursuit, are expressed with that energy and grace so peculiarly his own, in his letter to Linnaeus, published in the Journal de Paris, the original of which I preserve as an inestimable relic'.[149]

Exceptionally, Smith was perhaps the only English-speaking botanist of his time to refer to Rousseau's herbaria. The next reference to this aspect of Rousseau's botanical work in English had to wait until 1817 when the purported text on herbaria by Rousseau that was relayed by Bulliard, Lebreton and Richard was published by Amos Eaton in the United States in a translation of Bulliard's *Dictionnaire*, edited by Louis Claude Marie Richard. A central element of Rousseau's legacy as botanist

composent la bibliothèque du savant M. de Malesherbes, on trouve deux petits herbiers de Jean-Jacques, faits avec tout le soin et tout l'art possible; l'en est de *format in*-8°, et ne renferme que des cryptogames; et l'autre de *format in*-4°, est composé de plantes à fleurs distinctes'. Bulliard, *Dictionnaire élémentaire*, p.99.

148. Paul White, 'The purchase of knowledge: James Edward Smith and the Linnean collections', *Endeavour* 23:3 (1999), p.126-29.

149. James Edward Smith, *A Sketch of a tour on the continent*, 2nd edn, 3 vols, (London, 1807), vol.1, p.116. Smith seems to have been something of an admirer of Rousseau: 'the best and most religious people of my acquaintance are among his [Rousseau's] warmest admirers'; disregarding Rousseau's notorious personality, Smith writes: 'Truth is truth, wherever it comes from. No imperfections of humanity can discredit a noble cause.' Smith, *A Sketch of a tour*, p.117.

that was hardly known within the English-speaking world was preserved for a time, at least, in the French- and German-speaking one.

vii. The German reception

Rousseau enjoyed great popularity among German *Aufklärer* in the late eighteenth century; in the German-speaking regions his writings were readily available in the original and soon after their initial publication in translation; for example, *Du contrat social* was published in German in 1765. This was true in Swiss towns such as Zürich, where his works garnered considerable contemporary interest, despite being censored or forbidden to be sold.[150] He found admirers even among Swiss pastors who questioned his religious ideas.[151]

In this already receptive environment Rousseau's botany was welcomed by the *Göttingische Anzeigen von Gelehrten Sachen* in November 1781. Remarkably, in the same year, the letters to Mme Delessert appeared in translation as *J. J. Rousseaus Botanik für Frauenzimmer, in Briefen an die Frau von L***.[152] In late eighteenth-century Germany botanising was 'eine gesellschaftsfähige Beschäftigung'; for Alexander von Humboldt it was 'zur subjektven "Bildung des Verstandes" nicht fremd'.[153]

Following in Rousseau's footsteps, August Johann Georg Carl Batsch (1761-1802), professor of medicine and natural history at the University of Jena, wrote *Botanik fur Frauenzimmer und Pflanzenliebhaber, welche keine Gelehrten sind.*[154] Batsch, a collaborator of Goethe, concurred with the latter's interest in the natural method and may have organised the Jena botanical garden along these lines in the 1790s.[155] In an early work, his opinion of Rousseau's botanical competence was reserved:

> Ich weiss es, dass Rousseau, dieser seltne, und in vieler Rücksicht grosse Mann, die Pflanzenkenntnis eifrig liebte, und selbst eine Botanik für Frauenzimmer schrieb. Sie ist seiner würdig. Aber ein Mann wie er, dessen überwiegende Stärke auf einer andern Seite zu suchen war, konnte nicht so mit den Formen der Pflanzen und ihren Verhältnissen vertraut seyn, als ein

150. Speerli, 'Rousseau und Zürich', p.15ff.

151. For example Leonhard Usteri.

152. Rousseau, *J. J. Rousseaus Botanik für Frauenzimmer* (Mannheim, 1781).

153. Bettina Hey'l, *Das Ganze der Natur und die Differenzierung des Wissens: Alexander von Humboldt als Schriftsteller* (Berlin, 2007), p.65.

154. August Johann Georg Carl Batsch, *Botanik fur Frauenzimmer und Pflanzenliebhaber, welche keine Gelehrten sind* (Weimar, Verlag des Industrie Comptoirs, 1795); on works in natural history directed at female audiences, see Schiebinger, 'The philosopher's beard', p.287, n.5.

155. Nicolas Robin and Frank H. Hellwig, 'Plant systematics at Jena during the early nineteenth century: Fr. S. Voigt's treatment of the "méthode naturelle"', *Annals of the history and philosophy of biology* 10 (2005), p.117-42 (122-25).

andrer, der einen grossen Theil seines Lebens bei ihrer Untersuchung zubrachte.[156]

Batsch dedicated this work to women, as the sex responsible for the raising of children and for society; he claimed that a natural sympathy draws them to the observation and care for flowers, and that their pure souls delight in plants, which are innocent beings.[157]

Later Batsch identified himself as a follower of Rousseau, indeed as standing on Rousseau's shoulders:

> Ein Mann, der weit über ihn [Batsch] erhaben ist, der *unvergessliche Bürger von Genf*, war sein Vorgänger; diese entfernte Beziehung gereicht ihm zu einer Ehre, die er nur dem Zufall verdankt; nur darum, weil er später schrieb, als jener liebenswürdige und grosse Mensch [Rousseau], wurde es ihm möglich, mehr zu erzählen. Er [Batsch] nimmt es von dem Schicksal dankbar an.[158]

Rousseau's contributions to botany made perhaps their significant impression in the German world on Johann Wolfgang von Goethe (1749-1832), poet and statesman. Rousseau was also important for Adelbert von Chamisso (1781-1838), the Franco-German novelist and botanist. Rousseau's views on education, and its relation to nature, influenced German pedagogy and a significant work on Rousseau's botany was published in German in Berlin in 1885.[159] One of his herbaria was kept at the Berlin botanical garden until its destruction in WWII (see Appendix 3 and the discussion below).

Goethe read Rousseau's posthumous works soon after their appearance in [1780-]1782, as he was starting his career as Privy Councillor in the Duchy of Saxe-Weimar; forestry – a matter of vital interest to eighteenth-century governments – was one of his portfolios. His official duties set him on the path of nature study, and he credited Rousseau with playing a pivotal role in making that path accessible and attractive. In the history of his botanical studies Goethe spent several pages discussing Rousseau's role as a critical precursor, one who 'versenkt sich in dieses

156. 'I know that Rousseau, that rare, and in many respects great man, keenly loved learning about plants, and himself wrote a botanical work for women. It is worthy of him. But a man like him, whose greatness was overwhelmingly to be found on another side, could not be as knowledgeable about the forms of plants and their relationships, as another who spent a great part of his life investigating them.' August Johann Georg Carl Batsch, *Botanische Unterhaltungen für Naturfreunde zu eigner Belehrung über die Verhältnisse der Pflanzenbilding entworfen*, part 1 (Jena, Cröcker, 1793), p.xii-xiii.

157. Batsch, *Botanische Unterhaltungen*, p.xiii.

158. 'A man, who is greatly superior to him, the unforgettable citizen of Geneva, was his predecessor; this distant relation confers an honour on him, which he owes to chance; only because he wrote later than that charming and great man, was it possible for him to say more. He accepts this thankfully from destiny'. Batsch, *Botanik für Frauenzimmer*, p.174.

159. Jansen, *Jean-Jacques Rousseau als Botaniker*.

Reich [der Pflanzen], nimmt es ernstlich in sich auf, fühlt, dass ein gewisser methodischer Gang durch das Ganze möglich sei, getraut sich aber nicht damit hervorzutreten'.[160] As we have seen, Goethe attributed his awareness of the natural method to Rousseau's influence:[161] 'Seine Methode das Pflanzenreich ins Engere zu bringen, neigt sich [...] offenbar zur Einteilung nach *Familien*; und da ich in jener Zeit auch schon zu Betrachtungen dieser Art hingeleitet war, so machte sein Vortrag auf mich einen desto größeren Eindruck'.[162]

Goethe appreciated Rousseau's commitment to studying indigenous plants, even though it was at variance not only with the late eighteenth-century craze for exotic botany, but also with his own enthusiasm for the Italian flora.[163] Regarding the letters on botany, Goethe writes: 'Bei deren Überblick bemerkt man mit Vergnügen wie einheimisch ländlich er bei seinen Studien verfahren, indem nur Pflanzen vorgestellt sind, welche er auf seinen Spaziergängen unmittelbar konnte gewahr werden'.[164]

Goethe averred that Rousseau's botanical pedagogy complemented Goethe's stage of development in the 1780's: 'Und so wie die jungen Studierenden sich auch am liebsten an junge Lehrer halten, so mag der Dilettant gern vom Dilettanten lernen. Dieses wäre freilich [...] bedenklich, wenn nicht die Erfahrung gäbe, daß Dilettanten zum Vorteil der Wissenschaft vieles beigetragen [haben]'.[165] Goethe's assessment of

160. 'immerses himself in the [plant] realm, takes it seriously, feels, that a certain methodical way through the whole is possible, but does not trust himself to move forward with it', Johann Wolfgang von Goethe, 'Der Verfasser teilt die Geschichte seiner botanischen Studien mit', p.157.

161. From Linnaeus, on the other hand, Goethe claimed to have learned 'unendlich viel [...] nur nicht Botanik [an incredible amount [...] only not botany]'. Goethe to Zelter, 7 November 1816; Goethe, *Briefe*, ed. Bodo Morawe (Hamburg, 1965), vol.3, p.376.

162. My emphasis; 'His method of narrowing down the plant world lends itself to the classification of plants according to *families* [...] and since I too at that time had been led to conclusions of this kind, I was all the more impressed by his presentation'. Goethe, 'Der Verfasser teilt die Geschichte seiner botanischen Studien mit', p.159.

163. Goethe developed his concept of the *Urpflanze* while visiting the Botanical Garden of Padua, which contained many species he had not previously seen: 'Es ist erfreuend und belehrend, unter einer Vegetation umherzugehen, die uns fremd ist. Bei gewohnten Pflanzen sowie bei andern längst bekannten Gegenständen, denken wir gar nichts [...] Hier in dieser neu mir entgegentretenden Mannigfaltigkeit wird jener Gedanke immer lebendiger, daß man sich alle Pflanzengestalten vielleicht aus einer entwickeln könne'. Johann Wolfgang von Goethe, 'Italienische Reise', Erich Trunz and Herbert von Einem (eds) in *Goethes Werke*, Erich Trunz, *et al.* (eds) (Hamburg, 1950), vol.11, p.60.

164. 'One notes with pleasure how he proceeds in his studies in an indigenous, rural way, in which he introduces only those plants, that he could observe on his walks.' Goethe, 'Der Verfasser teilt die Geschichte seiner botanischen Studien mit', p.159.

165. 'And just as young students prefer young teachers, so also does the dilettante like to learn from the dilettante. This would of course be questionable [...] if experience did not show

Rousseau as a botanical dilettante might have gone too far, but he recommended Rousseau's botanical letters to Duke Karl August of Saxe-Weimar: 'In Rousseaus Wercken finden sich ganz allerliebste Briefe über die Botanick, worinn er diese Wissenschafft auf dass fasslichste und zierlichste [Weise] einer Dame vorträgt. Es ist recht ein Muster wie man unterrichten soll und eine Beylage zum Emil'.[166]

Goethe was also aware of Rousseau's herbaria. Rousseau showed 'eine sehr anmutige Sorgfalt für das Trocknen der Pflanzen und Anlegen von Herbarien beweist, und den Verlust desselben innigst bedauert, wenn irgendeins zugrunde geht'.[167] Yet Goethe concluded (erroneously) that Rousseau 'weder Geschick noch anhaltende Sorgsamkeit haben mochte, um besonders bei seinen vielfachen Wanderungen auf Erhaltung [der Pflanzen] zu achten'.[168]

A relatively neglected figure in the transmission of Rousseau's botany in Germany is Adelbert von Chamisso (1781-1838), a self-made naturalist and writer who, like Rousseau, had little formal education; his parents had fled France during the Revolution, settling in Berlin, where there had been a long-standing Huguenot French community dating from the revocation of the Edict of Nantes. Chamisso served as an infantry officer in the Prussian army, but left that occupation in 1808, joining the circle of Mme de Staël at Coppet, near Geneva, Switzerland, in 1810. Mme de Staël was sensitive to Rousseau's interest in botany,[169] and her son, Auguste de Staël (1790-1827), 'avait un peu le goût de la botanique', which he shared with Chamisso.[170] Under the influence of Mme de Staël's circle, Chamisso pursued botanical research for two years. In 1815 he joined a round-the-world voyage as botanist on the Russian ship *Rurik*, and published the fascinating account of his travels in his *Tagebuch* (1821). He is probably most well-known for *Peter Schlemihls wundersame Geschichte* (1813), the story of a man who, in a version of the classic tale of

that dilettantes contribute a great deal to science.' Goethe, 'Der Verfasser teilt die Geschichte seiner botanischen Studien mit', p.159.

166. 'In Rousseau's works there are the most delightful letters concerning botany, in which he explains this science to a lady in the most understandable and elegant way. It is truly a model of how to teach and a supplement to Emile'. Goethe to Duke Karl August of Saxe-Weimar, 17 June 1782; Johann Wolfgang von Goethe, *Briefe*, vol.1, p.398.

167. 'very delightful care for the drying of plants and the making of herbaria, and deeply regretted their loss, whenever any were ruined', Goethe, 'Der Verfasser teilt die Geschichte seiner botanischen Studien mit', p.159.

168. 'neither the skill nor the perseverance to give much attention to the preservation of plants on his many excursions'. Goethe, 'Der Verfasser teilt die Geschichte seiner botanischen Studien mit', p.159.

169. Anne Louise Germaine Necker de Staël-Holstein, *Œuvres complètes de Madame la baronne de Staël-Holstein* (Paris 1861), vol.1, p.18.

170. Candolle, *Mémoires et souvenirs*, p.479.

the devil's bargain, sold his shadow in return for endless wealth. In Schlemihl's case, his pact with the devil causes him to find social relations unbearable, and he turns (much as Rousseau did) to botany; he buys a pair of boots that allows him to cover seven leagues at a time. Traversing the old and new worlds in his wonderful boots, he develops plant geography and draws up a *Historia stirpium* for the entire earth, with the exception of Australia. This exercise 'lui permet de retrouver ses semblables par la médiation de la géographie botanique'.[171]

Upon his return to Germany in 1818 Chamisso was appointed curator of the Berlin botanical garden; his descriptions of Mexican trees, plants of the San Francisco Bay area, and north Germany in *Übersicht der nutzbarsten und schädlichsten Gewächse in Norddeutschland* (1829) are esteemed for their careful treatment of their subjects. The genus *Chamissoa* Kunth (Amaranthaceae) and many species were named in his honour. Plants he named are denoted by the abbreviation, 'Cham.' It was possibly Chamisso who brought Rousseau's small ('Berlin') herbarium to Berlin – perhaps as a gift or purchase from Mme Delessert's son, Benjamin Delessert;[172] unfortunately it perished during the bombardment of Berlin in World War II, and the only record of it comes from Jansen's *Jean-Jacques Rousseau als Botaniker* (Berlin, 1885), which provides a catalogue of its contents.[173]

Alexander von Humboldt (1769-1859), son of the Prussian king's chamberlain, grew up in comfortable circumstances, enticed to exotic flora from a young age by the dragon tree of the Berlin botanical garden and his father's collection of exotic trees at his boyhood home of Schloss Tegel. He explored South America on a five-year journey (1799-1804), penetrating many areas of the continent that were practically unknown to any Europeans, even the Spanish or Portuguese. The theory of plant geography that Humboldt formulated on the basis of his discoveries forged a whole new way of looking at plants, even though he himself claimed a long ancestry for this idea.[174]

Rousseau's influence may have acted on various aspects of Humboldt's life and work: (1) his desire, expressed circa 1795, to lead a peaceful life in

171. Drouin, *L'Herbier*, p.243.

172. Ruth Schneebeli-Graf, *Botanisieren mit Jean-Jacques Rousseau*, p.138. Unfortunately, Schneebeli-Graf does not explain the 'Recherchen' that led her to conclude that Chamisso may have acquired the herbarium from Benjamin Delessert. The herbarium is classified as 'Kriegsverlust'. Personal communication, H. W. Lack, Director, Berlin Botanical Garden.

173. Jansen, *Jean-Jacques Rousseau als Botaniker*, p.277-92.

174. Alexander von Humboldt, *Sur les lois que l'on observe dans la distribution des formes végétales*, excerpt from the Mémoire read before the Institut de France, 29 January 1816 (Paris, 1816), p.1-2.

a naturally idyllic setting such as that of Lake Lucerne;[175] (2) his emotional response to the beauty and magnificence of nature and (3) his advocacy of studying plants in their native habitat.

Humboldt considered Rousseau to have evoked the beauty of nature for his readers in a peculiarly effective way, although he regarded Rousseau merely as an inspiring author, who loved botany, but who was 'peu instruit':

> Si je rappelle ici l'éloquence entraînante de Rousseau, les descriptions pittoresques de Clarens et de la Meillerie sur les bords du lac de Genève, c'est que dans les principaux écrits de cet herborisateur peu instruit mais zélé, écrits qui ont devancé de vingt années les *Epoques de la nature* de Buffon, l'enthousiasme déborde aussi bien que dans les immortelles poésies de Klopstock, de Schiller, de Goethe, de Byron, et se manifeste surtout par la précision et l'originalité du langage. Un écrivain peut, sans avoir eu en vue les résultats directs de la science, inspirer un goût vif pour l'étude de la nature, par l'attrait de ses descriptions poétiques, alors même qu'elles portent sur des lieux très circonscrits et bien connus.[176]

Yet common concerns united Humboldt and Rousseau. Humboldt shared Rousseau's deep appreciation of magnificent, abundant nature. As Rousseau wrote: 'Je sens des *extases*, des *raptures* inexprimables à me fondre pour ainsi dire dans le système des êtres, à m'identifier avec la nature entière'.[177] He rejected the useful side of botany, especially for medicine; botany is beautiful rather than useful, or rather only useful to the extent that it is beautiful. In a similar vein, Humboldt wrote, 'en parlant des formes végétales, je songe à *l'émotion* que leur aspect peut produire, nullement au *secours* que l'on en peut tirer pour l'étude de la botanique'.[178] During his voyage in South America Humboldt commented:

> Plusieurs Européens ont exagéré l'influence des climats sur l'esprit, et affirment qu'il est impossible de s'y livrer à des travaux intellectuels, mais nous devons publier le contraire et dire d'après notre propre expérience que jamais nous n'avons joui d'autant de forces qu'en contemplant les beautés et la magnificence qu'offre ici la nature. Sa *grandeur*, ses *productions*

175. Charles Minguet, *Alexandre de Humboldt: historien et géographe de l'Amérique espagnole* (Paris, 1969), p.96.
176. Alexander von Humboldt, *Cosmos: essai d'une description physique du monde*, ed. J. Grange (Paris, 2000), vol.1, p.399. There is no evidence that Humboldt had seen any of Rousseau's herbaria; these were all in private hands at the time, and in contrast with Bulliard and Lebreton, Humboldt did not indicate that he was aware of their existence. He could have seen the herbier for Madelon Delessert in Benjamin Delessert's Paris museum, although I have no concrete evidence for this speculation.
177. My emphasis; Rousseau, *Rêveries*, *OC*, i.1065-66.
178. My emphasis; Humboldt, *Cosmos*, vol.1, p.423-24.

infinies et nouvelles nous *électrisaient*, elles nous *transportaient de joie* et nous rendait pour ainsi dire invulnérables.[179]

More specifically, three features of Rousseau's botanical thought foreshadow Humboldt's geographical perspective on plants: (1) the observation (admittedly not unique to Rousseau) that climate and habitat in general play a crucial role in whether plants flourish or languish; (2) that transplantation leads to changes, or even degeneration, a phenomenon already well-established in the botanical literature of Rousseau's day; and (3) the conclusion that the best way to study nature is in nature itself unlike Linnaeus, who 'l'a trop étudiée dans des herbiers et dans des jardins et pas assez dans la nature elle-même'.[180]

Humboldt, perhaps more than any naturalist of his era, acted on Rousseau's criticism of Linnaeus, although his orientation was toward the exotic rather than the local. Inspired from a young age by tropical vegetation (the Berlin botanical garden's dragon tree and his father's collection of exotic trees), the paintings of the Ganges by William Hodges (who accompanied Cook's second world voyage and traveled in India and Russia), and Bernardin de St Pierre's *Paul et Virginie*, he set out to explore nature as no European before him had ever done.[181]

Humboldt and Rousseau both argued that plants should be studied in their native habitats where they are free, and therefore vigorous: 'Les idées de vigueur et de liberté sont inséparables aussi dans les productions de la nature'.[182] A plant removed from its native habitat has been changed or diminished by transplantation to a foreign clime. Moreover, as beneficiaries of Darwin's investigations, we know that organisms adapt themselves to their habitats through generations of natural selection. Even before Darwin, however, Rousseau and Humboldt were aware of the intimate connection between geography (understood as climate and habitat) and plant life.

Their reflections on the European treatment of exotic flora likewise show affinities. Humboldt shared Rousseau's dislike of transplantation;[183] he agreed with Rousseau that hothouses only show us degenerate plants, exiled and denatured from their native climes.[184] Hothouses, according to Humboldt, are 'hôpitaux à l'usage des plantes languissantes', offering

179. My emphasis; Humboldt to Don Antonio Joseph Cavanilles, 22 April 1803; Alexander von Humboldt, *Correspondance scientifique et littéraire d'Alexandre de Humboldt*, ed. J.-B.-M.-A. Dezos de La Roquette, (Paris, 1869), vol.1, p.162. Cavanilles (1745-1804) was a Spanish ecclesiastic and botanist.

180. Rousseau, *Confessions*, *OC*, i.645.

181. Humboldt, *Cosmos*, vol.1, p.346

182. Humboldt, *Cosmos*, vol.1, p.425.

183. Cook, 'Jean-Jacques Rousseau and exotic botany'.

184. Rousseau, *Emile*, *OC*, iv.480.

'un reflet affaibli' of 'ce spectacle, auquel nous ne pouvons songer sans soupirer après les contrées ou la force de vie coule avec plus d'abondance'.[185] Herbaria specimens, in Humboldt's view, are preferable to such weakened transplants: 'aux yeux du botaniste zélé qui a parcouru le monde, des plantes cueillies sur les Cordillères ou dans les plaines de l'Inde et séchées dans un herbier ont souvent plus de prix que les mêmes espèces vivantes qui ont grandi dans une de nos serres d'Europe'.[186] Rousseau might well have agreed.

viii. The Delessert connection

One of the most important legacies of Rousseau's botany was its impact on Madeleine-Catherine Delessert's son, Jules Paul Benjamin Delessert (1773-1847). Delessert, a wealthy banker, industrialist, philanthropist and avid botanist, was the younger brother of Madelon Delessert, for whom Rousseau wrote the eight letters on botany and made the herbarium now at the Musée Jean-Jacques Rousseau (Montmorency).[187] Benjamin Delessert's elder brother, Etienne (1769-1794), a member of the natural history society of Edinburgh, began a herbarium of plants from his voyages in France, Holland, Switzerland, Germany, Denmark, the United States, England and Scotland. It should be noted that neither son was in a position to be instructed by Rousseau's letters as Madelon was; Benjamin was only born in 1773, near the end of the series, while his older brother, born in 1769, was only two years old when it began. However, Rousseau made it clear that he wrote not just for Madelon or exclusively for female children, but for all of Mme Delessert's children – 'vos enfants'.[188]

The older son inspired his younger brother, Benjamin, to botanise. Benjamin accompanied his elder brother on collecting expeditions, and at the latter's early death inherited his plant collections, which formed the basis for the famous 'Musée botanique' in Paris that made Benjamin Delessert one of the great nineteenth-century patrons of science.[189] Through many important acquisitions, Delessert's 'Musée botanique' became an obligatory destination for every botanist, and was open to any legitimate researcher.[190]

185. Humboldt, *Cosmos*, vol.1, p.424.
186. Humboldt, *Cosmos*, vol. i, p.425.
187. The herbarium Rousseau made for Madelon Delessert was kept separately from the rest of the herbarium, 'ce qui tendait à minorer la fonction qu'il joua dans la constitution réelle de la collection'. This herbarium was acquired by the Musée Jean-Jacques Rousseau (Montmorency, France) in 2001. Hoquet, 'La bibliothèque botanique de Benjamin Delessert', p.113, n.1.
188. Rousseau to Mme Delessert, 30 August 1773; *CC*, xxxix.190.
189. Stafleu and Cowan, *Taxonomic literature*, under 'Delessert, Jules Paul Benjamin'.
190. The museum was open from noon to 4.00 p.m. every day except Monday (and presumably Sunday).

Rousseau's influence on Benjamin Delessert is attested by Delessert's curator, Antoine Lasègue (1793-1873), in his comprehensive account of Delessert's collections:

> M. Benjamin Delessert avait puisé le goût qui le portait vers l'étude des plantes dans les lettres de Jean-Jacques Rousseau sur la botanique, dans ces lettres charmantes ou l'aridité de la science disparait sous les agréments du style, et qu'on croirait, tant l'auteur a su se renfermer dans les choses les plus fondamentales, écrites d'hier, quoique 70 années au moins nous séparent de l'époque où elles ont été rédigées. Un motif touchant ramenait sans cesse M. Delessert vers leur lecture.[191]

Rousseau's influence on Delessert was even more far-reaching, since Delessert constituted his immense herbarium following the natural method, which had likewise formed the basis for Rousseau's famous eight letters to Mme Delessert.

Hoquet argues that Delessert's choice of natural family classification may have also been reinforced by his close friendship with Augustin-Pyramus de Candolle (1778-1841), a major figure in the development of modern taxonomy, who appreciated Jussieu's influence on Rousseau.[192] 'L'attachement rousseauiste aux formes spontanées, intactes encore de la main de l'homme, apparait aussi dans la recherche d'une classification naturelle, qui anime les travaux d'Augustin-Pyramus de Candolle, à la suite d'Antoine-Laurent de Jussieu. Or, Benjamin épouse totalement l'entreprise candollienne et consacre une partie de sa fortune au triomphe de la méthode naturelle de son ami.'[193]

Delessert's collection contained 250,000 specimens, representing 86,000 species from the approximately 95,000 known at the time. Delessert acquired some of the most important herbaria in existence, such as those of Louis-Guillaume Le Monnier (1717-1799),[194] Nicolaas Laurens Burman (1734-1793) and his father, Johannes Burman (1707-1779); the Burman herbarium alone contained 29,000 specimens, including rich finds from Ceylon, the Cape of Good Hope and India; it served as the basis for Linnaeus's *Flora zeylanica*.[195] With the Burman

191. Lasègue, *Musée botanique de Benjamin Delessert*, p.43-44.
192. Candolle, *Histoire de la botanique genevoise*, p.20.
193. Hoquet, 'La bibliothèque botanique de Benjamin Delessert', p.115.
194. Le Monnier was personal physician to Louis XV and professor of botany at the Jardin du roi. His role in promoting botanical exploration enabled him (somewhat unethically) to acquire specimens from a number of important collectors funded by the royal purse. Since these expeditions had been supported by the royal household, they should by right have become part of the Cabinet du roi.
195. Linnaeus's *Flora zeylanica* (Stockholm, Laurentius Salvius, 1747) was based on Paul Hermann's Ceylonese specimens contained in the Burman collections; the Burmans also forwarded information to Linnaeus about the *Garcinia mangostana* L. collected by

collections, Delessert likewise acquired the Lapland herbarium of Linnaeus, the basis for Linnaeus's *Flora lapponica*, the first attempt to catalogue the flora of Lapland.[196] Candolle consulted Delessert's collections, as did many other scientists of note.[197]

Candolle's recollections of the Delessert family and of Benjamin in particular provide unparalleled insight into the importance that his friend placed on his twin interests of botany and philanthropy. The two men were acquainted from boyhood; they not only corresponded and visited each other in Paris, but also exchanged botanical specimens; Candolle recounts evenings spent at the home of Benjamin's sister, Marguerite Madeleine (Madelon) Gautier, née Delessert, during which he and Benjamin examined specimens from the latter's herbier, while other family members read poetry: 'Tantôt ils nous interrompaient pour nous faire remarquer quelque idée sublime, tantôt nous les appelions pour leur montrer quelque plante remarquable.'[198] Candolle received gifts of valuable botanical specimens from Delessert that he reciprocated upon obtaining the collections of L'Héritier de Brutelle after the latter's shocking murder.[199]

With Candolle, Delessert was active in philanthropy; he established twenty soup kitchens in two years in all the quarters of Paris. A governor of the Banque de France, he founded the first credit unions (caisses d'épargne). This philanthropy followed Rousseau's credo that no man should be so poor that he must sell himself, and none so rich that he could buy another.[200]

ix. Conclusion

Rousseau's popularisation of botany spread across Europe to Russia and into both Latin and North America. His works were so popular that a

---

Laurent Garcin. Linnaeus was well-acquainted with Johannes Burman, having lodged with him in Holland in 1735. Nicolaas, in turn, studied with Linnaeus in Uppsala in 1760.

196. The Lapland herbarium, the primary material for Linnaeus's *Flora lapponica* (Amsterdam, S. Schouten, 1737) was bound in a book, and therefore formed part of the Delessert library that was bequeathed by François Delessert's heirs to the Institut de France (Paris) in 1868. Hoquet, 'La bibliothèque botanique de Benjamin Delessert', p.110. Because the Delessert heirs disapproved of the conservation practices of the Museum d'histoire naturelle (Paris), they bequeathed the Delessert herbarium to the city of Geneva, which in turn donated it to its present home, the Conservatoire et jardin botaniques of Geneva.

197. Lasègue, *Musée botanique de Benjamin Delessert*, p.44; Candolle, *Mémoires*, p.112, 192.

198. Candolle, *Mémoires*, p.112.

199. Candolle, *Mémoires*, p.191. The practice was to make gifts of 'doubles', that is, duplicate specimens. Charles L'Héritier de Brutelle (1746-1800) collected widely in the Paris area, and had herbaria by other botanists containing collections from Cayenne, the Antilles, Guadeloupe and Sierra Leone. Notoriously, he was murdered.

200. Candolle, *Mémoires*, p.176-78; Rousseau, *Du contrat social*, *OC*, iii.391-92.

variety of editors and authors, especially proponents of the sexual system, used them to retail their own view of botany. Whether Rousseau, who did not expound the sexual system in the *Lettres* to Mme Delessert, would have approved of the result did not matter to these translators and editors; the mere fact of a work having a connection with him gave it caché. Hence Martyn, who wrote twenty-four letters to complement Rousseau's eight, would always play second fiddle to his notorious contemporary. Similarly, Rousseau's executors published as Rousseau's work of dubious provenance, exploiting his fame for various ends, of which financial gain was only one; at the same time, they gave their deceased friend an even greater reputation in death than he had had in life. Bulliard, Lebreton and Richard relayed a text to readers that they claimed Rousseau had written; perhaps they had no way to authenticate the text, and its original remains unknown.

Others, such as Goethe, Mme Roland and Benjamin Delessert, took Rousseau's works as they were; they did not seek to manipulate them to serve their own ends, but rather, to derive inspiration from them. Their testimonials to Rousseau's influence as a botanist repay attention today, when the plant's habit awaits rediscovery by Rousseau's readers.

10. Reading Rousseau on botany at the end of nature[1]

> Mon imagination qui se refuse aux objets de peine laissait mes sens se livrer aux impressions légères mais douces des objets environnants.[2]

For Rousseau botany was first and foremost a contemplative activity of 'une âme expansive' seeking serenity: to study the plant as it grew, to learn the differences among plant habits and hence, natural kinds; this activity focused his mind so that he forgot his presumed tormenters. This is a major teaching of the septième promenade of the *Rêveries*. In the first letter on botany to Mme Delessert Rousseau extended this askesis to a generally applicable principle: nature study calms the passions and nourishes the soul.[3] Being in nature studying plants is *good* for us; it is, to use his word, 'salutaire'.

This brings us to a critical question: what was Rousseau's stance toward the natural world, and what are the ramifications of this stance today? To put this question into modern terms: were Rousseau alive today would he be a 'shallow', mainstream, anthropocentric environmentalist interested only in what nature can do for humanity such as provide foods, medicines and other material as well as intangible benefits?[4] In line with this position, would he have considered environmental preservation compatible with capitalism and the existing political order? Or would he have leaned toward 'deep', biocentrism or ecophilosophy, supporting the view that nature has value in *itself*, independent of its usefulness for humanity?[5] Might he have adopted a third, hybrid position, perhaps? It is, of course, impossible to answer this question with certainty; yet however anachronistic the question may be, it is important to entertain a possible answer if we are to assess critically the value Rousseau's botanical studies may have for readers today.

1. The phrase is drawn from Bill McKibben, *The End of nature* (New York, 1989). Many critics have questioned what they see as an essentialist view of nature that freezes 'nature' in time and place like a museum piece. They point to the social, cultural and political construction of nature in particular times and places. See for example Andrew Biro, *Denaturalizing ecological politics: alienation from nature from Rousseau to the Frankfurt School and beyond* (Toronto, 2005), ch.1.
2. Rousseau, *Rêveries*, *OC*, i.1063.
3. Rousseau to Mme Delessert, 22 August 1771; *CC*, xxxviii.252.
4. See for example Wilson, *The Diversity of life*.
5. See for example Arne Naess, 'The shallow and the deep, long-range ecology movement: a summary', *Inquiry* 16 (1973), p.95-100.

Given Rousseau's view of botany as therapeutic, and hence in a way *useful* to him, our immediate reaction to the questions posed above might be that he would adopt shallow ecology, interested only in nature's use for him. Such a view would be bolstered by his advocacy of agriculture as the backbone of the national character and economy, epitomised in the utopia at Clarens. In *Projet de constitution pour la Corse* he went even further, advocating the maximum possible exploitation of the land; reading his text, one wonders if the approach he espoused would not quickly exhaust the land.

However, this would sell Rousseau's vision short. He did indeed depend on nature to perform a service – to salve his wounded and troubled soul, but nature can only perform this service for him if left to its own devices. This broaches the question of natural value that has exercised many environmental thinkers. Rousseau presages their discourse by asserting that when nature is altered and transformed by human beings it cannot play a salutary role for them; nature's degraded condition is observable in gardens, charged with *amour-propre* and the social passions:

> L'homme a dénaturé beaucoup de choses pour les mieux convertir à son usage, en cela il n'est point à blâmer; mais il n'en est pas moins vrai qu'il les a souvent défigurées et que quand dans les œuvres de ses mains il croit étudier vraiment la nature, il se trompe. Cette erreur a lieu surtout dans la société civile, elle a lieu de même dans les jardins. Ces fleurs doubles qu'on admire dans les parterres, sont des monstres dépourvus de la faculté de produire leur semblable dont la nature a doué tous les êtres organisés.[6]

These plants are themselves deformed and denatured products of the social passions; they can provide no succour to the soul battered by those very passions. To seek succour among such products of human desire is to seek the cure or palliative in the disease itself; this of course was precisely Rousseau's strategy in drawing botany as a salutary science from the bosom of the sciences, also the products of our passions. Unlike botany, however, which can point us towards the benevolent 'art divin' of the Author of nature, the horticultural monstrosity points only toward human art, which is neither divine, nor benevolent.

It seems evident, therefore, that Rousseau's position coheres with those philosophers who argue for (1) the objective existence of nature beyond a human evaluator (a necessary assumption to avoid the solipsistic objection to intrinsic value), and (2) nature's intrinsic value independent of human needs, desires and manipulations.[7] Were he alive

6. Rousseau to Mme Delessert, 22 August 1771; *CC*, xxxix.234.

7. See for example Holmes Rolston III, *Conserving natural value* (New York, 1994), p.159, 186, 192ff.; Robert E. Goodin, *Green political theory* (Cambridge, 1992).

today, it seems likely Rousseau would find ecophilosophy appealing, even if he might take issue with some aspects of its programme, such as population reduction.[8] In Rousseau's view, intrinsic value is the foundation of nature's value as a therapeutic resource.

By contrast, plants considered as mere use values leaves him cold. Nature does of course serve human and animal needs as a 'magasin d'aliments',[9] and needs being met in this way suggested to Rousseau the beneficence of nature's Author, but this cannot compare to 'cette chaîne de rapports et de combinaisons qui accable de ses merveilles l'esprit de l'observateur'.[10] Rousseau made it clear early on in his botanical studies that plants' primary purpose for him was *edification*; the study of nature detached him from himself and brought him closer to the 'Auteur' of nature; in this context, he was not interested in poultices, enemas or foods to feed him or flavour his meals. Plants' uses as medicines ingested or otherwise taken into the physical body held no interest for him, and indeed; he rejected the knowledge possessed by Adam – 'le premier Apothicaire' – who considered plants from a purely instrumental perspective.[11] On these practical points, Rousseau possessed considerable knowledge, evident in his annotations to the *La Botanique* of the Regnaults, but this is the knowledge of the apothecary, not the botanist[12] who seeks an 'usage pour la Sagesse et pour la vertu'.[13] Indeed, he was content with his 'foin pour toute nourriture';[14] in other words, the 'hay' he studied while botanising nourished what most needs nourishing: the soul, not the body.

Plants thus took on a particular therapeutic role for Rousseau as a 'pharmacie', purging negative passions from his soul when he studied them for themselves, independent of human needs or desires: the paradigm of nature study diametrically opposed to his own was that of chemistry, pharmacy or herbalism which distils, grinds and dissolves, rather than studying plants whole or at worst, dissected in order to learn about their structure.

No mortar and pestle is required to transform plants into medicine for the soul. The only processing they require is in the ordering mind of the human observer, who uses, if with reservations, the mediations of language and schemes of classification in order to acquire 'quelque idée du système végétal', and the 'chaîne' that holds it together.[15] Rousseau

8. Naess, 'The shallow and the deep', p.96.
9. Rousseau, *Rêveries OC*, i.1064.
10. Rousseau, *Confessions*, *OC*, i.641.
11. Rousseau, *Rêveries OC*, i.1064; see also 1063, 1065.
12. See BdAN [Munch MS], *passim*.
13. Rousseau to the Duchess of Portland, 3 September 1766; *CC*, xxx.314.
14. Rousseau, *Rêveries OC*, i.1060.
15. Rousseau, *Confessions*, *OC*, i.641.

sensed that contemplating the 'ensemble' of nature entailed more than 'une admiration stupide et monotone', and he was prepared to learn about various systems of classification, not simply look stupidly at plants; he first discovered the wonderfully neat sexual system of Linnaeus that he later found 'vide'. Offered taxonomic choices from the outset of his botanical studies, he sought out other systems such as that of the seventeenth-century English botanist, John Ray, a father of the natural method of classification.

Rousseau's preferred objects of botanical study were indigenous plants that have not been degraded by human action.[16] In approaching the plant realm in and for itself, Rousseau's central idea was that of the habit or 'port': the resemblance among members of a given plant family. While Rousseau seems to have been something of a taxonomic pluralist, he was nonetheless dedicated to the proposition that there are kinds in nature with knowable attributes. The habit inheres in wild plants that are allowed, like Julie's, to flourish according to their kinds. Wild plants serve as a *pharmakon* of the soul without having to do or be anything but *themselves*, unencumbered by human concerns. This kind-ness of organisms, perceivable 'au premier coup d'œil', moves the contemplator of nature who values it for itself, but it does not move Kleinjogg, the thrifty peasant: 'la botanique [...] est une étude de pure curiosité et qui n'a d'autre utilité réelle que celle que peut tirer un être pensant et sensible de l'observation de la nature et des merveilles de l'univers'.[17]

Another question that might vex the modern reader is whether Rousseau demands a wilderness in which to practise botany. Again the question may appear anachronistic, if we assume that because Europe was less densely populated than it is today, it had more wilderness areas. However, it would be mistaken to make this assumption, for lower population density does not necessarily mean land preservation. In Switzerland Rousseau was confronted with human activity in seemingly isolated places, and found it challenging to locate the untrammelled areas and plants for which he longed: 'Il n'y a que la Suisse au monde qui présente ce mélange que la nature sauvage et de l'industrie humaine'.[18] Ostervald's survey from 1764 found that the dispersed population of the Suisse romande had produced wasteland in many areas.[19] Where human penetration of the landscape is so advanced, nature 'fuit les lieux fréquentés; c'est au sommet des

16. Rousseau to the Duchess of Portland, 12 February 1767; *CC*, xxxii.135.
17. Rousseau to Mme Delessert, March/April 1774; *CC*, xxxix.234.
18. Rousseau, *Rêveries*, *OC*, i.1071-72.
19. Ostervald, *Description*, p.49-50.

montagnes, au fond des forêts, dans des Iles désertes qu'elle étale ses charmes les plus touchants'.[20]

Given Rousseau's desire for forests and wild plants to study, he has been accused of contributing to the 'trouble with wilderness', the idea that wilderness is the 'locus of pure, hence truly valuable nature'.[21] In response to this criticism of Rousseau as wilderness advocate 'avant la lettre', I suggest that while he undoubtedly prized untrammelled places and organisms, Rousseau seems to have been content with relatively modest venues for his plant studies: what Edward O. Wilson has called 'micro-wildernesses'; these 'are not a trivial part of wild Nature'.[22] Long ignored in cities and other unlikely places, these areas are teeming with life, including life we have yet to document; they are abundant in a world irrevocably altered by human activity. Rousseau acknowledges that he investigated the flora of the eighteenth-century micro-wilderness by botanising at his birds' cage[23] and collecting mosses in the Bois de Boulogne: 'J'ai fait cet hiver une seule herborisation dans le bois de Boulogne, et j'en ai rapporté quelques mousses que je ne crois pas vous avoir envoyées; elles augmenteront d'autant votre petit Moussier'.[24]

The 'micro-wilderness' offers satisfaction and edification to the careful investigator. Like Switzerland, where Rousseau commenced his botanical studies, the micro-wilderness cannot be understood as either first or second nature – it is heterogeneous, hybrid and impossible to define by any binary opposition. While a cultivated field may offer little or nothing to the naturalist, micro-wilderness studies are compatible with significant levels of human intervention.[25]

While Rousseau might not require a wilderness for his plant studies, he would nonetheless like to visit one from time to time. The central message of Rousseau's botanical teachings is that there is no substitute for nature study *in* nature, actively, on foot, carefully observing, noting, comparing and appreciating what one finds; the study of nature is above all an accessible, enjoyable *practice* that facilitates contemplation of nature and its Author. Even a six-year old child[26] can begin 'un herbier

20. Rousseau, *Julie*; *OC*, ii.479.
21. William Cronon, 'The trouble with wilderness; or getting back to the wrong nature' quoted in Zev Trachtenberg, 'Did we inherit our trouble with wilderness from Rousseau?', unpublished MS. Wilderness is problematic for Cronon, J. Baird Callicott and others as an ethnocentric construct that does not and has never existed, given the use and alteration of the environment by indigenous peoples. See Rolston, *Conserving natural value*, p.187, n. A wilderness is usually taken to be about 10,000 square kilometers.
22. Edward O. Wilson, *The Creation: an appeal to save life on earth* (New York, 2006), p.18.
23. Rousseau, *Rêveries*, *OC*, i.1061.
24. Rousseau to Malesherbes, 17 April 1772; *CC*, xxxix.38; see also Rousseau to Malesherbes, 19 December 1771; *CC*, xxxviii.300-03.
25. Lichens, for example, abound in the country parks of Hong Kong.
26. Madelon Delessert was four when the first letter on botany was composed in August 1771.

[...] qui s'augmentera *sans cesse* avec [se]s connaissances, et contiendra enfin l'histoire de *toute la végétation* du pays'.[27] Making the herbarium will help the child bond with his or her country. Botany can therefore be not only edifying, but also produce biodiversity surveys and foster patriotism.

Rousseau's contemporaries might have chosen to encounter the plant kingdom in the scientific botanical garden full of exotics, in the seemingly unplanned Elysium of Julie, or down on the farm with the Swiss peasant, Kleinjogg, tilling the fields with his numerous progeny. Rousseau had nothing against the Kleinjoggs of this world; indeed, he praised them, but he found his own way back to nature – not through the artifice of the garden, or through agriculture – but through contemplation of the plants that he believed the Author of nature has bestowed on humanity. If a wilderness does not offer its 'productions' for investigation, a micro-wilderness might have to suffice; this venue suits the modest botanist whose goal is not fame, but tranquility.

The way is available to anyone with the desire and patience to pursue it, but Rousseau did not expect everyone to follow him down this path because he believed only sensitive souls would be prepared to regard the plant kingdom from his contemplative perspective.[28] For him a plant could be just a plant and nothing more – that sufficed for it to do the work that he required: 'Je suis l'observateur et non moraliste. Je suis le Botaniste qui décrit la plante. C'est au médecin qu'il appartient d'en règler usage'.[29]

27. The emphasis is added; Rousseau to Mme Delessert, 11 April 1773; *CC*, xxxix.142. He may have even been thinking about this pedagogical problem at the time, since he started the botanical instruction course for Mme Delessert and her children a couple of months after he finished *Considérations sur le gouvernement de Pologne*. See *OC*, iii.966.
28. Rousseau, *Rêveries*, *OC*, i.1062-63.
29. Rousseau, *Mon portrait*, *OC*, i.1120. This work is dated by Rousseau's editors as c.1764 when he was starting his botanical studies.

Appendix 1: Rousseau's botanical sources[1]

Methodology

A core argument of this book is that Jean Jacques Rousseau studied and practised botany on the basis of a solid grounding in the science, including the works of the most important botanical authorities of time in addition to many older, illustrated works. However, there is no one document that sets forth the contents of Rousseau's botanical library or other sources that he used, so the list presented here has been reconstructed from a variety of sources; even after consulting these sources, it is often difficult to be certain which works he owned, merely consulted or knew about, or perhaps requested, but did not receive. In the sources we have consulted, Rousseau mentioned ninety specific works and eighty-two authors; twenty-two works in this list comprise two or more volumes.[2] We therefore believe that a conservative estimate of the number of works in his botanical library would be about 100. A possible corroboration of this estimate may be found in Rousseau's statement that to reject nomenclature would be to throw *300 volumes* (i.e. about 100 works) of botanical 'figures et descriptions' on the fire.[3]

This appendix therefore sets forth our best understanding to date of Rousseau's botanical sources. In preparing this document, we have striven for the highest possible degree of accuracy, carefully examining the following sources: (1) the *Correspondance complète*, (2) the so-called 'Introduction' to the *Fragmens pour un dictionnaire des termes d'usage en botanique*, (3) notes from botanical books in MsR80 (BPUN), 'Extraits de livres de botanique', and (4) the *Fragments de botanique*. Given the doubts raised in Chapter 9 of the present work concerning the authorship of *Fragmens pour un dictionnaire des termes d'usage en botanique*, I do not rely upon this text for arriving at any conclusions concerning Rousseau's knowledge of botany.

While we would like to claim that this document presents the entirety of works on botany that Rousseau owned and consulted, further books

1. Commentary by Alexandra Cook.
2. In some cases Rousseau discusses an author without specifying a work or works. It therefore seems likely that he owned and/or was acquainted with more than the ninety works that we have been able to identify with certainty.
3. The emphasis is added; Rousseau, 'Introduction', *OC*, iv.1209.

from his library may turn up in the future. As recently as 2007, Rousseau's copy of Dominique Chabrey's *Omnium stirpium sciagraphia et icones* (Geneva, 1678) was discovered in the holdings of the Lloyd Library, Cincinnati, Ohio.

A short history of Rousseau's botanical library

Rousseau's collection of botanical sources went through at least three major phases. He began to acquire his botanical library in 1764, acting on recommendations from Du Peyrou and possibly others, such as his teachers, d'Ivernois and Gagnebin. Du Peyrou shipped this collection by boat from Yverdon in May 1766,[4] and it was held in England by His Majesty's Customs until Rousseau paid for its release.[5] He promptly added a variety of works to his libary, including works of English botany such as Ray's *Synopsis*, given to him by one of his most important botanical correspondents, the Duchess of Portland. His acquisitions of texts continued throughout the decade, and drew on all eras of botany: antiquity, the Renaissance and the seventeenth and eighteenth centuries. It is clear from the prices reported in the correspondence that he went to considerable expense and trouble to obtain a useful and reasonably complete library that should remain intact even if he sold it.[6]

There are indications in Rousseau's correspondence that in late 1769 to early 1770 he was planning to sell his botanical library and that he had given first refusal to Du Peyrou, as well as letting Gouan and other interested parties look at it.[7] Rousseau also said he would return books that he had borrowed or claimed to have borrowed, for example Haller's *Historia stirpium*, which he had probably received as a gift from Du Peyrou. There were many reasons for Rousseau to sell his botany books, including a lessening interest in botany (that subsequently revived), and the burden of transporting his library during his many moves.

4. Du Peyrou consigned this precious cargo, together with Rousseau's dried plants, guitar and other items 'au bateau de Thuillard, Mandrot et Eternod parti d'ici le 24 du mois dernier'. Du Peyrou to Rousseau, 1 June 1766; *CC*, xxix.242. This boat company served the river route between Yverdon and Holland.
5. 'Je suis occupé à faire vendre mes livres et mes estampes qui m'ont été envoyés de Suisse bien malgré moi. Excepté ceux de *botanique* que je *garde*, je vends tout le reste pour en payer le port et la douane. Cette douane seule monte à quinze louis [...] Ce n'est pas à *Alger* que cela se fait, c'est à Londres.' My emphasis; Rousseau to Guy, 20 January 1767; *CC*, xxxii.56. In his letter of 1 June 1766 (cited in n.3) Du Peyrou had offered to pay the carriage and customs charges via his London bankers.
6. Further details of his acquisitions, especially illustrated works from the Renaissance, are discussed in chapter 7.
7. Rousseau to Gouan, 26 December 1769; *CC*, xxxvii.196; Rousseau to Latourrette, 26 January 1770; *CC*, xxxvii.213.

However, according to his own testimony in the *Dialogues*,[8] Rousseau did not in fact sell his botany books until 1775, and then to his British botanical correspondent, Daniel Malthus, who in turn willed most of them to his cousin and ward, Jane Dalton (c.1742-1817), also a botanist.[9] Having disposed of his library, but regaining an interest in botany, Rousseau borrowed books from which to take notes.[10] He also acquired more books, either by purchase and/or as gifts. By the time of his death on 2 July 1778, he had a library of fourteen works (two in-folio, the most expensive editions), comprising about twenty volumes.[11] These included well-known works by Linnaeus, Adanson, and Haller, as well as less well-known ones by Amman, Chabrey, and Scopoli. Persual of the list below reveals the wide-ranging, catholic character of Rousseau's botanical sources, and sheds much-needed light on a neglected feature of his formation as a botanist.

Please note that early-modern botanical authors were often known by more than one vernacular name (e.g. Lobel versus L'Obel) plus a Latin name (e.g. Lobelius). In most cases we have opted to use the *vernacular* name the author himself used, and have indicated the other names once in parentheses; for example, we have preferred 'Dalibard' to 'D'Alibard' (the version Rousseau used) and Caspar to Gaspard Bauhin. Some works mentioned here (e.g. by Magnol) do not appear in the main bibliography as they are not discussed in the main text.

8. Rousseau, *Dialogues*, *OC*, i.832 However, the marquis de Girardin claimed that Rousseau sold his books and herbarium to Vincent Louis Dutens (1730-1812). 'Notices bibliographiques: Herbiers', *OC*, iv.1888 (see n.6 above).
9. Malthus described Jane Dalton as '[l]a petite *cousine*, qui est botaniste à toute outrance'. Malthus to Rousseau, *CC* xxxv.54. Dalton's bookplate is present in two books that Rousseau autographed and annotated:. Haller, *Historia stirpium* and Vaillant, *Botanicon parisiense*. Rousseau's copies of Sauvages, *Methodus foliorum* and Ray, *Synopsis* are in the Thomas Malthus collection donated to Jesus College, Cambridge in 1949. These facts argue for Rousseau having sold the library to Daniel Malthus just as he claimed.
10. Rousseau, *Rêveries*, *OC*, i.1061
11. 'Quant à tous ses livres [de] botaniques [...] quelques-uns qui lui restaient encore, avec un grand herbier. Il n'avait ici qu'une Vingtaine de Volumes des Livres les plus Courants.' Girardin to Du Peyrou, 19 September 1778; *CC*, xli.289. See also Girardin to Du Peyrou, 4 October 1778; *CC*, xlii.11.

Detailed enumeration of Jean Jacques Rousseau's botanical sources[12]

* The specific work is not mentioned by Rousseau.

** Notes.

_______ Authors and/or works under discussion.

Author(s)	Works	Means of access[i]	Source(s)[ii]	Supporting quotes from letters[iii] / Informational notes
Adanson, Michel (1727-1806)	*Familles des plantes* (1763) 2 vols	Owned	Letter No. 3664, 20 November 1764; Guy to Rousseau	'[A]u moyen de ce que vous souhaitez, j'ay Differé jusqu'à ce jour pour ramasser Le peu que je vous envoye. [...] Donc, mémoire: [...] *La famille des plantes*, 2 vols 8°'.
			Letter No. 4359, 30 April 1765; Du Peyrou to Rousseau	'Adanson n'est pas grand partisan de ce Suédois [Linnaeus], et peut être a-t-il raison dans quelques points.'
			Letter No. 5681, 20 January 1767; Rousseau to Guy	'Entre autres les feuilles des plantes de M. Adanson où la feuille h du tome premier manque, au lieu de laquelle g est redoublé.'
		Requested new pages	Letter No. 5721, 9 February 1767; Coindet to Rousseau	'J'ay profité de l'occasion d'envoyer un Paquet à Londres pour vous faire parvenir Sous Couvert de Mr Davenport Le défet que vous demandez de la famille des Plantes d'Adanson.'
		Owned	Letter No. 6937, 11 May 1772; Rousseau to Malesherbes	'[J]e sais bien que si quelqu'un traitoit M. Adanson avec autant d'indignité et même de mauvaise foi qu'il traite M. Linnaeus [...] je m'échaufferois pour sa juste défense comme je fais pour celle de Linnaeus'.
			BdAN [Munch MS], art. 218, L'Euphraise. See also BdAN [Munch MS], art. 10, 246, 250	'(47) Ce n'est pas le sentiment de M. Adanson, à beaucoup près. Voyez ce qu'il en dit, Familles des plantes, Tome 2, p.205'.
			Letter No.: 7313, 4 October 1778; Girardin to Du Peyrou	'Catalogue des Livres que Mr Rousseau avoit apporté a Ermenonville [...] Adamson famille des plantes 2 v. Paris 1763'.
	General	Knew	*Introduction*, *OC*, iv, 1208	'M. Adanson fait encore plus, il prend une nomenclature toute nouvelle et ne fournit aucun renseignement pour y rapporter celle de M. Linnaeus. [...] mais M. Adanson n'en cite jamais ni genre ni phrases. [...]. Ainsi il faut opter entre M. Linnaeus et M. Adanson [...], et jetter tous les livres de l'un ou de l'autre au feu.'

i By 'means of access' we mean the various routes by which Rousseau became aware of the work of a particular author for example by ownership, loan, gift or a mere reference in a letter. In some cases the means of access and/or the specific work he refers to is unknown, so we can only state that he 'knew' an author, but not a specific work or works.

ii Due to space constraints, we use the *CC* letter number instead of volume and page number. The full names of correspondents are found in Appendix 2.

iii The original orthography has been retained.

12. Prepared by Rinske Kuiper under the direction of Alexandra Cook; background research was conducted by Michelle Ha Ming Wai.

Author(s)	**Works**	**Means of access**[i]	**Source(s)**[ii]	**Supporting quotes from letters**[iii] / **Informational notes**
			Fragments de botanique, OC, iv, 1253	'M. Adanson Blame M. Linnaeus d'avoir critiqué cette singuliére phrase de Tournefort: *Vicia flora vicia sepium*, comme s'il ignorait, ajoute M. Adanson, que des plantes de diverses famillles peuvent avoir de la ressemblance dans quelques-unes de leurs parties: mais ce n'est pas de cela ce me semble qu'il s'agit ici.'
Amman, Johann (1707-1741)	*Stirpium rariorum* (1739)	Owned	Letter No. 7313, 4 October 1778; Girardin to Du Peyrou	'Catalogue des Livres que Mr. Rousseau avoit apporté a Ermenonville [...] Ammani (Joannis) Stirpium rariorum (Petropolis fig.1739.)'.
Barbeu Dubourg, Jacques (1709-1799)	*Le Botaniste françois* (1767), 2 vols	Heard about	Letter No. 6367, 11 June 1768; Rousseau to Guy	'[O]n m'en annonce un nouveau intitulé le Botaniste français : le connaissez-vous? En dit-on du bien?'
		Knew	Letter No.6598, 8 August 1769; Rousseau to abbé Baurin	'Je crois que vous ferez bien de prendre votre Barbeu; du moins le second volume. Je prendrai peu de livres, parce qu'ils sont embarrassans.'
		Consulted	*Fragments de botanique, OC*, iv, 1253	*Botan. françois*. T.I. p. 154. 'Le public qui n'a point de méthode et ne connoit tout au plus que des espéces ne doit pas donner au Cerisier le nom de prunier; mais quiconque connoît et admet les genres placera sans surprise et sans peine le Cerisier dans le genre des pruniers'.
Barrelier, Jacques (1606-1673)	*Plantae per Galliam, Hispaniam et Italiam* (1714)	Owned	Letter No. 4622, 29 August 1765; Guy to Rousseau	'Je vous ay enfin adressé le petit envoy qui contient: [...] Barrelieri icones plantarum, fol. (rare)'.
Bauhin, Caspar [Gaspard] (1560-1624)	*Prodromos theatri botanici* (1620)	Requested	Letter No. 5681, 20 January 1767; Rousseau to Guy	'Si vous avez quelque occasion de m'envoyer le livre de M. Guettard [...]. Joignez-y le prodromus de Gaspard Bauhin, S'il vous tombe sous la main.'.
		Owned	Letter No. 7313, 4 October 1778; Girardin to Du Peyrou	'Catalogue des Livres que Mr Rousseau avoit apporté a Ermenonville [...] Prodromus H. admen. 1620'.
	Pinax theatri botanici (1623)	Requested	Letter No. 4994, 20 January 1766; Rousseau to Guy	'J'avois eu l'intention de prendre à Strasbourg chez M. Koenig quatre livres de botanique, savoir: Mathiolus, Michelius, Valerius Cordus et le Pinax de Gaspard Bauhin [...] Je vous prie, quand vous écrirez à M. Koenig, [...] de le prier d'envoyer ces quatre livres'.
		Knew	Letter No. 5482, 20 October 1766; Rousseau to the Duchess of Portland	'Je ne connois point les noms Anglois; ceux que je connois sont tous du Pinax de Gaspard Bauhin ou du Species plantarum de M. Linnaeus [...] Si par hasard, Madame la Duchesse, vous aviez aussi le Species plantarum ou le Pinax, ce point de réunion nous Seroit très commode pour nous entendre.'
			Letter No. 6185, 4 January 1768; Rousseau to the Duchess of Portland	'[E]t je suis fort trompé si ce n'est *l'Anthericum ossifragum* de Linnaeus, appellé par Gaspard Bauhin Pseudo-Asphodelus anglicus ou scoticus.'

Author(s)	Works	Means of access[i]	Source(s)[ii]	Supporting quotes from letters[iii] / Informational notes
			Letter No. 6933, 17 April 1772; Rousseau to Malesherbes	'[I]l ne faut très souvent que recourir à l'Auteur commun que tous deux citent assez constamment savoir Gaspard Bauhin. C'est dans son Pinax que je cherche leur concordance'.
				'[P]resque toute la nomenclature de Tournefort soit tirée mot à mot du Pinax'
		Owned	Letter No. 7313, 4 October 1778; Girardin to Du Peyrou	'Catalogue des Livres que Mr. Rousseau avoit apporté à Ermenonville [...] Bauhini (Gaspari) Pinax Theatri Botanici Basilea helv. 1623'.
		Knew	*Introduction, OC*, iv, 1204	'[L]e *Pinax*, fruit de quarante ans de travail est encore aujourd'hui le guide de tous ceux qui veulent travailler sur cette matière et consulter les anciens auteurs.'
	General	Knew	*Introduction, OC*, iv, 1207	'[I]l vaut cent fois mieux adopter celle-là [la nomenclature Linnéene] que de n'en voir aucune, ou retomber dans les phrases de Tournefort et de Gaspard Bauhin.'
			BdAN [Munch MS], art. 190 - La Sauge des bois. See also BdAN [Munch MS], arts. 217, 244, 347	'(38) Ce qu'il y a de plaisant est que les phrases de Tournefort étant presque toutes tirées de Gaspard Bauhin, tout l'honneur qu'en cela les François veulent faire à leur compatriote remonte à un Suisse en toute équité.'
Bauhin, Jean (1541-1612)	*Historia Plantarum* (1650-1651) 3 vols	Knew	BdAN [Munch MS], art. 15 - Le Safran	'(7) Jean Bauhin, tout sage et savant qu'il étoit, a pris ces stigmates pour des étamines.'
			BdAN [Munch MS], art. 31 - Le Grand Plantain. See also BdAN [Munch MS]: arts. 61, 116, 160, 244, 305	'(3) Je croirois que ce qu'on doit appeller Plantain à Bouquet est le Plantago rosea de Jean Bauhin, variété de celui-ci très commune en Allemagne, mais que je ne me rappelle pas l'avoir jamais vue en France.'
Blackwell, Elizabeth (1700-1758)	*A curious herbal, containing 500 cuts of the most useful plants* (1737-1739) 2 vols	Heard about	Letter No. 3549, 7 October 1764; Du Peyrou to Rousseau	'Quand à l'Herbier d'Elisabeth Blackwell, cet ouvrage est fait sans aucune Methode'.
			Letter No. 6218, 24 January 1768; Malthus to Rousseau	'J'attends vos ordres touchant ces difficultés. Nous avons dans le genre des planches : [...] Blackwell à 12 guinées**: 500, celles dont on se sert dans la médecine.' ** 12 guineas = £802
Buc'hoz, Pierre-Joseph (1731-1807)	*Traité historique des plantes qui croissent dans la Lorraine et les trois évêchés* (1762) 10 vols	Requested	Letter No. 3504, 15 September 1764; Rousseau to Duchesne	'J'apprends dans les journaux qu'on souscrit chez Durand pour un ouvrage intitulé Traité historique des plantes qui croissent dans la Lorraine, etc. Je voudrais bien souscrire pour cet ouvrage'.
		Discussed	Letter No. 3554, 10 October 1764; Rousseau to Du Peyrou	'Cet ouvrage, dont deux volumes ont déjà paru, en aura 20 in 8°, avec des Planches gravées qui se distribuent séparément pour pouvoir être reliées à part. L'Ouvrage ne se distribue que par Souscription.'

Author(s)	Works	Means of access[i]	Source(s)[ii]	Supporting quotes from letters[iii] / Informational notes
		Requested	Letter No. 3664, 20 November 1764; Guy to Rousseau	'[A]u moyen de ce que vous souhaitez, j'ay Differé jusqu'à ce jour pour ramasser Le peu que je vous envoye. [...] Donc, Mémoire: [...] La Souscription des plantes de Lorraine, avec 3 vol.'
		Discussed	Letter No. 3939, 27 January 1765; Rousseau to Duchesne	'Quoique l'Auteur du traité des plantes de la Lorraine soit mort, il n'est pas juste qu'on nous ait vendu ses trois volumes quinze francs sans aucune planche. Il est juste qu'on donne au moins les planches qui étoient faites, ou qu'on rende les quinze francs.'
			Letter No. 4465, 4 June 1765; Guy to Rousseau	'L'auteur des Plantes de la Lorraine n'est point mort comme on l'a annoncé; on promet de donner La suite de ce que vous avez et Suivre exactement ce qui a Eté annoncé par le prospectus.'
		Owned	Letter No. 4622, 29 August 1765; Guy to Rousseau	'Je vous ay enfin adressé le petit envoy qui contient: Les tomes 4 Et 5 des plantes de Lorraine, on promet de donner le[s] planches incessamment'.
		Requested	Letter No. 4577, 11 August 1765; Rousseau to Guy	'Je vois dans l'Avant coureur que le cinquième Tome des Plantes de la Lorraine paroit; vous savez que je n'ai reçu ni le quatriéme ni aucune Planche. Voyez je vous prie S'il y auroit moyen de mettre cette affaire en régle, et si enfin vous retirez les planches, n'allez pas les faire plier dans les in 12.'
Buffon, Georges-Louis Leclerc, comte de (1707-1788)	*Histoire naturelle, générale et particulière: avec la description du cabinet du roi* (1749-1788) 36 vols	Owned in part	Letter No. 2537, 12 March 1763; Guy to Rousseau	'De sorte que j'adresse à Pontarlier à M. François Gresset une Caisse contenant [...] et Les tomes 1. 2. 3. 5. 6. 7. 8. Et 9 de *Buffon*'.
			Letter No. 2550, 20 March 1763; Rousseau to Duchesne	'Pourquoi donc le quatrième tome de M. de Buffon a-t-il manqué plutôt que les autres? Pourquoi le réimprime-t-on? C'est justement ce quatrième tome qui me fait le plus de besoin, étant celui où il est question de moi, et précisément à la suite des in-12 que j'ai.'
			Letter No. 2571, 27 March 1763; Guy to Rousseau	'Il est vrai qu'il paroit asséz singulier que précisément le volume qui vous intéresse se trouve manquer dans l'exemplaire de M. *Buffon* que je vous ay envoyé?'
		Requested	Letter No. 3255, May 1764; Panckoucke to Rousseau	'Je n'ai pu joindre à L'envoye les nouveaux volumes 4° de M. De *Buffon*, je vous les enverrai dans La 1ere balle, dites moi ceux qui vous manquent de cet ouvrage et je vous les ferai parvenir.'
		Knew	Letter No. 3273, 13 May 1764; Rousseau to Rey	'J'estime que la collection de tout ce qui est imprimé et manuscrit donnera six Volumes in quarto de la grosseur et du caractére à peu près de ceux de l'histoire naturelle de M. de Buffon.'
		Owned	Letter No. 3374, 1 July 1764; Panckoucke to Rousseau	'Vous me deviez pour ces deux volumes [les tomes x, xi, de M. de *Buffon*] reliés trente Livres que vous me payerez à votre volonté.'

Author(s)	Works	Means of access[i]	Source(s)[ii]	Supporting quotes from letters[iii] / Informational notes
		Knew	Letter No. 3396, 12 July 1764; Claude Carloman de Rulhière to Rousseau	'M. de *Buffon* a exposé dans le 1er volume de son Histoire naturelle un système sur la formation des terres nouvelles, et vous connoissés mieux que moi, monsieur, les vraisemblances et les difficultés de ce système.'
		Owned	Letter No. 3400, 15 July 1764; Rousseau to Rey	'Si l'envoi des Livres que je vous ai demandés n'est pas encore fait je vous prie d'en retrancher le Tome IV de Buffon; car on m'envoye de Paris ce Tome relié, avec les Tomes X et XI.'
		/	Letter No. 3432, late July 1764; Panckoucke to Rousseau	'M. Duchesne m'a renvoyé Le *Buffon*, c'est Sans doute par votre ordre'.
		Owned	Letter No. 3452, 15 August 1764; Guy to Rousseau	'Je vous adresse par Pontarlier une petite Caisse contenant les 3. volumes de *Buffon* que j'ai payé 48 Lt que je porte compte, puisque vous les voulez.'
		/	Letter No. 3620, 4 November 1764; Rousseau to Du Peyrou	'Je suis fort touché de ce que vous me marquez de la part de M. et Madame de Buffon [...] qu'ils Sachent, je vous en Supplie, que je les honore et les aime de tout mon cœur.'
		Owned in part	Letter No. 3664, 20 November 1764; Guy to Rousseau	'[A]u moyen de ce que vous souhaitez, j'ay Differé jusqu'à ce jour pour ramasser Le peu que je vous envoye. [...] Donc, Mémoire: [...] '3. vol. de *Buffon* 4°.'
			Letter No. 4455, 30 May 1765; Guy to Rousseau	'Dans le premier Envoy que je vous feray, il y aura outre les livres des Plantes il y aura les tomes 12 Et 13 de *Buffon*.'
		Requested illustrations	Letter No. 5164, 19 April 1766; Rousseau to Guy	'Dans les cahiers d'estampes enluminées pour l'Histoire naturelle que M. de Buffon fait imprimer, il doit y en avoir pour la botanique; je vous prie de prendre ces derniers chez M. Panckoucke à mesure qu'ils paroîtront, en les payant, et de les rassembler pour me les envoyer avec les bonnes feuilles ou autre chose.'
		Owned	Letter No. 6897, 16 October 1771; Rousseau to Panckoucke	Je n'aurois pas tant tardé, Monsieur, à vous faire les remerciemens que je vous dois et pour le beau présent des fables de La Fontaine et pour celui du Volume de M. de Buffon'.
Camerarius, Joachim (1534-1598)	*Hortus medicus et philosophicus* (1588)	Owned	Letter No. 6924, 19 January 1772; Rousseau to Malesherbes	'[J]e suis allé à votre bibliothèque, j'y ai parcouru les livres de Botanique qui me sont tombés sous la main, j'ai pris note de la pluspart, et j'en ai emporté les quatre dont voici les titres. Tous petit in 4°. [...] 2. Camerarius Hortus medicus'.
Cesalpino, Andrea (1519-1603)	*De plantis libri XVI* (1583)	Looked for	Letter No. 6924, 19 January 1772; Rousseau to Malesherbes	'Il faut que je n'aye pas su trouver les places de tous les livres de Botanique, car j'ai cherché vainement Cesalpin*'.
		Knew	*Introduction*, *OC*, iv, 1203	'Il commença pourtant à se former de vrais Botanistes, tels que Clusius, Cordus, Cesalpin*, Gesner, et à se faire de bons livres et instructifs sur cette matière, dans lesquels même on trouve déjà quelques traces de méthode.'

Author(s)	**Works**	**Means of access**[i]	**Source(s)**[ii]	**Supporting quotes from letters**[iii] / **Informational notes**
Chabrey, Dominique (1610-1669)	*Omnium stirpium sciagraphia et icones* (1678)	Owned	Letter No.7313, 4 October 1778; Girardin to Du Peyrou	'Catalogue des Livres que Mr. Rousseau avoit apporté a Ermenonville [...] Chabraei (Dominii) omnium Stirpium susgraphica** Genève 1678 in 4°.' ** This book is in the Lloyd Library, Cincinnati, Ohio. Found in 2007, see Kobayashi, 'Seven botanical books annotated by Roussseau'.
Clusius, Carolus (Charles de l'Ecluse) (1526-1609)	*Rariorum aliquot stirpium per Hispanias observatarum historia* (1576)	Knew	BdAN [Munch MS], art. 115 – Le Tamaris	'(24) Il devient [souvent] beaucoup plus grand, et Clusius assure en avoir vu en Espagne dont un homme avoit peine à embrasser le tronc.'
	Rariorum plantarum historia (1601)	Requested	Letter No. 6173, 28 December 1767; Rousseau to Rey	'Livres à acquerir C. Clusii rariorum plantarum historia.'
	General	Searched	Letter No. 6509, 19 December 1768; Rousseau to Du Peyrou	'[J]'ai été obligé de faire venir à grands frais de Hollande et d'Angleterre le peu que j'en ai, encore ai-je cherché par tout ceux de Clusius sans pouvoir les trouver.'
		Owned	Letter No.6635, 23 November 1769; Rousseau to Rey	'Les Livres de Botanique que vous m'avez envoyés m'ont fait grand plaisir et surtout le Clusius qu'on trouve difficilement.'
		Knew	*Introduction, OC*, iv, 1203	'Il commença pourtant à se former de vrais Botanistes, tels que Clusius, Cordus, Cesalpin, Gesner, et à se faire de bons livres et instructifs sur cette matière, dans lesquels même on trouve déjà quelques traces de méthode.'
Columna, Fabius (Fabio Colonna) (1567-1650)	*Minus cognitarum rariorumque nostro coelo orientium stirpium ekphrasis* (1616)	Requested	Letter No. 6620, 6 October 1769; Rousseau to Gouan	'Je suis bien reconnoissant des recherches que vous avez eu la bonté de faire sur l'Ecphrasis de Columna [...] pour l'acquérir à votre préjudice, n'ayant assurement ni le pouvoir ni la volonté d'en donner le prix que De Bure en demande, ainsi, Monsieur, qu'à moi ne tienne que vous n'en fassiez l'acquisition si cela vous convient.'
			Letter No. 6924, 19 January 1772; Rousseau to Malesherbes	'Il faut que je n'aye pas su trouver les places de tous les livres de Botanique, car j'ai cherché vainement [...] Columna*'.
Cordus, Valerius (1515-1544)	*Annotationes in pedacii Dioscorides Anazarbei de materia medica libros V* (1561)	Requested	Letter No.4994, 20 January 1766; Rousseau to Guy	'J'avois eu l'intention de prendre à Strasbourg chez M. Koenig quatre livres de botanique, savoir: Mathiolus, Michelius, Valerius Cordus* et le Pinax de Gaspard Bauhin [...] Je vous prie, quand vous écrirez à M. Koenig, de le saluer de ma part, et de le prier d'envoyer ces quatre livres'.
		Owned	Letter No.5681, 20 January 1767; see also Letter No.5681bis, 8 January 1767; Rousseau to Guy	'Note de M. Koenig. [...] Valerius *Cordus**.'
			Letter No.6655, 26 January 1770; Rousseau to Latourrette	'Je ne doute pas que vous n'ayez la plupart de mes livres [...] Le Valerius Cordus* est assez rare.'

Author(s)	**Works**	**Means of access**[i]	**Source(s)**[ii]	**Supporting quotes from letters**[iii] / **Informational notes**
	General	Knew	*Introduction*, *OC*, iv, 1203	'Il commença pourtant à se former de vrais Botanistes, tels que Clusius, Cordus, Cesalpin, Gesner, et à se faire de bons livres et instructifs sur cette matière, dans lesquels même on trouve déjà quelques traces de méthode.'
Crantz, Johann Nepomuk Freiherr von (1722-1797)	(1769) *Stirpium austriacarum*, 2nd edn, 2 vols and/or (1767) *Classis umbelliferarum* (the latter a supplement to the first)	Consulted / Knew	Letter No.6620, 6 October 1769; Rousseau to Gouan	'Je vois que dans la figure de Crantz* [du *selinum palustre*] il y a plusieurs feuilles caulinaires, il n'y en a qu'une dans le specimen, dans Crantz les feuilles radicales sont plus petites que les caulinaires dans le specimen elles sont plus grandes. [...] A l'égard de *l'Athamanta lib[a]notis* je n'ai rien du tout à en dire parce que je ne l'ai pas trouvé dans l'herbier [...] non plus que le *selinum carvifolia* de Crantz dont j'ai la figure aussi dans ses ombelliféres, à laquelle je ne trouve rien de semblable dans celles de l'herbier. ** *Selinum carvifolium* in *Classis umbelliferarum*, p.59.
			Letter No.6645, 26 December 1769; Rousseau to Gouan	'Je vous avois offert le *Selinum carvifolia* (non celui de *Crantz**, mais celui de Linnaeus) [...] Mais quoique celui-ci lui donne pour synonyme le Seseli bienne de Crantz, je douterois que ce fût lui, attendu que Crantz cite la figure de Vaillant [*Botanicon*, p.54, t.9, f.4]'. ** See *Classis umbelliferarum*, p.93; *Stirpium austriacarum*, par.1, p.162.
	Stirpium austriacarum (1769) 2nd edn, 2 vols	Consulted / Knew	Letter No.7001, 9 August 1773; Rousseau to Mme Delessert	'[C]ette disposition [...] a donné lieu à la dénomination de colomniféres, que M. Crantz* Botaniste Allemand donne à une Classe nombreuse qui comprend avec les Malvacées, les Géraniums et d'autres genres.' ** See *Stirpium austriacarum*, par.1, p.143-44. ** Crantz was born in Luxembourg, but spent his career as a physician, botanist, and mining entrepreneur in the Austrian empire.
	General	Knew	*Introduction*, *OC*, iv, 1208	'On a obligation à M. Crantz*, malgré sa passion contre M. Linnaeus, d'avoir, en rejettant son système, adopté sa nomenclature.' ** Crantz proposed a natural method of classification and polemicised against Linnaeus in several works.
		Knew	BdAN [Munch MS], art. 214 bis - Le Calament	'Les feuilles et les fleurs sont trop petites dans la figure, de sorte qu'elle ressemble au Melissa Nepeta plustôt qu'au vrai Calament. Au reste ces deux plantes varient si fort et sont [entre elles] si ressemblantes que Crantz* doute avec raison si l'on doit les séparer.'
Dalibard, Thomas François (1709-1779)	*Floræ parisiensis prodromus* (1749)	Owned	Letter No. 4670, 18 September 1765; Du Peyrou to Rousseau	'Je vous envoye donc d'Alibard*'.

Author(s)	Works	Means of access[i]	Source(s)[ii]	Supporting quotes from letters[iii] / Informational notes
		Gave away	Letter No. 6305, 24 March 1768; Rousseau to Marie Charlotte Hippolyte de Campet de Saujon, comtesse de Boufflers	Note 3, R. A. Leigh: 'Rousseau allait dîner chez le médecin Laubel, à Gisors. Rousseau lui offrit ce jour-là un livre de botanique (le Prodromus de Dalibard [...]) Cet exemplaire [...] porte la note suivante: '24 Mars 1768, Reçu ce livre de Monsieur Jean-Jacques Rousseau'.
		Knew	Letter No. 6916, 2 January 1772; Malesherbes to Rousseau	'[J]e comptois qu'il me Seroit épargné par M. D'Alibard qui a donné une énumeration des plantes des environs de Paris dans le Système et le langage de Linnaeus, mais il se trouve que depuis le livre de Mr D'Alibard, Mr Linnaeus a changé beaucoup de ces noms.'
			Letter No. 6933, 17 April 1772; Rousseau to Malesherbes	'A l'égard des plantes dont Tournefort n'a pas tiré les noms du Pinax, on en trouve aisément la concordance dans les Auteurs François Linnaeistes, tels que Sauvages, Gouan, Gerard Guettard et d'Alibard qui l'a presque toujours suivi.'
		Knew	*Fragments de botanique*, *OC*, iv, 1254	'M. d'Alibard dans son *Florae parisiensis prodomus* a confondu le receptacle avec le placenta.'
Dillenius, Johann Jakob (1687-1747)	*Historia muscorum* (1741)	Requested	Letter No. 6134, 25 November 1767; Rousseau to Guy	'Ce sont des Livres de botanique qui se trouvent chez Briasson et dont je vous prie de vouloir bien faire pour moi l'acquisition, ayant soin de choisir de bonnes épreuves, et non pas des exemplaires de rebut, et de faire relier ceux qui ne le seront pas. [...] Dilleni historia muscorum. 4°'.
			Letter No. 6145, 3 December 1767; Guy to Rousseau	'Si j'eusse trouvé Les Livres que vous demandez par votre Lettre du 25 du mois passé, Reliés, je vous les aurois tout de suite Envoyés; Mais M. Briasson ne vend Ses Livres qu'en feuille'.
		Owned	Letter No.6151, 10 December 1767; Rousseau to Guy	'[M]ais j'avoue que je voudrais bien pouvoir renvoyer le Dillenius, dont je ne suis point content du tout, et qui même me devient parfaitement inutile, faute des explications des figures qu'il doit contenir et du texte historique, qui n'y sont pas.'
			Letter No.6171, 26 December 1767; Guy to Rousseau	'Le Dillenius est imparfait, c'est que j'ay pensé que quelques jours de plus ou de moins ne porteroient aucun préjudice à cette affaire: tout ce que je puis vous assurer, c'est que M. Briasson a donné exactement ce que vous demandiez par votre Lettre.'
			Letter No.6218, 24 January 1768 Malthus to Rousseau	'Je n'ai pas pû la trouver, mais Mademoiselle Dalton qui l'a [...] veut positivement que vous l'acceptiez. Cette édition moderne de Dillenius, qui n'a que les estampes, & la table, & qui se vend ici at 25 shill'.
			Letter No.6258, 18 February 1768; Guy to Rousseau	'[Q]uestion de régler un Compte; parce que les Bons comptes font les bons amis, voicy donc ce que la Maison a déboursé à compte Sur l'année 1767. Pour Dillenii historiae muscorum'.

Author(s)	Works	Means of access[i]	Source(s)[ii]	Supporting quotes from letters[iii] / Informational notes
		Gave away	Letter No.6479, 7 November 1768; Rousseau to Liotard	'[J]e serais peut-être fort embarrassé pour les déterminer sans le secours de mon Dillenius, que je n'ai plus.' ** Rousseau may have given his copy to Gouan; see Chapter 6.
		Knew	Letter No.6536, February 1769; Rousseau to Samuel Auguste André David Tissot	'[J]e m'amuse à observer et déterminer les mousses et les lichens sur les troncs des arbres, et je soupçonne déjà que cette étude pourroit être encore perfectionnée, même après Michelius et *Dillenius*'.
			Letter No. 6913, 19 December 1771; Rousseau to Malesherbes	'[N]'ayant point l'ouvrage de *Dillenius*, guide indispensable dans ces recherches je ne suis parvenu qu'avec beaucoup d'effort et souvent avec doute à déterminer les espéces que je vous envoye.'
		Borrowed	Letter No. 6933, 17 April 1772; Rousseau to Malesherbes	'Excepté l'Histoire des Mousses de *Dillenius*, j'ai à moi les autres livres de Botanique que vous tenez à Malesherbes et dont vous m'envoyez la note.'
Dioscorides, Pedanius of Anazarbos (40-90)	*De materia medica libri quinque* (Edition unknown)	Knew / Owned?	Letter No. 5693, 31 January 1767; Rousseau to Davenport	'[J]'ai pensé qu'il vaudroit mieux qu'on y mit mon vieux Dictionnaire de Calepin in folio, [...] qui contient encore plus surement les noms Grecs de plantes tirés de Dioscoride* et de Theophraste'.
		Knew	Introduction, *OC*, iv, 1202-1203	'Au lieu d'étudier les plantes sur la terre, on ne les étudioit plus que dans Pline et Dioscoride*, et il n'y a rien si fréquent dans les Auteurs de ces tems-là, que d'y voir nier l'existence d'une plante par l'unique raison que Dioscoride n'en a pas parlé. [...] Toutes les autres n'étant pas celle dont Dioscoride avoit parlé doivent être proscrites de dessus la terre.'
		Knew	BdAN [Munch MS], art. 295 - L'Aster, ou œil de Christ; see also BdAN [Munch MS], art. 305	'Dioscoride* parle de l'Aster dans son livre des descriptions des plantes, qu'il composa sous le règne d'Auguste. (3) Note de JJR au bas de l'article: (3) Ceci me paroit dit bien affirmativement. Rien n'est plus douteux, ce me semble, que le tems où Dioscoride a vécu, et il n'est pas non plus très certain que l'Aster dont cet Auteur parle soit celui qu'on décrit ici.'
Dodoens, Rembert (1517-1585)	*Stirpium historiae pemptades sex* (1583)	Requested	Letter No.5960, 5 July 1767; Rousseau to Coindet	'[V]oyez dans quelqu'une de vos courses de la rue St Jacques si vous ne trouveriez point le Livre suivant petit in folio. *Remberti Dodonœi Pemptades*.'
			Letter No.6134, 25 November 1767; Rousseau to Guy	'Est-il donc absolument impossible de trouver à Paris le *Dodonaei Pemptades* &c. sur les figures de Lobel.'
			Letter No.6145, 3 December 1767; Guy to Rousseau	'J'ay donné à Nos Coureurs des Ventes Le titre du *Dodonaei Pemptades* &c. Si on le trouve, je vous l'enverray tout de suite.'
			Letter No. 6173, 28 December 1767; Rousseau to Rey	'Livres à acquerir Remberti Dodonaei – Stirpium historiœ Pemptades sex.'

Author(s)	Works	Means of access[i]	Source(s)[ii]	Supporting quotes from letters[iii] / Informational notes
		Consulted	BdAN [Munch MS], art. 305 - La Mille-feuille; see also BdAN [Munch MS] art. 295, 314	'Tournefort a fait de l'Achilla'a une millefeuille et non de la millefeuille un Achilla'a. Dodonée* est un des premiers parmi les modernes qui ait confondu ces deux noms.'
Duchesne, Antoine-Nicolas** (1747-1827)	*Manuel de botanique* (1764)	Heard about	Letter No. 3587, 20 October 1764; Du Peyrou to Rousseau	'Rien de nouveau dans mes découvertes que l'annonce d'un ouvrage intitulé *Manuel de Botanique* par Mr Duchesne fils à Paris'. ** Not to be confused with Nicolas-Bonaventure Duchesne (1711-1765), French publisher, bookseller.
			Letter No. 3769, 23 December 1764; Du Peyrou to Rousseau	'Ce Sera aussi à votre loisir que je vous prie, Monsieur, de me marquer le titre de l'ouvrage de Tournefort que vous me conseillez d'avoir, & celui du *Manuel de Botanique*** que j'ay vu chez vous'. ** It is unclear if Duchesne's work is being discussed; see Chapter 6.
Duhamel du Monceau, Henri Louis (1700-1782)	*Traité des arbres et arbustes qui se cultivent en France en pleine terre* (1755), 2 vols	Requested	Letter No. 4417, 19 May 1765; Rousseau to Duchesne	'[J]'aurai encore besoin du Traité des arbres et arbustes de M. Duhamel'.
			Letter No. 4465, 4 June 1765; Guy to Rousseau	'M. Guerin ne peut point Separer le Traité des arbre et arbustes d'avec les autres ouvrages de M. Duhamel, dont le tout compose vol. in 4° qu'il nous veut 102 Lt. Voyez si vous souhaitez toute la Collection.'
		Declined	Letter No. 4484, 15 June 1765; Rousseau to Duchesne	'Les Livres de M. Duhamel sont trop chers pour moi, et puisqu'on n'en peut séparer l'ouvrage dont j'ai besoin je m'en passerai.'
		Gift	Letter No. 4601, 23 August 1765; Tschudi to Rousseau	'Il y a prés de quinze jours que j'ai Remis au Courrier de Strasbourg les deux volumes in quarto du traité des arbres et arbustes de Mr. Duhamel'.
			Letter No.4930, 25 December 1765; Tschudi to Rousseau	Vous Exigez, Monsieur, que je vous Dise le prix de L'ouvrage de M. *Duhamel*. Assurement, je ne veux pas vous forcer d'avoir La genérosité de L'accepter.
Ellis, John (1710 -1776)	*Directions for bringing over seeds and plants from the East-Indies and other distant countries in a state of vegetation* (1771)	Gift	letter : 6713, 15 May 1770; William Constable to Rousseau	'J'ay pris La Liberté, Monsieur, de vous Envoyer une petite Brochure Anglaise**.' ** Leigh speculates that this is the work in question; see letter 6713, note a.
Fusée-Aublet, Jean Baptiste Christophe (1720-1778)	*Histoire des plantes de la Guiane françoise* (1775), 4 vols	Owned	Letter No.7313, 4 October 1778; Girardin to Du Peyrou	'Catalogue des Livres que Mr. Rousseau avoit apporté à Ermenonville [...] Aublet (Fusée) Histoire des plantes de la Guyane françoise 4 vol. fig.'
Galen (131-201 CE)	General	Knew	Introduction, OC, iv, 1202	'A la renaissance des Lettres tout disparut pour faire place aux anciens livres; il n'y eut plus rien de bon et de vrai que ce qui étoit dans Aristote et dans Galien.'

Author(s)	**Works**	**Means of access**[i]	**Source(s)**[ii]	**Supporting quotes from letters**[iii] **/ Informational notes**
Garidel, Pierre Joseph (1659-1737)	*Histoire des plantes qui naissent autour d'Aix* (1715)	Borrowed	Letter No. 6728, 7 June 1770; Rousseau to Du Peyrou	'Il y en a deux qui ne sont pas à moi, savoir le Haller qui vous appartient, et le Garidel* qui appartient à M. le marquis de Mirabeau.'
Garsault, Francois-Alexandre-Pierre de (1691-1778)	*Les Figures de plantes et animaux d'usage en médecine* (1764-1765), 6 vols	Borrowed	Letter No. 4359, 30 April 1765; Du Peyrou to Rousseau	'[I]l vous faut un Garsault. Si vous voulez le mien, je vous l'envoyerai'.
			Letter No. 4408, 17 May 1765; Du Peyrou to Rousseau	'Si parmi les livres que vous avez, il y en a qui puissent vous convenir, de les prendre avec vous á Môtiers pour le temps que vous voudrez, par exemple le Garsault'.
			Letter No.4417, 19 May 1765; Rousseau to Duchesne	'J'ai découvert avec une peine infinie les noms de botanique de plusieurs plantes de *Garsault*'.
		Requested	Letter No.4504bis, 30 June 1765; Rousseau to Duchesne	'Voici les articles de ce second envoi que je vous prie de m'expédier le plustot qu'il vous sera possible. 1. les quatre volumes de M. de Garsault avec la description que vous m'annoncez.'
		Owned	Letter No.4622, 29 August 1765; Guy to Rousseau	'Je vous ay enfin adressé le petit envoy qui contient: [...] Garsault cinq vol. in 8.'
Gerard, John (1545-1612)	*The Herball or general historie of plantes* (1636)	Knew	Letter No.5482 20 October 1766; Rousseau to the Duchess of Portland	'Je ne connois point les noms Anglois; ceux que je connois sont tous du Pinax de Gaspard Bauhin ou du Species plantarum de M. Linnaeus, et je ne puis en faire la Synonymie avec Gérard*.'
		Requested	Letter No. 5654 December 1766; Rousseau to Paul Vaillant	'Monsieur Vaillant est prié de vouloir bien remettre à Monsieur Davenport le Livre de Botanique qui lui a été addressé par M. Guy libraire à Paris pour J.-J. Rousseau.'
		Borrowed	Letter No. 5838, 29 April 1767; Rousseau to Bernard Granville	'Il** lui renvoye aussi son *Gérard**, et lui fait bien des excuses d'en avoir abusé si longtems.' (** il = Rousseau, letter written in third person).
		Requested	Letter No. 6218, 24 January 1768; Malthus to Rousseau	'Je ne sais pas si vous voulez dire par Gérard* emaculatus, le titre du livre, car vous l'avez sousligné comme les autres: nos libraires ne connoissent pas ce titre, & la seule bonne édition que nous avons du Gérard, est Johnson [Thomas, 1600-1644] sur Gérard.'
		Owned	Letter No.6655, 26 January 1770; Rousseau to Latourrette	'Si même la revue de mon herbier et de mes livres de botanique pouvoit vous amuser quelques momens, le tout pourroit être déposé chez vous et vous le visiteriez à votre aise. Je ne doute pas que vous n'ayez la plupart de mes livres. Il peut cependant s'en trouver d'Anglois comme Parkinson et le *Gérard** émaculé que peut être n'avez-vous pas.'
Gérard, Louis (1733-1819)	*Flora Gallo-provincialis* (1761)	Knew	Letter No.6933, 17 April 1772; Rousseau to Malesherbes	'A l'égard des plantes dont Tournefort n'a pas tiré les noms du Pinax, on en trouve aisément la concordance dans les Auteurs François Linnaeistes, tels que Sauvages, Gouan, Gerard* [...]. qui l'a presque toujours suivi.'

Author(s)	**Works**	**Means of access**[i]	**Source(s)**[ii]	**Supporting quotes from letters**[iii] **/ Informational notes**
Gessner, Konrad (1516-1565)	*De raris et adminrandis herbis* (1555)	Borrowed	Letter No.6924, 19 January 1772; Rousseau to Malesherbes	'[J]e suis allé à votre bibliothèque, j'y ai parcouru des livres de Botanique qui me sont tombés sous la main, j'ai pris note de la pluspart, et j'en ai emporté les quatre dont voici les titres. Tous petit in 4°. [...] 4. Gessner de Lunariis.'
	General	Knew	*Introduction, OC*, iv, 1203	'Il commença pourtant à se former de vrais Botanistes, tels que Clusius, Cordus, Cesalpin, Gesner, et à se faire de bons livres et instructifs sur cette matière, dans lesquels même on trouve déjà quelques traces de méthode.'
		Knew	BdAN [Munch MS], art. 230, L'Herbe aux cuillers, ou cochléaria	'(35) et même bien peu vraisemblable: car cette plante, comme l'ont remarqué Gesner et Lobel a très peu de rapport au Telephium de Pline, et moins encore au Britannica.'
Gouan, Antoine (1733-1821)	*Hortus regius monspeliensis* (1762)	Requestec	Letter No.6134, 25 November 1767; Rousseau to Guy	'Ce sont des Livres de botanique qui se trouvent chez Briasson et dont je vous prie de vouloir bien faire pour moi l'acquisition [...] Gouan Hortus monspeliensis. 8°.'
			Letter No.6145, 3 December 1767; Guy to Rousseau	'Si j'eusse trouvé Les Livres que vous demandez par votre Lettre du 25. du mois passé, Reliés, je vous les aurois tout de suite Envoyés; Mais M. Briasson ne vend Ses Livres qu'en feuille'.
		Owned	Letter No.6258, 18 February 1768; Guy to Rousseau	'[Q]uestion de régler un Compte; parce que les Bons comptes font les bons amis, voicy donc ce que la Maison a déboursé à compte Sur l'année 1767. [...] – Gouan hortus Monspeliensis.'
			Letter No.6620, 6 October 1769; Rousseau to Gouan	'[O]n est réduit à étudier seul. J'ai été forcé par la même raison de renoncer à l'Hortus Cliffortianus aux amoenitates academicae [...] et à beaucoup d'autres livres qui me seroient également nécessaires pour suppléer par leurs descriptions à la sécheresse du Species et ne vous en déplaise à celle de l'Hortus Monsp.'
	Flora monspeliaca (1765)	Requested	Letter No. 6022, 13 August 1767; Rousseau to Coindet	'Si vous pouvez sans embarras vous charger de quelques bagatelles à l'aide d'un porte manteau, voici encore des commissions. Deux ou trois feuilles de papier doré tout uni, c'est à dire sans dessein. [...] Libraire rue St. Jacues. *Noia Monspeliaca***. *Autore Ant: Gouan*.' ** This title is mispelled but it seems likely to be the *Flora monspelliaca.*
		Knew	Letter No. 6573, 26 May 1769; Rousseau to Clappier	'Je douterois au contraire que M. Gouan lui-même ait bien connu le Gentiana filiformis, puisqu'il lui donne (*Flora monsp*: p.469) un synonyme de Magnol.'
			Letter No.6620, 6 October 1769; Rousseau to Gouan	'On est réduit à étudier seul. J'ai été forcé par la même raison de renoncer à [...] celle de l'Hortus Monsp: et de [la] Flora Monspel'.

Author(s)	**Works**	**Means of access**[i]	**Source(s)**[ii]	**Supporting quotes from letters**[iii] **/ Informational notes**
			Letter No.6933, 17 April 1772; Rousseau to Malesherbes	'A l'égard des plantes dont Tournefort n'a pas tiré les noms du Pinax, on en trouve aisément la concordance dans les Auteurs François Linnaeistes, tels que Sauvages, Gouan, Gérard, Guettard et d'Alibard qui l'a presque toujours suivi.'
Gronovius, Johan Friedrik (1611-1671)	*Flora orientalis* (1755)	Owned	Letter No.7313, 4 October 1778; Girardin to Du Peyrou	'Catalogue des Livres que Mr Rousseau avoit apporté à Ermenonville [...] Rauwolfil (Leonhard) flora orientalis** (Lugd. 1755).' ** Gronovius assembled this work from the collection of Rauwolf, made in 1573 to 1575 in the Near East.
Guettard, Jean-Etienne (1715-1786)	*Observations sur les plantes* (1747), 2 vols	Requested	Letter No.5681, 20 January 1767 (also 5681 bis); Rousseau to Guy	'Si vous avez quelque occasion de m'envoyer le livre de M. Guettard sur les plantes, vous me ferez grand plaisir.'
			Letter No.5721, 9 February 1767; Coindet to Rousseau	'Je vous enverrai par une autre occasion le livre de Guettard.'
		Owned	Letter No.6620, 6 October 1769; Rousseau to Gouan	'M. Guettard dit que le bord des feuilles est légèrement crenelé, dans le specimen il ne l'est point du tout, il ajoute que le bout des feuilles est mousse, dans le specimen.'
	General	Knew	Letter No.6933, 17 April 1772; Rousseau to Malesherbes	'A l'égard des plantes dont Tournefort n'a pas tiré les noms du Pinax, on en trouve aisément la concordance dans les Auteurs François Linnaeistes, tels que Sauvages, Gouan, Gerard, Guettard et d'Alibard qui l'a presque toujours suivi.'
Haller, Albrecht von (1707-1777)	*Iter helveticum* (1739)	Requested	Letter No.4504bis, 30 June 1765; Rousseau to Duchesne	'Voici les articles de ce second envoi que je vous prie de m'expédier le plustôt qu'il vous sera possible. [...] 3. Les ouvrages de M. Haller intitulés: Iter Helveticum, et Enumeratio Stirpium Helvetiae sur tout le dernier, s'il n'est pas d'un prix exhorbitant.'
			Letter No.6173, 28 December 1767; Rousseau to Rey	'Livres à acquérir Haller, Iter helveticum. 4e.'
	Enumeratio methodica stirpium Helvetiae indigenarum (1742) 2 vols	Requested	Letter No.4504bis, 30 June 1765; Rousseau to Duchesne	'Voici les articles de ce second envoi que je vous prie de m'expédier le plustot qu'il vous sera possible. [...] 3. Les ouvrages de M. Haller intitulés: Iter Helveticum, et Enumeration Stirpium Helvetiae sur tout le dernier, s'il n'est pas d'un prix exhorbitant.'
		Owned	Letter No.7303, 19 September 1778; Girardin to Du Peyrou	'Quant à tous ses livres [de] botaniques [...] quelques uns qui lui restoient encor, avec un grand herbier. Il n'avoit icy qu'une Vingtaine de Volumes des Livres les plus Courants tels que Le Species de Linnaeus, Tournefort, La 1re Edition d'Haller et quelques autres de ce genre.'
			Letter No.7313, 4 October 1778; Girardin to Du Peyrou	'Catalogue des Livres que Mr. Rousseau avoit apporté à Ermenonville [...]. Halleri (Alberti) Enumeratio methodica stirpium helveticae indigenarum fol. Gothing 1742.'

Author(s)	**Works**	**Means of access**[i]	**Source(s)**[ii]	**Supporting quotes from letters**[iii] **/ Informational notes**
	Historia stirpium indigenarum Helvetiae inchoata (1768) 3 vols	Owned	Letter No.6509, 19 December 1768; Rousseau to Du Peyrou	'Je me suis ruiné en livres de Botanique, et j'avois bien résolu de n'en plus acheter; cependant je sens que m'affectionnant aux Plantes des Alpes, je ne puis me passer de celui de Haller.'
			Letter No.6643, 21 December 1769; Rousseau to the Duchess of Portland	'[E]lle est bien décrite dans la dernière édition des plantes de Suisse de M. Haller**'. ** Annotated by Rousseau. This work is at the Royal Horticultural Society, London. See: Cook, 'Jean-Jacques Rousseau's copy of Albrecht von Haller's Historia stirpium'
			Letter No. 6593, 21 July 1769; Rousseau to Du Peyrou	'Je n'aurois pas tardé si longtems, mon cher Hôte, à vous remercier du livre de M. *Haller*.'
			Letter No. 6728, 7 June 1770; Rousseau to Du Peyrou	'Il y en a deux qui ne sont pas à moi, savoir le Haller qui vous appartient.'
			Introduction, OC, iv, 1208	'Mais M. Haller dans son grand et excellent traité des plantes alpines, rejette à la fois l'un et l'autre [...]. M. Haller s'attache à une synonymie exacte, par laquelle, quand il n'y joint pas la phrase de M. Linnaeus, on peut du moins la trouver indirectement par le rapport des synonymies.'
Hermann, Paul (1646-1695)	General	Discussed	*Introduction, OC*, iv, 1205	'[L]es botanistes forcés de chercher un fil pour s'en tirer, s'attachèrent enfin sérieusement à la méthode ; Herman**.' ** Hermann wrote five descriptive botanical works; it is unclear which work(s) Rousseau has in mind.
Hildegard of Bingen (1098-1179)	General	Discussed	*Introduction, OC*, iv, 1202	'Voilà tout l'art que les Myrepsus, les Hildegardes*, [...] et les autres Docteurs de ces tems-là mettoient à l'étude des plantes dont ils ont parlé dans leurs livres, et il seroit difficile peut-être au peuple d'en reconnoître une seule sur leurs noms ou sur leurs descriptions.' ** See Chapter 5.
Hill, Sir John (1716-1775)	*The British herbal* (1756-1757)	Offered	Letter No.6218, 24 January 1768; Malthus to Rousseau	'Il y a aussi Hill's British Herbal, avec des planches, à dix guinées enluminées**, 45 shil.: simple – toutes les plantes de la Grande Bretagne, & les espèces étrangères. Il combat à chaque page le système de Linnée il l'a approuvé avant & depuis.' ** 10 guineas = £668 in today's currency.
	Flora Britanica (1760)	Gift	Letter No.5772, 10 February 1767; the Duchess of Portland to Rousseau	'I met with a book the other day which I wish to convey to you. It is Ray's Synopsis of British plants in Linnaeus's Method**.' ** Rousseau means Hill's *Flora Britannica*
		Owned	Letter No.5837, 29 April 1767; Rousseau to the Duchess of Portland	'Je reçois, Madame la Duchesse, avec une nouvelle reconnoissance, les nouveaux témoignages de vôtre souvenir et de vos bontés dans le livre que M. Granville m'a remis de votre part.'

Author(s)	**Works**	**Means of access**[i]	**Source(s)**[ii]	**Supporting quotes from letters**[iii] **/ Informational notes**
			Letter No.5971, 10 July 1767; Rousseau to the Duchess of Portland	'J'y ai apporté le dernier Livre que vous m'avez envoyé.'
			Letter No.6063, 12 September 1767; Rousseau to the Duchess of Portland	'[N]'ayant point encore ici mes livres de Botanique, excepté la *Flora Britannica*, qui ne m'a pas quitté un seul moment.'
			Letter No.6567, 17 April 1769; Rousseau to Clappier	'C'est le Synopsis de Ray et de Dillenius rangé par le docteur Hill selon le système de Linnaeus et gâté en ce que ces transpositions coupent souvent d'une manière ridicule le texte des premiers auteurs.'
Jacquin, Nikolaus Joseph, Freiherr von (1727-1817)	*Selectarum stirpium Americanarum historia* (1763)	Consulted	BPUN MSR 81	Extraits de livres de botanique.
Jussieu, Antoine-Laurent de (1748-1836)	*Genera plantarum* (1789)	/		** Rousseau botanised with him and knew his natural method; see Chapter 6.
Jussieu, Bernard de (1699-1777)	/	/	/	** Bernard de Jussieu's 'natural orders' from the Jardin du Trianon were published for the first time in Antoine Laurent de Jussieu, *Genera plantarum*; Rousseau botanised with Bernard de Jussieu, see Chapter 6.
Latourrette, Marc-Antoine Claret de (1729-1793)	*Démonstrations élémentaires de la botanique* (1766), 2 vols**	Knew	Letter No.6512, 23 December 1768; Rousseau to Clappier	'Le livre de M. de La Tourrette est très bon, mais il ne traite que des plantes usuelles; il suit les systèmes de Tournefort et de Linnaeus, au lieu qu'il faudroit un système élémentaire pour commencer l'étude des plantes, avant de prendre les grands systèmes pour les classer.' ** Rozier co-authored this work; see entry on Rozier (below).
Lémery, Nicolas (1645-1715)	*Pharmacopé'e universelle* (1763-1764), 2 vols.	Knew	BdAN [Munch MS], art. 281 - Le Chardon Hemorrhoïdal and 282 - Le Chardon-Marie, ou l'Artichaut sauvage	'Le Chardon-Marie contient beaucoup de sel & d'huile, selon Lémery (3) Note de JJR au bas de l'article: (3) C'est son calice et son receptacle qui peuvent se manger à la poivrade ainsi que ceux de la grande Carline, dont nos paysans Suisses sont souvent leur déjeuner.'
Linnaeus, Carolus (1707-1778)	*Systema naturae* (1735)	Owned	Letter No.3769, 23 December 1764; Du Peyrou to Rousseau	'[M]on Professeur en Médecine & en Botanique [Frédéric Samuel Neuhaus], m'a fournie, Linnœi Systema naturae'. ** See Chapter 4
		Requested	Letter No.4232, 2 April 1765; Du Peyrou to Rousseau	'Mr Panckouke m'a assez mal Servi dans l'envoy qu'il m'a fait d'une partie des Livres que je lui avois commis. Il n'a pas eu beaucoup d'égard au choix des Exemplaires, & m'a laissé plusieurs articles de côté, entr'autres le *Genera plantarum*, & le Systema naturae de Linnaeus.'
		Owned	Rêveries, *OC*, i, 1043	'[J]'allois une loupe à la main et mon *Systema Naturae* sous le bras.'

Author(s)	Works	Means of access[i]	Source(s)[ii]	Supporting quotes from letters[iii] / Informational notes
	Fundamenta botanica (1736)	Requested	Letter No.6173, 28 December 1767; Rousseau to Rey	'Livres à acquérir C. Linnaei Fundamenta botanica. 8°.'
	Hortus cliffortianus (1737)	Renounced	Letter No.6620, 6 October 1769; Rousseau to Gouan	'J'ai été forcé par la même raison de renoncer à l'Hortus Cliffortianus.'
	Genera plantarum (1737)	Knew	Letter No.3769, 23 December 1764; Du Peyrou to Rousseau	'Mon Professeur en Médecine & en Botanique, m'a fournie, Linnœi [...] Genera plantarum'.
		Requested	Letter No.4232, 2 April 1765; Du Peyrou to Rousseau	'Mr Panckouke m'a [...] laissé plusieurs articles de côté, entr'autres le Genera plantarum'.
			Letter No.4504bis, 30 June 1765; Rousseau to Duchesne	'Voici les articles de ce second envoi que je vous prie de m'expédier le plustôt qu'il vous sera possible. [...] 2. Linaei Genera Plantarum.'
		Discussed	Letter No.4622, 29 August 1765; Guy to Rousseau	'Je vous ay enfin adressé le petit envoy qui contient: [...] Genera Plantarum'.
			Letter No.6218, 24 January 1768; Malthus to Rousseau	'Je me suis servi de [...] Linnaei Genera.'
		Knew	Letter No.6937, 11 May 1772; Rousseau to Malesherbes	'[N]on seulement il le [Tournefort] cite assez souvent dans son *Species* et presque à chaque article dans son *Genera*'.
		Owned	Letter No.7313, 4 October 1778; Girardin to Du Peyrou	'Catalogue des Livres que Mr Rousseau avoit apporté à Ermenonville [...]. Linné (Caroli) genera plantarum 1, v.1764.'
	Critica botanica (1737)	Requested	Letter No.6173, 28 December 1767; Rousseau to Rey	'Livres à acquerir [...] Ejusdem critica botanica. 8°.'
	Öländska och gothländska resa (1745)	Requested	Letter No.6173, 28 December 1767; Rousseau to Rey	'Livres à acquerir [...] Ejusdem Iter.Oelandicum Ejusdem Iter.Gotlandicum.'
	Amœnitates academicae (1749-1769), 7 vols	Knew	Letter No.6620, 6 October 1769; Rousseau to Gouan	'J'ai été forcé par la même raison de renoncer [...] aux amoenitates academicae.'
	Philosophia botanica (1751)	Requested	Letter No.4504bis, 30 June 1765; Rousseau to Duchesne	'Voici les articles de ce second envoi que je vous prie de m'expédier le plustôt qu'il vous sera possible. [...] 2. Linaei Philosophia Botanica.'
		Owned	Letter No.4622, 29 August 1765; Guy to Rousseau	'Je vous ay enfin adressé le petit envoy qui contient: [...] Philosophia Botanica, 1 vol.'
			Letter No.6891, 21 September 1771; Rousseau to Carl Linnaeus	'[J]e tire un profit plus réel de votre Philosophia botanica que de tous les livres de morale.'
	Species plantarum (1762-1763), 2 vols	Owned	Letter No.3769, 23 December 1764; Du Peyrou to Rousseau	'[M]on Professeur en Médecine & en Botanique, m'a fournie, Linnœi [...] *Species plantarum* [...] me recommandant les dernières Editions de cés ouvrages.'

Author(s)	Works	Means of access[i]	Source(s)[ii]	Supporting quotes from letters[iii] / Informational notes
			Letter No.4232, 2 April 1765; Du Peyrou to Rousseau	'Mr Panckouke [...] m'a laissé plusieurs articles de côté, entr'autres le Genera plantarum, & le Systema naturae de Linnaeus, qu'il m'est important d'avoir pour tirer quelque fruit de l'ouvrage que j'[ai] qui est le Species plantarum.'
			Letter No.4408, 17 May 1765; Du Peyrou to Rousseau	'[I]l y en a qui puissent vous convenir, de les prendre avec vous à Môtiers pour le temps que vous voudrez, par exemple le Garsault et le Species plantarum.'
			Letter No.5482 20 October 1766; Rousseau to the Duchess of Portland	'Je ne connois point les noms Anglois; ceux que je connois sont tous du Pinax de Gaspard Bauhin ou du Species plantarum de M. Linnaeus.'
			Letter No.6479, 7 November 1768; Rousseau to Pierre Liotard	'Les dénominations de plusieurs d'entr'elles ne sont pas exactes ou du moins ne sont pas dans mon *Species* de l'édition de 1762.'
			Letter No.6620, 6 October 1769; Rousseau to Gouan	'J'ai été forcé [...] de renoncer à l'Hortus Cliffortianus aux amoenitates academicae [...] et à beaucoup d'autres livres [...] necessaires pour suppléer par leurs descriptions à la sécheresse du Species.'
			Letter No.6641, 17 December 1769; Rousseau to Latourrette	'Ce n'est point lui, et il n'en est fait aucune mention que je sache ni dans le *Species* ni dans aucun Auteur de Botanique.'
			Letter No.6937, 11 May 1772; Rousseau to Malesherbes	'[N]on seulement il le [Tournefort] cite assez souvent dans son Species.'
			Letter No.7303, 19 September 1778; Girardin to Du Peyrou	'Quant à tous ses livres [de] botaniques [...], quelques uns qui lui restoient encor, avec un grand herbier. Il n'avoit icy qu'une Vingtaine de Volumes des Livres les plus Courants tels que Le Species de Linnaeus.'
			Letter No.7313, 4 October 1778; Girardin to Du Peyrou	'Catalogue des Livres que Mr Rousseau avoit apporté à Ermenonville [...] Linné (Caroli) specie plantarum holm. 2 v. 1762.'
	Systema vegetabilium (1774) (13th ed.)	Owned	Letter No.7313, 4 October 1778; Girardin to Du Peyrou; Rêveries, *OC*, i, 1061	'Catalogue des Livres que Mr. Rousseau avoit apporté à Ermenonville et [...] Linné (Caroli) Systema vegetabilium** Editio XIII. A Paris accd. Murray Gothing 1774.' ** Rousseau refers to this work as 'Régnum végétabile de Murray'; see entry on Murray, below.
	General	Knew	*Introduction, OC*, iv, 1205-1206; see also 1206, 1207, 1208	'M. Linnaeus plein de son système sexuel [...], forma le projet d'une refonte générale dont tout le monde sentoit le besoin [...] Il fit plus, il l'exécuta, et après avoir préparé dans son Critica Botanica les règles sur lesquelles ce travail devoit etre conduit, il détermina dans son Genera plantarum ces genres des plantes, ensuite les espèces dans son Species [...] il établit enfin une nomenclature éclairée, fondée sur les vrais principes de l'art qu'il avoit lui-même exposés.

Author(s)	**Works**	**Means of access**[i]	**Source(s)**[ii]	**Supporting quotes from letters**[iii] / **Informational notes**
			Fragments de botanique, *OC*, iv, 1255-1256	'M. Linnaeus donne le même nom de receptacle à l'adhérence du fruit à la plante et à l'adhérence de la semence au fruit'.
Linnaeus, Carolus fil. (1741-1783)	*Plantarum rariorum horti upsaliensis* (1767)	Requested	Letter No.6134, 25 November 1767; Rousseau to Guy	'Ce sont des Livres de botanique qui se trouvent chez Briasson et dont je vous prie de vouloir bien faire pour moi l'acquisition [...] Caroli Linnaei filii plantae upsalienses.'
			Letter No.6145, 3 December 1767; Guy to Rousseau	'Si j'eusse trouvé Les Livres que vous demandez par votre Lettre du 25. du mois passé, Reliés, je vous les aurois tout de suite Envoyés; Mais M. Briasson ne vend Ses Livres qu'en feuille'.
		Owned	Letter No.6258, 18 February 1768; Guy to Rousseau	'[V]oicy donc ce que la Maison a déboursé à compte Sur l'année 1767. [...] - Caroli Linaei filii planta.'
Lobel, Matthias de (1538-1616)	*Plantarum seu stirpium icons* (1581)	Requested	Letter No.6134, 25 November 1767; Rousseau to Guy	'Est-il donc absolument impossible de trouver à Paris le *Dodonaei Pemptades* &c. sur les figures de Lobel.'
			Letter No.6145, 3 December 1767; Guy to Rousseau	'Si j'eusse trouvé Les Livres que vous demandez par votre Lettre du 25. du mois passé, Reliés, je vous les aurois tout de suite Envoyés; Mais M. Briasson ne vend Ses Livres qu'en feuille.'
			Letter No. 6173, 28 December 1767; Rousseau to Rey	'Livres à acquérir [...] Lobelii Plantarum icones.'
Loesel, Johannes (Loeselius) (1607-1655)	*Flora prussica* (1703)	Requested	Letter No. 6173, 28 December 1767; Rousseau to Rey	'Livres à acquérir [...] Flora Prussica.'
		Borrowed	Letter No. 6924, 19 January 1772; Rousseau to Malesherbes	'[J]e suis allé à votre bibliothèque, j'y ai parcouru les livres de Botanique qui me sont tombés sous la main, j'ai pris note de la pluspart, et j'en ai emporté les quatre dont voici les titres. Tous petit in 4°. [...] 3. Flora Prussica.'
			Letter No. 6991, 2 May 1773; Rousseau to Malesherbes	'J'ai aussi fait remettre à M. de St Remi les deux livres de votre bibliothèque qui me restoient, savoir le Flora Prussica.'
		Consulted		'[L]a transposition du Carduus crispus et du Carduus Acanthoides qu'on démontre au Jardin du Roi l'un pour l'autre, ce que le Botanicon m'avoit déjà fait sentir et que j'ai bien vérifié dans la flora prussica.'
Ludwig, Christian Gottlieb (1709-1773)	*Definitiones generum plantarum* (1737)	Heard about	Letter No. 3769, 23 December 1764; Du Peyrou to Rousseau	'[M]on Professeur en Médecine & en Botanique [Frédéric Samuel Neuhaus], m'a fournie [...] *Ludwig Genera plantarum*.'
		Owned	Letter No. 6038, 26 August 1767; Du Peyrou to Rousseau	'Je vous porte les Livres que vous m'avez demandés, Savoir l'Agrostographie de Scheuzer, et Ludwig.' ** See Chapter 4
	Observationes in methodum plantarum sexualem (1739)	Requested	Letter No. 6003, August 1767; Rousseau to Du Peyrou	'Si vous pouviez aussi trouver la méthode de Ludwig ou quelque autre bon Livre de Botanique, vous me feriez grand plaisir.'

Author(s)	Works	Means of access[i]	Source(s)[ii]	Supporting quotes from letters[iii] / Informational notes
		Owned	Letter No. 6052, 8 September 1767; Coindet to Rousseau	'Il [M. Du Peyrou] vous a apporté le Livre Sur les Gramen de Schutzer & un autre ouvrage de Botanique**qu'il me remettra pour vous le faire parvenir.' ** Coindet refers to Letter 6003.
	General	Knew	*Confessions*, *OC*, i, 643	'Ce grand observateur [Linnaeus] est à mon gré le seul avec Ludwig qui ait vu jusqu'ici la botanique en naturaliste et en philosophe.'
			Fragments de botanique, *OC*, iv, 1255	'[P]ar cette raison n'appeler polypétales comme a fait depuis Mr Ludwig que les fleurs qui ont plus de six pétales.'
Magnol, Pierre (1638-1715)	*Botanicum monspeliense* (1676)	Knew	Letter No.6573, 26 May 1769; Rousseau to Clappier	'Je douterois au contraire que M. Gouan lui-même ait bien connu le *Gentiana filiformis*, puisqu'il lui donne (*Flora monsp*: p.469) un synonyme de *Magnol*** qui ne lui convient en aucune sorte'. ** Gouan cites both works of Magnol on the reference page.
	Hortus regius monspeliense (1697)			
Matthioli, Pietro Andrea (1500-1577)	*Opera quae extant omnia* (1674)	Owned	Letter No.4994 20 January 1766; Rousseau to Guy	'J'avois eu l'intention de prendre à Strasbourg chez M. Koenig quatre livres de botanique, savoir: Mathiolus, Michelius, Valerius Cordus et le Pinax de Gaspard Bauhin [...] Je vous prie, quand vous écrirez à M. Koenig, de le saluer de ma part, et de le prier d'envoyer ces quatre livres'.
			Letter No. 5681, 20 January 1767; Rousseau to Guy	'Note de M. Koenig. *Matthiolus Ed. de Bâle*'.
	General	Knew	BdAN [Munch MS], art. 314 - Le satyrion	'La racine et toute la figure (a) appartiennent à l'Orchis mascula de Linna'us, ainsi que les synonymes cités de Dodonée de Matthiole* et de Gérard.'
Micheli, Pier Antonio (1679-1737)	*Nova plantarum genera*** (1729)	Requested	Letter No. 4994, 20 January 1766; Rousseau to Guy	'J'avois eu l'intention de prendre à Strasbourg chez M. Koenig quatre livres de botanique, savoir: Mathiolus, Michelius**, Valerius Cordus et le Pinax de Gaspard Bauhin [...] Je vous prie, quand vous écrirez à M. Koenig, de le saluer de ma part, et de le prier d'envoyer ces quatre livres.' ** See Chapter 6
		Owned	Letter No. 5681, 20 January 1767; Rousseau to Guy	'Note de M. Koenig. [...] *Michelius**.'
			Letter No. 6063, 12 September 1767; Rousseau to the Duchess of Portland	'Car ayant, dans l'intervalle reçu mes livres de Botanique, j'y ai trouvé à l'aide des figures que Michelius* avoit fait un genre de cette plante.'
			Letter No. 6641, 17 December 1769; Rousseau to Latourrette	'[C]e n'est point lui, et il n'en est fait aucune mention que je sache ni dans le Species ni dans aucun Auteur de Botanique, hors le seul Michelius*.'
			Letter No. 6643, 21 December 1769; Rousseau to the Duchess of Portland	'[C]'est un autre souchet que je ne trouve décrit ni cité par aucun autre botaniste que le seul Michelius*.'

Author(s)	Works	Means of access[i]	Source(s)[ii]	Supporting quotes from letters[iii] / Informational notes
		Knew	*Introduction, OC*, iv, 1206	'[C]es deux mots seuls choisis avec discernement et appliqués avec justesse, font souvent mieux connoître la plante que ne faisoient les longues phrases de Micheli* et de Plukenet.'
Miller, Philip (1691-1771)	*Figures of [...] plants [...] described in the gardeners [sic] dictionary* (1760), 2 vols	Offered	Letter No.6218, 24 January 1768; Malthus to Rousseau	'J'attends vos ordres touchant ces difficultés. Nous avons dans le genre des planches: Miller** à dix guinées**: 250 plantes, les plus rares ou les plus belles.' ** It seems most likely that Malthus refers to this work. ** 10 guineas = £668 in today's currency.
Morison, Robert (1620-1683)	*Plantarum historia universalis Oxoniensis* (1680-1699)	Requested	Letter No.6173, 28 December 1767; Rousseau to Rey	'Livres à acquérir [...] Morisson, de Plantis umbelliferis. 4°.'
		Owned	Letter No.6218, 24 January 1768; Malthus to Rousseau	'Je vous envoye [...] Morison de plantis umbelliferis, ce qui est assez rare; les deux derniers tomes ne le sont pas.'
	General	Searched	Letter No.6924, 19 January 1772; Rousseau to Malesherbes	'Il faut que je n'aye pas su trouver les places de tous les livres de Botanique, car j'ai cherché vainement Cesalpina Columna, et Morisson*.'
Murray, Johan Andreas (Anders) (1740-1791)	*Systema vegetabilium, 13th edn* (1774)	Owned	*Rêveries, OC*, i, 1061	'[M]e voilà sérieusement occupé du sage projet d'apprendre par cœur tout le Régnum végétabile de Murray.' ** This is Linnaeus' *Systema vegetabilium*, 13th edn; see entry on Linnaeus (above) and Letter 7313.
Myrepsus, Nicolaus (late 13th c.)	General	Discussed	*Introduction, OC*, iv, 1202	'Voilà tout l'art que les Myrepsus [...] et les autres Docteurs de ces tems-là mettoient à l'étude des plantes [...], et il seroit difficile peut-être au peuple d'en reconnoître une seule sur leurs noms ou sur leurs descriptions.'
Neuhaus, Frédéric-Samuel (1733-1802)	*Catalogue raisonné des plantes usuelles en Suisse* (not extant)	Heard about	Letter No.3662, 20 November 1764; Du Peyrou to Rousseau	'Avant de finir ma longue épître, il faut dire un mot sur un ouvrage de botanique qui va paroître, & qui sera justement notre fait. C'est une espéce de catalogue raisonné des plantes usuelles de la Suisse, avec leurs usages & leurs noms. J'ay dit à l'Auteur qui est notre Médecin & mon guide pour mes Etudes Elémentaires, de joindre à son ouvrage un article bien essentiel pour nous.'
Parkinson, John (1567-1650)	General	Owned	Letter No. 6655, 26 January 1770; Rousseau to Latourrette	'Je ne doute pas que vous n'ayez la plupart de mes livres. Il peut cependant s'en trouver d'Anglois comme Parkinson** et le Gérard émaculé que peut être n'avez-vous pas.' ** Parkinson wrote two works on botany; we do not know which book is meant.
Petiver, James (ca. 1663-1718)	*Opera, Historiam naturalem spectantia* (1764), 3 vols	Gift	Letter No. 5425, 10 September 1766; the Duchess of Portland to Rousseau	'I shall send you Ray's Synopsis by the first opportunity & with it Petiver*.'
			Letter No. 5725, 12 February 1767; Rousseau to the Duchess of Portland	'[L]e Petiver* m'aide beaucoup par ses figures qui prêtent à mon imagination.'

Author(s)	Works	Means of access[i]	Source(s)[ii]	Supporting quotes from letters[iii] / Informational notes
	Opera, Historiam naturalem spectantia (1767), 2 vols	Discussed	Letter No. 6218, 24 January 1768; Malthus to Rousseau	'On a donné depuis peu le Petiverus* complet en deux volumes in folio – le prix est de six guinées** [...]. [John] Millar [sic]** [...] a acheté tout ce qu'il y avoit de Petiverus*'. ** 6 guineas = about £400 in today's currency. ** This is most probably John Millan, who reissued the *Opera* in 1767.
Pliny the Elder (23-79 CE)	*Caii Plinii secundi historia naturalis libri XXXVII*** (1723)	Requested	Letter No.3768, 23 December 1764; Rousseau to Duchesne	'Pourriez-vous me trouver un *Pline in usum Delphini* qui ne fut pas trop cher?' **A famous edition of Pliny
		Owned	letter n° 6145, 3 December 1767; Guy to Rousseau	'Il me tomba ces jours passés un beau Pline* fol. Je l'ay joint au paquet, je Souhaite qu'il Vous fasse plaisir.'
			Letter No.6151, 10 December 1767; Rousseau to Guy	'Le Pline* est en effet très beau, mais c'est jeter les perles aux pourceaux, ces espèces de livres étant trop savants pour moi, et je ne puis plus me fatiguer à les lire. Je vous suis sensiblement obligé de votre attention; mais cependant ce *Pline* est absolument inutile dans mes mains, et j'ai grande envie de vous le renvoyer.'
			Letter No.6171, 26 December 1767; Guy to Rousseau	'A l'Egard du Pline*, je ne vous l'ay envoyé que parce que je l'ay eu En Echange contre quelques Brochures, et que voyant votre goût décidé pour l'Histoire Naturelle** je pensay qu'il pourroit vous faire quelque plaisir.'
	General	Knew	*Introduction, OC*, iv, 1202	'Au lieu d'étudier les plantes sur la terre, on ne les étudioit plus que dans Pline et Dioscoride.'
Plukenet, Leonard (1642-1706)	*Almagestum botanicum* (1696)	Knew	*Introduction, OC*, iv, 1205	'Les noms des plantes devenoient non-seulement des phrases mais des périodes. Je n'en citerai qu'un seul de Plukenet* qui prouvera que je n'exagère pas.'
			Introduction, OC, iv, 1206	'[C]es deux mots seuls choisis avec discernement et appliqués avec justesse, font souvent mieux connoître la plante que ne faisoient les longues phrases de Micheli et de Plukenet*.'
Pontedera, Guilio (1688-1757)	*Anthologia* (1720)	Borrowed	Letter No.6924, 19 January 1772; Rousseau to Malesherbes	'[J]e suis allé à votre bibliothèque, j'y ai parcouru les livres de Botanique qui me sont tombés sous la main, j'ai pris note de la pluspart, et j'en ai emporté les quatre dont voici les titres. Tous petit in 4°. 1. Pontedera l'Anthologie &c.'
		Knew	Letter No.6933, 17 April 1772; Rousseau to Malesherbes	'J'ai commencé de lire l'Anthologie de Pontedera, et j'y trouve contre le Système sexuel des objections qui me paroissent bien fortes et dont je ne sais pas comment Linnaeus S'est tiré.'
		Gave back	Letter No.6991, 2 May 1773; Rousseau to Malesherbes	'J'ai aussi fait remettre à M. de St Remi les deux livres de votre bibliothèque qui me restoient, savoir le Flora Prussica et l'Anthologie de Pontedera.'

Author(s)	Works	Means of access[i]	Source(s)[ii]	Supporting quotes from letters[iii] / Informational notes
Ray, John (1627-1705)	*Methodus plantarum nova* (1682)	Requested	Letter No.6134, 25 November 1767; Rousseau to Guy	'Ce sont des Livres de botanique qui se trouvent chez Briasson et dont je vous prie de vouloir bien faire pour moi l'acquisition [...] et de faire relier ceux qui ne le seront pas. Ray, Methodus emendata et aucta &c. 8°.'
			Letter No.6145, 3 December 1767; Guy to Rousseau	'Si j'eusse trouvé Les Livres que vous demandez par votre Lettre du 25. du mois passé, Reliés, je vous les aurois tout de suite Envoyés; Mais M. Briasson ne vend Ses Livres qu'en feuille.'
			Letter No.6163, 20 December 1767; Rousseau to Guy	'S'il est absolument impossible de trouver à Paris le livre intitulé Methodus emendata et aucta Auctore J. Raiio, Je vous serois bien obligé si vous pouviez me le faire venir d'Angleterre.'
		Owned	Letter No.6218, 24 January 1768; Malthus to Rousseau	'Je vous envoye Raii methodus emendata & aucta.'
	Synopsis methodica stirpium britannicarum (1690)	Gift	Letter No.5425, 10 September 1766; the Duchess of Portland to Rousseau	'I shall send you Ray's Synopsis by the first opportunity.'
		Knew	Letter No.5482 20 October 1766; Rousseau to the Duchess of Portland	'[C]eux que je connois sont tous du Pinax de Gaspard Bauhin ou du Species plantarum de M. Linnaeus, et je ne puis en faire la Synonymie avec Gérard qui leur est anterieur à l'un et à l'autre, ni avec le Synopsis, qui est antérieur au second et qui cite rarement le premier.'
			Letter No.5725, 12 February 1767; Rousseau to the Duchess of Portland	'Le Synopsis comprend des descriptions à ma portée et que je suis en état de Suivre sans m'arracher les yeux'.
		Knew	Letter No.6645, 26 December 1769; Rousseau to Gouan	'Je crois les méthodes des botanistes très bonnes pour classer les plantes déjà connues; [...] Ray me paraît avoir approché plus qu'aucun autre de cette méthode élémentaire.'.
	General	Knew	*Introduction, OC*, iv, 1205	'[L]es botanistes forcés de chercher un fil pour s'en tirer, s'attachèrent enfin sérieusement à la méthode ; [...] Ray*.'
			Fragments de botanique, OC, iv, 1254	'Selon Rey les fleurs appellées indéterminément polypétales sont celles dont la corolle est composée de plus de 5 pétales.'
		Knew	BdAN [Munch MS], art. 84 - La Bette; see also BdAN [Munch MS], arts. 101, 250, 337.	'(15 bis) Il est vrai que Ray ne fait qu'une espèce de la Bette maritime et de celle-ci, mais tous les autres Botanistes les distinguent.'
Regnault, Nicolas-François (1746-1810) and Geneviève de Nangis Regnault (1746-1802)	*La Botanique mise à la portée de tout le monde*** (1774)	Knew		** Rousseau annotated a copy in 1777-1778; see Chapter 7 and BdAN [Munch].
			BPUN MSR 81	Extraits de livres de botanique.

Author(s)	**Works**	**Means of access**[i]	**Source(s)**[ii]	**Supporting quotes from letters**[iii] / **Informational notes**
Reede tot Drakenstein, Hendrik Adriaan van (1637-1691)	*Hortus indicus malabaricus* (1678-1703), 12 vols	Knew	*Introduction, OC*, iv, 1209	'Pour les plantes exotiques découvertes récemment, il [Linnaeus] a cité un ou deux auteurs modernes et les figures de Rheedi, de Rhumphius et quelques autres, et s'en est tenu là.'
Rivinus, Augustus Quirinius (Bachmann) (1652-1723)	General	Discussed	*Introduction, OC*, iv, 1205	'[L]es botanistes forcés de chercher un fil pour s'en tirer, s'attachèrent enfin sérieusement à la méthode Herman, Rivin*.'
Rozier, Jean-Baptiste Francois (1734-1793)	*Démonstrations élémentaires de la botanique* (1766), 2 vols**	Knew	Letter No. 6512, 23 December 1768; Rousseau to Clappier	'Le livre de M. de La Tourrette est très bon, mais il ne traite que des plantes usuelles.' ** Rozier co-authored this work.
Rumphius, Georg Eberhard (Rumpf) (1627/28-1702)	*Herbarium amboinense* (1750), 6 vols	Rejected gift	Letter No. 7093, 11 July 1776; Rousseau to the Duchess of Portland	'[I]l en seroit de même et par la même raison de l'herbarium amboinense, et cette perte seroit regrettable à proportion du prix de ce livre et de l'envoi. Voila les raison[s] qui m'empêchent d'accepter ce superbe cadeau.'
	General	Knew	*Introduction, OC*, iv, 1209	'Pour les plantes exotiques découvertes récemment, il [Linnaeus] a cité un ou deux auteurs modernes et les figures de Rheedi, de Rhumphius* et quelques autres, et s'en est tenu là.'
Sauvages de La Croix, François Boissier de (1706-1767)	*Methodus foliorum* (1751)	Knew	Letter No. 4408, 17 May 1765; Du Peyrou to Rousseau	'Si parmi les livres que vous avez, il y en a qui puissent vous convenir, de les prendre avec vous á Môtiers pour le temps que vous voudrez, par exemple le Garsault et le *Species plantarum*, et même le Sauvages quoiqu'il ne m'appartienne pas.'
		Requested	Letter No. 4417, 19 May 1765; Rousseau to Duchesne	'[J]'aurai encore besoin [...] d'un in-12 intitulé *Methodus Foliorum*, par M. de Sauvages.'
		Owned	Letter No. 4577, 11 August 1765; Rousseau to Guy	'En me faisant cet envoi je vous prie de m'envoyer aussi la note exacte de ce que je vous dois jusqu'à ce moment, en y comprenant le Sauvages** que je prendrai le parti de garder pour moi.' ** annotated by Rousseau; his copy is at Jesus College, Cambridge; see Chapter 5.
			Letter No.4622, 29 August 1765; Guy to Rousseau	'Je vous ay enfin adressé le petit envoy qui contient: [...] Methodus foliorum, de Sauvages.'
		Knew	Letter No.6573, 26 May 1769; Rousseau to Clappier	'[L]e Gentiana filiformis n'étant point pourpre et ne pouvant guère être appellé rameux, puisque, comme le dit très bien Sauvages, il n'a communément que trois fleurs.'
			Letter No.6933, 17 April 1772; Rousseau to Malesherbes	'A l'égard des plantes dont Tournefort n'a pas tiré les noms du Pinax, on en trouve aisément la concordance dans les Auteurs François Linnaeistes, tels que Sauvages.'
Scheuchzer, Johannes (1684-1738)	*Agrostographia sive graminum* 1719)	Requested	Letter No.5960, 5 July 1767; Rousseau to Coindet	'S'il est impossible de trouver l'*Agrostographie* ou traité des gramens de Scheuzer, voyez dans quelqu'une de vos courses de la rue St Jacques.'

Author(s)	**Works**	**Means of access**[i]	**Source(s)**[ii]	**Supporting quotes from letters**[iii] **/ Informational notes**
			Letter No.6003, August 1767; Rousseau to Du Peyrou	'Ne pourriez-vous point trouver où vous êtes l'Agrostographia, ou Traité des Gramens de Scheuzer?'
		Owned	Letter No.6038, 26 August 1767; Du Peyrou to Rousseau	'Je vous porte les Livres que vous m'avez demandés, Savoir l'Agrostographie de Scheuzer, et Ludwig.'
			Letter No.6052, 8 September 1767; Coindet to Rousseau	'Il [M. Du Peyrou] vous à apporté le Livre Sur les Gramen de Scheuzer.'
			Letter No.6933, 17 April 1772; Rousseau to Malesherbes	'J'ai cependant aussi l'Agrostographie de Scheuzer.'
Scheuchzer, Johann Jakob (1672-1733)	*Iter alpina* (1723), 4 vols	Requested	Letter No.6173, 28 December 1767; Rousseau to Rey	'Livres à acquérir J. J. Scheuzer Iter alpinum. 4°.'
Scopoli, Giovanni-Antonio (Joannes Antonii) (1723-1788)	*Flora carniolica* (1760)	Owned	Letter No.7313, 4 October 1778; Girardin to Du Peyrou	'Catalogue des Livres que Mr. Rousseau avoit apporté à Ermenonville [...] Scopoli (Joann. Antoin.) flora Carniolica, Vienne 1760.'
Séguier, Jean-François (1703-1784)	*Plantae veronenses stirpium* (1745-1754), 3 vols	Discussed	Letter No.6546, 28 February 1769; Rousseau to Du Peyrou	'M. Séguier, célèbre par le Plantae Veronenses que vous avez peut-être ou que vous devriez avoir, vient de m'envoyer des plantes qui m'ont remis sur mon herbier et sur mes bouquins.'
Stillingfleet, Benjamin (1702-1771)	*Miscellaneous tracts relating to husbandry* (1759)	Owned	Letter No. 5655, 2 January 1767; Rousseau to Malthus	'J'ai acquis votre Stillingfleet et je n'en suis pas plus avancé.'
Suardus, Paulus (1479-1481)	General	Knew	*Introduction, OC*, iv, 1202	'Voilà tout l'art que les [...], les Suardus*, [...] et les autres Docteurs de ces tems-là mettoient à l'étude des plantes [...], et il seroit difficile peut-être au peuple d'en reconnoître une seule sur leurs noms ou sur leurs descriptions.'
Tabernae-montanus, Jakobus Theodorus (Bergzabern) (1520-1590)	*Eicones plantarum seu stirpium* (1590)	Requested	Letter No. 6173, 28 December 1767; Rousseau to Rey	'Livres à acquérir Tabernaemontani icones plantarum.'
Theophrastus of Eresos (c.370-285 BCE)	General	Knew	Letter No. 5693, 31 January 1767; Rousseau to Davenport	'[J]'ai pensé qu'il vaudroit mieux qu'on y mit mon vieux Dictionnaire de Calepin in folio [...] qui contient encore plus sûrement les noms Grecs de plantes tirés de Dioscoride et de Theophraste*.'
Tournefort, Joseph Pitton de (1656-1708)	*Institutiones rei herbariae* (1700), 3 vols.	Requested	Letter No. 4417, 19 May 1765; Rousseau to Duchesne	'Je vous suis fort obligé du soin que vous avez eu de prendre et de faire relier pour moi le *Tournefort* et le *Vaillant*.'
			Letter No. 3768, 23 December 1764; Rousseau to Duchesne	'Je suis bien impatient de Savoir si nous aurons les Planches du Livre de Botanique. En cas qu'oui voici une note d'autres livres que je vous prie d'y joindre pour m'en faire un envoi le plustôt qu'il se pourra. Tournefort. Insitutiones Rei Herbariae la meilleure édition.'

Author(s)	Works	Means of access[i]	Source(s)[ii]	Supporting quotes from letters[iii] / Informational notes
			Letter No.3769, 23 December 1764; Du Peyrou to Rousseau	'Ce Sera aussi à votre loisir que je vous prie, Monsieur, de me marquer le titre de l'ouvrage de *Tournefort* que vous me conseillez d'avoir.'
			Letter No.4321, 21 April 1765; Rousseau to Duchesne	'J'ai reçu enfin la petite caisse; elle avait été déballée et ouverte comme à l'ordinaire; elle contenait plusieurs articles auxquels je n'avais pas songé; mais en revanche elle n'en contenait aucun de ceux que je vous avais demandés, pas même le *Tournefort*, dont la privation me fâche beaucoup.'
			Letter No. 4351, 28 April 1765; Rousseau to Duchesne	'J'ai, Monsieur, un extrême besoin de deux livres de botanique, l'un de Tournefort, en trois volumes in-4°, intitulé Institutiones rei herboriae.'
		Bound for Rousseau	Letter No. 4404, 15 May 1765; Coindet to Rousseau	'[J]e n'ay rien vu encore qui puisse vous convenir excepté *Tournefort* que je vous aurois envoyé Si je n'avois pas trouvé l'autre jour Mr Duchesne occupé à vous le faire relier.'
		Owned	Letter No. 4417, 19 May 1765; Rousseau to Duchesne	'Je vous suis fort obligé du soin que vous avez eu de prendre et de faire relier pour moi le *Tournefort* et le *Vaillant*.'
			Letter No. 7313, 4 October 1778; Girardin to Du Peyrou	'Catalogue des Livres que Mr Rousseau avoit apporté à Ermenonville [...]. Tournefort (Joss. Pithon) Institutiones rei herbarie 3 vol. fig. Parisiis e Typog. Regia 1700.'
	Elemens de botanique (1694), 3 vols	Requested	Letter No. 4351, 28 April 1765; Rousseau to Duchesne	'J'ai, Monsieur, un extrême besoin de deux livres de botanique, l'un de Tournefort, ou, si vous ne trouvez pas celui-là, le livre du même auteur intitulé Elémens de botanique.'
	Relation d'un voyage du Levant (1717)	Knew	BPUN MSR 81	Extraits de livres de botanique.
	General	Knew	Letter No. 6933, 17 April 1772; Rousseau to Malesherbes	'[Q]uand Tournefort n'a fait que prendre la phrase de Bauhin, de citer l'auteur original et non pas celui qui l'a transcrit, comme on fait très injustement en France. De sorte que, quoique presque toute la nomenclature de *Tournefort* soit tirée mot à mot du Pinax on croiroit à lire les botanistes françois qu'il n'a jamais existé ni Bauhin ni Pinax.'
			Letter No.6916, 2 January 1772; Malesherbes to Rousseau	'Vos noms sont ceux de Linnaeus. Ceux Sous lesquels mon herbier est rangé Sont principalement ceux de Tournefort et de Vaillant, les Seuls qu'on connoissoit quand j'ai commencé à herboriser.'
			Letter No.6954, 16 July 1772; Rousseau to Mme Delessert	'De grands Botanistes très bien oculés ne les ont pas mieux vues. *Tournefort* lui-même n'en fait aucune mention.'

Author(s)	**Works**	**Means of access**[i]	**Source(s)**[ii]	**Supporting quotes from letters**[iii] **/ Informational notes**
			Letter No.7303, 19 September 1778; Girardin to Du Peyrou	'Quant à tous ses livres [de] botaniques [...], quelques uns qui lui restoient encor, avec un grand herbier. Il n'avoit icy qu'une Vingtaine de Volumes des Livres les plus Courants tels que Le Species de Linnaeus, Tournefort, La 1re Edition d'Haller et quelques autres de ce genre.'
		Knew	*Introduction, OC*, iv, 1203	'Enfin parurent ces deux illustres frères, qui ont plus fait eux seuls pour le progrès de la Botanique que tous les autres ensemble qui les ont précédés et même suivis jusqu'à Tournefort.'
			Introduction, OC, iv, 1205	'[L]'immortel Tournefort l'emporta sur eux tous; il rangea le premier systématiquement tout le règne végétal; et réformant en partie la nomenclature, la combina par ses nouveaux genres avec celle de Gaspard Bauhin.'
			Introduction, OC, iv, 1207	Ce n'est pas que cette nomenclature Linnéene n'ait encore ses défauts et ne laisse de grandes prises à la critique; mais [...] il vaut cent fois mieux adopter celle-là que de n'en voir aucune, ou retomber dans les phrases de Tournefort et de Gaspard Bauhin.'
Tragus, Hieronymous (Bock) (1498-1554)	*Kreütter Buch*** (1545)	Owned / Gave away	Letter No. 6655, 26 January 1770; Rousseau to Latourrette	'[J]'avois aussi Tragus**, mais je l'ai donné à M. Clappier.' ** See Chapter 7
Vaillant, Sébastian (1669-1722)	*Botanicon parisiense* (1743)	Requested	Letter No.3768, 23 December 1764; Rousseau to Duchesne	'[U]ne note d'autres livres que je vous prie d'y joindre pour m'en faire un envoi le plustôt qu'il se pourra. Vaillant. Botanicon Parisiense.'
			Letter No.4351, 28 April 1765; Rousseau to Duchesne	'J'ai, Monsieur, un extrême besoin de deux livres de botanique. L'un de *Tournefort* [...] l'autre livre est de M. Vaillant, intitulé *Botanicon Parisiense*.'
		Owned	Letter No.4417, 19 May 1765; Rousseau to Duchesne	'Je vous suis fort obligé du soin que vous avez eu de prendre et de faire relier pour moi le *Tournefort* et le *Vaillant*.'
			Letter No.6991, 2 May 1773; Rousseau to Malesherbes	'[L]a transposition du Carduus crispus et du Carduus Acanthoides qu'on démontre au Jardin du Roi l'un pour l'autre, ce que le Botanicon m'avoit déja fait sentir et que j'ai bien vérifié dans la flora prussica.' ** Rousseau's copy is at Harvard University; see Chapter 6.
Valmont de Bomare, Jacques-Christophe (1731-1807)	*Dictionnaire raisonné universel d'histoire naturelle* (1764) 5 vols	Owned	Letter No.3597, 24 October 1764; Guy to Rousseau	'J'ay souscrit pour l'histoire des plantes de La Lorraine, Et je vous Enverray les deux volumes qui paroissent, avec un Nouveau Dictionnaire d'histoire Naturelle En 5. vol. 8° qui paroît depuis quelques mois: on en dit du bien.'
			Letter No.3664, 20 November 1764; Guy to Rousseau	'[A]u moyen de ce que vous souhaitez, j'ay Différé jusqu'à ce jour pour ramasser Le peu que je vous envoye. [...] Donc, mémoire: [...] Le Nouveau dictionnaire d'Histoire naturelle, 5. vol. 8.'

Author(s)	**Works**	**Means of access**[i]	**Source(s)**[ii]	**Supporting quotes from letters**[iii] **/ Informational notes**
Villanova, Arnald(us) (1240-1311)	General	Knew	*Introduction*, *OC*, iv, 1202	'Voilà tout l'art que les [...] Villanova* [...] et les autres Docteurs de ces tems-là mettoient à l'étude des plantes dont ils ont parlé dans leurs livres, et il seroit difficile peut-être au peulpe d'en reconnoître une seule sur leurs noms ou sur leurs descriptions.'

Appendix 2[1]

Rousseau's botanical correspondents and contacts[2]

Baurin, abbé (fl.1769-1787), Catholic priest, archdeacon of Salmorenc (Rhône-Alpes); botanised with Rousseau on Mont Pilat (August 1769).

Becket, Thomas (?1722-1813), English publisher; business partner of Pieter Abraham De Hondt (see below). Published *Miscellaneous works of M. J.-J. Rousseau*, 5 vols, (1767), *Recueil de Lettres de M. J.-J. Rousseau [...] relatives à sa persécution et à sa défense* and Hume's account of his quarrel with Rousseau.

Bernardin de St Pierre, Jacques-Henri (1737-1814), French writer, military engineer and botanist; friend of Rousseau; wrote *Etudes de la nature* (1784) and the romantic novel, *Paul et Virginie* (1788), intendant of the Jardin des plantes and cabinet d'histoire naturelle, Paris (1792-1793); Professeur de morale at the Ecole normale supérieure (1794); member of the Institut de France (1798).

Bondeli, Julie von (1731-1778), Swiss author whose salon was a centre of Bern's intellectual life of in the 1750s and 1760s.

Boothby, Sir Brooke, 7th baronet (1743-1824), English gentleman, poet, author; met Rousseau at Wootton and had his *Dialogues* printed at Lichfield (1780).

Boufflers-Rouverel, Marie Charlotte Hippolyte Campet de Saujon, comtesse de (1724-1800), French aristocrat, mistress of the prince de Conti; her salon at the Temple was one of the most brilliant of the period.

Boy de La Tour, Julie Anne Marie (1715-1780), wife of a Swiss banker, Pierre Boy de La Tour (1706-1758); carried on his business after his death. Her daughters, Madeleine-Catherine Delessert and Julie Emélie Willading née Boy de la Tour, botanised with Rousseau.

Brémond d'Ars, Marie Madeleine de, marquise de Verdelin (1728-1810), French aristocrat known for her correspondence with Rousseau.

Buffon, Georges Louis Leclerc, comte de (1707-1788), French naturalist,

1. Based on background research by Michelle Ha Ming Wai, with the assistance of Brian Ho Kin Ting and Irwin Chan Yu Shing.
2. This list is limited to persons known directly to Rousseau.

best-selling author; elected to the Académie royale des Sciences (Paris) at the age of 26; intendant of the Jardin du roi (1739); co-author of *Histoire naturelle*; unanimously elected to the Académie française (1753).

Buttafoco, Mattéo, comte (1731-1806), Corsican patriot, general in French service; asked Rousseau to write a constitution for Corsica. Elected to the Etats généraux of France in 1789, but returned to Corsica in 1794.

Clappier, Pierre (1740-1818), French physician, botanist of Montpellier; friend of Pierre Liotard (see 'Liotard, Pierre' below).

Coindet, François (1734-1809), bank employee in Geneva; moved in high society, but neglected his work; assisted Rousseau with book acquisitions.

Constable, William (1721-1791), English gentleman, amateur botanist, Fellow of the Royal Society (1775).

Dalberg, Nils (1736-1820), Swedish; physician to the Prince Royal, who became Gustav III of Sweden (1771).

Davenport, Richard (d.1771), English gentleman; hosted Rousseau at Wootton Hall in Staffordshire (1766-1767). Davenport educated his grandson, Davies Davenport (1757-1837), according to Rousseau's method without success.

Delessert, Madeleine Catherine (1747-1816), daughter of Pierre Boy de La Tour (1708-1758) and Julie Anne Marie Boy de La Tour (see above); married Etienne Gabriel Delessert (1735-1816), a Swiss banker (1767). Recipient of Rousseau's *Lettres élémentaires sur la botanique*, written for her daughter and other children. He visited her frequently in Paris (1776-1778).

Deleyre, Alexandre (1721-1797), French philosopher, revolutionary politician, author; député at the Convention; voted for the execution of Louis XVI.

Deluc, Jean André (1727-1817), Swiss geologist, meteorologist; a leader of the Représentants in Geneva (1760s). Fellow of the Royal Society (1773).

Dewes, Mary (c.1746-1814), English gentlewoman; daughter of Anne Granville Dewes, and niece of the botanical artist, Mary Delany, née Granville (1780-1788), a close friend of the Duchess of Portland. Dewes's uncle was Bernard Granville, who befriended Rousseau in England (see 'Granville, Bernard', below).

Dombey, Joseph (1742-1794), French botanist; explored South America in search of rare plants and other curiosities; died in the service of the French revolutionary regime. Gave Rousseau an important herbarium in 1768 that Rousseau may have later returned c.1770; see Chronology and Appendix 3.

Donin de Champagneux, Luc-Antoine (1744-1807), French politician, journalist, lawyer; associate of Roland de La Platière, but escaped the guillotine.

Du Peyrou, Pierre Alexandre (1729-1794), Neuchâtel businessman and owner of plantations in Surinam; a long-standing friend of Rousseau and, as one of his literary executors, was closely involved with the complicated publication of the latter's *Collection complète* in Geneva (1780-1789).

Duchesne, Marie Antoinette (after 1713-1793), wife of Nicholas-Bonaventure Duchesne (see 'Duchesne, Nicolas-Bonaventure', below); carried on her husband's business after his death, as 'la veuve Duchesne'.

Duchesne, Nicolas-Bonaventure (1711-1765), French publisher, bookseller.

Duclos, Charles Pinot, (1704-1772), French philosophe, member of several academies; protégé of King Louis XV's powerful mistress, Mme de Pompadour.

Dutens, Vincent-Louis (1730-1812), French diplomat, author; moved to England as a youth; ordained an Anglican minister (1758); had a wide literary correspondence; Fellow of the Royal Society and the Académie des Inscriptions et Belles-Lettres.

Flamanville, Marie Bonaventure Jean Joseph Augustin Le Conte de Nonant de Raray, chevalier de (1751-1779), French army officer, chevalier of the Order of Malta.

Gagnebin, Abraham (1707-1800), physician in La Ferrière in the Swiss Jura, botanist; longtime associate of Albrecht von Haller; had encyclopedic knowledge of the local flora; Gagnebin served as guide to Rousseau's collège de botanique.

Girardin, René Louis de Girardin, marquis de (1735-1808), French aristocrat, landscape designer, author of *De la composition des paysages* (1777), in which he set out the theories of landscape design that he applied on his estate at Ermenonville, Oise, where Rousseau died in 1778. His ideas had a great influence in France; particularly important was Rousseau's tomb. Girardin drew up the list of Rousseau's last botanical library.

Gouan, Antoine (1735-1821), French physician, botanist; student of François Boissier de Sauvages de la Croix (1706-1767); professor of botany at the University of Montpellier (1767); member of the Société royale des sciences of Montpellier from 1757, author of *Hortus Regius Monspeliensis* (1762), *Flora Monspeliaca* (1765) and other botanical works.

Granville, Bernard (1699-1775), English gentleman related to the earls of Bath; lived at Calwich Abbey, near Wootton; Rousseau became acquainted with the Duchess of Portland through Granville, who was the brother of a close friend of the Duchess, the botanical cut paper artist, Mary Delany, née Granville (1780-1788).

Guy, Pierre (1715-1795), French printer, bookseller; his second wife was the widow of his business associate, Duchesne.

Guyot, Alexandre-Nicolas-Joseph, chevalier de Maître (1730-?), French aristocrat, amateur botanist; friend of Abraham Gagnebin. Created marquis de Maîche-Villecomte (1784).

Haller, Albrecht von (1707-1777), Swiss patrician of Bern; an enormously prolific author in botany, medicine, and literature; he is a founder of experimental physiology. He taught at Göttingen (1736-1753), but returned to Bern, where he served in a number of administrative posts. His *Historia stirpium indigenarum helvetiae* (1768) documented the flora of greater Switzerland for the first time, while his poem, 'Die Alpen', was an early expression of Helvetism.

Harcourt, George Simon Harcourt, Viscount Nuneham (1736-1809), English aristocrat, politician; created a Rousseau museum in his country house, Nuneham Courtenay.

Hess, Caspar (1727-1800), Swiss academic; held professorships in Hebrew, natural law and moral philosophy, logic and rhetoric, in Zürich. In 1762 he and his bride visited Môtiers, and invited Rousseau to Zürich.

Hirzel, Johann Kaspar (1725-1803), Swiss physician, author; first physician of Zürich; co-founder and first president of the Helvetische Gesellschaft (1762); author of *Die Wirthschaft eines philosophischen Bauers* (1761), widely read in its French translation, *Le Socrate rustique* (1762).

Hondt, Pieter Abraham de (fl. 1760-1776), a Dutch printer; worked with Thomas Becket in London (1760-1776).

Hume, David (1711-1776), Scottish historian and philosopher, famous for his *Treatise of human nature* (1739-1740), and *Enquiry concerning human understanding* (1748), a simplification of the former; his *History of England* (1754-1762) made his reputation; his efforts to settle Rousseau in England (1766) ended acrimoniously.

D'Ivernois, François-Henri (1722-1778), French-born Swiss merchant, politician. Bourgeois of Geneva (1748); a leader of Geneva's Représentants, which advocated political reforms; member of the Conseil des Deux Cents; suspected of collusion with Louis Mandrin, an outlaw, bandit and smuggler.

D'Ivernois, Jean Antoine (1703-1765), honorary physician to the King of Prussia, member of the Academy of Sciences of St Petersburg, author of the first flora of Neuchâtel, containing 1,227 species, and correspondent of Haller, who used his work.

Jussieu, Antoine Laurent de (1748-1836), French botanist; nephew of Bernard de Jussieu, was appointed to the Académie royale des sciences at the age of 25, arranged the Jardin du roi according to his own natural method and published his uncle's natural method in *Genera plantarum* (1789).

Jussieu, Bernard de (1699-1777), French botanist; studied medicine at Montpellier and held a series of posts at the Jardin du roi; organised a scientific garden at the Petit Trianon, Versailles, according to his natural method.

Keith, George (1692/3?-1778), Scottish Jacobite, hereditary Earl Marischal and chief of the clan Keith; punished for loyalty to the Stuarts by attainder and confiscation of his Scottish estates. Served the King of Prussia in various military and other posts; governor of Neuchâtel (departed 1763), where he befriended Rousseau.

Laliaud, Henri (1718-1780), French apothecary, tradesman; helped Rousseau find accommodation in the late 1760s.

Latourrette, Marc Antoine Louis Claret de (1729-1793), French magistrate and botanist; had an especially rich herbarium; helped establish the botanic garden of the Ecole vétérinaire, Lyon; co-author with abbé Rozier (see 'Rozier, abbé Jean Baptiste François') of *Démonstrations élémentaires de botanique* (1766), and author of *Voyage au Mont Pilat* (1770).

Laurencin, Jean Baptiste Espérance Blandine de Laurencin de Chanzé, comte de (1740-1812), French army officer and novelist.

Lenieps, Toussaint Pierre (1697-1774), Master jeweller banished from Geneva for his association with the opposition figure, Micheli du Crest (imprisoned 1731). Banker in Paris, leader of the Swiss opposition abroad.

Linnaeus, Carolus (knighted 1761; henceforth: Carl von Linné) (1707-1778), Prolific Swedish physician, botanist and systematiser; published his first major work of classification, *Systema naturae*, in 1735; professor at Uppsala from 1741. His *Species plantarum* (1753) established modern binomial nomenclature.

Liotard, Pierre (?1729-1796), Weaver, soldier and botanist; taught himself Latin, and was appointed director of the Grenoble botanic garden (1782).

Luze Warney, Marianne Françoise de (1728-1796), Swiss aristocrat and relative of the Roguin-Delesserts; married the businessman, Jean Jacques de Luze (1728-1779).

Malesherbes, Chrétien Guillaume de Lamoignon de (1721-1794), French statesman, botanist, member of the Académie des Sciences (1750). Guillotined for serving as Louis XVI's defence lawyer.

Malthus, Daniel (1730-1800), English gentleman, botanist and father of the economist, Thomas Robert Malthus (1766-1834); acquired Rousseau's botanical books in 1775, and left most of them to his cousin, Jane Dalton (c.1742-1817).

Mirabeau, Victor Riqueti, marquis de (1715-1789), French aristocrat and physiocrat; author of *L'Ami des hommes ou Traité sur la population* (1756) and a letter appended to the French translation of Hirzel's *Socrate rustique.*

Montmorency Luxembourg, Charles François Frédéric de, duc de Luxembourg (1702-1764), French aristocrat and career soldier; protector of Rousseau; maréchal of France (1757). Rousseau authored two letters to him containing reflections on Neuchâtel and Switzerland (1763).

Moultou, Paul Claude (1731-1787), French Huguenot pastor trained in Geneva; ordained minister (1754); resigned the ministry (1773); entrusted by Rousseau with his unpublished manuscripts before his death 1778; co-published Rousseau's works with Pierre Alexandre Du Peyrou (see 'Du Peyrou, Pierre Alexandre').

Orlow, Grigory Grigoryevitch, count (1734-1783), Russian aristocrat, lover and chief advisor of Catherine the Great; led the coup that put her on the throne (1762). After Catherine discarded him for Grigory Potemkin, he lived abroad.

Panckoucke, Charles-Joseph (1736-1798), important French publisher; published Buffon, Voltaire, the supplement to the *Encyclopédie*, 5 vols (1776-1777), the *Encyclopédie méthodique* (starting 1782). He asked Rousseau to write a work on mosses, and Rousseau is supposed to have started a herbarium for him c.1771; its whereabouts are unknown.

Portland, Duchess of (Margaret Cavendish Harley Bentinck) (1715-1785), wealthy English aristocrat, avid botanist and collector; married William Bentinck, Duke of Portland (1734); botanised with Queen Charlotte and Mary Delany (see 'Granville, Bernard' above). Rousseau met the Duchess through Bernard Granville in England in 1766-1767.

Pramont, abbé de (d.1786), Canon of the cathedral of Vannes and vicar-general of the diocese of Viviers; interested in botany and music. At Pramont's request, Rousseau annotated the Regnaults' *La Botanique mise à la portée de tout le monde.*

Pury, Lieutenant-colonel Abraham de (1724-1807), former officer in the Sardinian service, militia officer and member of the Conseil d'état (principality of Neuchâtel), ardent Helvetic patriot and author, member of Rousseau's 'collège botanique', and father-in-law of Pierre Alexandre Du Peyrou.

Rey, Marc Michel (1720-1780), Swiss publisher and bookseller domiciled in Amsterdam; principal publisher of Rousseau.

Richard, Claude, (1705-1784), French horticulturalist; head gardener of the Trianon garden at Versailles and a close collaborator of Bernard de Jussieu.

Rozier, abbé Jean Baptiste François (1734-1793), French cleric and botanist; collaborated with Latourrette at the Ecole vétérinaire of Lyon and published an important journal, *Journal de physique.* Botanised with Rousseau in the Rhônes-Alpes region; his herbarium is in the Bibliothèque municipale of Lyon.

Thouin, André (1747-1824), French botanist and horticulturist, head gardener at the jardin du roi (Paris); professor of cultivation at the Jardin des Plantes; coordinated an immense botanical exchange and correspondence network that influenced both science and public policy. Rousseau botanised with him (1773-1778).

Tissot, Samuel Auguste André David (1728-1797), physician of Lausanne

(Vaud), medical author, and expert on nervous diseases. He promoted inoculation, opposed blood-letting and promoted the ancient idea of a 'regime' of diet and exercise; authored a treatise against masturbation, *L'Onanisme*, and *Avis au peuple sur sa santé* (1761), translated into seventeen languages.

Tscharner, Vincenz Bernhard (1728-1778), Swiss author, publisher, and politician; co-founder of the Société typographique de Berne; president of the Société économique de Berne and a founding member of the Helvetische Gesellschaft. Author of *Historische, geographische und physikalische Beschreibung des Schweizerlandes. In alphabetischer Ordnung abgehandelt* (1782-1783).

Tschudi, Jean Baptiste Louis Théodore, baron de (1734-1784), Swiss diplomat, military officer, author, horticulturist and a founder of the Académie of Metz (1761); wrote encyclopedia articles and a treatise on exotic botany, *De la transplantation, de la naturalisation, et du perfectionnement des végétaux* (London and Paris, 1778).

Appendix 3

Rousseau's herbaria: summary tables

Table 1: Rousseau's extant herbaria[1]

Place:	Carnavalet (Paris)	Montmorency	MdAD (Paris)	Neuchâtel[2]	Zürich
Institution:	Musée Carnavalet	Musée Jean-Jacques Rousseau	Musée des arts décoratifs	Bibliothèque publique et universitaire	Zentralbibliothek
Provenance:	Rousseau / Girardin?[3]	Rousseau	Rousseau	Rousseau / Dombey?	Rousseau
Contents:	8 specimens, incl. American species[4]	168 species	315 specimens, incl. mosses, algae and 11 lichens	400 + specimens[5]; 5 boxes / 'cartons'	100 plants, 1 alga; incl. all important families
Organisation:	Not clear	Alphabetical	Not clear	Linnaean?	Linnaean
Recipient:	Not known	Madelon Delessert[6]	Malesherbes?[7]	Malthus, Dombey? Stanislas Girardin	Julie Emélie Boy de la Tour[8]
Dimensions:	Quarto	23 × 16.4 cm	Portatif, in-12, 15.8 × 10 cm	40 cm × 30 cm × 9 cm	21.6 × 10.6 cm
Year, if known:		1774		1768	1772
Other information:	Doubts as to its authenticity	Letter to Mme Delessert, 28 May 1774[9]		Part of 'grand herbier'?[10] See Table 2	Letter to Julie Anne Marie Boy de la Tour, 16 April 1772[11]

1. Full details of these herbaria will be given in *Œuvres complètes de Jean Jacques Rousseau* (Paris, forthcoming).
2. See François Matthey, 'La dernière passion de Jean-Jacques Rousseau', *Revue neuchâteloise* 25:100 (Autumn 1982), p.27-40; Henry Cheyron, 'L'amour de la botanique: les annotations de Jean-Jacques Rousseau sur la Botanique de Regnault,' *Littératures* 4 (Autumn 1983), p.55-95; 'Ray et Sauvages annotés par Jean-Jacques Rousseau', *Littératures* 15 (Autumn 1986), p.83-99.
3. 'L'écriture est indiscutablement celle du philosophe'. B. de Montgolfier and M. Gallet, 'Souvenirs de Voltaire et de Rousseau au Musée Carnavalet', *Bulletin du Musée Carnavalet* 13:2 (November 1960), p.2-23 (14). On the other hand, 'Es wird vermutet, dass dieses Herbar nicht von Rousseau sondern von Marquis de Girardin zusammengestellt worden ist [It is suspected that this Herbarium is not from Rousseau, but was assembled by the Marquis de Girardin]'. R. Schneebeli-Graf, *Botanisieren mit Jean-Jacques Rousseau* (Thun, 2003), p.137. The incompleteness of the plant identifications and lack of strict Linnaean order might support the latter theory.
4. 'Species' refer to scientific classifications, while a 'specimen' exemplifies a species; a given herbarium might contain duplicates or varieties, and therefore more specimens than species, as in the case of the 'grand herbier'. See Table 2.
5. Henry Cheyron, 'L'amour de la botanique: les annotations de Jean-Jacques Rousseau sur la

Botanique de Regnault', *Littératures* 4 (1981), p.53-95 (88, n.79), suggests these are the 300 or so plants Rousseau added to the 'grand herbier' given to him by Joseph Dombey (1742-1794), a student of Antoine Gouan and explorer of South American flora. See Table 2.

6. Marguerite-Madeleine Delessert (1767-1839), daughter of Madeleine-Catherine Delessert, née Boy de la Tour (1747-1816), and recipient of Rousseau's course of instruction in *Lettres élémentaires sur la botanique*.
7. This herbarium was possibly one of the herbiers portatifs offered to Malesherbes. See Rousseau to Malesherbes, 21 October 1771 (*CC*, xxxviii.282-83) and 17 April 1772 (*CC*, xxxix.36-37).
8. Julie Emélie Boy de la Tour (1751-1826), younger sister of Mme Delessert, married Charles Emmanuel Willading of Bern in 1778; the herbarium was left by their daughter and her husband, Henri Faesy, to the central library of Zürich. See Théophile Dufour, 'Petit herbier pour Mademoiselle Julie Boy de la Tour', *Annales de la société Jean-Jacques Rousseau* 2 (1906), p.260-68; Ruth Schneebeli-Graf, *Das Zürcher Herbar von Jean-Jacques Rousseau* (Zürich, 1980).
9. Rousseau to Mme Delessert, 28 May 1774 (*CC*, xxxxi.252-54).
10. Dombey gave Rousseau 'un très grand nombre de plantes étrangères et rares parfaitement belles et bien conservées', inspiring him to make it his 'unique bibliothèque' (*CC*, xxxv.307).
11. Rousseau to Julie Anne Marie Boy de la Tour, 16 April 1772 (*CC*, xxxxi.34-36).

Table 2: Rousseau's herbaria: missing or of dubious authenticity

Place:	Berlin	Chaalis	Paris	N/A	N/A
Institution or designation:	Botanisches Museum	Musée Jacquemart-André	Muséum national d'histoire naturelle, Laboratoire de Phanérogamie	Duchess of Portland	'Grand herbier'
Provenance:	Rousseau	Fusée Aublet[12]	Fusée Aublet[13]	Rousseau	Dombey, Rousseau
Contents:	11 vols; catalogue by Jansen[14]; 153 species	13 specimens, 2 or 3 possibly from Rousseau	A few annotations in Rousseau's hand; rest from Fusée Aublet; 15 vols.[15]; 1500 specimens[16]	Unknown	Orig. nearly 2,000 specimens, 1,500 species; R added 300 specimens; catalogue at Berlin (destroyed)
Organisation:	Linnaean	Unknown	Linnaean	Unknown	Probably Linnaean[17]
Dimensions:	Quarto	Various	Quarto	Unknown; portatif?	Folio
Year:	1777	Probably 1778	1778 acquired by Rousseau	1773?	1768
Other information:	Destroyed during WW II	Probably acquired from Fusée Aublet	Probably acquired from Fusée Aublet	Letter to Duchess of Portland[18]	Given to Daniel Malthus;[19] a portion perhaps at BPUN

12. Jean-Baptiste-Christophe Fusée Aublet (1720-1778), first explorer of the flora of French Guyana and author of *Flore de la Guiane françoise*, 4 vols (Paris, 1775), moved in the same

'group of enlightened magistrates, courtiers and authors which included Malesherbes, Rousseau [...] and several others'. Frans A. Stafleu, *Linnaeus and the Linnaeans: the spreading of their ideas in systematic botany, 1735-1789* (Utrecht 1971), p.282. Rousseau acquired part of one of Fusée Aublet's herbaria in May 1778. See A. Cook, 'Jean-Jacques Rousseau and exotic botany', in *Eighteenth-Century Life* special issue, *Exoticism and the culture of exploration*, R. Maccubbin and C. Knellwolf (eds), 26:3 (Fall 2002), p.181-201 (188-89). There are, however, no known letters between Rousseau and Fusée Aublet.

13. This appears to be the partial herbarium of Fusée Aublet that Rousseau acquired in May 1778.
14. Albert Jansen, *Jean-Jacques Rousseau als Botaniker* (Berlin, 1885), p.277-92.
15. See J. Lanjouw and H. Uittien 'Un nouvel herbier de Fusée Aublet découvert en France', *Mededeelingen van het Botanisch Museum en Herbarium van de Rijksuniversiteit te Utrecht* 75 (1940), p.133-70 (140-41).
16. Personal communication, Cécile Aupic, Chargée de conservation des herbiers historiques, Herbier du Muséum national d'histoire naturelle, Paris, 5 November 2007.
17. The order would have been Linnaean since Dombey was a student of Antoine Gouan, a Linnaean botanist.
18. Rousseau to the Duchess of Portland, 22 October 1773 (*CC*, xxxix.203-204).
19. There are at least three accounts of the fate of this herbarium: (1) it was 'donné presque en entier à M. Malthus qui a acheté mes livres de botanique'. Rousseau, *Dialogues*, *OC*, i.832, n.**; (2) The marquis de Girardin claimed that Rousseau sold his books and herbarium to Vincent Louis Dutens (1730-1812). 'Notices bibliographiques: herbiers', *OC*, iv.1888; (3) Or Rousseau may have returned it to Joseph Dombey, as suggested in a letter of 7 June 1770 to Du Peyrou (*CC*, xxxviii.32), and in the seventh promenade (*OC*, i.1060); see also Cheyron, L'amour de la botanique', p.88, n.79, p.90, n.135, p.92, n.189; Matthey, 'La dernière passion, p.35ff.

Bibliography

Manuscripts and herbaria

France

Chaalis, Musée Jacquemart-André: Herbarium (purportedly of Jean-Jacques Rousseau).

Paris, Bibliothèque de l'Assemblée nationale, Rousseau's annotated copy of Nicolas François and Geneviève de Nangis Regnault, *La Botanique mise à la portée de tout le monde ou collection des plantes d'usage dans la médecine, dans les alimens et dans les arts: avec des notices instructives puisées dans les auteurs les plus célèbres, contenant la description, le climat, la culture, les propriétés et les vertus propres à chaque plante, précédé d'une introduction à la botanique, ou Dictionnaire abrégé des principaux termes emploiés dans cette science*, 2 vols (Paris, Chez l'auteur, 1774), cited as 'BdAN [Munch MS]', an unpublished MS of Rousseau's notes prepared by Valérie Munch. An imperfect transcription of these annotations was published by V. D. Musset-Pathay as 'Notes inédites de J.-J. Rousseau sur la botanique', in *Œuvres inédites de J.-J. Rousseau* (Paris, 1825), vol.1, p.283-372. Leigh published an extract from them in *CC*, xl.A666.

Paris, Musée Carnavalet: Herbarium (purportedly of Jean-Jacques Rousseau).

Paris, Muséum national d'histoire naturelle: Herbarium of Fusée Aublet containing annotations by Jean-Jacques Rousseau.

–, 'Catalogue des graines donnés par M. Jean-Jacques Rousseau pour le cabinet de M. le duc d'Orléans', Ms 324.

Lyon, Bibliothèque municipale, Fonds ancien, Herbarium of Abbé Jean-Baptiste François Rozier.

Lyon, Jardin botanique, Herbarium of Claret de la Tourette and General Herbarium.

Montmorency, Musée Jean-Jacques Rousseau: Herbarium of Madelon Delessert.

Great Britain

Natural History Museum, London, Banksian MS103, f.573.

Switzerland

Neuchâtel, *Bibliothèque publique et universitaire*

Herbarium of Rousseau

MsA147: Jean-Antoine d'Ivernois, *Catalogue méthodique des plantes qui croissent naturellement dans la souveraineté de Neufchâtel & Valangin.* 2 vols. Fait l'hiver de l'an 1745 à 1746.

MsR21: Jean-Jacques Rousseau, 'Caractères de botanique'.

MsR80: Jean-Jacques Rousseau, 'Extraits de livres de botanique' and 'Signes ou abréviations pour les descriptions et caractères des plantes'

MsR81: Jean-Jacques Rousseau, 'Catalogue de tous les Mémoires de la Chymie contenus dans tous les volumes de l'Histoire de l'Académie R[oyale] des sciences'.

Jean-Antoine d'Ivernois, *Catalogue méthodique des plantes qui croissent naturellement dans la souveraineté de Neuchâtel et Valangin, fait l'hiver de l'an 1745 à 1746*, 2 vols, BPUN MsA147.

Archives de l'Etat de Neuchâtel: 'Inventaire du mobilier de l'Hôtel Du Peyrou', *Plumitif civil de la Cour de justice de Neuchâtel*, novembre 1792-août 1798, cahier 58, p.154, no. 207.

Zürich, Zentralbibliothek: Herbarium of Julie Emélie Willading Boy-de la Tour.

Texts published before 1800

Adanson, Michel, *Familles des plantes*, 2 vols (Paris, Vincent, 1763-1764).

Altmann, Johann Georg, *L'Etat et les délices de la Suisse en forme de relation critique* (Amsterdam, Wetsteins & Smith, 1730).

Amman, Johann, *Stirpium rariorum in imperio Rutheno sponte provenientium* (St Petersburg, Imperial Academy of Sciences, 1739).

'Avantages de la médiocrité', *Journal helvétique* (June 1744), p.528ff.

Bachaumont, Louis P. de, *Mémoires secrets pour servir à l'histoire de la république des lettres en France depuis MDCCLXII jusqu'à nos jours ou journal d'un observateur*, vol.5 (London, John Adamson, 1777).

Barbeu Dubourg, Jacques, *Le Botaniste françois: comprenant toutes les plantes communes et usuelles, disposées suivant une nouvelle méthode, et décrites en langue vulgaire*, 2 vols (Paris, Chez Lacombe, 1767).

–, 'Réflexion sur ce qu'il a plu à Jean-Jacques Rousseau de dire à des médecins', *Gazette de médecine* 49 (19 June 1762), p.385-90.

Barrelier, Jacques, and Antoine de Jussieu, *Plantae per Galliam, Hispaniam et Italiam observatae: iconibus aeneis exhibitae* (Paris, Stéphane Ganeau, 1714).

Batsch, August Johann Georg Carl, *Botanik fur Frauenzimmer und Pflanzenliebhaber, welche keine Gelehrten sind* (Weimar, im Verlage des Industrie Comptoirs, 1795).

–, *Botanische Unterhaltungen für Naturfreunde zu eigner Belehrung über die Verhältnisse der Pflanzenbilding entworfen*, part1 (Jena, Cröcker, 1793).

Bauhin, Caspar, *Pinax theatri botanici* (Basel, Ludwig König,1623).

Bauhin, Jean, *Historia plantarum universalis*, 3 vols (Yverdon, n.p., 1650-1651).

Blackwell, Elizabeth, *A Curious herbal containing five hundred cuts of the most useful plants which are now used in the practice of physic: engraved on folio copper plates, after drawings taken from the life*, 2 vols (London, J. Nourse, 1737-1739).

Boulduc, Simon, 'Analise de l'ipecacuanha', *Histoire de l'Académie royale des sciences. Année MDCC. Avec les mémoires de mathématique et de physique, pour la même année, tirés des registres de cette Académie* (Paris, Charles Estienne Hochereau, 1719), p.1-5.

–, 'Essai d'analise des plantes', *Histoire de l'Académie royale des sciences. Année MDCCXXXIV. Avec les mémoires de mathématique et de physique, pour la même année, tirés des registres de cette académie* (Paris, de l'Imprimerie royale, 1736), p.101-106.

–, 'Observations analytiques de la coloquinthe', *Histoire de l'Académie royale des sciences. Année MDCCI. Avec les mémoires de mathématique et de physique, pour la même année, tirés des registres de cette Académie* (Paris, Charles Estienne Hochereau, 1719), p.12-17.

Buc'hoz, Pierre-Joseph, *Traité historique des plantes qui croissent dans la Lorraine et les trois Evêchés: contenant leur description, leur figure, leur noms, l'endroit où elles croissent, leur culture, leur analyse et leurs propriétés, tant pour la médecine, que pour les arts et métiers*, 10 vols (Paris, Fetil, 1770).

Buffon, Georges, Louis Leclerc, comte de, *Histoire naturelle, générale et particulière*, 36 vols (Paris, de Imprimerie royale, 1749-1788), vol.2.

Bulliard, Jean Baptiste François, *Dictionnaire élémentaire de la botanique, ou, exposition par ordre alphabétique des préceptes de la botanique, et de tous les termes, tant françois que latins, consacrés à l'étude de cette science* (Paris, chez l'auteur and Didot, Barrois and Belin, 1783).

–, *Histoire des champignons de la France, ou Traité élémentaire renfermant dans un ordre méthodique les descriptions et les figures des champignons qui croissent naturellement en France* (Paris, chez Pauteur, Barrois le jeune, Belin, Croullebois, Bazan, 1791).

Camerarius, Joachim, *Hortus medicus et philosophicus: in quo plurimarum stirpium breves descriptiones, novae icones non paucae, indicationes locorum natalium, observationes de cultura earum peculiares, atque insuper nonnulla remedia euporista, nec non philologica quaedam continentur*, ed. Joachim Camerarius *et al.* (Frankfurt, Johann Feyerabend, Sigmund Feyerabend, Heinrich Tack & Peter Fischer, 1588).

Cesalpino, Andrea, *De plantis libri XVI* (Florence, Apud Georgium Marescottum, 1583).

Chabrey, Dominique, *Omnium stirpium sciagraphia et icones, quibus plantarum et radicum tum in hortis cultarum, tum in urbium fossis et muris... spontè provenientium, nomina, figura, natura, natales, synonyma, usus et virtutes, docentur* (Geneva, Samuel de Tournes, 1678)

Clusius, Carolus, *Rariorum aliquot stirpium per Hispanias observatarum historia* (Antwerp, Christoph Plantin, 1576).

Colonna, Fabio, *Minus cognitarum rariorumque nostro coelo orientium stirpium ekphrasis: qua non paucae ab antiquioribus Theophrasto, Dioscoride, Plinio, Galeno aliisq. descriptae, praeter illas etiam in Phytobasano editas disquiruntur ac declarantur. Item de aquatilibus aliisque nonnullis animalibus libellus [...] Omnia fideliter ad vivum delineata, atque aeneis-typis expressa cum indice in calce voluminis locupletissimo* (Rome, Jacob Mascardum, 1616).

Cordus, Valerius, *Annotationes in Pedacii Dioscoridis Anazarbei de medica materia libros V*, ed. Conrad Gessner (Strasbourg, Josias Rihel, 1561).

'Cours de botanique à ceux qui veulent apprendre cette science', *Journal helvétique* (February 1747), p.143-46.

Crantz, Johann Nepomuk Freiherr von, *Classis umbelliferarum emendata cum generali seminum tabula et figuris aeneis in necessarium instit. rei herbar. Supplementum* (Leipzig, Johann Paul Kraus, 1767).

Daléchamps, Jacques, *Historia generalis plantarum*, 2 vols (Lyon, G. Rovillium, 1586-1587).

Dalibard, Thomas François, *Florae parisiensis prodromus* (Paris, Durand & Pissot, 1749).

D'Apples, Jean Benjamin, 'Mémoires sur le Faltranck', *Nouvelles de la république des lettres* (July 1709), p.17-25.

Darwin, Erasmus, *Botanic garden: a poem in two parts* (London, J. Johnson, 1791).

Dillenius, Johann Jacob, *Historia muscorum* (Oxford, The Sheldonian theatre, 1741).

Dodoens, Rembert, *Stirpium historiae pemptades sex sive libri XXX*

(Antwerp, Christoph Plantin, 1583).
Duchesne, Antoine-Nicolas, *Manuel de botanique, contenant les propriétés des plantes utiles pour la nourriture, l'usage de la médecine, employées dans les arts, d'ornemens pour les jardins et qu'on trouve à la campagne aux environs de Paris* (Paris, Didot le jeune and Panckoucke, 1764).
Duhamel du Monceau, Henri Louis, *Traité des arbres et arbustes qui se cultivent en France en pleine terre*, 2 vols (Paris, H. L. Guerin & L. F. Delatour, 1755).

Ellis, John, *Directions for bringing over seeds and plants, from the East-Indies and other distant countries in a state of vegetation* (London, 1771).

[Fontenelle, Bernard le Bovier de], 'Analises de la coloquinthe, du jalap, de la gomme gutte, et de l'ellébore noir', in *Histoire de l'Académie royale des sciences. Année MDCC. Avec les mémoires de mathématique et de physique, pour la même année, tirés des registres de cette académie* (Paris, Charles Estienne Hochereau, 1719), p.58-62.
–, 'Analise de l'ipecacuanha', in *Histoire de l'Académie royale des sciences. Année MDCC. Avec les mémoires de mathématique et de physique, pour la même année, tirés des registres de cette académie* (Paris, Charles Estienne Hochereau, 1719), p.46-48.
–, 'Sur le fer des plantes', *Histoire de l'Académie royale des sciences. Année MDCCVI. Avec les mémoires de mathématique et de physique, pour la même année, tirés des registres de cette académie* (Paris, Jean Boudot, 1707), p.38-40.
–, 'Sur les sels des plantes', in *Histoire de l'Académie royale des sciences. Année MDCXCIX. Avec les mémoires de mathématique & de physique, pour la même année, tirés des registres de cette académie* (Paris, la veuve de Jean Boudot, 1718), p.63-65.
Fusée Aublet, Jean Baptiste Christophe, *Histoire des plantes de la Guiane françoise*, 4 vols (Paris, Pierre-François Didot jeune, 1775).

Garcin, Laurent, 'A letter from Dr Laurence Garcin, of Neuchâtel, F. R. S. to Sir Hans Sloane Bart, late P. R. S. concerning the cyprus of the Ancients: done from the French by W. Watson, F. R. S., *Philosophical transactions* 45 (1748), p.564-78.
–, 'Lettre aux éditeurs', *Journal helvétique* (September 1744), p.257-79.
–, 'Lettre de M. Garcin sur la nature et l'usage des pilules madurines, avec des remarques intéressantes sur les effets du thé', *Journal helvétique* (May 1747), p.473-90.
–, 'Lettre de M. Garcin sur les œufs philosophiques, servant de réponse à celle de M. [Daniel] Gagnebin, insérée dans le journal de mars dernier', *Journal helvétique* (June 1745), p.517-29.
–, 'The settling of a new genus of plants, called after the Malayans, mangostans; by Laurentius Garcin, M. D. and F. R. S. Translated from the French by Mr. Zollman, F. R. S.', *Philosophical Transactions* 38 (1733-1734), p.232-42.
Garidel, Pierre-Joseph, *Histoire des plantes qui naissent aux environs d'Aix* (Aix, J. David, 1715).
Garsault, François-Alexandre-Pierre de, *Les Figures de plantes et animaux d'usage en médecine, décrites dans la 'Matière médicale' de M. Geoffroy [...] dessinés d'après nature par M. de Garsault*, 6 vols (Paris, Didot le jeune, 1764-1765).
Gerard, John, *The Herball or general historie of plantes* fol. (London, John Norton, 1597).

–, *The Herball or general historie of plantes*, ed. Thomas Johnson (London, Adam Filip, Joice Norton & Richard Whitakers, 1636).

'Genf', *Göttingische Anzeigen von gelehrten Sachen* 2:145 (29 November 1781), p.1162-67.

Gessner, Konrad, *De raris et admirandis herbis, quae, sive quod noctu luceant, sive alias ob causas, lunariae nominantur, et obiter de aliis etiam rebus, quae in tenebris lucent, commentariolus: cum iconibus quibusdam herbarum novis* (Zürich, Andreas and Jacob Gesner, 1555).

Gouan, Antoine, *Histoire des poissons, contenant la description anatomique de leurs parties externes et internes, et le caractère des divers genres rangés par classes et par ordres: avec un vocabulaire complet, des tables raisonnées en latin et en françois* (Strasbourg, A. König, 1770).

–, *Historia piscium: sistens ipsorum Anatomen externam, internam, atque Genera in Classes et Ordines redacta: accedunt Vocabularium locupletissimum, Indices latini ac gallici, Experimenta circa Motum natatorium et muscularem, Respirationis mechanismum, Auditus et Generationis organa* (Strasbourg, A. König, 1770).

–, *Hortus regius monspeliensis: sistens plantas tum indigenas tum exoticas no. MM.CC. ad genera relatas cum nominibus specificis, synonymis selectis, nominibus trivialibus, habitationibus indigenarum, hospitiis exoticarum, secundum sexualem methodum digestas* (Lyon, De Tournes frères 1762).

–, *Flora monspeliaca, sistens plantas No.1850 ad sua genera relatas, et hybrida methodo digestas; adjectis nominibus specificis, trivialibusque, synonymis selectis, habitatiombus plurium in agro Monspeliensi nuper detectarum, et earum quæ in usus medicos veniunt nominibus pharmaceuticis, virtutibusque probatissimis* (Lyon, Benédict Duplain, 1765).

Gronovius, Johannes Fredericus, and Leonhard Rauwolf, *Flora orientalis, sive, Recensio plantarum quas botanicorum coryphaeus Leonardus Rauwolffus, medicus augustanus, annis 1573, 1574, et 1575, in Syria, Arabia, Mesopotamia, Babylonia, Assyria, Armenia et Judaea crescentes observavit, et collegit: earumdemque ducenta specimina, quae in bibliotheca publica Lugduno-Batava adservantur, nitidissime exsiccata et chartae adglutinata in volumen retulit* (Leiden, Wilhelm de Groot, 1755).

Guettard, Jean Etienne, *Observations sur les plantes*, 2 vols (Paris, Durand, 1747).

Haller, Albrecht von, *Bibliotheca botanica*, 2 vols (Zürich, Orell, Gessner, Fuessli, 1771-1772), vol.2.

–, *Enumeratio methodica stirpium helvetiae indigenarum* (Göttingen, Abram Vandenhoek, 1742).

–, 'Extrait de deux pièces que Monsieur Haller, docteur en médecine, médecin du roi de la Grande-Bretagne, professeur en anatomie, en botanique et en chirurgie dans l'université de Gottingue et membre de la Société Roïale des sciences d'Angleterre et de Suède, a publiées depuis peu, sur la botanique', *Journal helvétique* (August 1740), p.172-210.

–, *Historia stirpium indigenarum Helvetiae inchoata* (Bern, Societatis typographicae, 1768).

–, *Iter Helveticum anni 1739* (Göttingen, Sumtu reg. univers. offic. librariae, 1740).

Hill, Sir John, *Flora britanica, sive, Synopsis methodica stirpium britanicarum sistens arbores et herbas, indigenas et in agricultas in classes et ordines, genera et species redactas secundum systema sexuale, additis nonnullis noviter detectis, cum classium, ordinum et generum characteristicis, specierum descriptionibus, et virium epitome, tabulis aeneis illustrata, post tertiam editionem synopseos rainae, opere Dillenii concinnatam, nunc primum ad celeberrimi Caroli Linnaei methodum disposita* (London, J. Waugh, 1760).

–, *The British herbal: an history of plants and trees, natives of Britain, cultivated for use, or raised for beauty* (London, T. Osborne & J. Shipton, J. Hodges, J. Newbery, B. Collins, S. Crowder & H. Woodgate, 1756).

Hirzel, Johann Kaspar, *Le Socrate rustique, ou Description de la conduite économique et morale d'un paysan philosophe*, trans. Jean Rodolphe Frey des Landres (Zürich, Heidegger, 1762; Lausanne, François Grasset, 1767).

Homberg, Guillaume, 'Essais pour examiner les sels des plantes', in *Histoire de l'Académie royale des sciences. Année MDCXCIX. Avec les mémoires de mathématique et de physique, pour la même année, tirés des registres de cette académie* (Paris, la veuve de Jean Boudot, 1718), p.69-74.

'Idée générale du grand ouvrage de botanique, que M. Haller, professeur à Göttingen, vient de publier sur les plantes naturelles de la Suisse', *Journal helvétique* (February 1743), p.174-83.

D'Ivernois, Jean-Antoine, 'Lettre à Monsieur Cartier ministre du St Evangile et pasteur de l'église de la Chaux du Milieu, dans cette souveraineté: servant d'apologie aux médecins-botanistes Suisses, contre ce qu'on a avancé dans le *Journal helvétique* de décembre 1741 p.1155 sur l'état de la botanique, en Suisse', *Journal helvétique* (January 1742), p.3-38.

Jacquin, Nikolaus Joseph Freiherr von, *Selectarum stirpium americanarum historia* (Vienna, ex officina Krausiana, 1763).

[Jacson, Maria Elizabeth], *Botanical dialogues between Hortensia and her four children* (London, J. Johnson, 1797).

Jaucourt, Louis, chevalier de, 'Neuchâtel', in *Encyclopédie, ou dictionnaire raisonné des sciences, des arts, et des métiers, par une société de gens de lettres*, ed. Denis Diderot and Jean Le Rond d'Alembert, CD-ROM (Paris, 2000).

Jussieu, Antoine-Laurent de, 'Examen de la famille des renoncules', *Histoire de l'Académie royale des sciences. Année MDCCLXXIII. Avec les mémoires de mathématique et de physique, pour la même année, tirés des registres de cette académie* (Paris, de l'imprimerie royale, 1777), p.214-40.

–, 'Exposition d'un nouvel ordre des plantes adopté dans les démonstrations du jardin royal', *Histoire de l'Académie royale des sciences. Année MDCCLXXIV. Avec les mémoires de mathématique et de physique, pour la même année, tirés des registres de cette académie* (Paris, de l'imprimerie royale, 1778), p.175-97.

–, *Genera plantarum secundum ordines naturales disposita, juxta methodum in horto regio parisiensi exartam, anno M.DCC.LXXIV* (Paris, Herissant and Barrois, 1789).

Laborde, Jean-Benjamin de, and Beat Fidel Baron de Zurlauben, *Tableaux de la Suisse ou Voyage pittoresque fait dans les XIII cantons du corps helvétique, représentant les divers phénomènes que la nature y rassemble, et les beautés dont l'art les a enrichis*, 13 vols (Paris, Clousier, 1780; Lamy, 1784-1786).

Lamarck, Jean Baptiste Pierre Antoine de Monet de, 'Herbier (Herbarium)', in *Encyclopédie méthodique: botanique* (Paris, Panckoucke, 1789), vol.3, p.111-16.

–, 'Jardin de botanique', in *Dictionnaire encyclopédique de botanique* (Paris, Laporte, 1789), vol.3, p.211-12, in *Encyclopédie méthodique*, 190 vols (Paris and Liège, Panckoucke, 1782-1830).

Latourrette, Marc Antoine Flerieu de Claret de, and Jean-Baptiste François Rozier, *Démonstrations élémentaires de botanique pour l'usage de l'école vétérinaire*, vol.1 (Lyon, Jean-Marie Bruyset, 1766).

Lebreton, François, *Manuel de botanique, à l'usage des amateurs et des voyageurs: contenant des principes de botanique, l'explication du système de Linné, un catalogue des différens végétaux étrangers, les moyens de transporter les arbres et les semences; la manière de former un herbier, etc.* (Paris, Prault, 1787).

–, *Traité sur les propriétés et les effets du sucre* (Paris, Royez, 1789).

Lee, James, *An Introduction to botany containing an explanation of the theory of that science, and an interpretation of its technical terms: extracted from the works of Dr Linnaeus[...] with twelve copper plates, and two explanatory tables: to which is added an appendix; containing upwards of two thousand English names of plants* (London, J. and R. Tonson, 1760).

Leigh, George, and John Sotheby, *A Catalogue of the botanical and natural history part of the library of the late John, Earl of Bute* (London, Leigh & Sotheby, 1794).

Lémery, Nicolas, *Dictionnaire universel des drogues simples* (Paris, L.-Ch. d'Houry, 1759).

–, *Pharmacopée universelle*, 2 vols (Paris, De Saint & Saillant, 1763-1764).

–, 'Réflexions physiques: sur le défaut et le peu d'utilité des analyses ordinaires des plantes et des animaux', in *Histoire de l'Académie royale des sciences. Année MDCCXIX. Avec les mémoires de mathématique et de physique, pour la même année, tirés des registres de cette Académie* (Paris, de l'Imprimerie royale, 1721), p.173-88.

Lestiboudois, François Joseph, *Botanographie belgique, ou Méthode pour connoître facilement toutes les plantes qui croissent naturellement: ou que l'on cultive communément dans les provinces septentrionales de la France* (Lille, Imprimerie J. B. Henry, 1781).

Lettsom, John Coakley, *The Natural history of the tea tree: with observations on the medical qualities of tea and effects of tea-drinking* (London, Edward & Charles Dilly, 1772).

–, *Le Voyageur-naturaliste, ou Instructions sur les moyens de ramasser les objets d'histoire naturelle et de les bien conserver* (Amsterdam & Paris, Lacombe, 1775).

Linnaeus, Carolus, *Amœnitates academicae*, 7 vols (Stockholm, Laurentius Salvius, 1749-1769).

–, *Bibliotheca botanica* (Amsterdam, Salomon Schouten, 1736).

–, *Critica botanica* (Leiden, C. Wishoff, 1737).

–, *A Dissertation on the sexes of plants*, trans. James Edward Smith (Dublin, L. White, 1786).

–, *Disquisitio de quaestione ab Academia Imperiali Scientiarum Petropol. in annum MDCCLIX pro praemio proposita: sexum plantarum argumentis experimentis nouis, praeter adhuc iam cognita, vel corroborare, vel impugnare, praemissa expositione historica et physica omnium plantae partium, quae aliquid ad foecundationem et perfectionem seminis et fructus conferre creduntur: ab eadem Academia die VI. Septembris MDCCLX in conuentu publico praemio ornata* (St Petersburg, Academy of Sciences 1760).

–, *The Families of plants with their natural characters*, trans. Botanical

Society of Lichfield, 2 vols (Lichfield, J. Johnson, 1787).
–, *Flora lapponica* (Amsterdam, S. Schouten, 1737).
–, *Flora zeylanica* (Stockholm, Laurentius Salvius, 1747).
–, *Genera plantarum*, 2nd edn (Leiden, C. Wishoff, 1742).
–, *Genera plantarum*, 6th edn (Stockholm, Laurentius Salvius, 1764).
–, *Hortus cliffortianus* (Amsterdam, n.p., 1737).
–, *Öländska och Gothländska Resa* (Stockholm & Uppsala, Gottfried Kiesewetter, 1745).
–, *Species plantarum*, 2nd edn, 2 vols (Stockholm, Laurentius Salvius, 1762-1763).
–, *System of vegetables, A*, trans. Botanical Society of Lichfield (Lichfield, J. Johnson, 1783).
–, *Systema vegetabilium secundum classes ordines genera species cum characteribus et differentiis*, 13th edn, ed. Johann Andreas Murray (Göttingen, Johann Christian Dietrich, 1774).
'Livres nouveaux: Göttingue' *Journal helvétique* (October 1742), p.83.
Lobel, Matthias de, *Plantarum seu stirpium icones* (Antwerp, Christoph Plantin, 1581).
Loesel, Johannes, *Flora prussica: sive plantae in regno Prussiae sponte nascentes* (Kaliningrad, Georgius, 1703).
Louis XIV, 'Règlement ordonné par le roi pour l'Académie royale des sciences', in Bernard le Bovier de Fontenelle, *Eloges des Académiciens, avec l'histoire de l'Académie royale des sciences*, facsimile edn (Brussels, 1969 [The Hague, Isaac van der Kloot, 1740]), p.5-17.
Ludwig, Christian Gottlieb, *Institutiones historico physicae regni vegetabilis* (Leipzig, Gleditsch, 1742; revised edn 1757).
Lusitanus, Amatus, *In Dioscoridis Anazarbei de medica materia libros quinque ennarationes* (Venice [Gualterum Scotum], 1553).

Mahon, [John Mac], 'Faltranck', *Encyclopédie méthodique: médecine* (Paris, Panckoucke, 1793).
Martin, François de Vitré, *Description du premier voyage faict aux Indes orientales par les Français: contenant les mœurs, loix, etc. des Indiens, une description et remarque des animaux, épiceries, et fruicts que se trouvent aux Indes* (Paris, Laurens Sonnius, 1609).
Martyn, John, *Historia plantarum rariorum* (London, Richard Reily, 1728).
–, *Methodus plantarum circa Cantabrigiam nascentium* (London, Richard Reily, 1727).
–, *Tabulae synopticae plantarum officinalium* (London, G. Strahan, W. & J. Innys, C. Rivington, 1726).
Martyn, Thomas, *Sketch of a tour through Swisserland: with an accurate map: a new edition, to which is added a short account of an expedition to the summit of Mont Blanc, by M. De Saussure, of Geneva* (London, G. Kearsley, 1788).
–, *Thirty-eight plates, with explanations intended to illustrate Linnaeus's system of vegetables, and particularly adapted to the letters of the elements of botany* (London, B. & J. White, 1794).
Matthioli, Pietro Andrea, *Opera quae extant omnia: hoc est, commentarii in vi. libros pedacii Dioscoridis Anazarbei de medica materia* (Basel, Johann Koenig, 1674).
Mercier, Louis Sébastien, *L'An deux mille quatre cent quarante: rêve s'il en fût jamais*, vol.1 (London [Paris], 1771).
Merian, Maria Sibylla, *Metamorphosis insectorum surinamensium* (Amsterdam, 1705).
Micheli(us), Pier Antonio, *Nova plantarum genera juxta Tournefortii methodum disposita* (Florence, Bernardo Paperini, 1729).

Morison, Robert, *Plantarum historia universalis Oxoniensis pars [secunda]-tertia, seu, Herbarum distributio nova: per tabulas cognationis et affinitatis ex libro natur observata et detecta* (Oxford, Sheldonian Theatre, 1680-1699).

Parkinson, John, *Paradisi in sole paradisus terrestris: or A garden of all sorts of pleasant flowers which our English ayre will permitt to be noursed up, with a kitchen garden of all manner of Herbes, Rootes, & fruites, for meate or sause used with us, and an orchard of all sorte of fruitbearing trees and shrubbes fit for our land: together with the right orderinge, planting and preserving of them and their uses and vertues collected by John Parkinson apothecary of London* (London, Humphrey Lownes & Robert Young, 1629).

Pliny, the Elder, *Caii Plinii secundi historiae naturalis libri XXXVII: quos interpretatione et notis illustravit Joannes Harduinus [...] in usum serenissimi Delphini*, 3 vols in 2 vols, folio (Paris, A.-U. Coustelier, 1723).

Plukenet, Leonard, *Almagestum botanicum* (London, the Author, 1696).

[Plutarch], Plutarque, *Les Vies des hommes illustres Grecs et Romains comparée l'une avec l'autre par Plutarque de Chaeronée*, trans. Simon Goulart, *et al.* (Lyon, Paul Frellon, 1611).

Pontedera, Guilio, *Anthologia, sive, de floris natura libri tres, plurimis inventis, observationibusque, ac aereis [sic] tabulis ornati* (Padua, Typis Seminarii, 1720).

Ray, John, *Methodus plantarum nova* (London, Henry Faitborne & John Kersey, 1682).

–, *Synopsis methodica stirpium britannicarum*, 3rd edn, ed. J. J. Dillenius (London, William & John Innys, 1724).

Rauwolf, Leonhard, *Aigentliche Beschreibung der Raiss inn die Morgenländer* (Laugingen, Leonhart Reinmichel, 1582).

Reede tot Drakenstein, Hendrik Adriaan van, *Hortus indicus malabaricus*, 12 vols (Amsterdam, Johan van Someren and Jan van Dyck, 1678-1703).

Regnault, Nicolas François, and Geneviève de Nangis Regnault, *La Botanique mise à la portée de tout le monde ou Collection des plantes d'usage dans la médecine, dans les alimens et dans les arts: avec des notices instructives puisées dans les auteurs les plus célèbres, contenant la description, le climat, la culture, les propriétés et les vertus propres à chaque plante, précedé d'une introduction à la botanique, ou dictionnaire abrégé des principaux termes employés dans cette science*, 2 vols (Paris, chez l'auteur, 1774).

Richard, Louis Claude Marie (ed.), *Dictionnaire élémentaire de botanique de Bulliard, revu et presqu'entièrement refondu par Louis Claude Richard* (Paris, A. J. Dugour & Durand, 1798).

Rousseau, Jean-Jacques, *Botanik for fruentimmere i breve til Frue de L*** [Botany for women in the letters to Mme L**], trans. Odin Wolff (Copenhagen, S. Poulsen, 1789).

–, *Collection complète des œuvres de J. J. Rousseau*, vol.7, in-4°; vol.14, in-8° and in-12°; vol.24, in-8° and in-12°, Pierre Alexandre Du Peyrou and Paul Claude Moultou (eds), (Geneva, [Société typographique de Genève], [1780-]1782).

–, *Inleyding tot de kruidkunde, een werk geschikt tot gebruik van hen, welke zig in deeze wetenschap willen oeffenen; introduction à la botanique*, trans. J. van Noorden, vol.1 (London, Vernor & Hood; Rotterdam, L. Bennet, 1798).

–, *J.-J. Rousseaus Botanik für Frauenzimmer, in Briefen an die Frau* *von L***, trans. anon. (Mannheim, in der schwanischen Buchhandlung, 1781).

–, *Letters on the elements of botany addressed to a lady by the celebrated J.-J. Rousseau; translated into English, with notes, and twenty-four additional letters, fully explaining the system of Linnaeus*, trans. and ed. Thomas Martyn (London, B. White, 1785).

–, *Letters on the elements of botany: addressed to a lady by the celebrated J.-J. Rousseau; translated into English, with notes, and twenty-four additional letters, explaining the system of Linnaeus*, trans. and ed. Thomas Martyn, 2nd edn (London, B. White, 1787).

–, *Letters on the elements of botany: addressed to a lady by the celebrated J.-J. Rousseau; translated into English, with notes, and twenty-four additional letters, explaining the system of Linnaeus*, trans. and ed. Thomas Martyn, 4th edn (London, B. and J. White, 1794).

–, *Œuvres complètes, de J.J. Rousseau. Nouv. éd., classée par ordre de matières, et ornée de quatre-vingt-dix gravures*, Louis Sébastien Mercier, *et al.* (eds), vol.5 (Paris, Poinçot, 1789).

–, *Supplément à la collection complète des œuvres de J. J. Rousseau*, vol.13, in-4°; vol.26, in-8° and in-12°, Pierre-Alexandre Du Peyrou, ed., (Geneva, [Société typographique de Genève], 1782).

Rozier, Jean Baptiste François, *Cours complet d'agriculture*, vol.6 (Paris, Delalain fils, 1787).

Ruchat, Abraham, *Les Délices de la Suisse, une des principales républiques de l'Europe, où l'on peut voir tout ce qu'il y a de plus remarquable dans son Pays et dans celui de ses Alliez* (Leiden, Pieter van der Aa, 1714).

Rumphius, Georg Eberhard (Rumpf), *Herbarium amboinense*, 6 vols, ed. Johannes Burman (Amsterdam, Fransicum Changuion, Joannes Catuffe, Hermann Uytwerf, 1741-1750).

Saussure, Horace-Bénédict de, *Voyages dans les Alpes précédés d'un essai sur l'histoire naturelle des environs de Genève*, vol.1 (Neuchâtel, Samuel Fauche, 1779).

Sauvages de la Croix, François Boissier de, *Methodus foliorum seu plantae florae monspeliensis* (The Hague, n.p., 1751).

–, *Nosologia methodica sistens morborum classes, genera et species, juxta Sydenhami mentem et botanicorum ordinem* (Amsterdam, Frères De Tournes, 1763).

Savary de Bruslons, Jacques, *Dictionnaire universel de commerce*, ed. Philémon Louis Savary, 2 vols (Geneva, chez les Héritiers Cramer & Frères Philibert, 1742).

Scheuchzer, Johannes, *Agrostographia sive graminum, juncorum, cyperorum, cyperoidum, usque affinium historia* (Zürich, Bodmer, 1719).

Scheuchzer, Johann Jacob, *Ouresphoitës helveticus, sive itinera per Helvetiae alpines regiones*, 4 vols in 2 (Leiden, 1702; London, Henry Clements, 1708; Leiden, Pieter van der Aa, 1723).

Scopoli, Giovanni Antonio, *Flora carniolica: exhibens plantas Carnioliae indigenas et distributas in classes, genera, species, varietates ordine linnaeano* (Vienna, Johann Paulus Krauss, 1772).

Séguier, Jean-François, *Plantae veronenses seu Stirpium quae in agro veronensi reperiuntur methodica synopsis* (Verona, Imprimeur libraire, 1745).

Smith, James Edward, *A Sketch of a tour on the continent*, 2nd edn, 3 vols (London, 1807), vol.1.

Southwell, Thomas, *Medical essays and observations* (London, J. Knox, 1764).

Stanyan, Abraham, *L'Etat de la Suisse, écrit en 1714*, trans. anon.

(Amsterdam, chez Jean Garrel/ chez les frères Wetsten, 1714).
Stillingfleet, Benjamin, 'Observations on grasses', in *Miscellaneous tracts relating to natural history, husbandry and physick* (London, R. & J. Dodsley, 1762).

Tabernaemontanus, *Eicones plantarum seu stirpium* (Frankfurt, Nikolaus Basse, 1590).
[Tollot, Jean-Baptiste], 'Réflexions sur l'étude des plantes et sur leur usage', *Journal helvétique* (December 1741), p.1142-57.
Toscan, Georges, *L'Ami de la nature, ou Choix d'observations sur divers objets de la nature et de l'art* (Paris, 1800).
–, *Histoire du lion: de la ménagerie du muséum national d'histoire naturelle et de son chien* (Paris, Cuchet, 1795).
Tournefort, Joseph Pitton de, *Elemens de botanique*, vol.1 (Paris, de Imprimerie royale, 1694).
–, *Institutiones rei herbariae* (Paris, de l'Imprimerie Royale, 1700).
–, *Relation d'un voyage du Levant: fait par ordre du Roy: contenant l'histoire ancienne et moderne de plusieurs isles de l'Archipel, de Constantinople, des Côtes de la Mer Noire, de l'Arménie, de la Géorgie, des frontières de Perse et de l'Asie Mineure: avec les plans des villes et des lieux considérables, le génie, les mœurs, le commerce et la religion des différens peuples qui les habitent, et l'explication des médailles et des monumens antiques: enrichie de descriptions et de figures d'un grande nombre de plantes rares, de divers animaux, et de plusieurs observations touchant l'histoire naturelle* (Lyon, Anisson et Posuel, 1717).
Tragus, Hieronymus, *Kreüter Buch* (Strasbourg, Josias Rihel, 1546).
–, *Kreüter Buch* (Strasbourg, Josias Rihel, 1560).
–, *New Kreütter Buch* (Strasbourg, Wendel Rihel, 1539).

Vaillant, Sébastien, *Botanicon parisiense, Operis majoris prodituri prodromus* (Leiden, Pieter van der Aa, 1723).
–, *Botanicon parisiense* (Paris, Briasson, 1743).
–, *Sermo de structura florum: discours sur la structure des fleurs* (Leiden, Pieter van der Aa, 1718).
Van den Spiegel, Adriaan, *Isagoges in rem herbariam libri duo* (Padua, Paolo Meietti, 1606).
Voltaire, *Lettres philosophiques* (Amsterdam, Lucas, 1734).

Wakefield, Priscilla, *An Introduction to botany, in a series of familiar letters, with illustrative engravings*, 2nd edn (London, E. Newbery, 1798).

Young, Arthur, *Rural œconomy* (London, T. Becket, 1770).
–, *Travels in France*, 2 vols (London, W. Richardson, 1792).

Texts published after 1800

Adanson, Michel, *Familles des plantes*, repr. edn with an introduction by Frans A. Stafleu (New York, 1966).
Aldridge, Alfred Owen, 'Jacques Barbeu Dubourg: a French disciple of Benjamin Franklin', *Proceedings of the American Philosophical Society* 95:4 (17 August 1951), p.331-92.
Amigues, Suzanne, *Etudes de botanique antique* (Paris, 2002).
Arber, Agnes, *Herbals: their origin and evolution, a chapter in the history of botany 1470-1670*, 3rd edn (Cambridge, 1986).
–, 'A seventeenth-century naturalist: John Ray', *Isis* 34:4 (Spring 1943), p.319-24.
Aristotle, *The Politics and the*

constitution of Athens, ed. Steven Everson (Cambridge, 1996).

Åsberg, Marie, and William Stearn, 'Linnaeus's Öland and Gotland journey 1741', *Biological journal of the Linnean Society* 5:1 (March 1973), p.1-107.

Atran, Scott, *Cognitive foundations of natural history: towards an anthropology of science* (Cambridge, 1990).

–, 'Covert fragmenta and the origins of the botanical family', *Man* 18 (1983), p.51-71.

Aymonin, Gérard G., 'L'herbier de Lamarck', *Revue d'histoire des sciences* 34:1 (1981), p.25-58.

Bacon, Francis, *Great instauration and new Atlantis*, ed. Jerry Weinberger (Arlington Heights, IL, 1989).

–, *The New organon*, ed. L. Jardine and M. Silverthorne (Cambridge, 2000).

Bandelier, André, *Des Suisses dans la république des lettres: un réseau savant au temps de Frédéric le Grand* (Geneva, 2007).

Barnes, Barry, *et al.*, *Scientific knowledge: a sociological analysis* (Chicago, IL, 1996).

Bartissol, Carlos, 'Sources des idées médicales de Rousseau', doctoral dissertation (Paris, 1914).

Bauer, Eddy, 'Le cercle du jardin – esquisse historique d'une société', in *La Société du jardin de Neuchâtel* (Neuchâtel, 1963).

De Beer, Gavin R., 'Haller's *Historia stirpium*', *Annals of science* 9:1 (1953), p.1-46.

–, 'Jean-Jacques Rousseau: botanist', *Annals of science* 10:3 (1954), p.189-223.

–, 'Rodolph Valltravers, F. R. S.', *Notes and records of the Royal Society of London* 4:2 (October 1946), p.216-26.

–, and Bernard Gagnebin, 'Abraham Gagnebin de la Ferrière d'après sa correspondance', *Bulletin de la société neuchâteloise des sciences naturelles* 80 (1957), p.45-78.

Benguigui, Isaac, *Genève et ses savants: physiciens, mathématiciens et chimistes au XVIIIe et XIXe siècles* (Geneva, 2006).

Bensaude-Vincent, Bernadette, 'La nature laboratoire', in *Rousseau et les sciences*, ed. Bernadette Bensaude-Vincent and Bruno Bernardi (Paris, 2003), p.155-74.

Bensaude-Vincent, Bernadette, and Bruno Bernardi, 'Pour situer les *Institutions chymiques*', *Corpus* 36 (1999), p.5-38.

–, 'Rousseau chimiste', in *Rousseau et les sciences*, ed. Bernadette Bensaude-Vincent and Bruno Bernardi (Paris, 2003), p.59-76.

Berchtold, Jacques 'Le framboisier nourricier: Rousseau et l'érotisme végétal de la tradition pastorale', *Etudes Jean-Jacques Rousseau* 14-15 (2003-2004), p.23-49.

Beretta, Marco, 'Sensiblerie vs. mécanisme: Jean-Jacques Rousseau et la chimie', *Corpus* 36 (1999), p.103-20.

Bernardin de St Pierre, Jacques Henri, 'Essai sur J.-J. Rousseau', *Œuvres complètes de Jacques-Henri Bernardin de St Pierre: mélanges*, ed. L. Aimé-Martin (Brussels, 1820).

–, *La Vie et les ouvrages de Jean-Jacques Rousseau*, ed. Maurice Souriau (Paris, 1907).

Bewell, Alan, '"On the banks of the South Sea": botany and sexual controversy in the late eighteenth century', in *Visions of empire: voyages, botany, and representations of nature*, ed. David Philip Miller and Peter Hans Reill (Cambridge, 1996).

Bird, Alexander, and Emma Tobin, 'Natural kinds', *Stanford encyclopedia of philosophy* (http://plato.stanford.edu/entries/natural-kinds/).

Birn, Raymond, *Forging Rousseau: print, commerce and cultural manipulation in the late Enlightenment* (Oxford, 2001).

Biro, Andrew, *Denaturalizing ecological politics: alienation from nature from Rousseau to the Frankfurt School and beyond* (Toronto, 2005).

Blake, S. F., 'Byways of nomenclature', *American journal of botany* 36:1 (January 1949), p.8-9.

Blum, Carol, '"Soyez toujours ce que vous êtes": J.-J. Rousseau, Genève, la Corse et le paradoxe de la population', in Michael Böhler, *et al.* (eds), *Republikanische Tugend: Aussbildung [sic] eines schweizer Nationalbewusstseins und Erziehung eines neuen Bürgers* (Geneva, 2000).

Blunt, Wilfrid, *The Compleat naturalist: a life of Linnaeus*, with the assistance of William T. Stearn (London, 1971).

Boom, Brian M., 'Societal and scientific information needs from plant collections', in *Sampling the green world: innovative concepts of collection, preservation, and storage of plant diversity*, ed. T. F. Stuessy and S. H. Sohmer (New York, 1996).

Bosher, J. F., 'Huguenot merchants and the Protestant International in the seventeenth century', *The William and Mary quarterly* 3rd Series, 52:1 (January 1995) p.77-102.

Boschung, Urs, *et al.* (eds), *Repertorium zu Albrecht von Hallers Korrespondenz 1724-1777*, vol.1 (Basel, 2002).

Bourzat, Jean Dominique, *Une Dynastie de jardiniers et de botanistes: les Richard de Louis XV à Napoléon III* (Paris, 2009).

Bowen, Elenore Smith [Laura Bohannan], *Return to laughter: anthropological novel* (Garden City, N.Y., 1964).

Bowker, Geoffrey, and Susan Leigh Star, *Sorting things out: classification and its consequences* (Cambridge, MA, 1999).

Brandt, H. (ed.), 'Lettres inédites adressées à Abraham Gagnebin', *Actes de la Société jurassienne d'émulation* (1931), p.215-23.

Bridel, Philippe Sirice, *Le Conservateur suisse ou Recueil complet des étrennes helvétiennes*, vol.13 (Lausanne, 1831).

Bridson, Diane, and Leonard Forman, *The Herbarium handbook*, revised edn (Kew, 1992).

Brockway, Lucille, *Science and colonial expansion: the role of the British botanic gardens* (New York, 1979).

Brosse, Jacques, *La Magie des plantes* (Paris, 1990).

Browne, Janet, 'Botany in the boudoir and garden: the Banksian context', in *Visions of empire: voyages, botany, and representations of nature*, ed. David Philip Miller and Peter Hans Reill (Cambridge, 1996).

Bryce, W. J., *A Botanist's paradise: the establishment of scientific botany in Russia in the eighteenth century* (Ystalyfera, 2008).

Brygoo, Edouard, 'Les médecins de Montpellier et le Jardin du roi à Paris', *Histoire et nature* 14 (1979), p.3-29.

Calatrava, Juan, 'Rousseau et l'architecture', *Annales de la Société Jean-Jacques Rousseau: Jean-Jacques Rousseau et les arts visuels*, Actes du colloque de Neuchâtel (20-22 septembre 2001) 45 (2003), p.81-111.

Callot, Emile, *La Philosophie biologique de Goethe* (Paris, 1971).

Cameron, Hector C., *Sir Joseph Banks K.B., F.R.S.: the autocrat of the philosophers* (London, 1952).

Candolle, Augustin Pyramus de, *Essai sur les propriétés médicales des plantes, comparées avec leurs formes extérieures et leur classification naturelle* (Paris, 1804).

–, *Histoire de la botanique genevoise* (Geneva & Paris, 1830).

–, *Mémoires et souvenirs*, ed. Jean-

Marc Drouin and Jean-Daniel Candaux (Geneva, 2004).
–, *Théorie élémentaire de la botanique, ou, Exposition des principes de la classification naturelle et de l'art de décrire et d'étudier les végétaux* (Paris, 1813).
Cantor, Paul A., 'The metaphysics of botany: Rousseau and the new criticism of plants', *Southwest review* 70:3 (1985), p.362-80.
Cash, Philip, *Dr Benjamin Waterhouse: a life in medicine and public service (1754-1846)* (Sagamore Beach, MA, 2006).
Cheneval, Francis, 'The reception of Rousseau's political thought by Zurich's "Patriots"', in Michael Böhler, *et al.* (eds), *Republikanische Tugend: Aussbildung [sic] eines schweizer Nationalbewusstseins und Erziehung eines neuen Bürgers* (Geneva, 2000), p.425-46.
Cheyron, Henry, 'L'amour de la botanique: les annotations de Jean-Jacques Rousseau sur la Botanique de Regnault', *Littératures* 4 (1981), p.53-95.
–, 'Fragments pour un dictionnaire des termes d'usage en botanique', in *Dictionnaire de Jean-Jacques Rousseau*, ed. R. Trousson and F. S. Eigeldinger (Paris, 2001), p.351.
–, 'Herbiers', *Dictionnaire de Rousseau*, ed. R. Trousson and F. S. Eigeldinger (Paris and Geneva, 2001), p.404-407.
–, 'Ray et Sauvages annotés par Jean-Jacques Rousseau', *Littératures* 15 (Autumn 1986), p.83-99.
–, 'Une herborisation de Rousseau avec le docteur Neuhaus: sur un document inédit', *Bulletin de l'information: étude et documents de l'Association des amis de Jean-Jacques Rousseau* 48 (1996), p.1-8.
Cohn, Ferdinand, 'Jean-Jacques Rousseau als Botaniker', *Deutsche Rundschau* 47 (April-May-June 1886), p.364-85.
Coleman, Patrick, '"Aimer la patrie et les lois": l'objet de l'éducation républicaine chez Rousseau', in *Republikanische Tugend: Aussbildung [sic] eines schweizer Nationalbewusstseins und Erziehung eines neuen Bürgers*, ed. Michael Böhler *et al.* (Geneva, 1998), p.459-70.
Conklin, Harold C., *Ethnographic atlas of Ifugao: a study of environment, culture, and society in northern Luzon* (New Haven, CT, 1980).
Considine, John, 'Ancient Greek among the eighteenth-century languages of science: Linnaeus, Dillenius, and the lexicographical record', *International journal of the classical tradition* 16:3-4 (September-December 2009), p.330-43.
Cook, Alexandra, 'Botanical exchanges: Jean-Jacques Rousseau and the Duchess of Portland', *History of European ideas* 33 (2007), p.142-56.
–, 'Le "Disciple" critique le "maître": Jean-Jacques Rousseau et la nomenclature linnéenne', in *Les Mots et les choses: la langue bien faite?*, ed. P. Selosse and D. Reynaud (Lyon, forthcoming).
–, 'Idées et pratiques scientifiques dans la correspondance botanique de Jean-Jacques Rousseau', *Annales de la société Jean-Jacques Rousseau* 47 (2007; distributed October 2008), p.265-85.
–, 'Jean-Jacques Rousseau and exotic botany', *Eighteenth-century life* special issue 26:3 (Fall 2002; distributed Spring 2003): *Exoticism and the culture of exploration*, p.181-201.
–, 'Jean-Jacques Rousseau, "terminator", and telos in nature', *Pensée libre* 8 (2001), p.311-25.
–, 'Jean-Jacques Rousseau's copy of Albrecht von Haller's *Historia stirpium indigenarum Helvetiae inchoata (1768)*', *Archives of natural history* 30:1 (April 2003), p.149-56.

–, 'Jean-Jacques Rousseau et les réseaux d'échange botanique', in *Rousseau et les sciences*, ed. Bernadette Bensaude-Vincent and Bruno Bernardi (Paris, 2003), p.93-114.

–, 'Linnaeus and Chinese plants: a test of the linguistic imperialism thesis', *Notes and records of the Royal Society*, 64 (June 2010), p.121-38.

–, 'Politics of nature and voyages of exploration: some purposes and results', in *Voyages and exploration in the north Atlantic from the Middle Ages to the 17th century*, ed. Anna Agnarsdottir, 2nd edn (Reykjavik, 2001).

–, 'Propagating botany: the case of Jean-Jacques Rousseau', in *The Transmission of culture in Western Europe, 1750-1850: papers celebrating the bi-centenary of the foundation in Geneva of the Bibliothèque britannique*, ed. D. Bickerton and J. Proud (Berne, 1999), p.69-94.

–, 'Rousseau and the languages of music and botany', in *Musique et langage chez Rousseau*, ed. Claude Dauphin, *SVEC* 2004:08, p.75-87.

–, 'The septième promenade of the *Rêveries*: a peculiar account of Rousseau's botany?' in *The Nature of Rousseau's Rêveries: physical, human, aesthetic*, ed. John C. O'Neal *SVEC* 2008:03, p.11-36.

–, and Christopher Kelly, 'Introduction', *The Collected writings of Rousseau* (Hanover, NH, 2000), vol.8, p.xi-xxviii.

Cook, Harold, *Matters of exchange: commerce, medicine and science in the Dutch Golden Age* (New Haven, CT, 2007).

Cooper, Alix, *Inventing the indigenous: local knowledge and natural history in early modern Europe* (Cambridge, 2007).

Corsi, Pietro, *Lamarck: genèse et enjeux du transformisme: 1770-1830*, translated by D. Menard (Paris, 2001).

Courtois, L. J., 'Chronologie critique de la vie et des œuvres de Jean-Jacques Rousseau', *Annales de la société Jean-Jacques Rousseau* 15 (1923).

Crombie, A. C., *Styles in scientific thinking* (London, 1994), vol.2.

Cunningham, Andrew, 'The culture of gardens', in *Cultures of natural history*, ed. N. Jardine, J. A. Secord and E. C. Spary (Cambridge, 1996), p.38-56.

Cunningham, Andrew, and Perry Williams, 'De-centring the "big picture": *The Origins of modern Science* and the modern origins of science', *British journal for the history of science* 26 (1993), p.407-32.

Danet, Frédéric, 'Rozier Jean-Baptiste François', in *Herbiers de la région Rhône-Alpes 2e partie: catalogue des herbiers*, ed. Andrine Faure (Lyon, 2006), p.300-301.

Dannenfeldt, Karl H., *Leonhard Rauwolf: sixteenth-century physician, botanist, and traveler* (Cambridge, MA, 1968).

Darnton, Robert, *The Business of Enlightenment: a publishing history of the Encyclopédie* (Cambridge, MA, 1979).

Darwin, Charles, *On the origin of species*, facsimile edn, introduced by Ernst Mayr (Cambridge, MA, 1964).

Daston, Lorraine, and Katharine Park, *Wonders and the order of nature: 1150-1750* (New York, 1998).

Daston, Lorraine, and Peter Galison, *Objectivity* (New York, 2007).

David, Thomas, *et al.*, *La Suisse et l'esclavage des noirs* (Lausanne, 2005).

Davis, Michael, *The Autobiography of philosophy: Rousseau's Reveries of the solitary walker* (Lanham, MD, 1999).

Dawson, Virginia P., *Nature's enigma:*

the problem of the polyp in the letters of Bonnet, Trembley and Réaumur (Philadelphia, PA, 1987).
–, 'Foundations of natural history in eighteenth-century Geneva', in Patrick Coleman, Anne Hofmann and Simone Zurbuchen (eds), *Reconceptualizing nature, science, and aesthetics: contribution à une nouvelle approche des Lumières helvétiques* (Geneva, 1998), p.31-42.
Dean, John, 'Controversy over classification: a case study from the history of botany', in *Natural order: historical studies of scientific culture*, ed. Barry Barnes and Steven Shapin (London and Beverly Hills, CA, 1979).
Delaporte, François, *Le Second Règne de la nature: essai sur les questions de la végétalité au XVIIIe siècle* (Paris, 1979).
–, *Nature's second kingdom: explorations of vegetality in the eighteenth century*, trans. A. Goldhammer (Cambridge, 1982).
Derrida, Jacques, *De la grammatologie* (Paris, 1967).
Desjardins, Gustave, *Le Petit Trianon: histoire et description* (Versailles, 1885).
Desmond, Ray, *The History of the royal botanic gardens Kew*, 2nd edn (Kew, 2007).
Dietz, Bettina, 'Mobile objects: the space of shells in eighteenth-century France', *British journal for the history of science* 39:142 (September 2006), part 3, p.363-82.
Doolittle, James, 'Jaucourt's use of source material in the *Encyclopédie*', *Modern language notes* 65:6 (June 1950), p.387-92.
Drouin, Jean-Marc, *L'Herbier des philosophes* (Paris, 2008).
–, 'Les herborisations d'un philosophe', in *Rousseau et les sciences*, ed. Bernadette Bensaude-Vincent and Bruno Bernardi (Paris, 2003), p.77-92.
–, 'Linné et la dénomination des vivants: portrait du naturaliste en législateur', in *Fondements de la botanique: Linné et la classification des plantes*, ed. Thierry Hoquet (Paris, 2005), p.37-55.
–, 'Rousseau lecteur de Linné', *Bulletin d'histoire et d'épistémologie des sciences de la vie* 7 (2000), p.39-57.
–, and Bernadette Bensaude-Vincent, 'Nature for the people', in N. Jardine, J. A. Secord and E. C. Spary (eds), *Cultures of natural history* (Cambridge, 1996), p.408-425.
Drouin, Jean-Marc, and Luc Lienhard, 'Botanik', in *Albrecht von Haller: Leben, Werk, Epoche*, ed. Hubert Steinke, *et al.* (Göttingen, 2008), p.292-314.
Duckert-Henriod, Marie-Marguerite, 'Jean-Antoine d'Ivernois, médecin et botaniste (1703-1765)' in Schlup (ed.) *Biographies neuchâteloises* (Hauterive, 1996), vol.1, p.143-47.
Ducourthial, Guy, *Flore magique et astrologique de l'antiquité* (Paris, 2003).
–, *La Botanique selon Jean-Jacques Rousseau* (Paris, 2009).
Dufour, Alfred, *Histoire de Genève*, 3rd edn (Paris, 2004).
Dufour, Théophile, *Les Institutions chymiques de Jean-Jacques Rousseau* (Geneva 1904, 1905).
–, 'Petit herbier pour Mademoiselle Julie Boy de la Tour', *Annales de la société Jean-Jacques Rousseau* 2 (1906), p.260-68.
Dugour, A. J., 'Notice sur la vie et les écrits de l'Abbé Rozier', in *Cours complet d'agriculture: théorique, pratique, économique, et de médecine rurale et vétérinaire ou dictionnaire universel d'agriculture*, vol.10 (Paris, 1800), p.i-xvi.
Duncan, Alistair, *Laws and order in eighteenth-century chemistry* (Oxford, 1996).
Duris, Pascal, *Linné et la France* (Geneva, 1993).

–, 'Le néolinnéisme en France au XIXe siècle', in *Les Fondements de la botanique: Linné et la classification des plantes*, ed. Thierry Hoquet (Paris, 2005), p.131-54.

Eamon, William, *Science and the secrets of nature: books of secrets in medieval and early-modern culture* (Princeton, NJ, 1994).

Egmond, Florike, *The World of Carolus Clusius: natural history in the making 1550-1610* (Brookfield, VT, 2010).

Eigeldinger, Frédéric S., *Des Pierres dans mon jardin: les années neuchâteloises de J.-J. Rousseau et la crise de 1765* (Geneva, 1992).

–, 'Les Montagnons: un archétype social', in *La Ville s'étend sur tout le pays, actes du colloque Rousseau, Neuchâtel et l'Europe*, ed. Frédéric S. Eigeldinger and Roland Kaehr (Neuchâtel, 1993), p.75-94.

Eklund, Jon, *The Incompleat chymist: being an essay on the eighteenth-century chemist in his laboratory, with a dictionary of obsolete chemical terms of the period* (Washington, DC, 1975).

Elvin, Mark, 'Curiosity: some examples from Xie Zhaozhe 謝肇淛 (1567-1624) and Gilbert White (1720-1793), and others, of the observation of nature in late-imperial China and early-modern Europe', unpublished MS.

Emberley, Peter, 'Rousseau versus the savoyard vicar: the profession of faith considered', *Interpretation* 14:2-3 (1986), p.299-329.

Ereshefsky, Marc, *The Poverty of the Linnaean hierarchy* (Cambridge and New York, 2001).

Escherny, François Louis D', 'Eloge de J.-J. Rousseau', in *Mélanges de littérature, d'histoire, de morale et de philosophie*, vol.3 (Paris, 1811), p.1-192.

Fara, Patricia, *Pandora's breeches: women, science and power in the Enlightenment* (London, 2004).

–, *Sex, botany and empire* (Duxford, 2003).

Ferrazzini, Arthur, *Béat de Muralt et Jean-Jacques Rousseau* (La Neuveville, 1951).

Findlen, Paula, *Possessing nature: museums, collecting, and scientific culture in early modern Italy* (Berkeley, CA, 1994).

Fontes da Costa, Palmira, 'Women and the popularization of botany in early nineteenth-century Portugal: the Marquise of Alornà's *Botanical recreations*', in *Popularizing science and technology in the European periphery, 1800-2000*, ed. Faidra Papanelopoulou, Agustí Nieto-Galan and Enrique Perdiguero (Farnham, 2009), p.43-63.

Foster, Elizabeth A., 'Le dernier séjour de Jean-Jacques Rousseau à Paris, 1770-1778', *Smith College studies in history* 6:1-2 (October 1920-January 1921).

Foucault, Michel, *Les Mots et les choses* (Paris, 1967).

–, *L'Herméneutique du sujet: cours au Collège de France (1981-1982)*, edn established by Frédéric Gros under the direction of François Ewald and Alessandro Fontana (Paris, 2001).

–, 'Technologies of the self', in *Technologies of the self: a seminar with Michel Foucault*, ed. Luther H. Martin, *et al.* (Amherst, MA, 1988).

Francillon, Roger, 'The Enlightenment in Switzerland', in Patrick Coleman, Anne Hofmann and Simone Zurbuchen (eds), *Reconceptualizing nature, science, and aesthetics: contribution à une nouvelle approche des Lumières helvétiques* (Geneva, 1998), p.13-27.

François, Alexis, *Lyrisme et géologie: le séjour de J.-J. Rousseau à Genève en 1754* (Geneva, 1941).

Fueter, Eduard, 'Geschichte der exakten Wissenschaften in der schweizerischen Aufklärung (1680-1780)', doctoral dissertation,

University of Zürich (Zürich, 1941).

Galiffe, Jacques-Augustin, *Notices généalogiques sur les familles genevoises*, vol. 7, ed. Louis Dufour-Vernes and Eugène Ritter (Geneva, 1895).

Galison, Peter L., *Image and logic: a material culture of microphysics* (Chicago, IL, 1997).

Gasbarrone, Lisa, 'Blindness or oversight? A closer look at Rousseau's *Dictionnaire de botanique*', *French forum* 11:1 (January 1986), p.5-17.

–, 'The book of nature: Rousseau's floral collections and the text', *L'Esprit créateur*, 28:1 (1988), p.27-41.

Gautier, Maurice, 'Introduction', in Jean-Jacques Rousseau, 'Les *Institutions chymiques* de Jean-Jacques Rousseau', ed. Maurice Gautier, in *Annales de la société Jean-Jacques Rousseau* 12 (1918-1921), p.vii-xxiii.

Gelzer, Florian, and Béla Kapossy, 'Roman, Staat und Gesellschaft', in *Albrecht von Haller: Leben, Werk, Epoche*, ed. Hubert Steinke, *et al.* (Göttingen, 2008), p.156-81.

George, Sam, *Botany, sexuality and women's writing 1760-1830: from modest shoot to forward plant* (Manchester, 2007).

Gilbert, E. J., 'Un esprit – une œuvre: Jean Baptiste François Bulliard, dit Pierre (1752-1793)', *Bulletin trimestriel de la société mycologique de France* 68:1 (1952), p.5-131.

Girre, Loïc, *Traditions et propriétés des plantes médicinales: histoire de la pharmacopée* (Paris, 1997).

Godet, Charles H., 'Enumération des végétaux vasculaires qui croissent dans le canton de Neuchâtel', in *Mémoires de la société des sciences naturelles de Neuchâtel*, vol.2 (Neuchâtel, 1839).

Goldgar, Anne, *Impolite learning: conduct and community in the Republic of Letters, 1680-1750* (New Haven, CT, 1995).

Goethe, Johann Wolfgang von, *Goethes Werke*, 14 vols, Erich Trunz, *et al.* (eds) (Hamburg, 1948-1960).

–, *Briefe*, ed. Karl Robert Mandelkow and Bodo Morawe (Hamburg, 1962-1967).

Goodin, Robert E., *Green political theory* (Cambridge, 1992).

Gorham, George Cornelius, *Memoirs of John Martyn, F.R.S. and of Thomas Martyn, B.D., F.R.S., F.L.S. professors of botany in the University of Cambridge* (London, 1830).

Goubier-Robert, Geneviève, 'L'Elysée: leçon de morale dans un parc', *Annales de la société Jean-Jacques Rousseau: Jean-Jacques Rousseau et les arts visuels* (2003), p.113-29.

Grasset, J.-J., 'Jean-Jacques Rousseau à Montpellier', *Académie des sciences et lettres de Montpellier: mémoires de la section des lettres* (1847-1854), p.553-85.

Greene, Edward Lee, *Landmarks of botanical history*, ed. Frank N. Egerton, with Robert P. McIntosh and Rogers McVaugh, part 2 (Stanford, CA, 1983).

Guédès, Michel, 'Duchesne, Buisson, Durande, early followers of the natural method of the Jussieus', *Taxon* 22:2-3 (May 1973), p.211-19.

–, 'Notules de bibliographie botanique XVI-XXI', *Archives of natural history* 11:2 (1983), p.195-97.

Guichet, Jean-Luc, *Rousseau, l'animal et l'homme: l'animalité dans l'horizon anthropologique des Lumières* (Paris, 2006).

Guillod, Nathalie, 'Une esquisse de l'élite culturelle neuchâteloise dans la seconde moitié du XVIIIe siècle', *Revue historique neuchâteloise* 2 (2007), p.107-24.

Guyot, Charly, *Un Ami et défenseur de Rousseau: Pierre-Alexandre Du Peyrou* (Neuchâtel, 1958).

–, *Le Rayonnement de l'Encyclopédie en Suisse française* (Neuchâtel, 1955).

Hamy, Ernest Théodore, *Joseph Dombey: médecin, naturaliste, archéologue, explorateur du Pérou, du Chili et du Brésil 1778-1785: sa vie, son œuvre, sa correspondance* (Paris, 1905).

Harten, Hans-Christian and Elke, *Die Versöhnung mit der Natur* (Reinbek, 1989).

Haüy, René Just, *Traité de minéralogie*, 5 vols (Paris, 1801).

Hawkins, Richmond Laurin, 'Manuscripts of Jean-Jacques Rousseau at Harvard University', *The Romantic review* 20 (July-September 1929), p.209-21.

Hayes, Richard, 'Biographical dictionary of Irishmen in France: part XI', *Studies: an Irish quarterly review* 33:131 (September 1944), p.367-77.

Heller, John Lewis, 'Linnaeus's *Bibliotheca botanica*', *Taxon* 19:3 (June 1970), p.363-411.

Henry, Philippe, 'L'évolution démographique' in *Histoire de pays de Neuchâtel 2: de la Réforme à 1815* (Hauterive, 1991), p.140-57.

–, 'Les relations politiques entre Neuchâtel et Berlin au XVIIIe siècle', in *Schweizer in Berlin des 18. Jahrhunderts*, ed. Martin Fontius and Helmut Holzhey (Berlin, 1996), p.33-44.

Hey'l, Bettina, *Das Ganze der Natur und die Differenzierung des Wissens: Alexander von Humboldt als Schriftsteller* (Berlin, 2007).

Hodacs, Hanna, 'Linnaeans outdoors: the transformative role of studying nature "on the move" and outside', *British journal for the history of science* 44 (June 2011), p.183-209.

Holmes, Frederic Lawrence, *Eighteenth-century chemistry as an investigative enterprise* (Berkeley, CA, 1989).

Hoquet, Thierry, 'La bibliothèque botanique de Benjamin Delessert', *Bulletin du bibliophile* 1 (2002), p.102-41.

–, *Buffon-Linné: éternels rivaux de la biologie?* (Paris, 2007).

–, 'Linné dans le texte', in *Les Fondements de la botanique: Linné et la classification des plantes*, ed. Thierry Hoquet (Paris, 2005), p.155-75.

Huard, Pierre, 'L'enseignement médico-chirurgical', in *Enseignement et diffusion des sciences en France au XVIIIe siècle*, ed. René Taton (Paris, 1964), p.171-236.

Hughes, J. Donald, *Pan's travail: environmental problems of the ancient Greeks and Romans* (Baltimore, MD, 1994).

Hulliung, Mark, *The Autocritique of Enlightenment: Rousseau and the philosophes* (Cambridge, MA, 1994).

Humboldt, Alexander von, *Correspondance scientifique et littéraire d'Alexandre de Humboldt*, ed. J.-B.-M.-A. Dezos de La Roquette, vol.1 (Paris, 1869).

–, *Cosmos: essai d'une description physique du monde*, ed. J. Grange, vol.1 (Paris, 2000).

–, *Sur les lois que l'on observe dans la distribution des formes végétales*, read before the Institut de France, 29 January 1816 (Paris, 1816).

Im Hof, Ulrich, *Mythos Schweiz: Identität, Nation, Geschichte, 1291-1991* (Zürich, 1991).

Incorvati, Giovanni, 'Seul avec la nature et vous: le promeneur solitaire rencontre Linné et ses disciples', *Annales de la société Jean-Jacques Rousseau* 42 (1999), p.281-326.

Jackson, Myles, 'Review article: labor, skills and practices in the scientific enterprise: recent works in the cultural history of science', *Journal of modern history* 71:4 (1999), p.902-13.

Jacquat, Marcel S., 'Abraham Gagnebin, médecin (1707-1800)', *Biographies neuchâteloises*, ed. Michel Schlup, vol.1 (Hauterive, 1996), p.97-102.

–, 'Laurent Garcin: médecin-chirurgien, naturaliste (1683-1752)', *Biographies neuchâteloises*, ed. Michel Schlup, vol.1 (Hauterive, 1996), p.103-109.

–, 'Une page régionale d'histoire des sciences relue récemment: 3. Abraham Gagnebin (1707-1800) et son cabinet de curiosités à la Ferrière', *Bulletin de la société neuchâteloise des sciences naturelles* 123 (2000), p.23-26.

[Jacson, Maria Elizabeth], *Botanical lectures by a lady: altered from 'Botanical dialogues for the use of schools', and adapted to the use of persons of all ages; by the same author* (London, 1804).

Jansen, Albert, *Jean-Jacques Rousseau als Botaniker* (Berlin, 1885).

Jarvis, Charles, *Order out of chaos: Linnaean plant names and their types* (London, 2007).

Jarvis, Charles, and Phillip Cribb, 'Review: Linnaean sources and concepts of orchids', *Annals of botany* (2009), p.1-12.

Jeanjaquet, Jules, 'L'auteur de la chronique des chanoines et de mémoires du chancelier de Montmollin', *Musée neuchâtelois* 38 (1951), p.3-14, 43-54.

Jeanneret, Frédéric-Alexandre-Marie, and James-Henri Bonhôte, 'Abraham Gagnebin', *Biographie neuchâteloise* (Le Locle, 1863), vol.1, p.357-61.

–, 'D'Escherny', *Biographie neuchâteloise* (Le Locle, 1863), vol.1, p.314-20.

–, 'Jean-Antoine d'Ivernois', *Biographie neuchâteloise* (Le Locle, 1863), vol.1, p.524-27.

–,'Laurent Garcin' and 'Jean-Laurent Garcin', *Biographie neuchâteloise* (Le Locle, 1863), vol.1, p.373-88.

Jelmini, Jean-Pierre, 'Rousseau neuchâtelois', *Bulletin de l'Association Jean-Jacques Rousseau* 72 (2012).

Johnstone, Paul H., 'The rural Socrates', *Journal of the history of ideas* 5:2 (April 1944), p.151-75.

Jones, Colin, *Paris: biography of a city* (London, 2004).

Jost, François, *Jean-Jacques Rousseau suisse, étude sur sa personnalité et sa pensée*, 2 vols (Fribourg, 1961).

Jussieu, Antoine-Laurent de, 'Sixième notice historique sur le muséum', *Annales du muséum d'histoire naturelle* 11 (1808), p.1-39.

Kapossy, Béla, 'Republican political economy', *History of European ideas* 33:4 (December 2007), p.377-89.

Kelly, Christopher, *Rousseau as author: consecrating one's life to the truth* (Chicago, IL, 2003).

Kesseler, Rob, and Madeline Harley, *Pollen: the hidden sexuality of flowers* (London, 2004).

Keudell, Elise von, *Goethe als Benutzer der Weimarer Bibliothek*, reprint edn (Leipzig, [1931] 1982).

Klein, Ursula, 'Shifting ontologies, changing classifications: plant materials from 1700 to 1830', *Studies in the history and philosophy of science* 36 (2005), p.261-329.

–, and Wolfgang Lefèvre, *Materials in eighteenth-century science: a historical ontology* (Cambridge, MA, 2007).

Knapp, Sandra, 'Introduction', in Nick Knight, *Flora* (New York, 2001).

Knight, Nick, *Flora* (New York, 2001).

Kobayashi, Takuya, '*Fragments pour un dictionnaire des termes d'usage en botanique*: processus de rédaction', *Bulletin de l'association J.-J. Rousseau* 62 (2003), p.3-24.

–, 'Seven botanical books annotated by Rousseau', *Lloydiana* 11 (2007), p.10-11.
Koerner, Lisbet, *Linnaeus: nature and nation* (Cambridge, MA, 1999).
Kuchler, Sina, and Volker Wissemann, '"Botanik für Frauenzimmer": Untersuchungen zur Popularisierung von Botanik im Raum Weimar/Jena um 1800', *Haussknechtia: Mitteilungen der thüringischen botanischen Gesellschaft* 10 (2004), p.299-330.
Kuhn, Bernhard, '"A chain of marvels": botany and autobiography in Rousseau', *European Romantic review*, 17:1 (January 2006), p.1-20.
–, 'Natural history and the history of the self: geology and autobiography in the works of Goethe and Rousseau', *Colloquium helveticum* 25 (1997), p.41-62.
Kunz, Joëlle, *Schweizer Geschichte – einmal anders* (Geneva, 2006).

Laissus, Yves, 'Le jardin du roi', in *Enseignement et diffusion des sciences en France au XVIIIe siècle*, ed. René Taton (Paris, 1964), p.287-317.
–, 'Les cabinets d'histoire naturelle', in *Enseignement et diffusion des sciences en France au XVIIIe siècle*, ed. R. Taton (Paris, 1964), p.659-70.
Landrin, A., *Correspondance inédite de Linné avec Claude Richard et Antoine Richard* (Versailles, 1863).
Lanjouw J., and H. Uittien, 'Un nouvel herbier de Fusée Aublet découvert en France,' *Mededeelingen van het Botanisch Museum en Herbarium van de Rijksuniversiteit te Utrecht* 75 (1940), p.133-70.
Larson, James L., *Reason and experience: the representation of natural order in the work of Carl von Linné* (Berkeley, CA, 1971).
Lasègue, Antoine, *Musée botanique de Benjamin Delessert* (Paris 1845).
Leigh, R. A., 'Jean-Jacques Rousseau and the myth of antiquity', in *Classical influences on western thought A.D. 1650-1870: proceedings of an international conference held at King's College, Cambridge March 1977*, ed. R. R. Bolgar (Cambridge, 1979), p.155-68.
–, 'Les manuscrits disparus de J.-J. Rousseau: quelques observations et quelques fragments retrouvés', *Annales de la société Jean-Jacques Rousseau* 34 (1956-1958), p.31-81.
Leong, Elaine, 'Making medicines in the early-modern household', *Bulletin of the history of medicine* 82 (2008), p.145-168.
Lévi-Strauss, Claude, 'Jean-Jacques Rousseau, fondateur des sciences de l'homme', in Samuel Baud-Bovy, *et al.*, *Jean-Jacques Rousseau* (Neuchâtel, 1962), p.239-48.
–, *La Pensée sauvage* (Paris, 1962).
Lienhard, Luc, '"La machine botanique": zur Entstehung von Hallers Flora der Schweiz', in *Hallers Netz*, Martin Stuber, Stefan Hächler and Luc Lienhard (eds) (Basel, 2005), p.371-410.
Lindley, John, *A Natural system of botany, or, A systematic view of the organisation, natural affinities, and geographical distribution, of the whole vegetable kingdom*, 2nd edn (London, 1836).
Linnaeus, Carolus, *The 'Critica botanica' of Linnaeus*, trans. A. F. Hort (London, 1938).
–, *Philosophia botanica*, reprint edn (New York [Coldicote], 1966 [1751]).
–, *Philosophia botanica*, trans. Stephen Freer (Oxford, 2003).
–, *Species plantarum: a facsimile of the first edition 1753*, with introductory essays by William T. Stearn, 2 vols (London, 1957-1959).
–, *Systema naturae 1735: facsimile of the first edition*, introduction and trans. M. S. J. Engel-Ledeboer and H. Engel (Nieuwkoop, 1964).

Livesey, James, 'Botany and provincial Enlightenment in Montpellier: Antoine Banal père et fils 1750-1800', *History of science* 43 (2006) p.57-76.

–, 'London by the light of Montpellier: scientific networks between northern Europe, Britain and the Languedoc 1680-1789', in *SVEC* 2010:04, p.85-102.

Lloyd, G. E. R. (ed.), *Hippocratic writings*, trans. J. Chadwick, *et al.* (London, 1978).

Locke, John, *An Essay concerning human understanding*, ed. R. Woolhouse (London, 1997).

Lorch, Jacob, 'The natural system in biology', *Philosophy of science* 28:3 (July 1961), p.282-95.

–, 'The discovery of nectar and nectaries and its relation to views on flowers and insects', *Isis* 69:4 (December 1978), p.514-33.

Louis, Armand, *Mathieu de L'Obel (1538-1616): épisode de l'histoire de la botanique* (Ghent, Louvain, 1980).

Lovejoy, Arthur O., *The Great chain of being: a study of the history of an idea* (Cambridge, MA, 1936).

Mackay, David, 'Agents of empire', in *Visions of empire: voyages, botany, and representations of nature*, ed. David Philip Miller and Peter Hans Reill (Cambridge, 1996).

Magnin, Antoine, *Claret de la Tourrette: sa vie, ses travaux, ses recherches sur les lichens du Lyonnais d'après ses ouvrages et les notes inédites de son herbier* (Paris, 1885).

Magnin-Gonze, Joëlle, *Histoire de la botanique*, revised edn (Paris, 2009).

Mantelkow, Mariette, and Kenneth Nyberg, 'Linnaeus's apostles and the development of the *Species plantarum*', *Symb. bot. Upsal.* 33 (2005), p.73-80.

Margez, Marlène, Cécile Aupic and Denis Lamy, 'La restauration de l'herbier Haller du Muséum national d'histoire naturelle', *Support-Tracé* 5 (2005), p.58-68.

Marouf, Abderrazak, and Jean Vallade, *Dictionnaire de botanique: les phanérogames* (Paris, 2000).

Martin, Alison, 'Frauenzimmerbotanik: unschuldiger Zeitvertreib und Mode?' *Frauen und Gärten um 1800: Weiblichkeit, Natur, Ästhetik*, ed. Christiane Holm and Holger Zaunstöck (Halle, 2009), p.37-47.

Matthey, François, 'La dernière passion de Jean-Jacques Rousseau', *Revue neuchâteloise* 25:100 (Autumn 1982), p.27-40.

Mauzi, Robert, *L'Idée du bonheur dans la littérature et la pensée françaises au XVIIIe siècle* (Paris, 1979).

McClellan, James E. III, 'The scientific press in transition: Rozier's journal and the scientific societies in the 1770s', *Annals of science* 36 (1979), p.425-49.

McKibben, Bill, *The End of nature* (New York, 1989).

McVaugh, Michael, 'The *Experimenta* of Arnald of Villanova', *The Journal of medieval and renaissance studies* 1:1 (1971), p.107-18.

Metsger, Deborah. A., and Sheila C. Byers, 'Introduction', in *Managing the modern herbarium: an interdisciplinary approach*, ed. D. A. Metsger and S. C. Byers (Washington, DC, 1999), p.1-18.

Meyer, Ernst H. F., *Geschichte der Botanik*, vol.4 (Königsberg, 1857).

Miller, Hortense S., 'The Herbarium of Aylmer Bourke Lambert: notes on its acquisition, dispersal, and present whereabouts', *Taxon*, 19:4 (August, 1970), p.489-553.

Miller, Philip, *The Gardener's and botanist's dictionary*, 9th edn, ed. Thomas Martyn, 4 vols (London, 1807).

Minguet, Charles, *Alexandre de Humboldt: historien et géographe de l'Amérique espagnole* (Paris, 1969).

Montgolfier, B. de, and M. Gallet,

'Souvenirs de Voltaire et de Rousseau au Musée Carnavalet', *Bulletin du Musée Carnavalet* 13:2 (November, 1960), p.2-23.

Mornet, Daniel, *Les Sciences de la nature en France au XVIIIe siècle* (Paris, 1911).

Morton, A. G., 'Marginalia to Andrea Cesalpino's work on botany', *Archives of natural history* 10:1 (1983), p.31-36.

Mukerji, Chandra, *Territorial ambitions and the gardens of Versailles* (Cambridge, 1997).

Müller-Wille, Staffan, *Botanik und weltweiter Handel: zur Begründung eines natürlichen Systems der Pflanzen durch Carl von Linné (1707-1778)* (Berlin, 1999).

–, 'Collection and collation: theory and practice of Linnaean botany', *Studies in history and philosophy of biological and biomedical sciences* 38 (2007), p.541-62.

–, 'Gardens of Paradise', *Endeavour* 25:2 (June 2001), p.49-54.

–, 'Nature as marketplace: the political economy of Linnaean botany', *History of political economy* 35:1 (2003), p.154-72.

–, and Karen Reeds, 'A translation of Carolus Linnaeus's introduction to *Genera plantarum* (1737)', *Studies in the history and philosophy of biological and biomedical sciences* 38 (2007), p.563-72.

Murgia, Camilla, 'Le mythe de Rousseau chez Zurlauben et Laborde', *Annales de la société Jean-Jacques Rousseau* 45 (2003), p.583-617.

Naess, Arne, 'The shallow and the deep, long-range ecology movement: a summary', *Inquiry* 16 (1973), p.95-100.

Nauert, Charles G., Jr., 'C. Plinius Secundus (*Naturalis Historia*)', in *Catalogus translationum et commentariorum: medieval and Renaissance translations and commentaries*, ed. F. Edward Cranz and Paul Oskar Kristeller, vol.4 (Washington, DC, 1980), p.297-422.

Needham, Joseph, with Lu Gwei-Djen and Huang Hsin-Tsung, *Science and civilisation in China*, vol.6 *Biology and biological technology*, part 1: *Botany* (Cambridge, 1986).

Neumann, Josef N., 'Rousseaus Kritik an der Heilkunde seiner Zeit: zur Frage nach der handlungstheoretischen und ethischen Begründung medizinischen Handelns', *Medizinhistorisches Journal* 26:3-4 (1991), p.195-213.

Ogilvie, Brian W., 'Image and text in natural history, 1500-1700', in *The Power of images in early modern science*, ed. Wolfgang Lefèvre *et al.* (Basel, 2003), p.141-66.

–, 'The many books of nature: Renaissance naturalists and information overload', *Journal of the history of ideas* 64:1 (January, 2003), p.29-40.

O'Neal, John C., 'Nature as refuge in Rousseau's *Rêveries du promeneur solitaire*', in *The Nature of Rousseau's 'Rêveries': physical, human, aesthetic*, ed. John C. O'Neal, *SVEC* 2008:03, p.59-69.

Ostervald, Samuel Frédéric, *Description des montagnes et vallées du pays de Neuchâtel en 1764*, ed. Victor Benoit (Neuchâtel, 1861).

Partington, J. R., *A History of chemistry*, vol.2 (London, 1961).

Penzig, Otto A. J., *Contribuzioni alla storia della botanica* (Milan, 1905).

Perrot, Alain, *Jean Calvin et la science genevoise* (Geneva, 1996).

Phelps, Almira Hart Lincoln, *Familiar lectures on botany*, 18th edn (New York, 1843).

Pignal, Marc, 'Techniques particulières', *Orchidées: culture et protection* 23 (1995), p.43-47.

Pinault Sørenson, Madeleine, 'Rousseau et l'art du paysage', *Annales de la société Jean-Jacques Rousseau:* 'Jean-Jacques Rousseau et les arts visuels', Actes du colloque de Neuchâtel (20-22 September 2001) 45 (2003), p.271-300.

Plotkin, Mark J., Brian M. Boom and Maloyre Allison, *The Ethnobotany of Aublet's Histoire des plantes de la Guiane françoise (1775)* (St Louis, MO, 1991).

Polianski, Igor J., *Die Kunst, die Natur vorzustellen: die Ästhetisierung der Pflanzenkunde um 1800* (Cologne, 2004).

Ponçet, Vincent, *L'Herbier Dominique Villars (1745-1814): témoin de la flore du Dauphiné*, preface by G. G. Aymonin (Grenoble, 1999).

Prest, John, *The Garden of Eden: the botanic garden and re-creation of paradise* (New Haven, CT, 1981).

[Pury, Abraham], *Mémoires sur le comté de Neuchâtel en Suisse par le chancelier de Montmollin* (Neuchâtel, 1831).

–, *Extraits des chroniques ou Annales autrefois successivement par les chanoines du chapitre de Notre-Dame de Neuchâtel* (Neuchâtel, 1839, 1884).

Racault, Jean-Michel, 'L'utopie narrative en France et en Angleterre, 1675-1761', *SVEC* 280 (1991).

Rader, Linda, and Celia Ison, 'Legacy of mercuric chloride', in *Managing the modern herbarium: an inter-disciplinary approach*, ed. Deborah A. Metsger and Sheila C. Byers (Washington, DC, 1999), p.353-54.

Raj, Kapil, *Relocating modern science* (Basingstoke & New York, 2007).

Rappaport, Rhoda, 'G.-F. Rouelle: an eighteenth-century chemist and teacher', *Chymia* 6 (1960), p.68-101.

–, 'Rouelle and Stahl: the phlogistic revolution in France', *Chymia* 7 (1961), p.73-102.

Raven, Charles E., *English naturalists from Neckam to Ray: a study of the making of the modern world* (Cambridge, 1947).

–, *John Ray, naturalist: his life and works* (Cambridge, 1950).

Reeds, Karen M., *Botany in medieval and renaissance universities* (New York and London, 1991).

Reichler, Claude, *La Découverte des Alpes et la question du paysage* (Geneva, 2002).

Richard, Louis Claude Marie, *A Botanical dictionary*, trans. and ed. Amos Eaton, *et al.* (New Haven, CT, 1816).

–, *Démonstrations botaniques ou Analyse du fruit, considéré en général* (Paris, 1808).

–, *Dictionnaire élémentaire de botanique de Bulliard, revu et presqu'entièrement refondu par Louis Claude Richard* (Paris, A. J. Dugour & Durand, 1798).

–, *Dictionnaire élémentaire de botanique par Bulliard. Revu et presqu' entièrement refondu par Louis-Claude Richard*, 2nd edn (Paris, 1802).

Richard-Pauchet, Odile, '"Voici des fruits, des fleurs, des feuilles et des branches": une lecture érotique des lettres élémentaires sur la botanique de Rousseau', *Revue de l'AIRE* 34 (2008), p.197-214.

Richebourg, Marguerite, *Essai sur les lectures de Rousseau* (Geneva, 1934).

Riddle, John M., 'Dioscorides', in *Catalogus translationum et commentariorum: medieval and Renaissance translations and commentaries*, ed. F. Edward Cranz and Paul Oskar Kristeller, vol.4 (Washington, DC, 1980), p.1-144.

Roberts, Lissa, 'Setting the table: the disciplinary development of eighteenth-century chemistry as

read through the changing structure of its tables', in *The Literary structure of scientific argument*, ed. Peter Dear (Philadelphia, PA, 1991), p.99-132.
Robin, Nicolas, and Frank H. Hellwig, 'Plant systematics at Jena during the early nineteenth century: Fr. S. Voigt's treatment of the "méthode naturelle"', *Annals of the history and philosophy of biology* 10 (2005), p.117-42.
Roche, Daniel, 'Natural history in the academies', in *Cultures of natural history*, ed. N. Jardine, *et al.* (Cambridge, 1996), p.127-144.
Roger, Jacques, *Buffon: a life in natural history*, trans. Sarah L. Bonnefoi, ed. L. Pearce Williams (Ithaca, NY, 1997).
–, 'Lamarck et Jean-Jacques Rousseau', *Gesnerus* 42 (1985), p.369-81.
Rolston, Holmes, III, *Conserving natural value* (New York, 1994).
Roule, Lucie, 'Les médecins du Jardin du roi: origine médicale du Muséum national d'histoire naturelle', doctoral dissertation, Faculty of Medicine, Paris (Paris, 1942).
Roulet, Louis-Edouard, 'Abraham Pury: militaire (1724-1897)', in *Biographies neuchâteloises*, vol.1, ed. Michel Schlup (Hauterive, 1996), p.225-28.
–, 'La société helvétique et les Neuchâtelois', *Musée neuchâtelois* (1991), p.211-23.
Rousseau, Jacques, 'Sébastien Vaillant: an outstanding 18th century botanist', *Essays in biohistory and other contributions presented by friends and colleagues to Frans Verdoorn*, ed. P. Smit and R. J. Ch. V. ter Laage (Utrecht, 1970), p.195-228.
Rousseau, Jean-Jacques, 'Botanical writings', in *Collected writings of Rousseau*, trans. and ed. Alexandra Cook, vol.8 (Hanover, NH, 2000), p.93-255, 298-333, 345-49.
–, *La Botanique de J.-J. Rousseau: contenant tout ce qu'il a écrit sur cette science; l'exposition de la méthode botanique de M. de Jussieu; la manière de former les herbiers par M. Haüy* (Paris, 1802).
–, *La Botanique de J.-J. Rousseau, contenant tout ce qu'il a écrit sur cette science, augmentée de l'exposition de la méthode de Tournefort, de celle du système de Linné, d'un nouveau dictionnaire de botanique et de notes historiques*, ed. Albéric Deville, 2nd edn (Paris, 1823).
–, *La Botanique de J.-J. Rousseau ornée de soixante-cinq planches, d'après les peintures de P. J. Redouté* (Paris, 1805).
–, *Botanisieren mit Jean-Jacques Rousseau: die Lehrbriefe für Madeleine; das Herbar für Julie*, trans. Ruth Schneebeli-Graf (Thun, 2003).
–, *Botany: a study of pure curiosity: botanical letters and notes towards a dictionary of botanical terms*, trans. Kate Ottevanger (London, 1979).
–, *Cartas sobre os elementos de botanica, com as notas e addiçôes de Thomaz Martyn; (quarta ediçaô em Londres)*, trans. by a lady of the court (Lisbon, 1801).
–, *Correspondance complète de Jean-Jacques Rousseau*, ed. R. A., Leigh, 52 vols (Geneva, Madison, WI, Oxford, 1965-1998).
–, *Correspondance de J.-J. Rousseau avec Leonard Usteri*, ed. Paul Usteri and Eugène Ritter (Zürich and Geneva, 1910).
–, *Deux lettres à M. le mareschal duc de Luxembourg (le 20 et 28 janvier 1763)*, ed. Frédéric S. Eigeldinger (Neuchâtel, 1977).
–, *Institutions chimiques*, ed. Christophe van Staen (Paris, 2009).
–, 'Les institutions chymiques de Jean-Jacques Rousseau', ed.

Maurice Gautier, *Annales de la société Jean-Jacques Rousseau* 12-13 (1918-1921).
–, *Lettres inédites de Jean-Jacques Rousseau à Mmes Boy de la Tour et Delessert, comprenant les lettres sur la botanique*, ed. Philippe Godet and Maurice Boy de la Tour (Paris and Geneva, 1911).
–, 'Notes inédites de J.-J. Rousseau sur la botanique', in *Œuvres inédites de J.-J. Rousseau*, ed. V. D. Musset-Pathay, vol.1 (Paris, 1825).
–, *Œuvres complètes*, ed. Bernard Gagnebin and Marcel Raymond, 5 vols (Paris, 1959-1995).
–, *Russovy pis'ma o Botanikie s dopolneniem ego Botanicheskago Slovania, s ob"iaseniem trekh luchshikh metod Turneforta, Linneia, i Zhus'e: I s Botanicheskami chasami, izobriretennymi bezsmertnym Linnem*, trans. Vladimir Vasil'evich Izmailow (Moscow, 1810).
Rowell, Margery, 'Linnaeus and botanists in eighteenth-century Russia', *Taxon* 29:1 (February 1980), p.15-26.
Rudolf, Gerhard, 'Jean-Jacques Rousseau (1712-1778) – Stammvater der Naturheilkunde?', *Hippokrates* 12 (1969), p.474-79.
–, 'Jean-Jacques Rousseau (1712-1778) und die Medizin', *Sudhoffs Archiv: Zeitschrift für Wissenschaftsgeschichte* 53:1 (1969), p.30-67.
Rueff, Martin, 'Apprendre à voir: l'optique dans la théorie de l'homme', in *Rousseau et les sciences*, ed. B. Bensaude-Vincent and B. Bernardi (Paris, 2003), p.193-214.
Ruiz [López], Hipólito, 'Travels of Ruiz, Pavón and Dombey in Peru and Chile (1777-1788)', trans. B. E. Dahlgren, with official documents and epilogue by A. J. Barreiro, *Botanical series, Field museum of natural history* 21:467 (28 March 1940), p.5-327.
Ruse, Michael, *Darwin and design: does evolution have a purpose?* (Cambridge, 2003).
Rychner, Jacques, and Michel Schlup, 'Frédéric Samuel Ostervald: homme politique et éditeur (1713-1795)', in *Biographies neuchâteloises*, ed. Michel Schlup (Hauterive, 1996), vol.1, p.197-201.

Sachs-Hombach, Klaus, *Wege zur Bildwissenschaft: interviews* (Cologne 2004).
Saint-Amand, Pierre, 'Rousseau contre la science: l'exemple de la botanique dans les textes autobiographiques', *SVEC* 219 (1983), p.159-67.
Saint-Lager, Jean-Baptiste, *Histoire des herbiers* (Paris, 1885).
Savage, Spencer, *A Catalogue of the Linnaean herbarium* (London, 1945).
Schaer, Jean-Paul, 'Louis Bourguet: philosophe et naturaliste (1678-1742)', in *Biographies neuchâteloises*, vol.1, ed. Michel Schlup (Hauterive, 1996), p.17-22.
Schiebinger, Londa, 'Gender and natural history', in *Cultures of natural history*, ed. N. Jardine, J. A. Secord and E. C. Spary (Cambridge, 1996), p.163-77.
–, *The Mind has no sex? Women in the origins of modern science* (Cambridge, MA, 1989).
–, *Nature's body: gender in the making of modern science* (Boston, MA, 1994).
–, 'The philosopher's beard: women and gender in science', *The Cambridge history of science*, 4: *Eighteenth-century science*, ed. Roy Porter (Cambridge, 2003).
–, *Plants and empire: colonial biosprospecting in the Atlantic world* (Cambridge, MA, 2004).
–, 'The private life of plants: sexual politics in Carl Linnaeus and Erasmus Darwin', in *Science and sensibility: gender and scientific enquiry, 1780-1945*, ed. Marina Benjamin (Oxford, 1991).
–, and Claudia Swan (eds), *Colonial*

botany: science, commerce and politics in the early modern world (Philadelphia, PA, 2005).

Schlup, Michel, 'Aperçu de l'imprimerie et de l'édition neuchâteloises avant 1769', in *L'Edition neuchâteloise au siècle des Lumières: la société typographique de Neuchâtel (1769 à 1789)*, ed. Michel Schlup (Hauterive, 2002), p.29-59.

–, 'Neuchâtel aux alentours de 1769', in *L'Edition neuchâteloise au siècle des Lumières: la société typographique de Neuchâtel (1769 à 1789)*, ed. Michel Schlup (Hauterive, 2002), p.15-28.

Schlup, Michel, *et al.* (eds), *Bibliothèques et mécènes: deux siècles de donations à la bibliothèque de Neuchâtel* (Neuchâtel, 2006).

Schneebeli-Graf, Ruth, *Botanisieren mit Jean-Jacques* (Thun, 2003).

–, *Das Zürcher Herbar von Jean-Jacques Rousseau* (Zürich, 1980).

School of Hippocrates, 'Airs, waters, places', in *Hippocratic writings*, ed. G. E. R. Lloyd, trans. J. Chadwick and W. N. Mann (London, 1983), p.148-69.

Schwartz, Joel, *The Sexual politics of Jean-Jacques Rousseau* (Chicago, IL, 1984).

Scott, David, 'Rousseau and flowers: the poetry of botany', *SVEC* 182 (1979), p.73-86.

Selosse, Philippe, 'The underlying pattern of the Renaissance botanical genre pinax', in *Opening windows on texts and discourses of the past*, ed. Janne Skaffari, *et al.* (Amsterdam; Philadelphia, PA, 2005), p.161-78.

–, 'Peut-on classer le vivant?', *Linné et la systématique aujourd'hui*, ed. D. Prat, *et al.* (Paris, 2008).

Seneca, *Letter* 92.3, LS 63F, in *The Hellenistic philosophers: translations of the principal sources with philosophical commentary*, vol.1, ed. A. A. Long and D. N. Sedley (Cambridge, 1987).

Sénelier, Jean, *Bibliographie générale des œuvres de J.-J. Rousseau* (Paris, 1950).

Shteir, Ann B., 'Botanical dialogues: Maria Jacson and women's botanical writing', *Eighteenth-century studies* 23:3 (1990), p.301-17.

–, *Cultivating women, cultivating science: Flora's daughters and botany in England 1760-1860* (Baltimore, MD, 1996).

–, 'Iconographies of Flora: the goddess of flowers in the cultural history of botany', in *Figuring it out: science, gender and visual culture*, ed. Ann B. Shteir and Bernard Lightman (Hanover, NH, 2006).

–, '"The pleasing objects of our present researches": women in botany', in *Women and history*, ed. Valerie Frith (Toronto, 1995).

Sigrist, René, *L'Essor de la science moderne à Genève* (Lausanne 2004).

Simon, Jonathan, 'L'homme de verre? Les trois règnes et la promiscuité de la nature', *Corpus* 36 (1999), p.65-80.

Siraisi, Nancy G., *Medieval and early renaissance medicine: an introduction to knowledge and practice* (Chicago, IL, 1990).

Slaughter, Mary M., *Universal languages and scientific taxonomy in the seventeenth century* (Cambridge, 1982).

Sloan, Phillip R., 'Buffon, German biology and the historical interpretation of biological species', *British journal for the history of science* 12:2 (July 1979), p.109-53.

Smith, Sir James Edward, *Exotic botany*, 2 vols (London, 1804-1805).

Smith, Pamela H., *The Body of the artisan* (Chicago, IL, 2004).

–, 'Art, science and visual culture in early modern Europe', *Isis* 97 (2006), p.83-100.

Société botanique de Lyon, 'The Origin of herbaria', in *Bulletin trimestriel de la société botanique de*

Lyon, translated in *Bulletin of the Torrey Botanical Club*, 12:17 (December 1885), p.129-31.

Sokoloff, Dimitry D., *et al.*, 'The history of botany in Moscow and Russia in the 18th and early 19th centuries in the context of the Linnaean collection at Moscow University (MW)', *Huntia* 11:2 (2002), p.129-91.

Sonenscher, Michael, 'French economists and Bernese agrarians: the marquis de Mirabeau and the economic society of Berne', *History of European ideas* 33:4 (December 2007), p.411-26.

Sonntag, Otto, 'Introduction', *The Correspondence between Albrecht von Haller and Horace-Bénédict de Saussure* (Bern, 1990).

Spary, Emma, 'Scientific symmetries', *History of science* 62 (2004), p.2-46.

Spector, Céline, 'Rousseau et la critique de l'économie politique', in *Rousseau et les sciences*, ed. Bernadette Bensaude-Vincent and Bruno Bernardi (Paris, 2003), p.237-56.

Speerli, Leonore, 'Rousseau und Zürich: vom Erscheinen des ersten Discours bis zum Ausbruch der Revolution in Frankreich', doctoral dissertation, University of Zürich (Zürich, 1941).

Staël-Holstein, Anne Louise Germaine Necker de, *Œuvres complètes de Madame la baronne de Staël-Holstein*, vol.1 (Paris, 1861).

Stafford, Barbara Maria, *Body criticism: imaging the unseen in Enlightenment art and medicine* (Cambridge, MA, 1991).

Stafleu, Frans A., *Linnaeus and the Linnaeans: the spreading of their ideas in systematic botany, 1735-1789* (Utrecht, 1971).

–, 'Review of *The Gardener's dictionary* by Philip Miller', *Taxon* 18:6 (December 1969), p.713-15.

–, and Richard S. Cowan, *Taxonomic literature*, 2nd edn, 7 vols (Utrecht, 1979-1988).

Star, Susan Leigh, and James Richard Griesemer, 'Institutional ecology, "translations" and boundary objects: amateurs and professionals in Berkeley's Museum of Vertebrate Zoology, 1907-1939', *Social studies of science* 19 (1989), p.387-420.

Starobinski, Jean, *Jean-Jacques Rousseau: la transparence et l'obstacle* (Paris, 1971).

–, *Le Remède dans le mal: critique et légitimation de l'artifice à l'âge des Lumières* (Paris, 1989).

Stearn, William T., *Botanical Latin: history, syntax, terminology and vocabulary*, 4th edn (Newton Abbot, 1992).

–, 'The background of Linnaeus's contributions to the nomenclature and methods of systematic biology', *Systematic zoology* 8 (1959), p.4-21.

–, 'Major botanical publications of Linnaeus', in Carolus Linnaeus, *Species plantarum: a facsimile of the first edition 1753*, vol.1 (London, 1957), p.17-23.

–, 'Preface', in John Ray, *Synopsis methodica stirpium Britannicarum editio tertia, 1724* [and] Carl Linnaeus, *Flora Anglica 1754 & 1759*, facsimile edns (London, 1973).

–, 'A Royal Society appointment with Venus in 1769: the voyage of Cook and Banks in the *Endeavour* in 1768-1771 and its botanical results', *Notes and records of the Royal Society of London* 24:1 (1969), p.64-90.

–, 'Signs used by Linnaeus', in Carolus Linnaeus, *Species plantarum: a facsimile of the first edition 1753*, with an introduction by William T. Stearn (London, 1953), vol.1, p.162-63.

Stevens, Peter F., *The Development of*

biological systematics: Antoine-Laurent de Jussieu, nature, and the natural system (New York, 1994).

Störi, Fritz, 'Der Helvetismus des "Mercure suisse" ("Journal helvétique"), 1732-1784', doctoral dissertation, University of Zürich (Bern, 1953).

Stroup, Alice, *A Company of scientists: botany, patronage, and community at the seventeenth-century Parisian Royal Academy of Sciences* (Berkeley, CA, 1990).

Stuber, Martin, and Regula Wyss, 'Der Magistrat und ökonomische Patriot', in *Albrecht von Haller: Leben, Werk, Epoche*, ed. Hubert Steinke, *et al.* (Göttingen, 2008), p.347-80.

Stutzer, Walter, 'Jean-Jacques Rousseau und die Schweiz: zur Geschichte des Helvetismus', doctoral dissertation, University of Zürich (Zürich, 1950).

Theophrastus, *Enquiry into plants*, trans. Sir Arthur Hort (London and Cambridge, 1916).

Thériault, Serge A., *Jean-Jacques Rousseau et la médecine naturelle* (Montréal, 1979).

Thorndike, Lynn, *A History of magic and experimental science during the first thirteen centuries of our era*, vol.2 (New York, 1923).

Thurmann, Jules, *Abraham Gagnebin de la Ferrière* (Porrentruy, 1851).

Tolila, P., 'Rousseau et le matérialisme des fleurs', *La Pensée* 202 (1978), p.116-31.

Trachtenberg, Zev, 'Did we inherit our trouble with wilderness from Rousseau?' unpublished MS.

Tutin, T. G., *et al.* (eds), *Flora europaea*, 5 vols (Cambridge, 1964-1983).

Uzgalis, William, 'John Locke', *The Stanford encyclopedia of philosophy* (Winter 2010 edn), Edward N. Zalta (ed.) (http://plato.stanford.edu/archives/win2010/entries/locke/).

Vallette, Gaspard *Jean-Jacques Rousseau genevois*, 2nd edn (Geneva, 1911).

Vallière, Capitaine [Paul Emanuel] de, *Honneur et fidelité: histoire des Suisses au service étranger* (Neuchâtel, n.d.).

Venel, Gabriel-François, 'Chimie', in Denis Diderot and Jean Le Rond d'Alembert (eds), *Encyclopédie, ou Dictionnaire raisonné des sciences, des arts et des métiers, par une société de gens de lettres*, CD-ROM (Paris, 2000).

Walters, S. M., *The Shaping of Cambridge botany: a short history of whole-plant botany in Cambridge from the time of Ray into the present century* (Cambridge, 1981).

Waterhouse, Benjamin, *The Botanist: being the botanical art of a course of lectures on natural history, delivered in the University of Cambridge [Harvard]: together with a discourse on the principle of vitality* (Boston, MA, 1811).

White, Paul, 'The purchase of knowledge: James Edward Smith and the Linnean collections', *Endeavour* 23:3 (1999), p.126-29.

Williams, Roger L., *Botanophilia in eighteenth-century France: the spirit of the Enlightenment* (Dordrecht, 2001).

–, *The Letters of Dominique Chaix, botaniste-curé, International archives of the history of ideas*151 (Dordrecht, 1997).

Wilson, Edward O., *The Creation: an appeal to save life on earth* (New York, 2006).

–, *The Diversity of life* (Cambridge, MA, 1992).

Winkler, Markus, 'The poetics of the Enlightenment and Salomon Gessner's Idylls', in Patrick Coleman, Anne Hofmann, and

Simone Zurbuchen (eds), *Reconceptualizing nature, science, and aesthetics: contribution à une nouvelle approche des Lumières helvétiques* (Geneva, 1998), p.185-97.

Winsor, Mary P., 'Linnaeus's biology was not essentialist', *Annals of the Missouri botanical garden* 93 (2006), p.2-7.

–, 'Setting up milestones: Sneath on Adanson and Mayr on Darwin', in *Milestones in systematics*, ed. David M. Williams and Peter L. Forey (Boca Raton, FL, London, New York, Washington, DC, 2001).

–, 'Taxonomy was the foundation of Darwin's evolution', *Taxon* 58:1 (February 2009), p.43-49.

Wyngaard, Amy, *From savage to citizen: the invention of the peasant in the French Enlightenment* (Newark, DE, 2004).

Zimmer, Oliver, *A Contested nation: history, memory and nationalism in Switzerland, 1761-1891* (Cambridge, 2003).

Zoller, Heinrich, 'Albrecht von Hallers Pflanzensammlungen in Göttingen, sein botanisches Werk und sein Verhältnis zu Carl von Linné', *Nachrichten der Akademie der Wissenschaften in Göttingen: mathematisch-physikalische Klasse* 10 (1958), p.217-51.

Zurbuchen, Simone, 'Berliner "Exil" und Schweizer "Heimat": Johann Georg Zimmermanns Reflexionen über die Rolle des Schweizer Gelehrten', in *Schweizer im Berlin des 18. Jahrhunderts*, ed. Martin Fontius and Helmut Holzhey (Berlin, 1996) p.57-68.

Index